Resonance Ionization Spectroscopy 1992

Sponsors

Financial support from the following organizations is gratefully acknowledged

United States Department of Energy
Office of Health and Environmental Research, Washington, DC, USA

Atom Sciences, Inc.
Oak Ridge, Tennessee, USA

Coherent Inc.
Palo Alto, California, USA

Continuum
Santa Clara, California, USA

Kratos Analytical
Ramsey, New Jersey, USA

Lumonics Inc.
Kanata (Ottawa), Ontario, Canada

Spectra-Physics Lasers Inc.
Mountain View, California, USA

The University of Tennessee
Knoxville, Tennessee, USA

Los Alamos National Laboratory
Los Alamos, New Mexico, USA

RIS-92 Advisory Group

Resonance Ionization Spectroscopy 1992

Proceedings of the Sixth International Symposium on
Resonance Ionization Spectroscopy and its Applications
held in Santa Fe, New Mexico, USA, 24–29 May 1992

Edited by Charles M Miller and James E Parks

Institute of Physics Conference Series Number 128
Institute of Physics Publishing, Bristol and Philadelphia

CODEN IPHSAC 128 1–357 (1992)

British Library Cataloguing in Publication Data

A catalogue record for this book is available from the British Library

ISBN 0-7503-0230-5

Library of Congress Cataloging-in-Publication Data are available

Published by IOP Publishing Ltd, a company wholly owned by the Institute of Physics, London
Techno House, Redcliffe Way, Bristol BS1 6NX, England
US Editorial Office: IOP Publishing Inc., The Public Ledger Buildings, Suite 1035, Independence Square, Philadelphia, PA 19106, USA

Printed in the UK by Galliard (Printers) Ltd, Great Yarmouth, Norfolk

Preface

The Sixth International Symposium on Resonance Ionization Spectroscopy and its Applications, RIS-92, was held in Santa Fe, New Mexico, USA, 24–29 May 1992. The symposium was organized by a Program Committee formed as a part of the International Advisory Committee, which provides leadership and guidance for the symposia. This symposium was jointly coordinated by the Institute of Resonance Ionization Spectroscopy of the University of Tennessee and the Los Alamos National Laboratory of the United States Department of Energy. The symposia are a major focal point for updating the current state of the art in research involving resonance ionization and related techniques. The RIS technique has a substantial impact on research in medicine, biology, material and nuclear sciences, industry, and the environment. RIS-92, the sixth symposium in this series, placed emphasis on technique development and applications, particularly in basic physics and chemical analysis where ultrahigh sensitivity and selectivity are needed. These proceedings report a wide selection of current topics addressing the needs of both researchers in the field and potential users.

RIS-92 was structured with a balance of invited talks, plenary talks, and contributed presentations, both oral and poster. Ample time was allotted for discussion and personal interaction. In these symposia, subjects and speakers are chosen to present the latest and best developments in the field of RIS. In addition, emerging and related work, sometimes on the far fringes, is presented to help stimulate new thoughts, techniques, approaches, and applications. This leads to a diversity of subjects being covered, and this diversity is reflected in these proceedings.

The symposium featured nine technical sessions, each addressing particular applications or aspects of the resonance ionization technique. Session 1, *Ultrasensitive Applications*, was devoted to measurements of extremely rare isotopes where RIS is particularly suited for selective detection. Sessions 2 and 3, *Atomic Spectroscopy* and *RIS Dynamics*, focused on the basic physics underlying the resonance ionization process. Comparisons of RIS with other sensitive laser-based methods were provided by Session 4, *Laser Techniques in Flames and Plasmas*. Session 5, *Molecular RIS*, addressed the rich field of molecular analysis and spectroscopy which parallels much of atomic resonance ionization. Sessions 6 and 7, *Analytical Applications* and *Surface and Bulk Analysis*, addressed the concept of sensitive and selective elemental/isotopic determinations. Session 8 was devoted to *Biological and Medical Applications* of resonance ionization spectroscopy. Session 10, *Sources and Techniques*, explored the latest developments in laser technology and mass spectrometry.

Several highlights of the program for this symposium should be mentioned. The program began with a Short Course on *Resonance Ionization Spectroscopy*, which attracted more than 30 participants. The short course, offered for the second time in the history of the symposia, included a new discussion on molecular RIS and represented a growing interest in this subject. The Advisory Committee continued to make a conscientious effort to make graduate students an important part of the symposia, and succeeded in attracting numerous students from the US, Europe, and Asia. It was encouraging to note that some of the best and most innovative work was reported by graduate students, thus giving RIS a bright future.

Another highlight of the symposium was the presentation of excellence awards for outstanding work in RIS performed by a graduate student and for an outstanding poster presentation. Ludwin Monz, a graduate student working under the direction of Dr H-Jürgen Kluge, was awarded a cash prize and a certificate representing the Excellence Award for RIS Research and Development Performed by a Graduate Student. His work, Collinear Resonance Ionization Spectroscopy for the Detection of Strontium-90 and Strontium-89 in Environmental Samples, is an excellent example of the usefulness that RIS can have in solving difficult problems in analytical applications. Giuseppe Petrucci, along with his co-workers, was awarded books on RIS and a certificate for the Excellence Award for RIS Research and Development Presented in a Poster Session. Their work, Resonance Detection of Photons, is a good example of the development of RIS techniques for simple and effective chemical analysis. The Awards Committee also recognized H Lauranto and his co-workers for honorable mention for their work and poster presentation of Noise Reduction Techniques in RIS Utilizing Real-Time Laser Pulse Spectra.

A final highlight to be mentioned was the noteworthy work and the participation of undergraduate students and their professor, Dr R C Estler, from Fort Lewis College. Their work was cited by the Poster Awards committee for honorable mention because of their excellent presentation and work in an undergraduate setting. While much work in RIS requires equipment in the millions of dollars price range, this group demonstrated that RIS can be performed on a 'shoestring' and that RIS research within the undergraduate institution environment can provide meaningful and valuable research experience for undergraduate students. Such work can be stimulating, exciting, and motivating to the student while teaching him/her much knowledge about physics and chemistry.

The local host for RIS-92, Los Alamos National Laboratory, is to be commended for the excellent planning and arrangements made in hosting this very successful symposium. Special thanks are due to Jan Hull, Conference Coordinator, Kim Nguyen and the staff of the Protocol Office of Los Alamos National Laboratory, and to Elaine Roybal of the Isotope and Nuclear Chemistry Division. The standards they set in arranging this meeting will be difficult to duplicate in the future. We are also very appreciative of the hard work provided by the session chairmen in helping with the review of the manuscripts for these proceedings, and we are especially thankful for the dedicated work in the editing and management of these proceedings, provided by Maureen Clarke, Senior Commissioning Editor of the Institute of Physics Publishing.

The support provided by the following sponsors is also gratefully acknowledged: US Department of Energy, Los Alamos National Laboratory, LANL's Isotope and Nuclear Chemistry Division, The University of Tennessee, Atom Sciences, Lumonics, Continuum, Spectra-Physics Lasers, Coherent, and Kratos. Without their support, a well-balanced program would not have been possible.

The advice and support of our colleagues on the Advisory Committee is acknowledged. The members of the Advisory Committee for RIS-92 were: E Arimondo,

G Bekov, P Benetti, P Camus, G Goldstein, G S Hurst, H-J Kluge, P Knight, Y Y Kuzyakov, B Lehmann, A L'Huillier, T B Lucatorto, J C Miller, K Niemax, N Omenetto, H Rubinsztein-Dunlop, T J Whitaker, and N Winograd. Their suggestions and guidance were important to the success of the symposium, and their work and collaboration is gratefully appreciated.

Finally, we wish to acknowledge all the authors for their participation in the symposium and for the contributions of their manuscripts. We especially wish to thank the keynote speaker and the plenary and invited speakers for providing leadership for the technical presentations and discussions. Without their dedication and interest, the symposia would not be possible. We, like these proceedings affirm, believe that resonance ionization spectroscopy is an interesting and stimulating field of study, and has much to offer in solving many interesting and difficult problems in a variety of applications.

Charles M Miller
James E Parks

Contents

Section 3: RIS Dynamics

Inst. Phys. Conf. Ser. No 128: Keynote Address
Paper presented at RIS 92, Santa Fe, NM, USA, 24–29 May 1992

Some trends in resonance ionization spectroscopy

G. Samuel Hurst

Atom Sciences, Inc., Oak Ridge, Tennessee USA

ABSTRACT: Some general trends in RIS, including its widespread use as a state selective and sensitive analytical method, are noted. One of these capabilities (counting noble gas atoms) has made possible a solar neutrino experiment using a bromine compound.

1. SOME TRENDS IN RIS

In this paper, I will try to pick out some general trends on RIS and will illustrate one of them in detail by discussing the solar neutrino problem which Ray Davis brought to us at our very first symposium. Substantial progress has been made on a Maxwell demon for counting Kr atoms. This RIS based demon makes possible a new solar neutrino experiment which could yet help in solving the riddle. This progress will be described later in the program by N. Thonnard.

Prior to our first symposium (RIS-81, Gatlinburg) saturated RIS had been abundantly demonstrated and sensitive detection had begun. One-atom detection of cesium was described in which laser beams were directed through proportional counters capable of detecting single electrons. Work with noble gases had already begun and a device which incorporated RIS with a small mass spectrometer for counting Kr atoms in a closed system was referred to as Maxwell's demon. A system using a time-of-flight mass spectrometer was used for detecting K and rare earth isotopes. Enthusiasm for applications of these early RIS devices was rampant. The list is familiar today: impurities in semiconductors, dating the earth, the oceans, the ice caps, and old groundwater, detection of solar neutrinos, and double beta- decay.

Whereas the reported work at the first symposium made reference to only two papers with mass spectrometers, there was enough work at the second symposium (RIS-84, Knoxville) to have an entire session called Resonance Ionization Mass Spectrometry (RIMS). The need for mass analysis to reduce interferences from other species, including isobars, also received early recognition at the Institute of Spectroscopy (IOS) at Troitzk. The opportunity to incorporate various methods of atomization in these RIMS systems, including ion-beam sputtering was called Sputter Initiated RIS (SIRIS). We saw the acceptance of SIRIS and other techniques based on RIS for the characterization of surfaces and solid materials, including semiconductors for the electronics industry. Likewise, we saw increased efforts on noble gas analysis, especially the development of four-wave mixing schemes for the generation of the necessary short wavelength radiations. Since the driving force for counting noble gas atoms (with isotopic selectivity) is the dating of groundwater and polar ice and the solar neutrino experiment, a lot of effort was reported on the separation a few atoms from huge samples of materials. Atom Sciences reported on efforts to develop a Krypton demon which uses a time-of-flight spectrometer to detect Kr-85 and Kr-81. Weak interaction and particle physics topics included double-beta decay and several solar neutrino experiments using either chlorine, bromine, gallium, or molybdenum. Reports were made on searches for free quarks, magnetic monopoles and "nuclearites" in general, as possible cosmic ray components. Several proposals were discussed for using RIS in new kinds of searches for these unusual particles.

C. Grey Morgan organized and hosted our first meeting outside of Tennessee. The Swansea meeting (RIS- 86) succeeded in bringing together a more representative international participation. The most important trend was captured by Keith Boyer in his excellent summary of the meeting when he said "papers presented centered on accomplishments rather than plans." A strong session on photophysics and spectroscopy certainly contributed to this perception of maturation, as did specific accomplishments in the analysis of noble gases and impurities in solids. For the first time, there was an entire session on small molecules and another on biological and medical application where both atoms and large molecules were discussed. Significant results on the RIMS technique, with applications ranging from nuclear and particle physics to the environment and to geophysics were reported. We saw the emergence of on-line accelerator physics for the study of short-lived isotopes. Professor Kluge and his group (Mainz), reported on the first RIMS application to study isotopic shift and hyperfine structure in some gold isotopes and isomers.

The fourth symposia (RIS-1988) was hosted by Dr. T.B. Lucatorto at the National Bureau of Standards and assisted by Dr. J.E. Parks. It is gratifying to note that the basic theoretical and experimental studies remained strong, with outstanding reports on isotopically selective RIS. We had interesting reports on noble gas analysis with applications of wider scope, including meteorite dating studies. The use of RIS for surface and bulk analysis of materials continued to be one of the most active applications of RIS. Molecular applications matured substantially. A trend emerged to discuss the technological developments, including active interest in RIS schemes as a data base at the National Institute of Standards and Technology (NIST).

Time and space do not permit me to elaborate on the technical developments at our last meeting (RIS-1990) in Varese which was so graciously hosted by our colleague N. Omenetto. In addition to our proceedings, we have a brief meeting report (Hurst 1991) which highlights some superb advances in RIS. These include theories of isotope effects due to hyperfine structure interactions, a marvelous summary of how the on-line methods are leading to better understanding of nuclei, collinear methods for high isotopic selectivity, use of the SIRIS method for dramatic reduction in matrix effects in depth profiling of solids, rapid methods of genome sequencing, and the maturation of RIS of molecules.

In the early history of RIS it was believed that the method would have applications in three major branches, namely: (1) excited state kinetics, (2) materials separation, and (3) analytical methods. RIS, having the powerful combination of selectivity and sensitivity, has been advanced considerably as a new analytical method. Analytical applications of RIS have far exceeded its use in excited state kinetics or in materials separation. A number of articles on RIS have appeared in the Analytical Chemistry Journal. For instance, the March 1, 1992 issue contained a cover page article on genome sequencing and two other articles on RIMS systems. While RIS has been developed as a new analytical method, it has not been widely adopted by the analytical chemists; further recognition and adoption of RIS is a challenge to those who develop RIS. The barriers appear to be complexity and cost. A technological trend has been toward more sophistication as new features are discovered and incorporated into complex systems. Initially, there were dreams about using simple lasers with ionization detectors like proportional counters and single ion detectors. But, mass spectrometers were added to have the advantages of a RIMS system. Further, atomization systems have become complex with the use of atomic beams and ion sputtering sources. Currently, the more advanced systems, especially those using the collinear methods and the on-line nuclear physics techniques have brought RIS into the arena of small tandem accelerators and other very specialized analytical methods.

2. A SOLAR NEUTRINO APPLICATION

For many years a remarkable experiment has been running deep underground in a gold mine in South Dakota to study neutrino production in the interior of the sun. In this experiment, Ray Davis and his collaborators use a large tank filled with about 100,000 gallons of dry cleaning solution, (C_2Cl_4), to capture a tiny fraction of the neutrinos passing through the tank. Neutrino interactions with Cl-37 produce Ar-37 atoms which decay by electron capture with a half-life of 35 days. This is a favorable situation for detection with proportional counters, even though only 2 or 3 atoms of Ar-37 are produced per week! The cross section for neutrino interaction with matter is exceedingly small, however occasionally a neutrino with energy greater than 0.814 MeV is captured in a Cl-37 nucleus and produces an Ar-37 atom. By recovering the accumulated Ar-37 and observing the radioactive decay of these, the Davis team has been able to determine the flux of one of the neutrino sources in the Sun. They have shown that the number of these neutrinos, primarily the neutrinos accompanying positron emission from B-8 and having energies from 0 to 14 MeV, are about 1/3 of the "expected" flux. More specifically, they find an average of 0.472 ± 0.037 neutrino captures per day over the time period from 1970 to 1985, while the standard model predicts a rate of 1.8 neutrino captures per day. Further, there is about a factor of two variation in the mean from one year to another and Davis has suggested a possible anti-correlation with sunspot activity. Much of the significance of these experimental results is due to the careful theoretical work on stellar physics, see the book on Neutrino Astrophysics (John N. Bahcall, 1989.) The very reason we can refer to the "expected" flux is due to this effort, which concentrates on a standard stellar model. General acceptance of these calculations now provides a basic framework to which the results of any solar neutrino experiment must be compared.

If it could have been done by decay counting, the bromine experiment could have been done long ago (Scott 1976). Neutrino interaction with Br-81, produces Kr-81 with a half-life of 210,000 years, thus the feasibility of a radiochemical experiment, using a large vat filled with a bromine compound, would depend on the ability to count a few hundred atoms of Kr-81, by a direct method, independent of radioactive decay. The development of RIS for counting a small number of Kr-81 atoms has now made the bromine experiment feasible and would appear to be a natural sequel to the chlorine experiment (Hurst et. al. 1984). The threshold for the reaction is 470 kev and thus would detect the neutrino associated with electron capture in Be-7 (0.86 MeV) which is nearly 1000 times more abundant than the B-8 neutrino. Further, the Be-7 neutrino, while sensitive to the temperature of the sun, is far less so than the B-8 neutrino. One way out of the neutrino dilemma would simply be to assume that the interior temperature of the sun is somewhat less than the 15.5 million degrees predicted from the standard stellar model. In spite of the recent stress on ideas involving neutrino physics and particle physics this "old" idea cannot be ruled out entirely. A Cl-consistent model could be built around a 10% decrease in the core temperature of the Sun. A test could be the bromine experiment.

In addition to the chlorine experiment, there are two solar neutrino experiments in operation and these have been reviewed in an excellent article by Schwarzschild (October 1990).

For several years a collaboration in Japan has been searching for proton decay in large tanks of water equipped with photomultiplier tubes to detect Cerenkov radiation. An interesting byproduct of the failure to detect proton decay was the detection of solar neutrinos, beginning in 1987 in the Kamiokande II detector, which uses 3000 tons of water. This became the second solar neutrino experiment, and like the Homestake experiment it monitors primarily the relatively weak B-8 neutrino source in the Sun. In fact, the energy threshold (7.3MeV) of the Kamiokande II experiment, which depends on electron scattering by neutrinos, is much higher than in the Homestake experiment. The measured flux of solar neutrinos is about one-half of that predicted by the standard solar model and from the beginning of 1987 to the present there is no evidence of a time variation of this flux.

The third type of solar neutrino experiment to produce interesting results is based on the use of 30 tons of liquid gallium metal deep inside the Caucasus mountains in USSR and operated as a joint scientific enterprise known as the Soviet-American Gallium Experiment (SAGE). Another gallium detector is operating in the Gran Sasso d'Italia. This experiment called Gallex is primarily a European collaboration and is under the general direction of Til Kirsten of the Max Planck Institute (Heidelberg).The idea behind these experiments is the fact that very low energy neutrinos can interact to produce Ge-71 by capture in Ga-71 with a low energy threshold (0.23 MeV) permitting the measurement of the neutrino source associated with the p-p reaction (the starting reaction in the Sun's fusion process). To a first approximation, the gallium experiment should provide a way to separate questions of solar physics from those of neutrino physics. The first very preliminary result from the SAGE experiment is now available. No neutrinos were detected above the experimental background. If this result holds in future measurement, it is a strong indication of something new in neutrino physics.

Schwarzschild speculated on ways in which the results of these three experiments could fit into a unified picture involving some recent theories in neutrino and particle physics. This discussion is a continuation of another Physics Today article, Wolfenstein and Beier (198), written before the results of the SAGE experiment were obtained. The crux of the matter described in both articles is that neutrinos can change flavors from electron to mu or tau neutrinos when the neutrinos are passing through the high density region of the Sun, according to the MSW theory. The mechanism for these flavor changes is described by Mikheyev and Smirnov who built onto the formulation of Wolfenstein. This process can only occur if all types of neutrinos have finite rest masses. Whether an electron neutrino makes it out of the Sun without a flavor change is a rather complex problem depending on several parameters in the MSW theory. The excess of the mu neutrino mass above the electron neutrino mass is one of the parameters of great interest. With some values of these parameters it is possible to predict nearly total suppression of the low energy electron neutrinos seen by the gallium detector, about a factor of three suppression of the mid- energy neutrinos seen by the chlorine experiment and the smaller loss at the high energy region seen by the Kamiokande detector.

However, Davis has shown that there is a possible anti-correlation of neutrino flux with Sunspot activity and this is another puzzle to be resolved. Some theories have been advanced for this anti-correlation, in which it is suggested that if the electron neutrino has a large magnetic moment it may flip from a state of left-handed helicity into a state of right-handed helicity in the strong magnetic fields associated with sunspots. Electron- neutrino detectors are not sensitive to the right-handed helicity state (antineutrinos). If these ideas are correct, it will be several years before the gallium experiments begin to see neutrinos from a quieter Sun. At this time it is not possible to know whether the flavor mixing mechanism in the interior of the sun or the helicity flip at the photosphere is the dominant mechanism in understanding the results of the three solar neutrino experiments. Of course both theories could be correct, but it would be highly unusual for two entirely new mechanisms to be discovered in efforts to understand a limited set of experimental observations.

A bromine solar neutrino experiment that would be much more sensitive than the chlorine experiment and would detect the lower energy (line source at 0.8 MeV) neutrinos has considerable virtue. First, it would give information on both the Be-7 and the B-8 neutrino sources. With good knowledge of the B-8 flux from the Davis experiment and reduction of the uncertainties of the cross sections, the bromine experiment could be analyzed to give just the Be-7 flux. Recent work, Chen and Cherry (1991), suggests that the MSW oscillation in the Sun's interior could even be working in concert with a higher core temperature in a way that is consistent with all of the existing solar neutrino experiments. Any experiment that would measure the Be-7 solar neutrino flux to further define the core temperature of the Sun would be of great value. In addition, a bromine detector could give a unique way to study the time correlation of neutrinos with sunspot activity by using a detector which is much more sensitive than the chlorine experiment. Since the period of the solar activity is about 11 years, the Kr-81 atoms produced in a large bromine tank could be accumulated for about one year, and the counting statistics on approximately 1000 atoms should be quite good. If it turns out that the sunspot modulation of the neutrino flux is not seen with the more sensitive detector, then attention focuses on the flavor mixing mechanism in the interior of the Sun.

Just as the gallium detector was designed to separate solar physics from neutrino physics, the bromine experiment is designed to further define the new ideas in neutrino physics, namely to distinguish flavor mixing in the Sun's interior from helicity flips in the photosphere of the Sun. And, of course, if the riddle involves a combination of solar physics and neutrino physics, who would argue that new experiments are unnecessary?

3. REMARKS ON MAXWELL'S DEMON

Last year Maxwell's demon was recognized for contributions to entropy concepts, information theory, and the computational sciences (Leff and Rex, 1991). We comment on the relationship of laser based demons with the subjects of entropy and information. Boltzmanns epitaph in Vienna reads, $S=k\ln\Omega$ where S is the entropy and Ω is the number of microstates or complexions corresponding to a thermodynamic state. Brillouin (1962) defines two kinds of information; bound and free. Bound information is the consequence of a measurement which reduces Ω in a true thermodynamic sense, while free information is a measurement which is momentary in nature and cannot be stored. A measurement which implants an atom into the detector, such as in the noble gas detector, is an example of a laser technique that acquires bound information. A laser fluorescence method in which an atom is observed, but not stored, is a nice example of Brillouin's free information.

We estimate the ratio of ΔS to ΔI, where ΔI is the gain in bound information for our RIS based demon. To a good approximation, ΔS (for one laser shot) is the wall plug energy Q over the ambient temperature, T. Assume that a single atom has the same probability of occupying any small volume v in a large volume V. Then, $\Delta I=k\ln(V/v)$, where v is identified as the volume in which the atom is located at the instant of detection. Thus, $\Delta S/\Delta I=nQ/kT\ln(V/v)$ where n is the number of laser shots required to find the atom. Present methods of generating short wavelength radiations are very ineffecient, and if $Q=1$joule, $\Delta S/\Delta I=2.5 \times 10^{20} n/\ln(V/v)$. Now suppose we witness an exceedingly rare fluctuation in which the single atom, known to be in the large volume V before the laser shot, is found to be in the detector volume v during the first laser pulse. How large could the ratio V/v be without (even mometarily) violating the second law? The ratio is $V/v=10^{1000000000000000000000000}$ before the information gained becomes equal to the entropy created. This emphasizes that, while the second law is a statistical statement, it is likely to be irrefutable when lasers are used for observations. We can safely predict that the generalized second law (Rodd 1964), in the form $\Delta(S-I) \geq 0$, will not limit the selectivity or sensitivity of RIS!

References:
Bahcall J N 1989 Neutrino Astrophysics Cambridge U P New York
Brillouin Leon 1962 Science and Information Theory Academic Press New York
Chen C X and Cherry M L 1991 Astrophysical J. 377, L105
Hurst G S 1991 J. Anal. Atom. Spec. 6, 406
Hurst G S et al 1984 Phys. Rev. Lett. 53, 1116
Leff and Rex 1990 Maxwell's Demon: Entropy, Information, Computing Princeton University Press Princeton, NJ
Rodd, P. 1964 AM. J. Phys. 32, 333
Schwarzschild Bertram 1990 Physics Today 43, 17 (October 1990)
Scott R D 1976 Nature (London) 264, 729
Wolfenstein L and Beier E W 1989 Physics Today 42, 28 (July 1989)

Inst. Phys. Conf. Ser. No 128: Section 1
Paper presented at RIS 92, Santa Fe, NM, USA, 24–29 May 1992

RIS measurements with pulsed laser ion sources

J.E. Crawford, for the COMPLIS Collaboration[*]

Foster Radiation Laboratory, McGill University, Montréal H3A 2B2, Canada

ABSTRACT: A system called COMPLIS (Collinear Measurements with a Pulsed Laser Ion Source) has been designed, and is under construction at the new Booster-ISOLDE facility at CERN. In this system, isotopes implanted in a graphite substrate are allowed to decay; their radioactive daughters are laser-desorbed and ionized by RIS beams that probe the desorbed cloud. For studies of hyperfine structure and isotope shift, two modes of operation are possible. In experiments requiring moderate resolution, conventional Resonant Ionization Mass Spectroscopy measurements may be made. For experiments at high resolution, a second set of RIS beams will be used to perform collinear spectroscopy on the fast pulsed beam. The system is well suited to studies of short-lived isotopes of refractory metals in the platinum region.

1. INTRODUCTION

Of the many techniques available for the measurement of atomic hyperfine components, collinear fast-beam laser spectroscopy (CS) is among the most sensitive and versatile. The principle is well-established: ions are produced in an on-line mass-separator source, extracted by a high voltage electrode and manipulated to form a collimated beam. This is then usually neutralized by a charge-exchange cell, and the atomic beam formed is made collinear with a laser beam that excites a transition from either the atomic ground state, or a metastable level populated in the charge exchange. Experimental linewidths, which depend not only on the natural linewidths, but on details of the particular system (e.g., ion source, voltage stability) are generally better than 50 MHz. With fluorescence detection, it has been possible to make measurements on atomic beams with intensities as low as 10^4 atoms/s (Ahmad 1983). The limit is usually photomultiplier background from such sources as dark current, stray light, radioactivity, or metastable states populated in the charge exchange. Techniques that rely on particle (rather than photon) detection are attractive,because of higher detection efficiency and insensitivity to stray light. An elegant example of this kind is a variant of CS that depends on charge exchange and collisional ionization. In one version of the technique used in Sr isotope studies, a transition populates a metastable level; the difference in neutralization probability between this metastable level and the ground state is reflected as a change in ion current in a charge-exchange, and this forms the spectroscopic signal (Silverans 1985). A similar principle has been used for measurements of rare gas isotopes (Neugart 1986). In specific cases measurements have been performed with this technique on beams with intensities as low as 400 atoms/s (Borchers 1989). CS in conjunction with coincident particle detection has been used by a group at Daresbury (Eastham 1986) on even weaker beams, with intensities down to 60 atoms/s. However, such sensitivities are only achievable with beams containing very low isobaric contamination. A review of these techniques has been presented by Kluge (1991).

With its wide applicability and high sensitivity, RIS was an obvious candidate for detection in on-line studies of radionuclides. Since pulsed beams must be used in some steps of the RIS ladder, pulsed sources—synchronized with the RIS pulses—can be used to maximize the beam-target interaction. The technique was tested off-line at McGill (Lee 1987) and implemented in the PILIS system at Orsay (Lee 1988b). Similar techniques have been independently developed by the Mainz group (Krönert 1987). The major success of the method so far has been in experiments on Pt isotopes. This element (and other refractory metals from tantalum to iridium) are not available in sufficient intensities from the ion sources of ISOLs for conventional collinear spectroscopy. In the PILIS method (Le Blanc 1990) an ISOL beam of a selected Au isotope is implanted, accumulated in a graphite substrate and allowed to β decay to the daughter Pt isotope. After an appropriate delay, it is desorbed by a Nd:YAG pulse (λ = 532 nm., τ = 10 ns). RIS beams tuned for selected Pt transitions produce ionization just in front of the desorption region; these ions are accelerated and identified by time-of-flight. The spectroscopic information is derived from the lowest RIS transition ($5d^96s\ ^3D_3 \rightarrow 5d^9\ 6p\ 7_4$, 266 nm). Using a similar technique, with Pt derived from the decay of implanted Hg isotopes, the Mainz group has succeeded in making measurements down to ^{183}Pt, which decays with $T_{1/2}$ = 6.5 m (Hilberath 1992). Fig. 1 shows the variation of <r^2> for Pt and neighbouring elements. This variation, interpreted with information on nuclear moments and nuclear spectroscopic studies suggests a progression of shapes: Pt appears to be oblate for A≥192, likely becomes triaxial for 186≤A<192, and assumes a prolate shape at A=185. Nuclear structure systematics (Wood 1982), and theoretical calculations (refs. in Hilberath 1992) suggest a possible second shape transformation below A=180. Evidence for this transition has also been seen in in-beam studies of high-spin states (Dracoulis 1986).

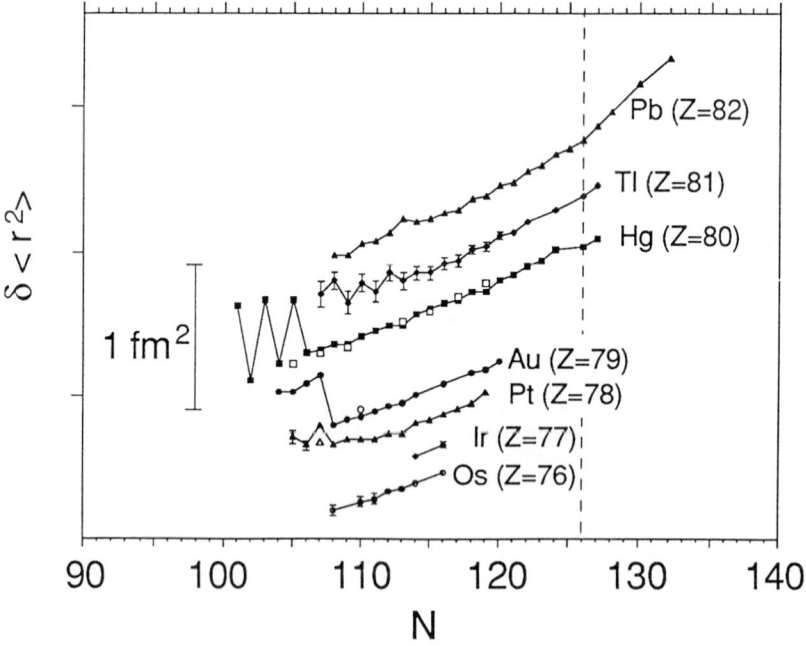

Fig. 1 Summary of δ<r^2> results for isotopes below the nuclear closed shells at Z=82, N=126. Isomeric nuclear states are shown by open markers

These studies have motivated a proposal for COMPLIS, a system of high sensitivity and resolution for investigation of the lighter Pt isotopes, and other elements in the refractory metal group.There is fortunately a very good chance of producing these isotopes at the new ISOLDE facility. Experience with previous ISOLDE beams suggests that Booster-ISOLDE sources should produce intense Hg isotope beams ($>10^7$ atoms/s for isotopes heavier than ^{182}Hg), and $^{A-4}$Pt isotopes may be produced by the ^{A}Hg alpha decay.

RIS detection on a continuous fast beam in collinear geometry has previously been used for studies of radioactive Yb isotopes (Schulz 1991a,b). The detection efficiency (10^{-5} ions per incoming atom) was comparable to that of conventional fluorescence detection, with relatively low background rates from collisional ionization (10^{-8} ions/ atom). This type of experiment suffers from two limitations. The background comes from not only the isotope studied, but from other unwanted isobars in the ISOL beam. Also, the duty cycle is low, because most of the atoms in the continuous beam are not in the interaction region during an RIS pulse. In this work, it was pointed out that substantial gains in sensitivity —perhaps 2 orders of magnitude—could be realized if collinear RIS detection were combined with a *synchronized pulsed* ion source. COMPLIS uses just this principle and contains elements that reduce background from isobaric contamination. If measurements at the highest resolution are not feasible, COMPLIS may be used for RIMS measurements using the same beam arrangement as the PILIS systems.

2. THE COMPLIS SYSTEM

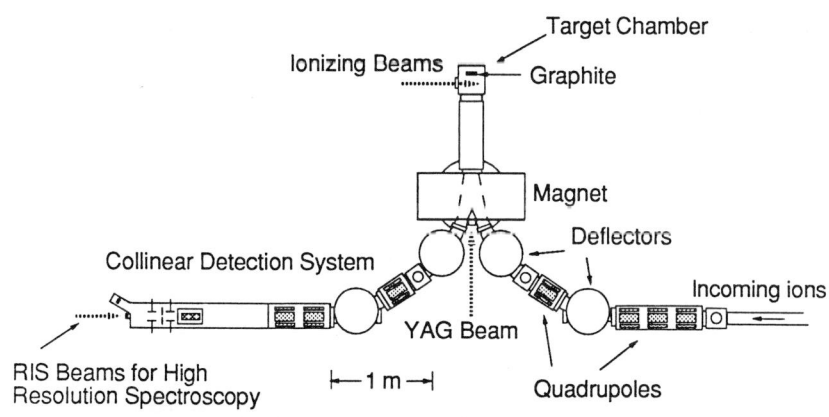

Fig. 2 Components of the COMPLIS system

The components of COMPLIS are shown in Fig. 2. Incoming 60 keV ions from the ISOL source are directed by electrostatic deflectors and by the magnet into the interaction chamber where they are retarded by a potential close to 60 kV; this ensures a shallow implantation in the graphite substrate. After nuclear decay produces a sufficient quantity of the isotope to be studied, those atoms are desorbed by a YAG pulse, ionized by RIS beams, and reaccelerated by the same potential to 60 keV. The magnet serves two purposes: it separates the incoming ISOL beam from the outgoing ions, and ensures that there is little contamination from unwanted masses desorbed and ionized by the YAG. The time of flight measured by a channelplate detector inserted at the end of the exit beam line then identifies the isotopic mass. The target chamber of the new system (Fig. 3) is similar to that of the ISOCELE PILIS systems, but has been redesigned to incorporate a number of beam monitors. A viewing tube permits observation of the YAG spot on a translucent screen that replaces one of the graphite targets; a second target position is left as open aperture so that a Faraday cup can be used to measure beam current passing through; a tape system can be inserted to collect the radioactive ion beam and to transport and monitor the source for α and γ radioactivity. In addition, a secondary electron monitor can be used to check current arriving at the tape. Laser spectroscopic measurements on short-lived isotopes may be done with implantations on a stationary graphite disk, with repeated desorption on the same spot. For longer-lived isotopes, it may be advantageous to rotate the target, implant over a larger area, and desorb with similar target displacement. A computer control system coordinates the various implantation, desorption and counting sequences.

If the system is used for ordinary RIMS measurements as in the PILIS experiments, only one set of RIS beams is used (the ionizing beams in Fig. 2). For experiments at moderate resolution (~1 GHz) , we obtain efficiencies (detected/implanted ions) ~10^{-5}. Of this, the main losses arise from the very small geometric overlap of the RIS beams with the desorbed atoms (~1%) and the low efficiency of the ionization step(<10%).High resolution may be achieved by injection-locking a pulsed dye laser to the frequency of a CW laser. This may be done by adding an excimer-pumped dye cell to the cavity of a tunable CW dye laser (Pinard 1977).

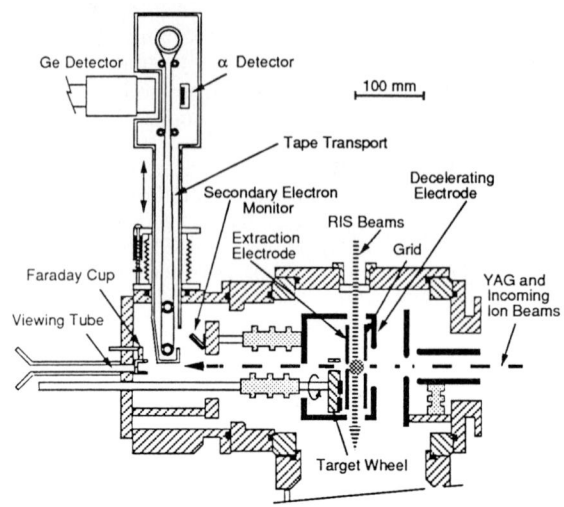

Fig. 3 Design of the COMPLIS desorption target

Although the inherent laser linewidth is Fourier-limited by the pulse width to 100 MHz for the frequency-doubled UV signal, the experimental resolution also depends on the angular divergence of the desorbed atomic beam.With tight collimation of the ionizing beams, it has been possible to obtain experimental linewidths of 170 MHz, but with a corresponding reduction in efficiency to about $10^{-6.}$ The PILIS mode of operation should be well suited for the study of I=0 nuclei in cases where the ISOL yield is greater than 10^5 atoms/s. Fig. 4 shows spectra for a series of even Pt isotopes obtained at ISOCELE, and in preliminary runs at the ISOLDE-3 facility.

3. COLLINEAR SPECTROSCOPY WITH A PULSED BEAM

With collinear detection, it should be possible to improve the system efficiency and achieve high resolution. In this COMPLIS arrangement, the beam geometry in the first ionization region is set to yield the highest possible intensity of the selected element. The ion bunches are neutralized in the charge-exchange cell. Residual ions are removed by an electrostatic deflector, and the atomic beam enters the laser-atom interaction region. The ions produced are then deflected and detected by the channelplate detector. In a COMPLIS-type test system at Orsay, a simple two-colour/three-step RIS excitation was used for tests on natural Pt ($5d^96s \rightarrow 6p \rightarrow 7s \rightarrow$ionization; 306.6, 505.9, 505.9 nm). About 1% of the ions entering the Cd vapour of the charge-exchange cell are neutralized to populate the atomic ground state, and are re-ionized by the pulsed dye laser in the collinear line. This efficiency can be improved by increasing the power available for the final ionization step. We believe that realistic aims for efficiencies of the system are as follows: desorption, 20%; first ionization, 10%, beam transport, 70%; charge exchange and reionization in the collinear section, 10%. The predicted overall efficiency of ~10^{-3} will be achievable only if lasers of sufficient power (~ 100 mJ cm^{-2}) are available for the photoionization step.

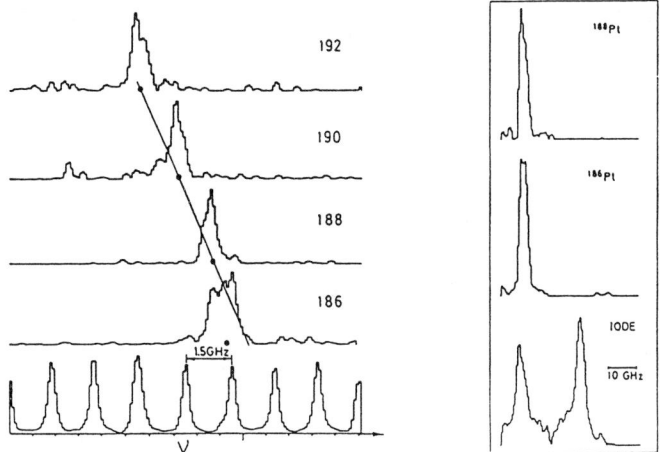

Fig. 4 Resolution obtained with RIS measurements on laser desorbed isotopes of Pt
with the ISOCELE PILIS system (left) and in tests at COMPLIS (right)

Elements of the design of the geometry and pulse timing of the first ionization region are
critical to minimize energy and time spreads in the later collinear section. Conventional ISOL
target-ion sources can produce ion beams with energy spreads $\delta E \sim 5$ eV. If the ions are
accelerated to energy E, the corresponding frequency linewidth that would be observed in a

spectroscopic measurement is $\delta v = \dfrac{v_0}{\sqrt{mc^2}} \cdot \dfrac{\delta E}{\sqrt{2E}}$. (For an isotope of mass 200 amu, and a

transition around 600 nm, this would be approximately 20 MHz.) However, in the laser
desorption source, δE may be much larger.

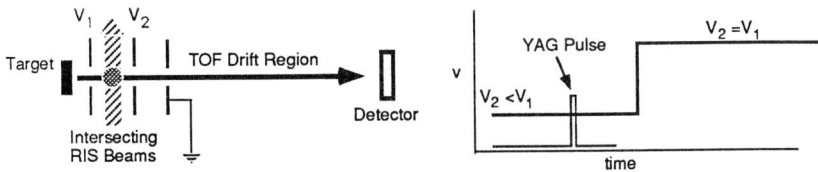

Fig. 5 A pulsed system to reduce energy spread in the desorption source

The simplified TOF system shown in Fig. 5 illustrates the problem. If the accelerating
electrodes of the desorption source are operated with constant DC voltages, the largest
contribution to the energy spread results from the potential gradient across the finite volume
in which the ions are formed by the RIS beams. A simple way to reduce this spread is to
pulse the electrode system so that the voltages V_1 and V_2 are equalized before any of the ions
produced enter the TOF drift region. In this way, each ion is accelerated by a force qE for the
same time, thus receiving the same impulse, and therefore the same final velocity. The
energy spread of a laser desorption source used in this geometry has been tested at McGill
(Ghalambor 1990) using a spherical electrostatic analyzer. With a constant 500V DC applied
across the initial accelerating region, the source energy spread was 35 eV; when the voltages

were equalized with the pulsing system, this value was reduced to about 7 eV, comparable to the spread in a conventional source. Calculations indicate that although the time-of-flight resolution will be poorer with this modification, it should be sufficiently high to identify neighbouring isotopes. This design is being implemented in the new COMPLIS source region.

4. REFERENCES

Ahmad S A *et al* 1983 *Phys. Lett.* **B 133** 47
Borchers W *et al* 1989 *Phys. Lett.* **B 216** 7
Eastham D A *et al*1986 *J. Phys. G: Nucl Phys.* **12** L205
Dracoulis G D *et al*1 1986 *J. Phys* **G12** L97
Ghalambor M 1990 M Sc Thesis (McGill) unpublished
Hilberath Th *et al* 1992 *Z. Phys.* **A342** 1
Kluge H-J 1991 *Resonant Ionization Spectroscopy 90* Inst. of Physics Conf. Ser. 114 ed J E Parks and N Omenetto (Bristol) p.1
Krönert U *et al* 1987 *Appl. Phys.* **A44** 339
Neugart R *et al* 1986 *Nucl. Inst. Meth.* **B17** 354
Schulz Ch *et al*1991a *Resonant Ionization Spectroscopy 90* Inst. of Physics Conf. Ser. 114 ed J E Parks and N Omenetto (Bristol) p.27
Schulz Ch *et al*1991b *J. Phys.* **B24** 4831
Le Blanc F *et al* 1991 *Resonant Ionization Spectroscopy 90* Inst. of Physics Conf. Ser. 114 ed J E Parks and N Omenetto (Bristol) p.439
Lee J KP *et al* 1988a *Proc. of 5th Int. Conf. on Nuclei far from Stability*, 1987, ed. I.S. Towner (A.I.P., New York) p.205
Lee J K P *et al* 1988b *Nucl. Inst. Meth.* **B26** 444
Pinard J and Liberman S 1977 *Opt. Comm* **20** 344
Silverans R F *et al* 1985 *Hyp. Int.* **24** 181
Wood J L 1982 Lasers in Nuclear Physics ed. C E Bemis Jr and H K Carter (Nucl. Sci. Res. Conf. Ser. vol. 3) (New York: Harwood) p 481

*Current members of the COMPLIS Collaboration are: F. Buchinger, J.E. Crawford, J.K.P. Lee (McGill University, Montréal); J. Arianer, F. Ibrahim, P. Kilcher, F. Le Blanc, J. Obert, J. Oms, J.C. Putaux, B. Roussière, J. Sauvage (IPN, Orsay); H.T. Duong, J. Pinard (Lab. Aimé Cotton, Orsay); I. Deloncle, J. Libert, P. Quentin (CSNSM, Orsay); N. Boos, G. Huber, M. Krieg, R. Neugart (Inst. für Physik, Universität Mainz); P. Juncar (Université de Paris); M. Pellerin, J.L. Vialle (Université de Lyon); T. Kühl (GSI Darmstadt).

Inst. Phys. Conf. Ser. No 128: Section 1
Paper presented at RIS 92, Santa Fe, NM, USA, 24–29 May 1992

13

Resonance ionization spectroscopy of ^{242m}Am fission isomers

H. Backe, P. Graffé, D. Habs[1], M. Hies, Ch. Illgner, H. Kunz, W. Lauth, H. Schöpe,
P. Schwamb, W. Theobald, P. Thörle[2], N. Trautmann[2] and R. Zahn

Institut für Physik der Universität Mainz Postfach 3980, D-6500 Mainz, Germany
[1]Max-Planck-Institut für Kernphysik Postfach 103980, D-6900 Heidelberg, Germany
[2]Institut für Kernchemie der Universität Mainz Postfach 3980, D-6500 Mainz, Germany

ABSTRACT: Optical spectroscopy with ^{242m}Am fission isomers is in progress
at the low target production rate of 10/s. The experimental method employed is
based on resonance ionization spectroscopy in a buffer gas cell with detection of the
ionization process by means of the fission decay of the isomers. The resonance ionization has been performed in two steps utilizing an excimer dye laser combination
with a repetition rate of 300 Hz. The first resonant step proceeds through terms
which correspond to wavelengths of 466.28 nm, 468.17 nm or 426.56 nm; the second
non resonant step is achieved with the 351 nm radiation of the excimer laser itself,
running with XeF. The frequency scans of the tuneable dye laser at 466.28 nm and
468.17 nm exhibit broad resonance ionization signals, the latter with a large isotope shift between ^{242m}Am and ^{243}Am which is in accord with the large quadrupole
moment of the ^{242m}Am fission isomer. For the 426.56 nm transition, no resonance
ionization signal was observed. This fact can be understood from our cross-section
measurements on ^{243}Am for the second non resonant step, being (14.1, 18.2 and
3.9)$\cdot 10^{-17}$cm^2 for the three excitation schemes, respectively.

1. INTRODUCTION

The optical hyperfine spectroscopy of fission isomers has been a challenge since Bemis et al.
[1, 2] measured the isomer shift of the optical $^{10}P_{7/2} - ^8S_{7/2}$ ($\lambda = 6405\text{Å}$) transition in ^{240m}Am
($T_{1/2} = 0.9$ ms). The final goal of our optical spectroscopy experiment on ^{242m}Am is the
observation of a resolved hyperfine pattern from which the nuclear spin, the nuclear g-factor and
the intrinsic quadrupole moment can be determined model-independently. Such data should
be of great value to understand nuclear matter in the state of high deformation. However
such experiments are extremely difficult. The production rate of fission isomers is very low,
typically only in the order of a few per second. The half lives of the fission isomers are very
short ($\leq$ 14 ms). Well-established in-beam methods [3] therefore can not be simply adapted
for the optical spectroscopy of fission isomers. We, therefore, developed in our laboratory an
ultrasensitive method which is based on the radioactive decay detected resonance ionization
spectroscopy (RADRIS) in a buffer gas cell [4]. As described in detail in ref. [5] this method is
well suited for the optical spectroscopy of fission isomers.

*) Work supported by the BMFT under contract 06 MZ 188 I

2. THE EXPERIMENTAL SET-UP AND PROCEDURE

The fission isomers were produced through the ^{242}Pu(d,2n)^{242m}Am reaction by using a pulsed (20 ms on, 20 ms off) 12 MeV deuteron beam supplied by the MP tandem Van de Graaff accelerator of the Max-Planck-Institut für Kernphysik in Heidelberg. The recoiling fission isomers leave the target with a primary energy of $\leq$100 keV. Conversion electron transitions and succeeding Auger cascades result in large nonequilibrium ionic charge states, typically between 10^+ and 35^+ [6]. After post-acceleration in an appropriate electrical potential difference of 90 kV the energy of the fission isomers is high enough to penetrate a 50 μg/cm^2 thick entrance window of the optical cell, shown in Fig. 1.

The cell, filled with 30 mbar argon and 0.3 mbar nitrogen, the latter serving as a quenching gas, is loaded with fission isomers in the beam on periods. At a mean beam current of 5 μA at the Pu - target, the fission isomer rate at the entrance of the optical cell amounted to 6/s. The pressure of the inert gas and the energy of the fission isomers are matched in such a way that as many as possible are stopped in the buffer gas. A certain fraction (about 13%) of the recoiling ions neutralize during the slowing down process [7]. The remaining ions are sucked into an appropriate electrical field onto an electrode and are in that way excluded from detection. The gas acts at the same time as a storage medium for the fission isomers. The diffusion time to the cell walls, being 1 cm apart, is in the order of 30 ms. Resonance ionization is performed in the beam off periods. The atoms are irradiated by two laser beams in order to accomplish resonance ionization via the atomic excitation schemes shown in Fig. 2.

To produce the laser beams we have been using an excimer laser EMG 104 MSC from Lambda Physik, lasing with XeF at a wavelength of 351 nm, in combination with one dye laser FL 2001. The excimer laser supplies every 40 ms, in the deuteron beam off periods, a burst of four pulses with an interval of 3 ms. The pulse energy of the excimer laser amounts to 35 mJ, the pulse width to 15 ns. An energy fraction of 30% has been used to pump the dye laser, the remaining energy is guided by dielectric mirrors into the optical cell. At the position of the experiment we measured a pulse energy per unit area of typically 15 mJ/cm^2.

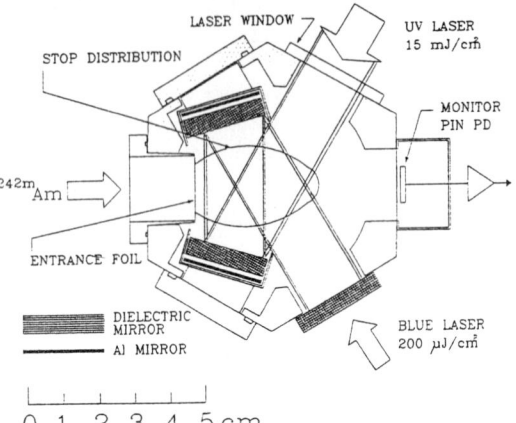

Figure 1: Optical cell. The irradiation of the stopping region in the optical cell is improved by installation of laser mirrors. Behind the dielectric laser mirrors for reflection of the UV an aluminium coated quartz plate is placed for reflection of the blue laser light. The UV laser radiation leaves the optical cell after five reflections, the blue dye-laser after two reflections.

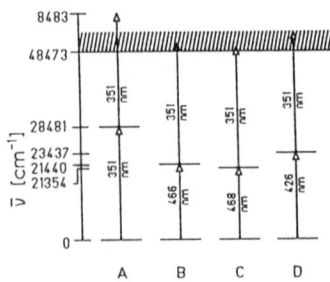

Figure 2: Excitation schemes for resonance ionization spectroscopy for ^{242m}Am fission isomers. At the excitation schemes B,C and D the resonance has been searched for by scanning the dye laser beam around the levels at 21440.350, 21353.863 and 23437.108 cm^{-1}. A constant background level results from the ladder A representing the quasi - resonant two step excitation with the 351 nm excimer beam alone through an intermediate level at 28480.856 cm^{-1}.

The dye laser supplies a beam with the same temporal structure as the pump laser. The dye laser output has been guided by a 10 m long and 600 μm thick quartz fibre to the chamber. The spectral width amounts to 7 GHz, and the pulse energy per unit area to 0.2 mJ/cm^2 at the position of the optical cell. The wave length has been calibrated with an accuracy of approximately 0.5 GHz from the absorption spectrum of tellurium utilizing the wave length tables of ref. [8].

In the experiment the excimer laser beam is delayed by about 15 ns with respect to the dye laser beam avoiding by this means a time overlap between the laser pulses. This delay was chosen to prevent depletion of the ground state population by the excimer laser alone. The resonantly ionized fission isomers are transported by an ion electrode system to a semiconductor detector, where they are detected by the fission. To separate true fission events from electrical breakdown pulses, the signal of the fission detector is recorded by an storage oscilloscope.

3. MEASUREMENTS AND RESULTS

First we measured the background without laser radiation. No fission events were observed with a beam charge of 94 mC, corresponding to a count rate of (0.000 ± 0.01) events per mC. This high background rejection is produced by the 90 ^{0} deflection geometry of the collection region, by which no direct path occurs from the stopping region to the fission fragment detector. With irradiation from the excimer laser, according to the excitation scheme A in Fig. 2, 33 RI events in 221 mC collected beam charge were observed, corresponding to $N_{RI}(\text{excimer}) = (0.15 \pm 0.03)/$ mC.

In Fig. 3 is shown the time distribution of the fission events. Through the transport time delay, the count rate increases 6 ms after the first laser pulse. In the range 20 to 40 ms and 0 to 10 ms one recognizes the exponential decay of the fission isomers. In Fig. 4 the energy distribution of fission fragments in the semiconductor detector is compared to a calibration spectrum, showing real fission fragments were observed.

In a further experiment the fission isomers were resonantly ionized according to the excitation scheme B of Fig. 2. The measured resonance is shown in the Fig. 5. The level at 21440.35 cm^{-1} was selected, because of the small isotope shift of $\delta\nu = 13.3$ mK the search region for the resonance is very small. The width of the resonance is 25 - 40 GHz. A hyperfine splitting could not be observed.

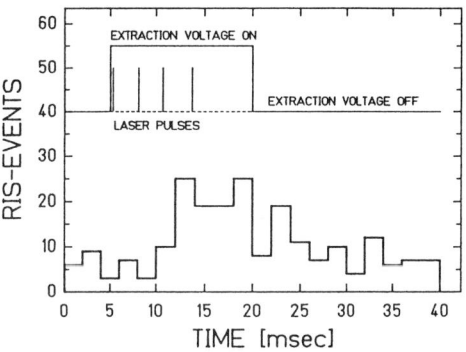

Figure 3: Time development of the fission fragment signal in the semiconductor detector. After a delay time of 6 ms following the transport time a sharp rise in the event rate is observed.

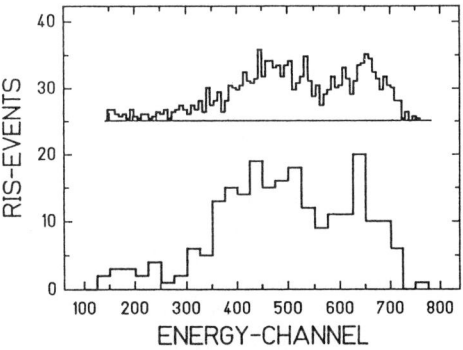

Figure 4: Energy distribution spectra of fission fragments in the semiconductor detector. The bottom part shows the energy distribution of fission fragments from resonantly ionized fission isomers, the top part the energy calibration spectrum taken with fission isomers which come to rest in the buffer gas as positively charged ions. The double humped energy distribution can clearly be recognized.

It is not clear whether the width of the structure is caused by saturation broadening or the superposition of many hyperfine components in combination with saturation broadening. The measurements according to excitation scheme C via an americium level at 21353.863 cm^{-1} (468.17 nm) was carried out to determine the isomer shift caused by the quadrupole deformation of the ^{242m}Am fission isomers. The isotope shift for this state was measured to be 67(5.5) mK [9], so that a large isotope shift due to the large quadrupole moment of the fission isomer could be expected. As can be seen from Fig. 5, a red shifted resonance has been found which is a factor X = 18.7(2.5) larger than the isotope shift between ^{243}Am and ^{241}Am. This factor is somewhat smaller than X = 26.8(2.0) measured by Bemis et al. for the 640.5 nm transition at ^{240m}Am which results in a quadrupole moment Q = 29.0(1.3) eb. Since the magnetic hyperfine splitting of the level at 21353.863 cm^{-1} (468.17 nm) may be rather large, not all hyperfine components may have been observed, resulting in a shift of the observed resonance with respect to the center of gravity. This would affect the accuracy of our isotope shift measurement.

Therefore, we have performed in addition a scan according to excitation scheme D in Fig. 2. The level at 23437cm^{-1} has a very small A - factor of only -2.2 mK (^{241}Am) but about a factor of two larger isotope shift with opposite sign $\delta\nu_{241-243}$ = -128 mK as the 468 nm transition. The result of an extensive search of a resonance is shown in Fig. 5. No statistically significant resonance was observed.

In order to understand the strengths of the observed resonances and the absence of the signal for the excitation scheme D we performed cross-section measurements for the excitation into the continuum. For that purpose a buffer gas cell has been built in which the resonance ionization process can be studied at the α-active isotope ^{243}Am (T$_{1/2}$=7370 a). A sample of $5\cdot10^{14}$ AmO$_2$ molecules has been prepared by electro-deposition on a 10 μm rhenium foil.

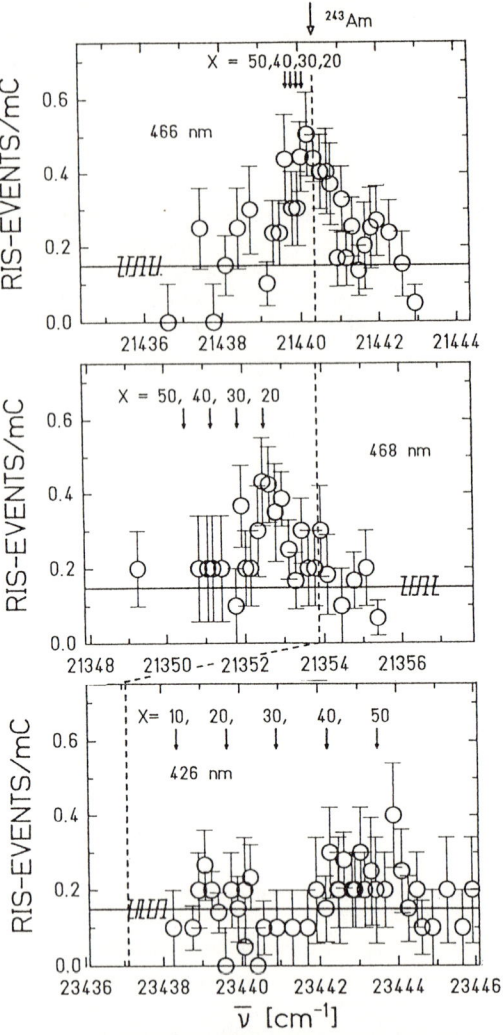

Figure 5: Resonance ionization signals for the ^{242m}Am fission isomers for the three excitation schemes B,C and D shown in Fig. 2. The number of RI events normalized to the deuteron beam charge is shown versus the frequency detuning of the first excitation step. The dashed vertical line corresponds to the position of the ^{243}Am resonance. The following numbers correspond to the spectra (N$_{RI}$ = RI events, T = data acquisition time, collected d-beam charge).
466 nm : N$_{RI}$ = 271, T = 56 h, Q = 1 C;
468 nm : N$_{RI}$ = 169, T = 35 h, Q = 0.63 C;
426 nm : N$_{RI}$ = 136, T = 43 h, Q = 0.77 C.

The filament has been electrically heated. At a temperature of 1300 °C Am atoms evaporate into the argon buffer-gas (30 mbar). The resonance ionization is performed with the same laser system as described in section 2. The resulting ions are transported with an electric field to a platinum wire. The charge is integrated with a charge sensitive pre-amplifier. About $1.5 \cdot 10^3$ ions are needed to obtain a peak to noise ratio of 1:1 in a single laser shot. The measured RI signals are shown in Fig. 6 as a function of the excimer laser photon flux per pulse Φ. The saturation cross-sections σ_{II} were deduced by a best fit procedure with a function proportional to $(1 - e^{-\frac{\sigma_{II} \cdot \Phi}{2}})$. The derived cross-sections for the different excitation schemes are listed in table 1.

To understand the resonance ionization signal measurements on the ^{242m}Am fission isomer described above we assume first of all that in all excitation schemes the first excitation step is saturated for all hyperfine components. This is a reasonable assumption for the excitation schemes B,C and D from the known photon intensities and atomic lifetimes [10]. It is not obvious for the ladder A since the strong emission lines of the excimer laser do not coincide exactly with the americium level at 23437.108 cm^{-1}. However, a very weak satellite line with a relative intensity of only about 10^{-4} is sufficient to saturate the transition. Secondly, we assume for ladder D that a signal of $2 \cdot 10^{-4}/\mu$C would not have been seen with statistical significance, c.f. Fig. 5. The count rate per collected beam charge N_{RIS} is given by

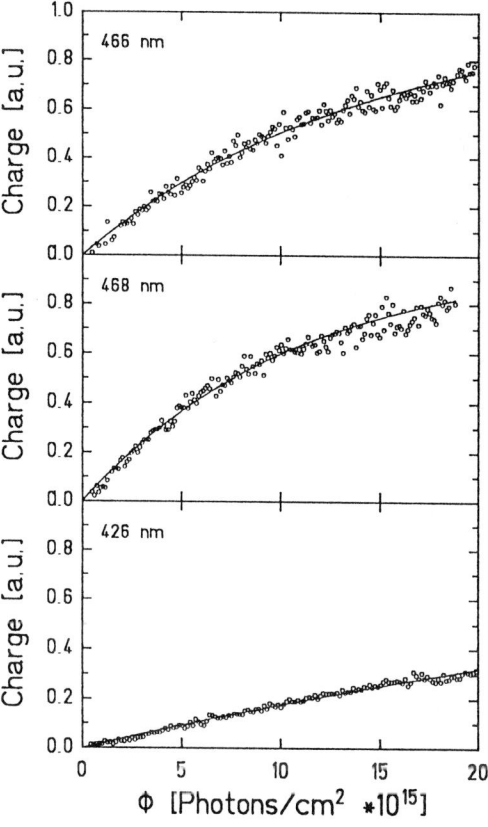

Figure 6: Cross-section measurements for the excitation schemes B, C, and D in Fig. 2. The background signal resulting from the 351 nm radiation of the excimer laser allone has been subtracted.

$$N_{RIS} = \frac{1}{2}(1 - e^{-\frac{\sigma_{II} \cdot \Phi}{2}}) \cdot \epsilon_{ovlp} \cdot n \cdot \frac{\epsilon_{neut}}{1 - \epsilon_{neut}} \cdot [\epsilon_{trans} \cdot \epsilon_{det} \cdot N^+]. \qquad (1)$$

Here $n \approx 1.2$ is the effective number of laser shots in the burst, $\epsilon_{neut} \approx 0.13$ is the fraction of americium fission isomers which have become neutral after the slowing down process, $\epsilon_{trans} \approx 0.35$ is the transport efficiency of fission isomers from the interaction domain to the detector, $\epsilon_{det} \approx 1.0$ is the detection efficiency of the photodiode for a fission event, and N^+ is the number of fission isomers per collected beam charge coming to rest as positively charged ions. The quantity in square brackets in equation (1) has been determined experimentally to be 45/mC. The experimental results can be reproduced with a pulse photon flux $\Phi \approx 2 \cdot 10^{16}/\text{cm}^2$ and an overlap efficiency between laser beams and fission isomer range distribution $\epsilon_{ovlp} \approx 0.15$. Both numbers are in reasonable accord with our experimental measurements.

Table 1: Saturation cross section for the second excitation step into the continuum with a wavelength of 351 nm.

First step [cm^{-1}]	23437.108	21440.35	21353.863	20031.427	19993.836
$\sigma_s[10^{-17}$ cm$^2]$	3.9(1.4)	14.1(4.0)	18.2(4.8)	13.7(4.1)	19.8(5.9)

The non resonant background induced by the excimer laser beam alone can be avoided if the Am atoms are ionized via autoionizing states. For that purpose we have searched for autoionizing states at ^{243}Am. The resonance ionization signal, scanning the wavelength of the second step, around the ionization limit and using the relay state at 23437.108 cm^{-1} is shown in Fig. 7. The states below the ionization limit are ionized due to collisions with the buffergas. Also for the relay states of 23307.526 cm^{-1}, 21440.35 cm^{-1}, and 21353.836 cm^{-1} many autoionizing states have been observed. With this information measurements at the fission isomer ^{242m}Am are planed with, hopefully, an improved peak to background ratio.

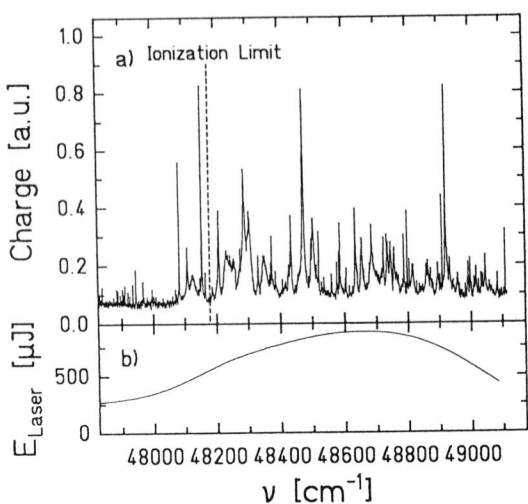

Figure 7: a) Scan of the second laser step around the ionization limit indicated by the dashed line [11]. The laser beam has a beam diameter of 8 mm and a spectral width of 7 GHz. Due to the large scanning range of 1200 cm^{-1}, the dye-laser pulse energy varied as shown in part b). The first step was kept on resonance at 23437.108 cm^{-1} with a pulse energy of 30 μJ.

References

[1] C.E. Bemis et al., Phys. Rev. Lett. 43, (1979) 1854.

[2] M.W.Johnson et al., Phys. Lett. B 161, (1985) 75.

[3] E. W. Otten, Treatise on Heavy-Ion Science, Vol.8, p.517, ed. D. A. Bromley, Plenum Publishing Corporation, 1989.

[4] W. Lauth et al., Phys. Rev. Lett. 68, (1992) 1675 .

[5] H. Backe et al., in press in Hyperfine Interactions.

[6] V. Metag et al., Phys. Rep. 65, (1992) 1.

[7] H. Backe et al., to be published in Nucl. Inst. Meth. B.

[8] J. Cariou et al., Atlas du Spectre d'Absorption de la Molecule Tellure Lab. Aime Cotton, Orsay, France, 1980.

[9] J.G. Conway, private communication, 1984,1991.

[10] Th. Arndt et al., Phys. Rev. A 38, (1988) 5084.

[11] F. Ames, private communication, 1992.

Inst. Phys. Conf. Ser. No 128: Section 1
Paper presented at RIS 92, Santa Fe, NM, USA, 24–29 May 1992

19

An ultrasensitive resonance ionization mass spectrometer for xenon

J D Gilmour, I C Lyon, I K Perera, S M Hewett, W A Johnston and G Turner
Geology Department, Manchester University, Manchester, UK

Abstract The sensitivity of our time-of-flight resonance ionization mass spectrometer has been increased by a factor of 100 by the addition of a cryogenic sample concentrator Evaluation of two-photon resonance schemes using wavelengths of 249.62nm, 252.50nm and 224.30nm shows the former to be the most sensitive.

1. Introduction

In a previous paper (Gilmour et al 1991) we described the development of a resonance ionization, time-of-flight mass spectrometer for the analysis of meteoritic xenon. Here we give a brief account of some more recent work including an investigation of different two-photon resonance ionization schemes and the development of a sample concentrator similar to that of Hurst (1984) that has increased the instrument's efficiency by a factor of approximately 100.

The resonance ionization scheme routinely employed for the two-photon excitation, one-photon ionization of xenon requires generation of light at 249.62nm. This is currently achieved using the doubled output of a Nd:YAG laser to pump a dye laser producing light at 652.265nm. This is then doubled and mixed with residual 1064nm light from the Nd:YAG laser. The resulting pulses have a duration of 6ns and energy around 1.5mJ.

The instrument consists of a two stage Wiley-McLaren ion source (Wiley and McLaren 1955) modified by the inclusion of a curved back plate (Werner 1974). Ions are focused by an accelerating einzel lens down a 65cm flight tube onto a pair of chevron mounted microchannel plates operated in analogue mode. The output signal is amplified and recorded on a LeCroy 9400 digital oscilloscope using both channels and a 5ns delay line to produce an effective sampling interval of 5ns. Averages of the signal generated by 200 successive laser shots are made and transferred to the computer where peaks are fitted to extract abundances. More detailed descriptions of the basic time-of-flight machine have been given elsewhere (Gilmour 1990, Gilmour et al 1991).

2. Two-Photon Resonance Schemes

To qualify as the upper state of the two-photon resonance from the even parity, $J=0$ ground state a level must be of even parity and have $J=0$ or 2. 5 such states are available: $^2P_{3/2}6p[5/2]_2$ (256.02nm), $^2P_{3/2}6p[3/2]_2$ (252.50nm), $^2P_{3/2}6p[1/2]_0$ (249.62nm), $^2P_{1/2}6p[3/2]_2$ (224.30nm) and $^2P_{1/2}6p[1/2]_0$ (222.58nm) where the jl coupling scheme has been used We routinely use the 249.62nm transition but in the course of our development we investigated the efficiency of resonance ionization processes based on two of the others, namely those at 252.50nm and 224.30nm.

The dye laser wavelength necessary to produce output light at 249.62nm from the laser system described above is 652.265nm. The dye used (DCM in methanol) allows a tuning range between 607 and 672nm. This allows us to reach the dye laser wavelength of 662.132nm necessary for generation of 252.50nm and the resulting doubled (331.066nm) and mixed wavelengths also lie conveniently within the range of our (KD*P) doubling and mixing crystals. Comparison of signal heights produced from identical samples with equal ionization laser pulse energies suggests that resonance ionization via the lower energy state is only 20% as effective as that using 249.62nm light. This result is consistent with that of about 25% previously reported (Schneider 1987) given the uncertainty in our pulse energy measurement (10%). Furthermore, the peak achievable pulse energy at 252.50nm was only 30% that at 249.62nm. Clearly using this wavelength would not improve the spectrometer's efficiency.

To attempt to reach the two higher energy candidates it was necessary to change the dye to a solution of rhodamine-590 in methanol since the necessary dye laser wavelengths (568.430nm 562.918nm) lie outside the tuning range of DCM. For similar reasons the mixing and doubling crystals were changed to KDP. It still proved impossible to generate pulses at 222.58nm but 1mJ pulses were achieved at 224.63nm. Comparison of signal heights with those routinely achieved at 249.6nm suggests that this process is only about one tenth as efficient. Although the higher energy $J=0$ state might be expected to be more efficient it seems likely that the routinely used resonant process is the most suitable candidate for use in a mass spectrometer.

3 The Sample Concentrator

A useful measure of the sensitivity of a static mass spectrometer is the lifetime against detection (Turner 1987) defined in $t_{det}=N/(dN_+/dt)$ where N is the number of sample atoms in the spectrometer and dN_+/dt is the number of sample atoms detected per unit time at the collector. In the basic time-of-flight instrument previously described the lifetime against detection of a xenon atom was measured by pulse counting at mass 128 for a sample of 10^9 xenon atoms to be close to 50 hours. In order to improve on this it was decided to construct a cryogenic sample concentrator similar to that previously described by Hurst et al (1984).

Such a device consists of a localised cold spot created in the back plate of the ion source of the mass spectrometer onto which the xenon sample condenses. 1 microsecond before the ionizing laser pulse another laser is fired to heat the cold spot. The evaporated sample is concentrated in the ionizing region of the spectrometer during the ionizing laser pulse, thus enhancing the sensitivity by two to three orders of magnitude. Details of the construction of our sample concentrator will be given in a forthcoming publication (Gilmour et al, in prep).

3.2 Sensitivity

Factors affecting the improvement in sensitivity of a time-of-flight mass spectrometer when operated with a sample concentrator include the interval between the heating and ionizing laser pulses, the diameter of the heating laser beam and the distance of the ionizing region above the condenser's surface. However, when tuned to peak performance a lifetime against detection of 20 minutes has been measured by pulse counting at mass 124 from a sample of 10^7 total xenon atoms. This corresponds to a count rate of 1 cps from a sample of 10^3 atoms.

The sensitivity can be broken down into a combination of two factors: the rate at which sample atoms condense on the active area of the sample concentrator and the efficiency at which those that are condensed are ionized after their subsequent release. The condensation rate can be characterised by the return time which is defined by t_{ret} in

$$dN_C/dt=(N-N_C)/t_{ret}=(N-N_C)Av/4V \qquad (1)$$

where after time t, N_C of the sample atoms have condensed. The second equality follows from the rate at which $(N-N_C)$ atoms of mean speed v in a volume V hit area A (Loeb 1961). It follows that

$$\ln((N-N_C)/N)=-t/t_{ret} \qquad (2)$$

which allows the return time to be experimentally deduced.

The heating laser shutter was closed for intervals of between one and sixty seconds and the magnitude of the signal measured from the first subsequent heating of the sample condenser recorded. A graph of $\ln((N-N_C)/N)$ against time is linear with a gradient corresponding to a return time of 10 seconds.

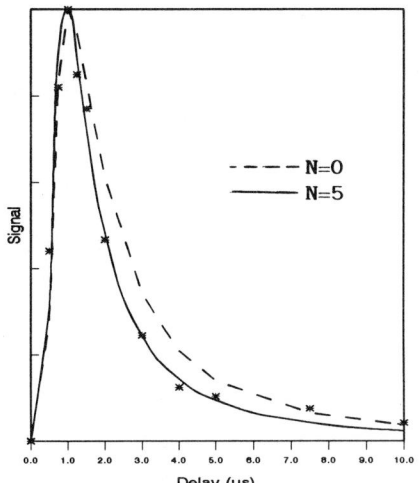

Figure 1: Signal vs Delay

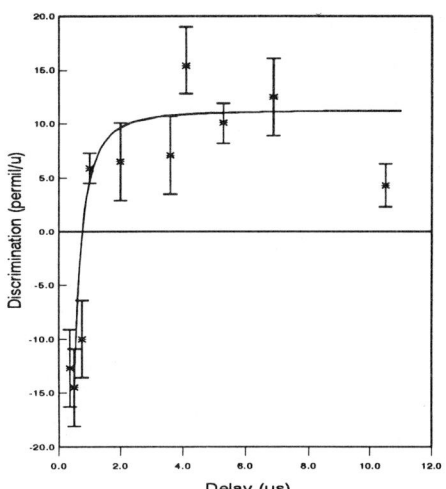

Figure 2: Discrimination vs Delay

Since the volume of our spectrometer is approximately 400cc and the mean speed of xenon atoms at room temperature is 220ms⁻¹, the return time indicates that the diameter of our cold spot is of the order of 1mm. This is consistent with the observed minimum diameter of the heating laser spot that prevents the sample building up outside the scope of the heating laser.

3.3 Velocity Spectrum

Figure 1 shows a graph of the signal recorded at mass 132 as a function of the delay between the firing of the two lasers. The distance of the ionizing region from the cold spot surface was approximately 0.2mm. and modelling of the data using a Maxwellian speed distribution modified by a $\cos^N(\Theta)$ angular distribution suggests that sample atoms leave the surface with a characteristic temperature of 350K and trajectories close to the normal.

Theoretically (Hurst et al 1984) the maximum temperature a stainless steel surface of area A cm² reaches when hit by a laser pulse of energy EmJ and duration t_h seconds is

$$T = 29.4[(E/A)^2/t_h]^{1/6} \qquad (3)$$

(where the constants in the original reference have been evaluated and combined). The velocity spectrum data were acquired using a heating laser with a beam diameter of about 1mm and intensity 25mJ, of which approximately 3mJ is estimated to have been absorbed by the surface. Thus the maximum temperature reached by the surface is in the region of 350K, which is in close agreement with the characteristic temperature of the released atoms velocity spectrum. This suggests that atoms remain in thermal equilibrium with the spots surface throughout the laser pulse.

3.4 Discrimination

The sample concentrator can be expected to give rise to a new source of discrimination in the spectrometer. This occurs because the ionising laser is fired a fixed interval after the heating laser and thus selects the detected ions for velocity. If a Maxwell speed distribution is assumed it can be shown that for atoms of isotopes with masses around m the discrimination caused by ionizing only those with velocity v is 3/2m -

$1.66 \times 10^{-27} v^2 / 2kT$ permil/u where the factor 1.66×10^{-27} arises in the conversion from kg to u.

Figure 2 shows a graph of measured fractionation against the delay between the heating and ionizing laser pulses. The solid line corresponds to the theoretical prediction with T=350K and velocities determined using the measured distance between the ionising laser's focus and the spot surface of 0.2mm and it can be seen that agreement is good.

3.5 Interferences

The wavelength necessary to achieve the two-photon excitation of xenon is in the UV and is thus capable of causing non-resonant ionization of hydrocarbons. This effect can be seen when the unfocused beam is passed through the source region but is significantly exacerbated by the focusing necessary to saturate the two-photon resonant transition.

Non-resonant ionization of hydrocarbons adversely affects the performance of the instrument in two ways. Firstly, the major peaks at masses 12, 24 and 36 arrive at the microchannel plates before the xenon peaks and discharge channels that have insufficient time to recharge before the arrival of the xenon ions. They are thus responsible for lowering the sensitivity of the instrument. Secondly, there is a minor group of peaks in the xenon region which produce interferences that the mass resolution of 200 is insufficient to resolve from neighbouring xenon peaks.

Although the effects of non-resonant ionization can be compensated for by recording spectra with the laser tuned away from resonance (usually to 249.903nm), the dramatic increase in sensitivity caused by the addition of the sample concentrator might be expected to accentuate the problem. However, as well as reducing the detection limit for xenon the sample concentrator decreases the size of the non-resonant contribution.

This reduction is due to two factors. Firstly, although the unilluminated portions of the sample concentrator are insufficiently cold to condense xenon, they are still capable of cryopumping hydrocarbons out of the spectrometer. Secondly, those hydrocarbon molecules that do condense on the active part of the sample concentrator leave the surface at the same temperature as the xenon atoms, but since they are of a different mass, they tend to arrive in the ionization region at a different time and not be ionized. Investigation of the velocity spectra of the hydrocarbons as they leave the surface of the cold spot suggests that the interferences at masses 12, 24 and 36 have a single parent with a mass around 40u.

4. Conclusion

Of the two-photon resonance, one-photon ionization schemes tested, that which uses a resonant wavelength of 249.62nm is the most efficient. To further improve ionization efficiency would seem to require development of a four-wave mixing technique.

Installation of a cryogenic sample concentrator has increased the sensitivity of our time-of-flight, resonance ionization mass spectrometer by a factor of 100; a xenon atom in the instrument now has a lifetime against detection of 20 minutes, which is significantly superior to those in coventional electron impact mass spectrometers. Furthermore, our instrument continually collects ions from all 9 xenon isotopes and so has no requirement for peak switching.

5. References

Gilmour J D 1990 *PhD Thesis* University of Sheffield
Gilmour J D, Hewett S M, Lyon I C, Stringer M and Turner G 1991 *Meas. Sci Technol.* 2,589-595.
Hurst G S, Payne M G, Phillips R C, Dabbs J W T and Lehmann B E 1984 *J App Phys* 55 1278-1284
Loeb L B 1961 *The Kinetic theory of gases*
Schneider K 1986 *Proc. 3rd Int. Conf. on Resonance Ionization Spectroscopy* 67-73
Wiley W C and McLaren I H 1955 *Rev Sci Inst* 26 1150-1157
Werner H W 1974 *Int. J. Mass. Spec. Ion. Proc.* 14 189-204

Inst. Phys. Conf. Ser. No 128: Section 1
Paper presented at RIS 92, Santa Fe, NM, USA, 24–29 May 1992

Laser-induced radiative electron–ion recombination in a storage ring

R Neumann*, M Grieser, D Habs, U Schramm, T Schüßler, D Schwalm and A Wolf

Physikalisches Insitut der Universität Heidelberg and
Max-Planck-Institut für Kernphysik, Heidelberg, Germany

* Gesellschaft für Schwerionenforschung (GSI), Darmstadt, Germany

ABSTRACT: Laser-stimulated radiative recombination in a storage ring was performed with merged beams of protons and electrons, continuing previous work with improved statistics, and was applied also to C^{6+}, thus extending the technique for the first time to a heavy ion. Both gain curves reveal a shoulder on the slope towards lower laser photon energy. For C^{6+}, a finite level above the spontaneous recombination background far into the region below the field-free ionization threshold is not excluded.

1. INTRODUCTION

The interaction between positive ions and free electrons can result in spontaneous radiative recombination, representing the reverse process of photoionization. The sum of the kinetic energy and the binding energy of a captured electron is emitted via a photon. The idea of enhancing spontaneous recombination by the presence of an intense resonant radiation field was proposed by Rivlin (1979) for positronium formation and by Neumann *et al* (1983) for the production of antihydrogen. The latter proposal already involved a storage ring configuration, including parallel overlapped beams of antiprotons and positrons with equal average velocity. A similar situation exists in the electron cooling section of an ion storage ring, providing ideal conditions for merging both particle beams parallel or antiparallel with a powerful laser beam.

The laser-induced rate of charge-changed ions as a function of laser photon energy provides access to various aims, e.g. the measurement of the electron velocity distribution, the study of external influences such as electric fields on the recombination process, and the selective population of certain states. Laser-induced recombination was first verified experimentally by Schramm *et al* (1991) at the Heidelberg Test Storage Ring TSR (Krämer *et al* 1990) and by Jousif *et al* (1991) in a single-pass experiment. The present paper reports on new results of proton-electron and C^{6+}-electron recombination at the TSR.

2. EXPERIMENTAL PROCEDURE

Similar to our first experiment (Schramm *et al* 1991) a 21 MeV proton beam (~ $6 \cdot 10^9$ stored protons) was merged with the intense electron beam ($4.7 \cdot 10^7$ electrons per cm^3) of the TSR. Light pulses (width ~ 20 nsec, power ~ 2 MW) from a dye laser, pumped by an excimer laser, were shot in antiparallel to the protons and electrons, and the hydrogen atoms formed by laser-stimulated recombination were detected in a microchannel plate behind the first bending magnet in coincidence with the laser pulses. Since the hydrogen beam leaves the ring on a straight trajectory, it was used to extrapolate the proton beam position inside the electron cooler by blocking the atoms with a scraper immediately behind the cooler section, and by measuring their impact positions on the channel-plate detector.

For C^{6+} recombination, the TSR was first operated with deuterons, again defining the ion beam position via the emerging neutral atomic beam. Switching from deuterons to the storage of $^{12}C^{6+}$, which has almost the same mass-to-charge ratio, ensured an identical beam position without changing the ring parameters except the ion velocity.

3. RESULTS AND DISCUSSION

The experimental signature of laser-induced recombination into an electronic substate of principal and orbital quantum numbers n and l is given by the ratio G of the laser-induced to the total spontaneous recombination rate within the time interval of laser interaction. This ratio, representing a gain, is described by

$$G_{nl}^0(\varepsilon) = G_{nl}^0(0) \frac{\sqrt{\pi}}{2} \left(\frac{k_B T_{\parallel}}{\varepsilon} \right)^{1/2} \exp\left(-\frac{\varepsilon}{k_B T_{\perp}} \right) \mathrm{erf}\left[\left(\frac{\varepsilon}{k_B T_{\parallel}} \right)^{1/2} \right].$$
(1)

Here, ε is the kinetic energy of the electron prior to recombination – eq. (1) being defined only for $\varepsilon > 0 -$, and $T_{\perp}$ and $T_{\parallel}$ are the transverse and longitudinal temperatures of the cooling electron beam. With regard to the laser-stimulated free-bound transition the kinetic energy can also be expressed by $\varepsilon = E_{\gamma} - E_0$, where E_{γ} is the photon energy and E_o is the field-free binding energy of the final atomic state. For electrons of zero kinetic energy, the gain reaches its maximum value, given by

$$G_{nl}^0(0) = \frac{Ic^2}{8\pi v^3} \frac{2}{\pi k_B (T_{\perp} T_{\parallel})^{1/2}} \frac{g_0(nl)}{n} \left(\sum_{n'=1}^{n_{cr}} \frac{g_0(n')}{n'} \right)^{-1},$$
(2)

where v and I are the laser frequency and intensity in the ion rest frame. The Gaunt factors $g_0(n,l)$ and $g_0(n')$, representing the continuum-bound state transition oscillator strengths for the given (sub)shells at $\varepsilon = 0$, were tabulated by Omidvar and Guimaraes (1990). The inverse sum in eq. (2) accounts for the fact that only ions in states below a critical n_{cr} contribute to G. Ions in states above n_{cr} are assumed to be reionized by motional electric fields, when passing the magnetic fields of various storage ring components downstream of the cooler, and thus should not reach the detector. For a more detailed discussion of G see Neumann *et al* (1983), Schramm *et al* (1991), and in particular Wolf (1992).

The measured gain curves of proton-electron and C^{6+}-electron recombination as a function of the relative energy ε are displayed in Fig. 1a and b. The data points were taken with numerous sweeps of the laser frequency, to average out possible laser intensity drifts. The two curves have essentially the same shape for $\varepsilon > 0$, and exhibit similar features in the region below the field-free ionization limit. For protons, the gain is dominantly due to recombination into the 2p sublevel of hydrogen, which decays rapidly to the ground state. Recombination into the 2s sublevel is almost negligible, since in this long-lived state the atoms are reionized by the laser field via photon absorption. In the case of C^{6+} the n=14 state of C^{5+} was populated. In contrast to the proton experiment a large number of fine structure substates are involved in the recombination process, requiring the summation of eq.(1) over all l values. The question to which extent the different sublevels contribute to the total gain was discussed in detail by Wolf (1992), with special regard to laser saturation. Owing to the different values of v, n, and the applied laser intensity I, the measurements yielded a much larger gain for protons than for C^{6+}, in accordance with eq.(2).

In the common rest frame of protons and electrons, the lower limit of the electron energy distribution coincides with the ionization limit of the hydrogen atom. Therefore, the gain curve should reach its maximum at the threshold ($\varepsilon = 0$), and immediately fall steeply to zero for $\varepsilon <$ 0, as described by eqs. (1) and (2). However, in contrast to this expectation, a finite gain occurs for negative values of ε. This behaviour can be interpreted in terms of the Stark effect. Due to space charge, the electron beam possesses an internal radial electric field, which is zero exactly in the centre and increases linearly to the outer beam edge. An electric field at the site of a proton deforms its potential and leads to the formation of a saddle point which, for an estimated field strength of ~5 V/cm, lies at a distance of $q^{1/2} \cdot 1.7$ μm and an energy $E_{sp} \approx -q^{1/2}$ ·1.7 meV, q being the nuclear charge. The main sources of the given average electric field are the variation of the space charge of the electron beam over the proton beam cross section and the motional electric field due to angular fluctuations of the solenoidal magnetic field which guides the electron beam. Free electrons moving in the deformed ion potential at initial energies below the field-free ionization theshold can recombine by radiative transitions into low-lying states, and give rise to a sub-threshold contribution to the laser-induced recombination spectrum.

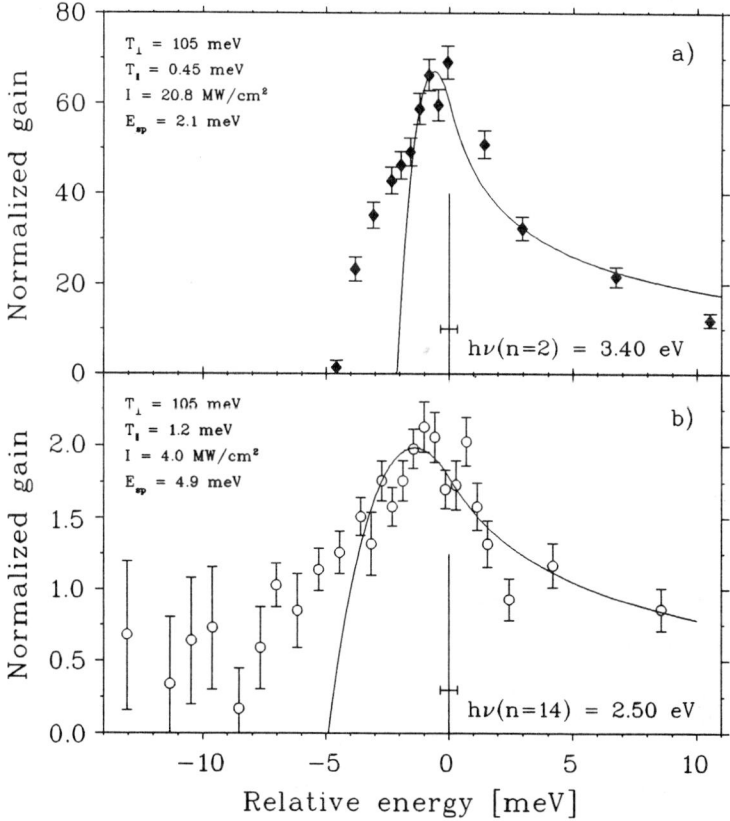

Fig. 1. Gain of laser-stimulated radiative recombination as a function of the difference between c.m. photon energy and field-free binding energy: (a) proton-electron recombination into the n=2 state of hydrogen, (b) C^{6+} - electron recombination into the n=14 state of C^{5+}.

In both cases, the gain diagram in Fig. 1 is supplemented by a theoretical curve, representing a modification (Schramm *et al* 1991) of eq.(1). In the range $\varepsilon <$ o, the fit includes only data points with at least 70% of the gain maximum. The fit yields reasonable electron temperatures, saddle point energies, and laser intensities, as listed in Fig. 1. The fit curve reproduces the lowering of the gain maximum and its shift to values of $\varepsilon < 0$ as well as the occurrence of a finite gain left to the maximum. The solid line reaches zero at the saddle point energy $E_{sp} = 2.1$ meV and close to $q^{1/2} \cdot 2.1$ meV for protons and C^{6+}, respectively, in agreement with the expected $q^{1/2}$ dependence. For small electron energies, the data points show an additional yield which cannot be explained by our present model including the electric field lowering of the ionization threshold. The total additional yield represents (7.6 ± 0.9) % and (13.8 ± 3.0) % of the integrated theoretical curve, for protons and C^{6+}, respectively. In addition, also a fit of these shoulders yields abscissa values compatible with the assumption of a $q^{1/2}$ dependence. Finally, within the present error bars, it cannot be excluded that the gain remains on a finite level above the spontaneous recombination background far into the region of negative ε, so that the additional yield so far observed for C^{6+} only represents a lower bound. The region of negative ε will be investigated in more detail in a forthcoming beamtime.

4. OUTLOOK

A significantly larger average laser-induced signal is envisaged through the concept of two successive transitions from the continuum to a Rydberg state, and from there further down to a low-lying state (safe against field ionization) by simultaneous irradiation of infrared and visible laser light (Wolf 1992), offering the possibility to use much higher laser pulse repetition rates at lower intensities.In the near future, we plan further experiments using ions with higher nuclear charge. The goals of these measurements will include the q dependence of the gain curve, the role of saturation, and the influence of l sublevel mixing due to static electric fields. The ions and the cooling electrons form an exotic plasma, subjects of further investigations being in particular the electron energy and density distribution in the vicinity of ions as a function of increasing ion charge, and whether there occurs a population of states below the ionization threshold of the recombined system, e.g. caused by three-body collisions (Beyer *et al* 1989). This also motivates the investigation of laser-stimulated recombination with highly charged ions at the Experimental Storage Ring (ESR) of GSI (Borneis *et al* 1992).

ACKNOWLEDGMENT.This work was supported by the German Federal Minister for Research and Technology (Bundesminister für Forschung und Technologie) under Contract No. 06HD133I.

REFERENCES

Beyer H F, Liesen D and Guzman O 1989 *Part. Acc..* **24** 163-75
Borneis S, Bosch F, Greten G, Jung M, Klaft I, Klepper O, Koenig W, Kozhuharov C, Kühl T, Marx D, Neumann R, Tan J, Franzke B and the ESR group 1992 *GSI Scientific Report* 1991, in press
Jousif F B, Van der Donk P, Kucherovsky Z, Reis J, Brannen E, Mitchell J B A and Morgan I J 1991 *Phys. Rev. Lett.* **67** 26-9
Krämer D, Bisoffi G, Blum M, Friedrich A, Geyer Ch Holzer B, Heyng H W, Habs D, Jaeschke E, Jung M, Ott W, Pollock R E, Repnow R, Schmitt F and Steck *Nucl. Instrum. Methods Phys. Res. A* **287** 268-72
Neumann R, Poth H, Winnacker A and Wolf A 1983 *Z. Physik A* **313** 253-62
Rivlin L A 1979, *Sov. J. Quantum Electron.* **9** 353-5
Schramm U, Berger J, Grieser M, Habs D, Jaeschke E, Kilgus G, Schwalm D, Wolf A, Neumann R and Schuch R 1991 *Phys. Rev. Lett.* **67** 22-5
Schramm U, Grieser M, Habs D, Kilgus G, Schüßler T, Schwalm D, Wolf A and Neumann R 1992 *GSI Scientific Report* 1991, in press
Wolf A, in: *Proc. NATO Advanced Research Workshop on Recombination of Atomic Ions,* Newcastle, Northern Ireland, 1991 Ed. W G Graham, Plenum 1992, in press
Omidvar K and Guimaraes P T 1990, *Astrophys. J. Suppl. Ser.* **73** 555-602

Inst. Phys. Conf. Ser. No 128: Section 1
Paper presented at RIS 92, Santa Fe, NM, USA, 24–29 May 1992

27

The second-generation RIS–TOF noble gas detector: detection limits below 100 atoms in less than 5 minutes

N Thonnard, M C Wright, W A Davis*, and R D Willis[+]

Atom Sciences, Inc., Oak Ridge, Tennessee 37830, USA

ABSTRACT: A second-generation RIS-TOF noble gas detector is described having more than an order of magnitude greater sensitivity than previously reported instrumentation. The system volume has been reduced to 1.8 liters, which when coupled to the factor of ~100 reduction in outgassing from the mass spectrometer walls, mass resolution of ~700, improvements to the atom buncher, and improvements in the RIS laser set-up, have yielded detection limits of ~100 atoms for the rare ^{81}Kr and ^{85}Kr isotopes.

1. INTRODUCTION

The detection and quantitation of minute noble gas samples has been of significant interest to a wide range of researchers for many years. The chemical inertness of the noble gases makes them good tracers in a wide variety of applications, while also making it feasible to efficiently and reproducibly separate a few noble gas atoms from a very large reservoir. The ultimate example of this is the solar neutrino detector pioneered by Davis, et al (1983), in which ~100 ^{37}Ar atoms are separated from 100,000 gallons of perchloroethylene (2×10^{30} molecules!). As ^{37}Ar is radioactive with a 35 day half-life, it can be efficiently measured using low-level decay counting in a small proportional counter. The 213,000 year half-life of ^{81}Kr, whose only known source is from cosmic ray interactions in the upper atmosphere, is well-matched to the dating requirements of old aquifers and deep ice cores. But, as there are only ~1,300 ^{81}Kr atoms per kg of modern water or ice, measurement of this valuable isotope has not been possible by conventional methods. Another important application of sensitive noble gas isotope measurements is of gas extracted from meteorites and minerals, where changes in isotopic ratios are indicators of cosmic ray exposure ages and thermal history. Static noble gas mass spectrometers using electron impact ionization typically require at least 10^5 atoms of each isotope for reliable measurements, although detection limits as low as 10^4 have been achieved.

In the late 1970's, G S Hurst and co-workers embarked on a program to develop an ultrasensitive noble gas atom counting system utilizing resonance ionization spectroscopy (RIS). In 1983, Atom Sciences started its own development of a noble gas atom counting system along the same principles charted by Hurst et al (1985, and references therein). The two main differences over the Hurst design were the use of a time-of-flight mass spectrometer and a temperature controlled closed-cycle He refrigerator for the "atom buncher" cold finger (Thonnard et al 1984). In spite of the 6 liter mass spectrometer volume and ~3×10^5 Kr min^{-1} outgassing rate, detection limits of ~ 1000 Kr atoms were achieved. The usefulness of the system

Current Addresses
*108 Evans Lane, Oak Ridge, TN 37830, USA
[+]Mantech Environmental Technology, Inc., Research Triangle Park, NC 27709, USA

was demonstrated with ^{81}Kr measurements of old ground water (Willis et al 1989) and Kr isotope ratio measurements (including ^{81}Kr) from meteorite samples that were much smaller than possible with conventional mass spectrometers (Willis et al 1991).

2. THE NEW RIS-TOF SYSTEM

Two key ingredients in the Hurst et al (1985) noble gas atom counting scheme are the generation of VUV radiation (116.5 nm for Kr) to enable saturation of the RIS process by using single-photon excitations for all steps, and the use of an "atom buncher" to spatially and temporally concentrate the noble gas atoms in the RIS laser beams. In the new system, generation of 116.5 nm photons by four-wave mixing in an Ar/Xe mixture remains, but generation of the 252 nm and 1507 nm photons needed for four-wave mixing has improved significantly. Substituting difference mixing in LiNbO$_3$ for second Stokes Raman shifting in generating 1507 nm photons, and doubling 504 nm in BBO instead of doubling and mixing in KDP, has resulted in very reliable operation of the RIS laser system, with saturation of the first transition achievable on a daily basis in a 2 mm diameter beam with minimal adjustments.

The time-of-flight (TOF) mass spectrometer was completely redesigned to increase sensitivity, reduce out-gassing, and improve resolution. Maintaining sufficient volume in the mass spectrometer (1.8 liters), enabled use of either a microchannel plate or modified Daly-type detector, provided very uniform acceleration fields, and reduced ion scattering in the flight tube, thereby leading to improved mass resolution and abundance sensitivity. The cold finger of the "atom buncher", which departs from the Hurst et al (1985) design by providing a very steep temperature gradient to minimize hydrocarbon build-up and krypton loss on inactive surfaces, was redesigned to improve electric field uniformity and flexibility during spectrometer baking.

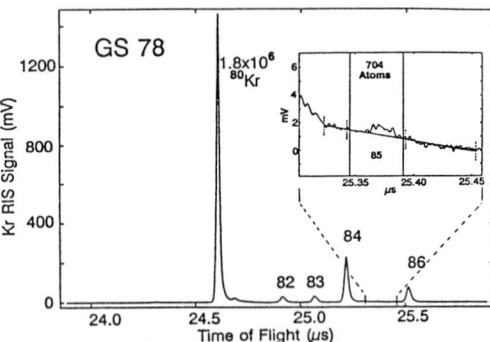

Figure 1. RIS-TOF signal from 704 ^{85}Kr atoms in 1.8 liter mass spectrometer.

3. SYSTEM PERFORMANCE

The performance of the new RIS-TOF system has met all design goals. In Figures 1 and 2, RIS-TOF mass spectra are shown containing very few ^{85}Kr atoms. By bleeding ~ 2-4x10^{-9} torr of a 10^5:1 Ar/Kr mixture into the static RIS-TOF system, ~10^6 Kr atoms are introduced for calibration, to which the unknown signal is compared; all results are referred to the number of atoms initially in the system. The sample becomes depleted as the atoms are ionized and implanted in the detector, which is typicaly operated until the signal has dropped below 10%. Mass resolution, m/Δm at half-maximum, is $\geq$ 700, while the high-mass tail one mass unit beyond a large peak is $\leq$ 1:500. From Figures 1 and 2, the regularity and low noise of the baseline permits a dynamic range of 10^4 to be achieved adjacent to a large peak. A 3σ

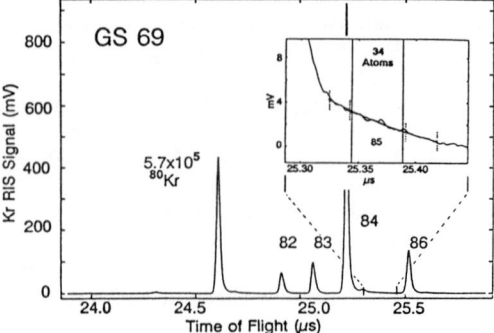

Figure 2. RIS-TOF detection limit showing noise in ^{85}Kr mass region.

detection limit of ~100 atoms in the mass 85 region was determined from the RMS noise in the baseline between the peak integration regions (vertical lines in expanded plot of Figure 2).

The RIS-TOF sensitivity is illustrated in Figure 3, in which half of the sample is ionized and detected in 2½ minutes. This compares to 15 minutes in the first RIS-TOF system, shown in Figure 4, and adapted from Willis et al (1989), and more than 60 minutes for a modern electron-impact system. Of equal importance is out-gassing, which was measured by blocking the RIS lasers for 5 and 10 minutes intervals and observing any rise in krypton signal (Figure 5).

4. CAUTIONS AND SOLUTIONS

As has been recently pointed out by several researchers (e g Fairbank et al 1989) isotopic anomalies, especially odd/even effects, are seen with RIS, even when laser bandwidths are

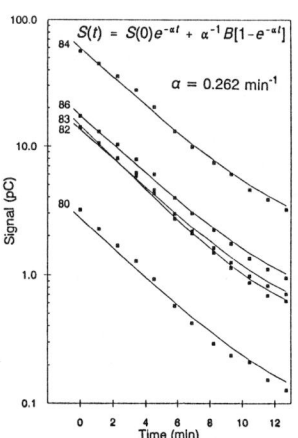

Figure 3. Sensitivity of RIS-TOF system. Note short detection half-life of sample.

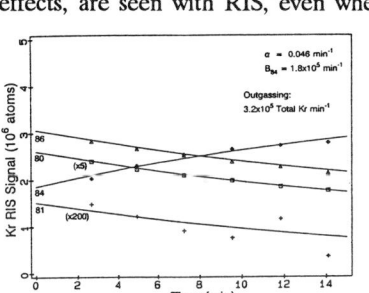

Figure 4. Sensitivity and out-gassing of first RIS-TOF system.

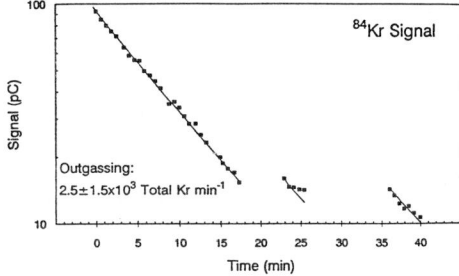

Figure 5. ^{84}Kr out-gassing test. RIS lasers were turned off at 18 to 23, and at 26 to 36 minutes.

large compared to isotope shifts. This can be seen dramatically in Figure 6, where the first few integrations show a higher ^{83}Kr than ^{82}Kr signal, even though the ^{82}Kr abundance is slightly higher. But, as ^{83}Kr is being pumped faster, the signal quickly drops below ^{82}Kr. Note that the same is true for ^{85}Kr. This effect is due to the relative polarization of the RIS laser beams (Figure 7); choosing the proper polarization can minimize the effect.

Another difficulty can arise due to saturation of the detector when too much sample is present in the system, as shown in Figure 8, where the ^{84}Kr and ^{86}Kr signals are below the expected pump-out line. (Also note that with a properly adjusted laser, the ^{82}Kr and ^{83}Kr pump-out rates are identical.) Though these errors should be avoided by proper adjustment of the system parameters, valuable data from one-of-a kind samples can be recovered when the system sensitivity is high (fast pump-out), out-gassing rates are low, and data are taken until most of the sample is depleted. Data analysis with

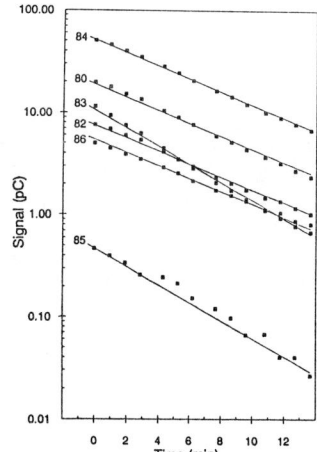

Figure 6. Odd-Even isotope effect with incorrect polarization.

conventional systems requires extrapolation of the signal from several integrations to zero time (i.e., the time of sample introduction). Isotopic biases due to sensitivity differences are corrected by calibrating with isotopic standards, but saturation effects may not be evident. Noting that the area of the pump-out curve (assuming equal detector efficiency) is proportional to the total initial number of atoms in the system, integrating the pump-out curve can significantly improve the accuracy of measurements in which biases exist. For example, using "biased" data, such as in Figures 6 or 8, the mean error in isotopic abundance measurements was reduced from $11\pm 14\%$ using extrapolated values, to $1.3 \pm 4.7\%$ using the pump-out fit. For "good" data using $\sim 10^6$ Kr atoms, the mean isotopic abundance error for masses 80 through 86 is $0.5\pm 1.4\%$. This latter error is random, and somewhat larger than what was observed previously when the system mass resolution was less, and hence possibly due to the severe undersampling introduced by our present 10 nsec sampling interval in the data acquisition system.

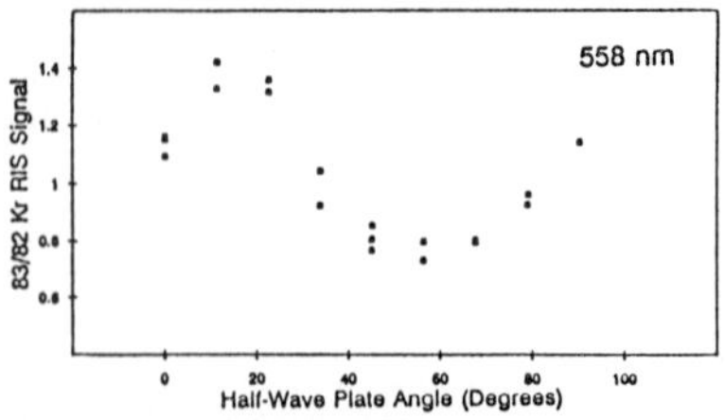

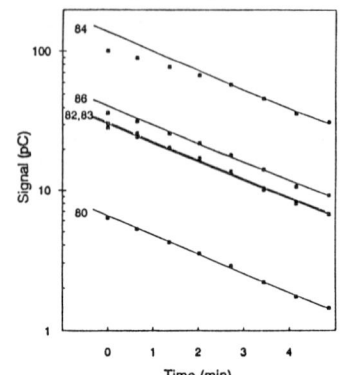

Figure 7. $^{83/82}$Kr isotope ratio as a function of RIS laser polarization.

Figure 8. Saturation of first few ^{84}Kr and ^{86}Kr integrations.

ACKNOWLEDGEMENTS

The work reported here was made possible with support from the U.S. Department of Energy, the National Science Foundation, BP R&D, and Atom Sciences Internal R&D funds. The assistance of D Ashburn, C Joyner and R Sangsingkeow, and the long-term encouragement and suggestions of GS Hurst and BE Lehmann is gratefully acknowledged.

REFERENCES

Davis R Jr, Cleveland BT and Rowley JK 1983 *Science Underground* ed MM Nieto *et al* (AIP Conf. Proc. 96) p2

Fairbank WM Jr, Spaar MT, Parks JE and Hutchinson JMR 1989 *Phys. Rev. A* **40** 2195

Hurst GS, Payne MG, Kramer SD, Chen CH, Phillips RC, Allman SL, Alton GD, Dabbs JWT, Willis RD and Lehmann BE 1985 *Rep. Prog. Phys.* **48** 1333

Thonnard N, Payne MG Wright MC and Schmitt HW 1984 *Resonance Ionization Spectroscopy 1984* ed GS Hurst and MG Payne (Bristol: Institute of Physics) pp 227-234

Willis RD, Thonnard N, Eugster O, Michel Th and Lehmann BE 1991 *Resonance Ionization Spectroscopy 1990* ed JE Parks and N Omenetto (Bristol: Institute of Physics) pp 275-278

Willis RD, Thonnard N, Wright MC, Lehmann BE and Rauber D 1989 *Resonance Ionization Spectroscopy 1988* ed TB Lucatorto and JE Parks (Bristol: Institute of Physics) pp 213-216

Inst. Phys. Conf. Ser. No 128: Section 1
Paper presented at RIS 92, Santa Fe, NM, USA, 24–29 May 1992

Attogram measurement of rare isotopes by CW resonance ionization mass spectrometry

B A Bushaw

Pacific Northwest Laboratory, Richland, WA 99352

ABSTRACT: Three-color double-resonance ionization mass spectrometry, using two single-frequency cw dye lasers and a cw carbon dioxide laser, has been applied to the detection of attogram quantities of rare radionuclides. ^{210}Pb has been measured in human hair and brain tissue samples to assess indoor radon exposure. Measurements on ^{90}Sr have shown overall isotopic selectivity of greater than 10^9 despite unfavorable isotope shifts relative to the major stable isotope, ^{88}Sr.

1. INTRODUCTION

The development of laser-based ultratrace isotopic measurements at Pacific Northwest Laboratory, based on three-color double-resonance excitation of atoms to Rydberg states followed by ionization with a cw CO_2 laser and mass spectrometric analysis, has been described at prior RIS symposia (Bushaw *et al* 1987, 1988, 1991). This approach has been shown to be capable of routinely achieving detection limits in the attogram range and relative isotopic selectivities of 10^{10} and greater. In the last two years we have progressed from spectroscopic characterization and methods development to applying these techniques to the measurement of isotopes relevant to "real world" problems for which currently existing measurement technology is not adequate to meet analytical needs. These studies have concentrated on the two radioisotopes Lead-210 and Strontium-90.

2. MEASUREMENT OF LEAD-210 FOR RADON EXPOSURE ASSESSMENT

The radioisotope ^{210}Pb is the first long-lived progeny in the decay of ^{222}Rn. It has therefore been suggested that accumulation of ^{210}Pb in biological tissues and the environment may be used to assess long-term integrated radon exposure. However, the measurement of ^{210}Pb at levels expected in small biological samples is generally beyond the capabilities of nuclear decay counting. The capabilities of laser induced ionization for measuring ^{210}Pb at subfemtogram levels have been described previously (Bushaw and Munley 1991) and the approach is only summarized here. Aqueous solutions containing the ^{210}Pb to be measured are evaporated in a graphite crucible, which is then placed within a vacuum system and heated electrothermally. As the temperature increases, the Pb salts in the sample are reduced to neutral metallic atoms on the graphite surface, which then vaporize and are ejected from the crucible. Just above the exit aperture of the crucible, the neutral metal atoms are subjected to three overlapped laser beams which cause specific ionization of the isotope of interest via the double-resonance excitation:

$$6s^2\,{}^3P_0 \xrightarrow{\lambda_1} 6p7s^3P_1 \xrightarrow{\lambda_2} 6p16p(1/2,3/2),J{=}2 \xrightarrow{\lambda_3} Pb^+ + e^- \qquad (1)$$

where the resonance excitations (λ_1 = 283.3 nm, 0.5mW and λ_2 = 420.9 nm, 200mW) are accomplished with single-frequency cw dye lasers (the first excitation step laser is frequency doubled), and the infrared photons for the ionization step (λ_3 = 10.6μm) are provided by a 10 watt cw CO_2 laser. The Pb^+ ions produced in this manner are then electrostatically extracted into a quadrupole mass filter which provides additional isotopic discrimination. Previously we reported ^{210}Pb detection limits of 0.3 femtograms using this approach. Since then, using pyrolytic coatings on the graphite crucible and increased experience have lowered the 3σ detection limits to 0.06 femtograms, as derived from statistical analysis of the data shown in Figure 1A.

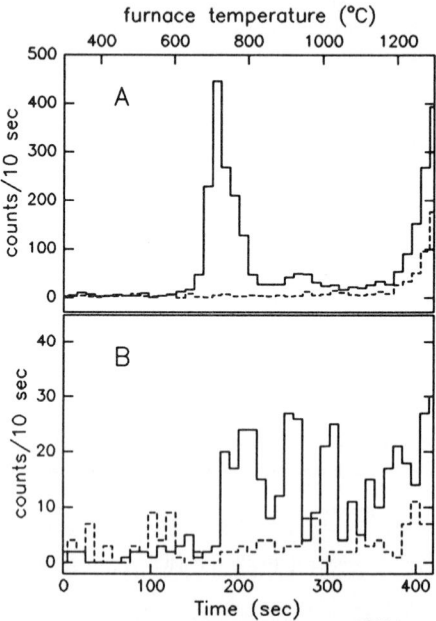

Fig. 1. Cw RIMS Determination of ^{210}Pb in A) a reference sample containing 4.4 femtograms and B) a human hair sample.

To test the measurement capabilities for "real world" biological samples, hair samples were obtained from subjects with known bedroom radon levels, as determined by standard charcoal canister measurements. These samples were obtained from persons in the Grand Forks, North Dakota area where elevated radon levels in homes are known to occur. These samples can not be loaded directly into the graphite crucible, but rather must be subjected to chemical pretreatment before the analysis is performed. The chemical procedure used began with wet ashing the hair (typically 0.3 - 0.5 grams of starting material) with nitric acid and hydrogen peroxide. The resulting Pb^{++} ions in the digestate were converted to the $Pb(OH)_4^=$ anion by treatment with concentrated ammonia solution. The Pb in this anionic form was then purified and concentrated by a two step ion exchange procedure using Dowex 1x4 anion exchange resin, with the final elution evaporated to a volume suitable for loading into the atomization crucible. A temperature evolution curve, typical for these samples, is shown in Figure 1B. Compared to the reference standard solutions (Fig 1A), the evolution of the ^{210}Pb is spread out into several peaks. This is attributed to matrix effects caused by residual materials that are co-eluted from the ion exchange columns and delay the reduction of the Pb ions to the neutral atomic form. Standard addition techniques using samples spiked with the stable isotopes showed that ion exchange recoveries and/or lower atomization efficiency reduced overall measurement sensitivity by a factor of 2 to 3. However, the standard additions also served as an internal calibration and it was still possible to obtain semiquantitative results by integration over the multiple evolution peaks. To assure that the observed signals were indeed derived from ^{210}Pb, samples were divided into two aliquots, with the second aliquot used as a "blank" and measured identical to the first except that the first excitation laser was detuned by 100 MHz from the ^{210}Pb resonance (dashed line in Figure 1B). Table 1 tabulates the measured bedroom radon levels and the corresponding ^{210}Pb content found in four different test hair samples. Even with this rather limited data set, it is apparent that there is a strong correlation between the radon exposure levels and the ^{210}Pb

Table I. ^{210}Pb concentrations in the hair of persons with known bedroom radon levels.

radon (pCi/L)	^{210}Pb (fg/g)[1]
0.3	1.2
10	5
65	22
80	18

(1)Semiquantitative measurement, estimated uncertainty is $\pm$ 30%.

concentrations. Continuing work in the area will focus on 1) improving the chemical procedures to reduce the matrix effects, and 2) making measurements on a larger number of hair samples, particularly in the 0 - 20 pCi/L expose range, to obtain a quantitative understanding of the correlation between radon exposure and ^{210}Pb levels in human hair.

While the predominant health risk associated with radon exposure has conventionally been thought to be an increased probability of lung cancer, recently it has been postulated that there may be additional risks (Henshaw *et al* 1990). In particular, because of the high solubility of radon in fat (Pohl and Pohl-Ruling 1967), it has been suggested that elevated radon expose may be a contributory factor in a number of fatty tissue diseases including brain tumors, Parkinson's and Alzheimer's diseases, and certain types of bone marrow related leukemias. We have begun testing these hypotheses by performing ^{210}Pb determinations on brain sections from Alzheimer's victims and comparing them to age-matched control brain sections from unafflicted subjects. These samples were obtained by Dr. G. I. Lykken from the Ramsey Brain Bank in St. Paul, Minnesota. Initial chemical preparation involved microwave digestion with mineral acid (2 - 3 grams starting material) at the Human Nutrition Research Center at the University of North Dakota. The resulting solutions were then transported to Pacific Northwest Laboratory where they were freeze dried to reduce the liquid volume, and then subjected to the same ion exchange procedures described above for hair samples. The results of the measurements on these samples are summarized in Table II. In all cases, the ^{210}Pb levels determined are extremely low, near the detection limits of the measurement system. Further, there appears to be no significant difference between the Alzheimer's and control samples. Thus we conclude that these measurements provide no evidence to support the hypothesis that radon exposure is a contributory factor in the development of Alzheimer's disease. Conversely, it should be realized that the limited number of samples and results very near the detection limits are not sufficient evidence to discount the hypothesis. Further efforts in this area

Table II. ^{210}Pb concentrations found in brain samples.

SAMPLE	TYPE[a]	fg ^{210}Pb/g tissue
BR-07	N	0.12 $\pm$ 0.04
BR-30	N	0.10 $\pm$ 0.07
BR-18	N	0.00 $\pm$ 0.02
BR-25	A	0.08 $\pm$ 0.04
BR-13	A	0.03 $\pm$ 0.03
BR-26	A	0.05 $\pm$ 0.04
BR-06	A	0.04 $\pm$ 0.04
BR-36	B	1.15 $\pm$ 0.10

(a) N: Normal brain, A: Alzheimer's victim, B: Bovine liver.

will concentrate on improving the chemical procedures, using larger initial sample sizes, and extending the measurements to bone marrow sections from Leukemia victims.

3. STRONTIUM-90 MEASUREMENTS

Methods for the rapid determination of low levels of the radioisotope ^{90}Sr are important for several reasons. It is a high-yield fission product in nuclear reactors (and weapons), with approximately 11 Kg produced for every Gigawatt-year of reactor operation. It has a high potential for biological damage because it can chemically substitute for calcium in bone material and the radioactive half-life of 28.5 years is comparable to the retention time within the bone. However, it is very difficult to measure by conventional radiochemical methods because of the low-energy pure β decay, without any associated gamma-ray emissions. Usual radiochemical methods involve a chemical separation of the strontium from a sample, followed by approximately one week for grow in and equilibration of the ^{90}Y progeny, and then a measurement of the ^{90}Y is performed. Even at moderately high ^{90}Sr levels, one to two weeks are required after the acquisition of a sample before analytical results can be delivered. Recent emphasis on DOE site remediation activities has prompted the investigation of cw RIMS techniques for providing near real time ^{90}Sr measurements to support and evaluate these activities.

The isotope shift for ^{90}Sr (as well as the stable isotopes) in the $5s^2$ $^1S_0$ - $5s5p$ $^1P_1$ (λ = 460.7 nm) resonance transition is known (Buchinger *et al* 1985), and the shifts for the even massed stable isotopes in the $5sns$ $^1S_0$ and $5snd$ $^{1,3}D_2$ Rydberg series have been studied in detail (Lorenzen *et al* 1983). However, no information is available for the ^{90}Sr shifts in Rydberg states, other than to expect that they will be anomolous because of the large nuclear volume changes that occur after filling the $N=50$ neutron shell at ^{88}Sr (Silverans *et al* 1988). Thus, the first task was to measure ^{90}Sr shifts in the Rydberg states. These measurements were performed with samples containing 30 -100 femtograms ^{90}Sr enriched to a relative abundance of 10^{-6} with respect to the stable isotopes. First, at low atomization temperatures, the spectrum was recorded with the first dye laser and mass spectometer tuned for ^{88}Sr while the second dye laser was scanned over the Rydberg resonance yielding a reference peak position for ^{88}Sr. Then the first laser and mass spectrometer were retuned for ^{90}Sr, the atomization temperature was increased and the scan of the second laser repeated. Comparison of the two spectra yielded the ^{90}Sr - ^{88}Sr isotope shifts listed in Table III. While the 86-84 and 88-86 isotope shifts are on the order of 250 MHz (primarily normal mass shift) in all these states, the observed 90-88 shifts were all less than 10 MHz. This is attributed to a rather unfortuitous couterbalancing of the mass and volume shifts. Further, the apparent lack of a significant difference between the S and D Rydberg series is not consistent with the those shifts observed for the stable isotopes and conventional models of isotope shifts (Heilig and Steudel 1974),

Table III. ^{90}Sr - ^{88}Sr isotope shifts in Rydberg states. Estimated uncertainties are $\pm$ 2 MHz.

State	Shift(MHz)
$5s14s$ $^1S_0$	-5.0
$5s16s$ $^1S_0$	-4.6
$5s12d$ $^1D_2$	-5.0
$5s13d$ $^1D_2$	-9.5
$5s14d$ $^1D_2$	-7.0
$5s15d$ $^1D_2$	-8.1
$5s16d$ $^1D_2$	-4.8
$5s17d$ $^1D_2$	-6.6
$5s19d$ $^1D_2$	-4.8
$5s16d$ $^3D_2$	-6.7

wherein the *D* series exhibit residual specific mass shifts while the *S* series does not. These inconsistencies are not yet fully understood and are currently being adressed by repeating existing measurements and extending the range of Rydberg states studied.

Despite these unfortunately small isotope shifts, which, in conjunction with residual Doppler broadening, limit optical isotope selectivty to approximately 10^2, efforts were still made to evaluate analytical capabilities. The detection limits were determined using a prepared sample containing 2.7×10^{-17} grams of ^{90}Sr in the presence of 4.7×10^{-9} grams of the stable strontium isotopes as shown in Figure 2. This was compared to a blank which contained an equivalent amount of the stable isotopes, but had no ^{90}Sr present, and is represented by the dashed line. The 27 attograms of ^{90}Sr present in the sample clearly produced a signal above the background of ca. 0.2 counts/sec that was observed with the blank. Integration and statistical analysis of the signals over the first 600 seconds of sample evolution evolution yields a 3σ detection limit of 2.1×10^{-18} grams, corresponding to 14,000 ^{90}Sr atoms or one radioactive decay every 20 hours. This experiment was then repeated with increased loading of the stable isotopes. When the load was increased to 38 nanograms, a 1.4×10^9

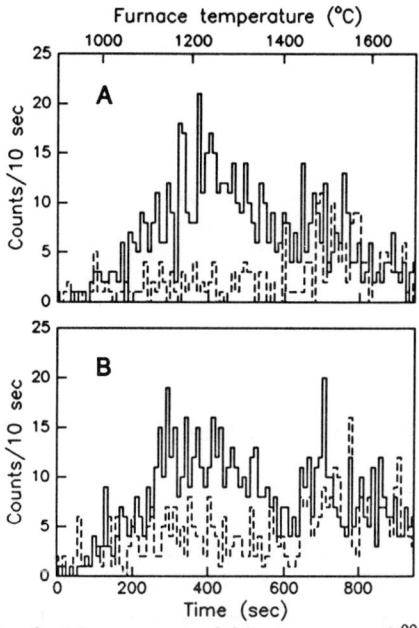

Fig. 2. Meaurement of 27 attograms of ^{90}Sr in the presence of A) 4.7 nanograms and B) 37 nanograms of the stable isotopes. Dashed lines are blank measurements without ^{90}Sr present.

excess above the 27 atograms of ^{90}Sr, the leakage of ^{88}Sr through the mass spectrometer becomes apparent as shown in Figure 2B. Analysis of the integrated data showed that the overall measurement selectivity for ^{90}Sr against ^{88}Sr was 3×10^9. Of this, only a factor of 70 was due to optical selectivity while the remaining 4×10^7 was provided by the mass spectrometer. While this overall selectivity is not sufficient to measure ^{90}Sr at background levels, it may be capable of performing useful screening measurements. For example, measurements on groundwater samples at the drinking water standard of 8 pCi/liter and (typical) stable isotope concentrations of 400 ppb requires a selectivity of 6×10^9, and thus the current measurement capabilities would only provide a signal to background ratio of 0.5. Obviously, it would be desirable to improve the selectivity by an order of magnitude or more. Current efforts are addressing these improvements through understanding the Rydberg state isotope shifts and improvements in sample atomization geometry to reduce residual Doppler broadening.

4. CONCLUSIONS

Three-color, double resonance ionization mass spectrometry using high resolution cw lasers has been shown to be capable of performing rare isotopes measurements in the attogram range. Measurements of ^{210}Pb have been successfully applied to human tissue samples for the assessment of radon exposure. Measurements of ^{90}Sr have shown excetionally low detection limits; however, extremely small isotope shifts observed in the Rydberg states studied thus far have limited overall isotopic selectivity to 3×10^9. If this selectivity can be improved moderately, there will be a wide range of applications for these measurements in environmental monitoring and remediation activities.

5. ACKNOWLEDGEMENT

This work is supported by the Office of Health and Environmental Research of the US Department of Energy (DOE) under contract DE-AC06-76RLO 1830. Special thanks to Dr. Glenn I. Lykken of the Univeristy of North Dakota for his acquisition of human tissue samples and assistance in developing chemical procedures for the separation of Pb.

6. REFERENCES

Buchinger F, R Corriveau, E B Ramsey, D Berdichevsky and D W L Sprung 1985 *Phys Rev C* **32** 2058.
Bushaw B A, B D Cannon, G K Gerke and T J Whitaker 1987 *Inst. Phys. Conf. Ser.* **84** 103.
Bushaw B A and G K Gerke 1988 *Inst. Phys. Conf. Ser.* **94** 277.
Bushaw B A and J T Munley 1991 *Inst. Phys. Conf. Ser.* **114** 387.
Henshaw D L, J P Eatough and R B Richardson 1990 *Lancet* **28** 1008.
Heilig K and A Stuedel 1974 *At. Data Nucl. Data Tables* **14** 613.
Lorenzen C-J, K Niemax and L R Pendrill 1983 *Phys. Rev. A* **28** 2051.
Pohl E and J Pohl-Ruling, 1967 *anais da Academia Brasileira de Ciencias* **39** 393.
Silverans R E, P Lievens, L Vermeeren, E Arnold, W Neu, K Wendt, F Buchinger, E B Ramsey and G Ulm 1988 *Phys. Rev. Lett.* **60** 2607.

Inst. Phys. Conf. Ser. No 128: Section 1
Paper presented at RIS 92, Santa Fe, NM, USA, 24–29 May 1992

Particle detection after state-selective collisional ionization in collinear fast beam laser spectroscopy of thallium

H. A. Schuessler[a], E. C. Benck[a], and F. Buchinger[b]

[a]Department of Physics, Texas A&M University, College Station, TX 77843 USA
[b]Foster Radiation Laboratory, McGill University, Montreal, Canada H3A 2B2

ABSTRACT: Collisional ionization of a fast beam of thallium atoms in a dilute argon atmosphere was used to observe the laser resonance between the $6s^2 6p\ ^2P_{3/2}$ metastable and the $6s^2 7s\ ^2S_{1/2}$ excited states at $\lambda=535$ nm.

I. INTRODUCTION

Improving the detection efficiency is of crucial importance in on-line laser spectroscopy of short-lived isotopes. To this end recently various methods based on monitoring the laser resonance by observing the changes in the ion beam current have been developed. They have the advantage that they are not limited by the solid angle of observation as is conventional fluorescence photon detection. These particle detection schemes employ either resonance ionization (Hurst 1988), or stepwise-excitation followed by field-ionization for fast atom beams (Kudryavtsev 1988), and laser optical pumping followed by near resonant charge neutralization for fast ion beams (Silverans 1985). Another particularly effective and simple method which is the inverse of state selective charge neutralization is laser optical pumping of a fast atom beam followed by state-selective collisional ionization. This method has already been applied successfully to the ultra sensitive detection of the noble gases radon, xenon (Neugart 1988), and krypton (Neugart 1986). The present paper describes the extension of collisional ionization to the detection of thallium. So far, on-line laser spectroscopy of thallium reached the very neutron-deficient isotopes and isomers down to A=187 (Schuessler 1992). In these experiments, purely optical detection of the laser induced fluorescence from radioactive ion beams of about 10^5 particles/ sec was used. Now we report on initial test measurements which aim to increase this sensitivity even further using state selective collisional ionization. On the one hand, in the noble gas experiments, the levels for state selective collisional ionization are about 10 eV apart and this energy difference is the major cause for the collisional selectivity. On the other hand, in thallium, the levels of interest are separated by only about 1 eV, but even then state selective collisional ionization works.

II. BASIC DETECTION PRINCIPLE AND EXPERIMENTAL ARRANGEMENT

The experimental setup, shown in Fig. 1, is a modified version of our collinear fast laser spectroscopy beam apparatus to which a collisional

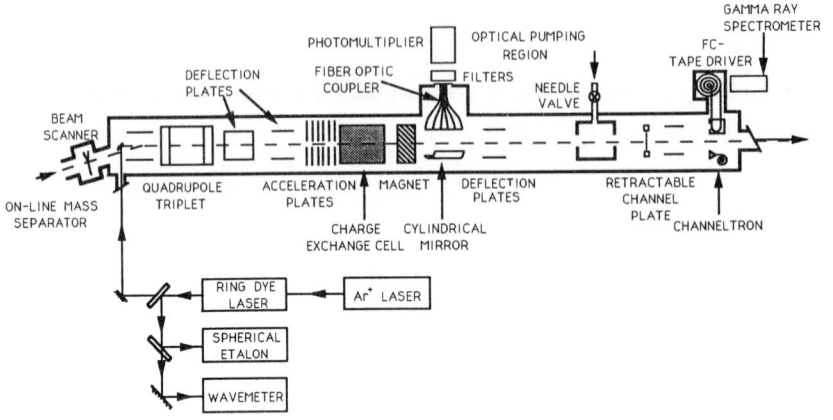

Fig. 1.   Collinear fast beam laser spectroscopy apparatus with
collisional ionization detection.

ionization cell has been added. A 50 KeV beam of thallium ions is mass
separated at the UNISOR on-line mass separator. The ion beam is then
converted to a fast atom beam predominantly populated in the $^2P_{3/2}$ metastable state using the near resonant charge exchange reaction with sodium
atoms. The remaining $T\ell^+$ ions are electrostatically deflected in the optical pumping region and only a pure neutral atom beam impinges into the
collisional ionization cell. A cw dye laser beam is copropagating with
the fast atom beam. Fig. 2 depicts in a partial energy level diagram of
$T\ell$, the energy levels involved. Spectroscopy is carried out in the $T\ell$
line at $\lambda=535$ nm between $^2P_{3/2}$ metastable and $^2S_{1/2}$ excited states.
Since the $^2S_{1/2}$ state can also decay to the $P_{1/2}$ groundstate by $\lambda=377$ nm
radiation, optical pumping to the groundstate takes place when the laser
is at resonance. This condition is obtained by leaving the laser frequency fixed and by Doppler tuning the ion beam velocity into resonance
in a post acceleration region. The laser power is more than 20 mW and
saturates the transition. Finally, the optical pumping is detected by
reionizing the fast thallium beam, employing state selective collisional
ionization in argon. A decrease in the total ion current is observed at
resonance. This flop-out signal is presently about 10^{-4} of the total ion
beam and has the potential to be more sensitive than the purely optical
signal provided that the background ion beam is minimal. For comparison,
the sensitivity of our purely optical detection scheme is also at the
10^{-4} photons/ion level and limited by the background of a few hundred
counts/second produced by the nuclear decay radiation of the short lived
isotopes. The apertures of the collisional ionization cell are differentially pumped to establish good vacuum in the entire apparatus and to
guaranty that ions are mainly produced in the collisional ionization cell
located after the optical pumping zone. The pressure in the collisional
ionization cell is adjusted by a needle valve to the optimal value of
about 50 mtorr and is monitored by a Baratron gauge.

III.   MEASUREMENTS AND RESULTS

So far we have tried only two gases, sulphurhexafluoride and argon. SF_6
is known to have an electron affinity of (1.05 eV) and is a readily

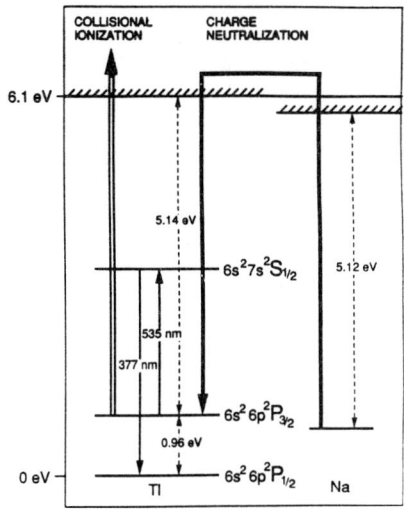

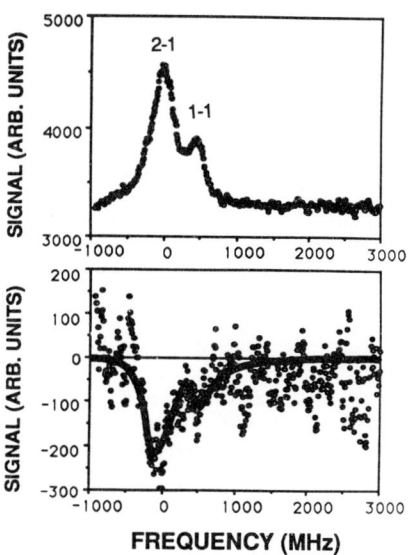

FREQUENCY (MHz)

Fig. 2. Partial energy-level diagram of Tℓ. The near resonant charge neutralization reaction with Na into the metastable 6s $^2P_{3/2}$-state and the collisional ionization with argon are indicated. Laser excitation at λ=535 nm and subsequent decay at λ=377 nm optically pump the Tℓ-atoms from the metastable $6s^26p\ ^2P_{3/2}$ to the ground state.

Fig. 3. Collinear fast beam signals of ^{203}Tℓ. The labelling of the HFS components is (F ↔ F'). Both signals, namely the optical signal (upper trace) and the collisional signal (lower case) were recorded simultaneously.

available candidate for the case that electron transfer is the dominant ionization process. Argon was chosen as a collision partner assuming that with it an electron is preferentially stripped from the excited metastable 6p $P_{3/2}$-state of Tℓ. The measurements showed that for a relative collision energy of 50 KeV the stripping process with argon has a much higher state selectivity than SF$_6$ even though the total ionization yield was lower than with SF$_6$. Fig. 3 displays a typical resonance curve taken for Tℓ. Both the collisional ionization signal as well as the fluorescence photon signal were taken simultaneously. An ion beam intensity of about 1x10^7 ions/sec was employed. However, the charge exchange cell was barely heated (T ≈60 °C), so that only a small number of atoms are produced by charge neutralization. Most of the primary ion beam is not used in the experiment and is swept out of the beam in the optical pumping region. In this way, an on-line situation is simulated where less than 10^4 ions/sec are expected. The optical pumping process was saturated as is apparent by the broad observed optical signals. The line width is about 300 MHz and three times as wide as obtained at lower laser

powers when fluorescence detection is used.   Comparing the two signals
it can be seen that in these initial measurements the absolute size of
the ion signal is about three times smaller than the optical signal.
Also, the signal to noise ratio is lower.   This is due to the large
background of isobaric ions other than thallium produced mainly in the
ion source.   Obvious improvements of the technique such as a better mass
separation of the primary ion beam and/or using a thermal ion source for
Tℓ should reduce this problem.   It is also possible to energy label the
ions produced during collisional ionization by putting the collision cell
at a known high potential and by using additional energy selective ion
detection in a second mass spectrometer.

This work is supported by DOE, UNISOR, the Teledyne Research Assistance
Program, the Center for Energy and Mineral Resources at Texas A&M
University, and the Canadian National Science and Engineering Research
Council.   UNISOR is a consortium of Universities and Oak Ridge National
Laboratory and is partially supported by them and DOE under Contract No.
DE-AC05-76OR00033.

REFERENCES

Hurst G S and Payne M G (1988), in Principles and Applications of
    Resonance Ionization Spectroscopy, Bristol, Adam Hilger Ltd.
Kudryavtsev, Y A and Petrunin V S (1988), Sov. Phys. JEPT 67 691.
Neugart R, Arnold E, Borchers W, Neu W, Ulm G, and Wendt K (1988), in
    Nuclei for from Stability, AIP Conference Proc. 164, 126.
Neugart R, Klempt W, and Wendt K (1986) Nuc. Instr. and Methods B17, 354.
Schuessler H A, Benck E C, Buchinger F, Iimura H, Li Y F, Bingham C, and
    Carter H K (1992), Hyperfine Interactions.
Silverans R E, Borghs P, De Bisschop and Van Hove M, (1985) Hyperfine
    Interactions 24-26 181.

Inst. Phys. Conf. Ser. No 128: Section 1
Paper presented at RIS 92, Santa Fe, NM, USA, 24–29 May 1992

41

^{81}Kr-detection by collinear beam spectroscopy with tunable infrared diode lasers: concept and preliminary experiments

B.E.Lehmann and A.Ludin

Physics Institute, University of Bern, Sidlerstr.5
CH-3012 Bern, Switzerland

ABSTRACT : A small Kr gas sample of $3 \cdot 10^{-3}$ cm^3 STP is continuously recycled through a plasma ion source. Ions are accelerated and focussed through a velocity filter into an alkali vapour charge exchange cell. The neutral beam is colinearly irradiated with cw IR diode laser light at 811 nm to excite the $1s_5$-$2p_9$ transition. Laser induced fluorescence will be used to study critical parameters of the system. The combined expected mass-spectrometric and optical isotope selectivity of the system is calculated.

1. INTRODUCTION

The RIS-technique for noble gases (Hurst et al. 1985) has successfully been used in recent years to measure ^{81}Kr concentrations in groundwater (Lehmann et al. 1991), in polar ice samples (Willis et al. 1990a) and in meteorites (Willis et al. 1990b). Krypton atoms are thereby excited from their ground state by 116.5 nm laser pulses with a bandwidth which is larger than isotope effects in the absorption spectrum of Kr. Any isotope selectivity in these schemes is achieved with non-optical methods by combining element-selective laser ionization with conventional mass filters. Elaborate isotope enrichment procedures have to be used (Lehmann et al. 1987, Thonnard et al. 1987) to preenrich environmental Kr samples where ^{81}Kr-concentrations are at a level of 10^{-12} or lower relative to total Kr.

Various authors have discussed possibilities of using narrowband cw lasers to excite noble gas atoms from excited metastable states in order to add optical isotope selectivity to the detection schemes (Fairbank 1987, Hardis et al. 1988, Whitaker and Cannon 1988, Cannon and Janik 1988). For Kr it should be possible to use tunable IR diode lasers to excite the $1s_5$-$2p_9$ transition at 811.5 nm.

2. CONCEPT

We use a modified arrangement of the apparatus that was used in the first preenrichment step for ^{81}Kr-analysis in groundwater samples by RIS (Lehmann et al. 1987). A small Kr gas sample of $3 \cdot 10^{-3}$ cm^3 STP of Kr is pumped in a closed cycle through a plasma ion source by a small turbomolecular pump. Kr ions are continuously extracted and accelerated to 10 kV, pass a 15 cm velocity filter (COLUTRON) and are focussed through a small

aperture approximately 150 cm down the beam line where individual Kr isotopes are spatially separated by about 2 mm. A total
Kr current of 150 nA equivalent to a flux of 0.5 ^{81}Kr atoms per
second can be extracted for several hours from such a small
(modern atmospheric) Kr sample. When tuned to mass 81 the
stable Kr isotopes of mass 78, 80, 82, 83, 84 and 86 are
attenuated by factors of at least 1500, 250, 1500, 4000, 6000
and 8000 respectively. Abundance sensitivities of >10^5 between
adjacent masses have been achieved with an improved 40 cm
velocity filter at Atom Sciences (Davis and Thonnard, 1988).

The fast Kr ion beam then passes through an alkali vapour
charge exchange cell where a fraction of the ions is converted
to neutral atoms in the metastable 1s$_5$ state. The beam of fast
metastable Kr atoms enters the detector where it is collinearly
irradiated by infrared light from a cw diode laser.

Hyperfine constants and isotope shifts for all Kr isotopes
including the rare isotopes ^{81}Kr and ^{85}Kr have been reported by
Cannon and Janik (1990) and Cannon (1992). The Doppler effect
in collinear beam spectroscopy causes a shift of the absorption
lines which depends on the velocity of the atoms.

Using the natural abundances
of the various Kr isotopes,
the given mass attenuation
factors of our Wien filter
(tuned to mass 81), the hyperfine constants and isotope
shifts of all Kr isotopes and
assuming a Lorentzian absorption line profile with a FWHM
of 5.3 MHz (the natural linewidth of the 1s$_5$-2p$_9$ transition), we calculate the combined optical and mass spectrometric selectivity of our
system as a function of beam
energy (Fig.1). The plotted
selectivity is relative to
the absorption at the strongest ^{81}Kr hyperfine transition
(13/2-11/2). The figure shows
that at energies of 10 keV,
800-1200 eV or 200-400 eV
^{81}Kr atoms with an abundance
of 5·10^{-9} in a natural Kr
sample would cause a weak
absorption peak. The detection of ^{81}Kr in environmental
samples will require another
four orders of magnitude of
isotope selectivity. Either

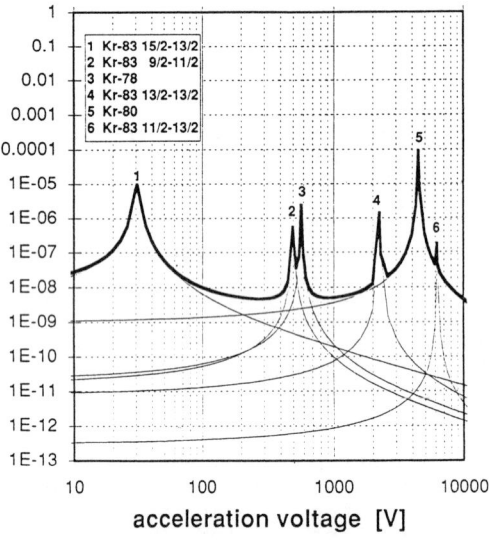

Fig.1 : Calculated absorption
line profiles vs beam energy
after passing the velocity
filter (tuned to mass 81).

additional selective IR excitation and ionization or photon
burst detection (Fairbank et al. 1991) have this potential. Use
of a more advanced velocity filter system (Davis and Thonnard
1988) will of course be an advantage. The ultimate apparatus
may even be based on photon/ion coincidence measurements.

3. DETECTOR FOR LASER INDUCED FLUORESCENCE

In order to investigate important parameters such as ion source performance (energy spread and stability), charge exchange efficiency, absorption line profiles and overall system stability, we have built a detector to observe laser induced fluorescence from the more abundant Kr isotopes (Fig.2).

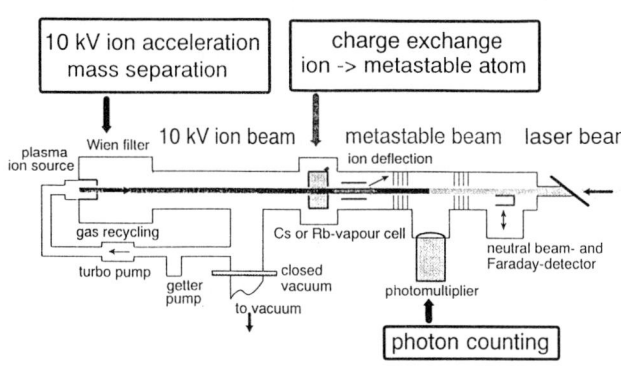

Fig 2. : Experimental Set-up

The charge exchange cell consists of a 6 cm diameter water-cooled stainless steel cylinder with 3 mm beam apertures. It is mounted on a CF63 flange, contains 1 gram of Cs (or Rb) and is heated by an external heater cartridge. Ions, which leave the cell without being neutralized, can be deflected. A Faraday detector can be moved into the beam. By changing the bias on the cup secondary electrons emitted from a Ni plate are either collected or rejected in order to separate charged and neutral fractions of the beam. Initial tests with Cs and Rb indicate very high beam neutralization at moderate temperatures (Fig.3).

The relative population of the desired $Kr^*(4p^5 5s,J=2)$ state in the emerging beam is not known yet and should be determined in a next phase of experiments.

The beam of neutral Kr atoms enters the interaction region in front of a RCA 8852 photomultiplier with an efficiency of 2% at 811 nm. In the present set-up the solid angle photon collection efficiency is about $3 \cdot 10^{-3}$, therefore, one out of every 20,000 photons emitted in the interaction region is detected.

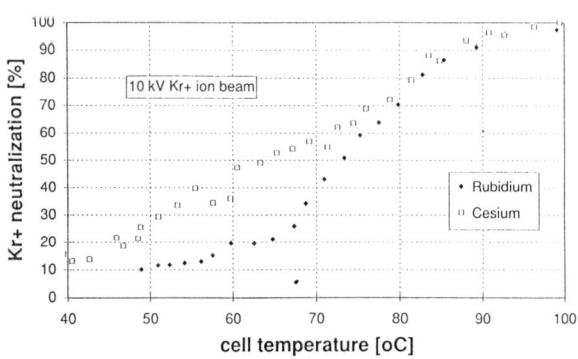

Fig. 3 : Kr ion beam neutralization in Cs and Rb vs temperature.

The tunable infrared diode laser beam (1 mW, 2 mm diameter) enters the interaction region through a Brewster angle window. Tuning and frequency stabilization is achieved with two small confocal Fabry-Perot interferometers as described by Lawrenz and Niemax (1989).

Beam skimmers before and after the interaction region are mounted to minimize stray light to the photomultiplier. We have currently achieved a background countrate (with laser, ion beam and charge exchange cell in operation) of 1 MHz (+/- 20 kHz). The detection of a fluorescence signal on the order of 50 kHz per channel (5 MHz IR scanning steps) is possible with the present geometry.

4. SUMMARY AND OUTLOOK

Collinear beam spectroscopy using tunable IR diode laser excitation from metastable states has the potential for very sensitive isotope analyses in noble gases. The laser system would be relatively inexpensive when compared to pulsed systems with their multiple frequency conversion steps used so far. The key advantage, however, would be that isotope enrichment procedures for environmental noble gas samples would no longer be necessary. With the present experiments we intend to determine important parameters of this new approach.

Our activities in the near future will focus on ion beam deceleration, further stray light reduction, improvements in light collection efficiency and beam alignment capabilities.

5. REFERENCES

Cannon B D and Janik G R, 1988, Inst Phys Conf Ser 94, 217
Cannon B D and Janik G R, 1990, Phys Rev A, 42, 1, 397
Cannon B D, 1992, Phys.Rev.A, in preparation
Davis W A and Thonnard N, 1988, Inst Phys Conf Ser 94, 233
Fairbank W M Jr, 1987, Nucl Instr Meth B29, 407
Fairbank W M, Hansen C S, LaBelle R D, Pan X J, Chamberlin E P,
    Fearey B L, Gritzo R E, Keller R A, Miller C M, Oona H, 1991
    SPIE Vol 1435, 86
Hardis J E, Peifer W R, Cromer C L, Migdall A L, Parr A C
    1988, Inst Phys Conf Ser 94, 237
Hurst G S, Payne M G, Kramer S D, Chen C H, Phillips R C,
    Allman S L, Alton G D, Dabbs J W T, Willis R D, Lehmann B E,
    1985, Rep Prog Phys 48, 1333
Lawrenz J and Niemax K, 1989, Spectrochimica Acta, 44B, 2, 155
Lehmann B E, Rauber D F, Thonnard N, Willis R D, 1987, Nucl
    Instr Meth B28, 571
Lehmann B E, Loosli H H, Rauber D, Thonnard N, Willis R D,
    1991, Appl Geochemistry 6, 419
Thonnard N, Willis R D, Wright M C, Davis W A, Lehmann B E,
    1987, Nucl Instr Meth B29, 398
Whitaker T J and Cannon B D, 1988, Inst Phys Conf Ser 94, 225
Willis R D, Thonnard N, Davis W A, Wright M C, Joyner C F,
    Craig H, 1990a, EOS 71, 1829
Willis R D, Thonnard N, Eugster O, Michel Th, Lehmann B E,
    1990b, Inst Phys Conf Ser 114, 275

Inst. Phys. Conf. Ser. No 128: Section 2
Paper presented at RIS 92, Santa Fe, NM, USA, 24–29 May 1992

Resonance ionization of metastable rare gas atoms

H Hotop, D Klar and S Schohl

Fachbereich Physik, Universität Kaiserslautern, W-6750 Kaiserslautern, FRG

ABSTRACT: In this paper, we present a survey of recent basic studies and applications involving resonant two-photon ionization of metastable rare gas atoms Rg(ms $^3P_{2,0}$) by cw lasers. Odd Rg(ns',nd') autoionization resonances, as excited from intermediate levels Rg(mp, J_i = 1,2), are discussed in some detail with emphasis on the dependence of their widths on orbital and total angular momentum and on atomic number. Two applications, based on efficient ionization via the Rg(mp, J_i = 3) level, are described: i) Absolute detection of metastable rare gas atoms; ii) Laser photoelectron attachment to molecules at sub-meV resolution.

1. INTRODUCTION

Metastable rare gas atoms play an important role in gaseous discharges and laser media; therefore, their reactions with photons, electrons, atoms, molecules and surfaces are actively studied. Sensitive detection of metastable rare gas atoms is both of fundamental and applied interest (Hurst et al. 1979, 1985, Whitaker and Cannon 1989, Hardis et al. 1989, Schohl et al. 1991) and multiphoton ionization schemes for this purpose have been demonstrated (Stebbings et al. 1975, Hurst et al. 1979, Ganz et al. 1982, Cannon and Janik 1989, Schohl et al. 1991). In this connection, a detailed knowledge of the photoionization dynamics (cross sections, auto-ionization structure etc.) is desirable. For about a decade, our group has actively investigated resonant two-photon ionization of the heavier metastable rare gas atoms Rg(ms $^3P_{2,0}$) (Rg = Ne-Xe; m = 3-6), and several parts of our results, notably those for Ne, were published (Ganz et al. 1982, 1983, 1984; Siegel et al. 1983; Harth et al. 1985, 1987; Schohl et al. 1991; Klar et al. 1991, 1992b). Like some other groups (e.g. Miller et al. 1985, Bushaw et al., 1987, Cannon and Janik 1989) we have pursued the route of cw-RIS, i.e. we use continuous wave lasers for ultra-sensitive resonance ionization spectroscopy, by which we can easily detect metastable rare gas atoms at densities of order 1 cm^{-3}. In this paper, we discuss in some detail the properties of the odd Rg(ns',nd',ng') autoionization resonances, as studied by resonant excitation via the Rg(mp, J_i = 1,2) intermediate levels; moreover, we briefly present two applications involving efficient two-photon ionization of metastable Rg(ms $^3P_2$) atoms via the closed transition to the Rg(mp $^3D_3$) level: i) Determination of the absolute Rg(ms $^3P_2$) flux (Schohl et al., 1991); ii) Laser photo-electron attachment to molecules at an energy resolution around 0.2 meV (Klar et al., 1992a).

2. RESONANT TWO-PHOTON IONIZATION OF METASTABLE RARE GAS ATOMS

2.1 Principle and experimental procedure

Figure 1 illustrates several options for two-photon RIS of metastable Rg atoms (Rg=Ar). For brevity we subsequently use Paschen's level notation; the levels of the $3p^54s/3p^54p$ configuration are represented by $1s_x$(x=2-5)/$2p_x$(x=1-10). Among the various transitions from $1s_3$ and $1s_5$ to $2p_x$, only the $1s_5$-$2p_9$ transition is closed, i.e. the $2p_9$ level can only decay back to $1s_5$. In the absence of hyperfine structure (as relevant for the most abundant isotopes ^{20}Ne, ^{40}Ar, ^{84}Kr, ^{132}Xe, normally used in our work), the $1s_5$-$2p_9$ represents a two-level transition, which can be easily saturated with cw lasers. Therefore, an average excitation probability in the $2p_9$ levels close to 50% and a clean polarization of both $2p_9$ and $1s_5$ can be achieved (Siegel et al. 1983). Consequently, excitation of $2p_9$ is ideally-suited for an efficient, loss-free two-step photoionization scheme involving cw lasers (Ganz et al. 1982, Schohl et al. 1991) and for the determination of the dynamical parameters (reduced dipole matrix elements, phase differences) of the photoionization step through polarization dependent ion and electron measurements (Siegel et al. 1983). For an investigation of the $Rg(ns',nd')$ autoionization resonances, which are Rydberg states of the $Rg^+(^2P_{1/2})$ ion core ($j_c=1/2$) and therefore embedded in the $Rg^+(^2P_{3/2}) + e^-$ continuum for sufficiently high n, it is preferable to choose an intermediate level with $J_i = 1,2$. In contrast to $2p_9$ ($j_c=3/2$), all these levels possess a more or less substantial fraction with $j_c = 1/2$ core, from which the $Rg(ns',nd')$ levels are efficiently reached by excitation of the valence p-electron. In two-photon ionization via the $J_i = 1,2$ levels with cw lasers, care has to be taken with the spatial arrangement of the two laser beams as a result of the losses, which occur by spontaneous emission to the $(1s_2, 1s_4, J=1)$ levels, followed by decay to the Rg ground level. The metastable atom should already interact with the ionizing laser 2 when it enters the exciting laser 1 (Ganz et al. 1983).

Figure 2 shows the experimental setup used for ion measurements. Metastable rare gas atoms are produced in a differentially-pumped supersonic dc discharge source. The collimated beam is crossed by two (anti-)collinear cw laser beams in the reaction chamber (metastable density about 10^6 cm^{-3}); photoions are extracted by weak electric fields (5-20 V/cm) and counted behind a quadrupole mass spectrometer, which serves to reject unwanted ions (e.g. due to Penning ionization of background gas). The lasers are linearly polarized, and the relative directions of the respective electric vectors $\vec{E}_1$ and $\vec{E}_2$ are normally chosen parallel or perpendicular such that the desired ionization path is optimized. A stabilized single mode laser excites the (Doppler-free) transition to the intermediate level, and a multimode or a single mode laser are used for the ionizing step. Fabry Perot interferometers serve as marker cavities to establish the frequency scale of the ionizing laser.

2.2 Experimental results and discussion

A comprehensive study of the resonant two-photon ionization of $Ne(1s_3, 1s_5)$ atoms (Ganz et al. 1982, 1983, 1984; Siegel et al. 1983, Harth et al. 1985, 1987) yielded the following main results:

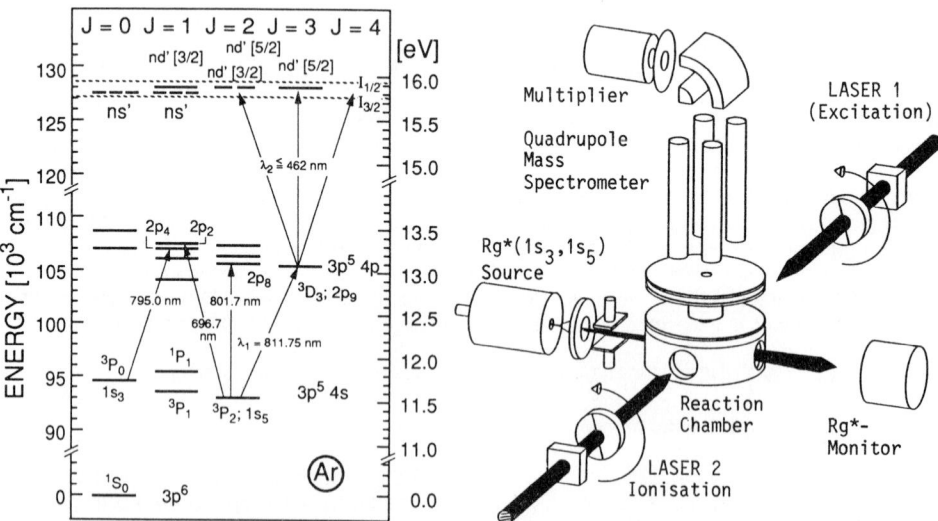

Fig. 1. Energy levels and transitions in Ar, relevant for two-photon ionization of metastable Ar($1s_5$,$1s_3$) atoms

Fig. 2. Experimental setup for resonant two-photon ionzation of metastable rare gas atoms with ion detection

i) The ionization process can be viewed as a removal of the valence p-electron with the ion core acting as spectator. In Ne(2p$_9$) ionization, electron spectrometry showed core-switching transitions (j$_c$ = 3/2 → 1/2), signalled by a detection of Ne$^+$(^{2}P$_{1/2}$) photoelectrons, to be negligible ($\lesssim$ 10^{-3}, Ganz et al. 1982). This observation justified a simplified analysis of the ion and electron experiments in the photoionization of polarized Ne(2p$_9$) atoms, from which the ratio σ_d /σ_s of the partial cross sections for ionization into the Ed and Es wave continuum and the respective phase difference Δ = δ_d - δ_s were determined in the electron energy range 0 < E < 0.7 eV (Siegel et al. 1983). Satisfactory agreement with theory (Chang and Kim 1982) was observed.

ii) Both the ns' <u>and</u> the nd' resonances are very narrow (Ganz et al. 1983, 1984, Harth et al. 1985, Ganz 1985, Klar et al. 1990, 1992b) in basic (although not quantitative) agreement with the theoretical prediction by Johnson and Le Dourneuf (1980). The reduced autoionization widths $\Gamma_r = \Gamma_n \cdot$ n*^{3} (n* = effective principle quantum number) fall in the range (80-370) cm^{-1} and thereby prevent a resolution of the Ne(ns',nd') resonances by conventional light sources or broadband multimode lasers (Ganz et al. 1983). The reduced width for Ne(ns', J=1) is 3 times larger than that for Ne(ns', J=0) and due mainly to the influence of the (ns' - Es) channel, which does not exist for the two channel case (ns', J=0 - j$_c$ = 3/2, Ed$_{3/2}$).

iii) All the observed Ne(ns', nd') resonances could be described very well by Beutler-Fano profiles (including a constant or slowly varying background σ_{nc} due to non-interfering continua), as given by

$$\sigma(E) = \sigma_{nc} + \sigma_{rc} \cdot (q + \epsilon)^2 / (1 + \epsilon^2) \qquad (1)$$

with ϵ = 2(E-E$_0$)/Γ, E$_0$ = resonance energy. q is the profile index or shape parameter, which can be expressed as q = M$_{ri}$ /(π V$_{cr}$ M$_{ci}$) (Fano 1961); M$_{ri}$ and M$_{ci}$ are the electric dipole matrix elements from the bound (intermediate) state i to the resonance state r and to the continuum c, respectively, and V$_{cr}$ is the autoionization amplitude connected to the width by Γ = 2π|V$_{cr}$|^2. Ganz et al. (1984) demonstrated both experimentally and by a MQDT analysis that the profile index q of the Ne(ns', J=1) resonance varies with the character of the Ne(3p, J$_i$ = 1,2) intermediate level, whereas the resonance width remains constant.

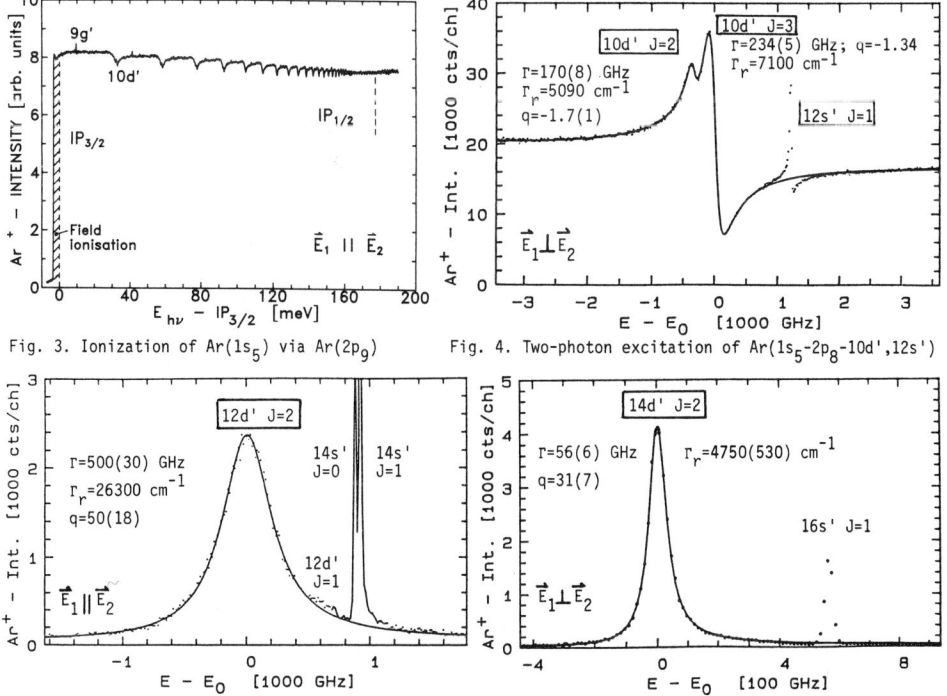

Fig. 3. Ionization of Ar(1s$_5$) via Ar(2p$_9$)

Fig. 4. Two-photon excitation of Ar(1s$_5$-2p$_8$-10d',12s')

Fig. 5. Two-photon ionization Ar(1s$_5$-2p$_2$-12d',14s')

Fig. 6. Two-photon ionization Ar(1s$_3$-2p$_4$-14d',16s')

Subsequent work by Wang and Knight (1986), who investigated pulsed-laser, two-photon excitation of Xe($1s_3$, $1s_5$), showed a strong dependence of the nd', J=1,2,3 resonances on J and another interesting feature for the nd', J=2 resonances: two distinct series exist, described in $j_c \ell$-coupling by nd' $[K=3/2]_2$ and nd'$[K=5/2]_2$; in Xe the reduced width for the former exceeds that for the latter by a factor of three. Wang and Knight (1986) also discussed the intensity propensities in the excitation of the ns', nd' resonances from several J_i=1,2 intermediate levels on the basis of $j_c \ell$-coupling calculations and pointed out the prominence of the mp'$[3/2]_1 \rightarrow$ nd'$[5/2]_2$, mp'$[3/2]_2 \rightarrow$ nd'$[5/2]_3$, and mp'$[1/2]_1 \rightarrow$ nd'$[3/2]_2$ transitions, also reflected in our data (see below).

For illustration of the experimental results, we present characteristic examples of cross sections, obtained in our resonant two-photon ionization studies of Ar($1s_5$, $1s_3$) via the Ar($2p_9$, $2p_8$, $2p_4$, $2p_2$) intermediate levels. Fig. 3 shows the data for Ar($2p_9$), as measured with a multimode dye laser 2 (width about 30 GHz = 1 cm^{-1}) with parallel linear polarizations ($\vec{E}_1 \| \vec{E}_2$). A nearly constant cross section is observed; the nd' and sharp ng' resonances (the latter excited due to configuration interaction) are weak features only, in qualitative agreement with the previous results for Ne($2p_9$), which did not even indicate the presence of resonances (Ganz et al. 1982). For perpendicular polarizations ($\vec{E}_1 \perp \vec{E}_2$) the continuum cross section decreases by (20-25) %, yielding cross section ratios σ_d / σ_s around 4 in the energy range $0 < E < 0.9$ eV, quite similar to the findings for Ne($2p_9$) (Siegel et al. 1983). We note that for $\vec{E}_1 \perp \vec{E}_2$, the nd'-resonance structure is more prominent and changes from a window-like appearance in Figure 3 to a dispersion-type feature. Figure 4 presents the results for Ar(10d', J=2,3; 12s', J=1), as excited from Ar($1s_5$) via Ar($2p_8[5/2]_2$) with $\vec{E}_1 \perp \vec{E}_2$. A strong continuum cross section and asymmetric resonances with small q-values are observed. The smooth curve through the data points represents a fitted cross section, using an extended expression (1) with two independent Fano profiles for the nd'$[5/2]_{2,3}$ resonances; the much narrower 12s', J=1 resonance (Γ = 17 GHz, Klar et al. 1992b) is somewhat broadened by the experimental resolution.

For excitation via intermediate levels with dominant j_c = 1/2 character the ionization spectra are qualitatively different and characterized by low continuum cross sections and high values for the profile index. As an example, Figure 5 shows the (12d', 14s') resonance region, reached by excitation from Ar($1s_5$) via Ar($2p_2[1/2]_1$) ($\vec{E}_1 \| \vec{E}_2$). The reduced width of the 12d'$[3/2]_2$ resonance in Figure 5 is about 5 times larger than that of the 10d'$[5/2]_2$ resonance in Figure 4. To corroborate this finding the nd'$[5/2]_2$ resonance was also excited from Ar($1s_3$) via Ar($2p_4[3/2]_1$); Figure 6 displays the data for the (14d',16s') region ($\vec{E}_1 \perp \vec{E}_2$). Again, nearly symmetric (Lorentzian) profiles (high $|q|$) on a very low continuum cross section are observed. The reduced width of the 14d', J=2 resonance, deduced from this spectrum, agrees with that of the 12d', J=2 resonance in Figure 4. We note that the reduced widths are practically independent of n in Ne and Ar and depend only weakly on n for Kr and Xe.

As mentioned above, the ns' and ng' resonances cannot be resolved at photon bandwidths around 1 cm^{-1}. Therefore, they were investigated in more detail with a single mode ionization laser (overall resolution < 50 MHz). Two examples are presented in Figures 7 and 8.

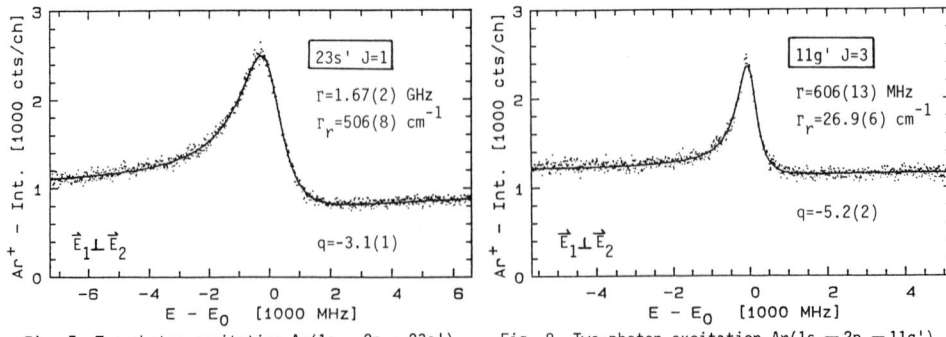

Fig. 7. Two-photon excitation Ar($1s_5$− $2p_8$− 23s')   Fig. 8. Two-photon excitation Ar($1s_5$− $2p_8$− 11g')

Figure 7 displays the profile of the Ar(23s', J=1) resonance; analogous data for the Ar(ns', J=0) series demonstrate that the reduced width of ns', J=0 is about 55 % <u>larger</u> than that of ns', J=1, very much in contrast to the situation in Ne. Fig. 8 shows the Ar(11g', J=3) resonance, which exhibits a very low reduced width (Γ_r = 27 cm^{-1}); this is not surprising, however, in view of the large orbital angular momentum and the near-zero quantum defect, which we determined as 0.004(3).

Our measured reduced widths for the Rg(ns',J=0) and Rg(ns', J=1) resonances are summarized in Figure 9. Numerical values and a comparison with other experimental and with theoretical work were presented by Klar et al. (1991, 1992b). The following major trends are observed:
i) the reduced widths of the Rg(ns', J) resonances do not vary monotonically with atomic size; they are largest for Kr, both for J=0 and J=1, and smallest for Ne, especially so for J=0.
ii) For Ne(ns'), the J=0 resonances are about three times narrower than the J=1 resonances; for Ar,Kr, and Xe, on the other hand, the J=0 resonances are (1.3-2) times broader than the J=1 resonances.
The variation of the Rg(ns') widths with atomic size can be qualitatively understood on the basis of the quantum defects μ_ℓ for the bound $\ell = 0$ electrons and of the phase shifts ($\delta_\ell \approx \pi\mu_\ell$) for the respective continuum electrons in combination with the variation of the size of the Rg-core from Ne to Xe. The quantum defect aspect tells us that Ne and Xe should have smaller widths than Ar and Kr, whereas the monotonically rising core radius favours a substantial increase of the widths from Ne to Xe (see also Klar et al., 1992b). The overall variation of the widths with a maximum for Kr reflects the combined influence of the two different trends.

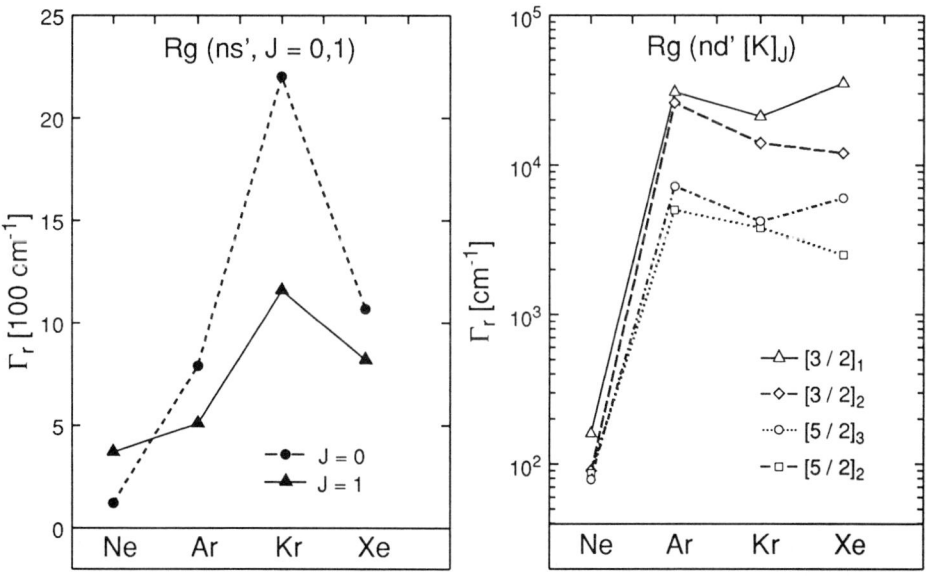

Fig. 9.   Reduced widths Γ_r of the   Fig. 10.   Reduced widths Γ_r of the
            Rg(ns', J=0,1) resonances   Rg(nd', J=1,2,3) resonances

The J-dependence ii) of the ns'-widths for Ar, Kr, and Xe is quite unexpected in view of the fact that the Rg(ns', J=1) resonance possesses three decay channels (Es, Ed$_{3/2}$, Ed$_{5/2}$) instead of only one (Ed$_{3/2}$) for Rg(ns', J=0). This problem was recently clarified by many body perturbation theory (MBPT) calculations, carried out by Tsemekhman et al. (Klar et al. 1992b). For Rg (ns', J=0), a first order calculation of the s'-d coupling amplitude provides a reasonable width for Ne and the correct trend for the dependence on atomic size. Electron correlation effects are important for quantitative results in Ar, Kr, and Xe, both for J=0 and J=1. For Rg(ns', J=1), the additional amplitudes due to s'-d exchange and due to s'-s decay are found to be of varying importance with atomic number. Destructive interference between the different s'-d amplitudes and the influence of Rg(ns', J=1) - Rg(nd', J=1) coupling are responsible for the fact that the (ns', J=1) resonances are narrower than the (ns', J=0) resonances for Ar, Kr, and Xe.

Figure 10 summarizes the experimental reduced widths for the Rg(nd') resonances (Klar et al. 1991 and references herein, Wang and Knight 1986). The widths for Ne are about two orders of magnitude lower than those for Ar, Kr, and Xe, for which similar values are observed for a given $[K]_J$ series. Relativistic MQDT calculations (Johnson et al. 1980, Johnson, Le Dourneuf 1980) for J=1 and MBPT calculations (Amusia et al. 1992) for J=1,2 show satisfactory, near-quantitative agreement with the experimental values for the nd'-widths. We note that so far, no ab initio calculations for the shape parameters q of the nd'- or ns'-resonances, excited from the Rg($2p_x$) levels, have been reported.

3. APPLICATIONS OF cw PHOTOIONISATION OF LASER-EXCITED Rg($2p_9$)

In this section we briefly present two applications of efficient cw two-photon ionization of Rg($1s_5$) atoms via the Rg($2p_9$) level, where ionization occurs inside the cavity of an Argon ion laser or of a tunable dye laser. The ionization probability of the laser-excited Rg($2p_9$) atoms near ionization threshold, as estimated from the relevant cross sections (Chang and Kim 1982) for an effective laser-atom interaction time of 1 μs and an ionizing laser flux density of 10^{20}/mm^2 (i.e. 50 W/mm^2 at a photon energy of 3 eV), amounts to values of 6 % to 18 % for Ne-Xe. We first discuss the use of this scheme for the absolute detection of metastable Rg($1s_5$, $^3P_2$) atoms, as outlined in Figure 11. The photoionization of Rg($1s_5$) atoms leads to a photoelectron current ΔI_p and - as a result of the Rg($1s_5$) depletion - to a decrease ΔI_s in the electron current, ejected from a conductive surface at the metastable atom detector. The coefficient γ for electron emission upon impact of Rg($1s_5$) is then directly and simply determined by $\gamma = \Delta I_s / \Delta I_p$; knowledge of γ yields the Rg($1s_5$) flux F($1s_5$) = I_s /eγ, where I_s is understood to be the fraction of the detector current due to Rg($1s_5$). This method was introduced and discussed in detail by Schohl et al. (1991). In the left part of Figure 11, a measurement example for Ar* impinging on a CuBe surface is shown for illustration: comparison of ΔI_s and ΔI_p yields γ = 0.22. It was found that the values of γ vary considerably with the detector surface, depending on the material, surface preparation, surface temperature and the time, over which the surface is exposed to the reactive metastable rare gas beam (Schohl et al. 1991). In situ determination of γ values is therefore mandatory in quantitative work; we plan to take advantage of the described method in measurements of absolute collision cross sections involving metastable rare gas atoms.

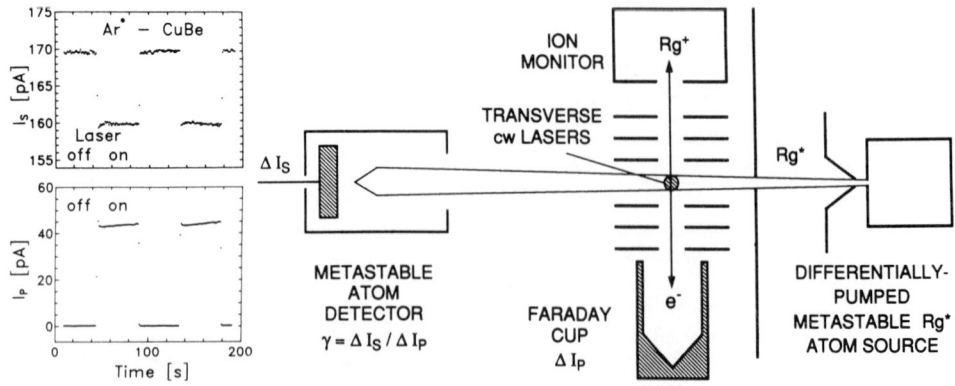

Fig. 11. Illustration of the cw laser photoionization method for the absolute detection of metastable rare gas atoms

As a second example, we sketch the Laser Photoelectron Attachment (LPA) method (Klar et al. 1992a), illustrated in Figure 12. Monoenergetic electrons of variable, well-known energy are produced by cw photoionization of laser-excited $Ar(2p_9)$ atoms with an intracavity tunable dye laser. The photoelectrons interact with a static target gas, and the negative ions, formed in electron attachment processes, are mass-analyzed and detected with an electron multiplier. To avoid electric fields during electron production and attachment, the experiment is pulsed: the exciting laser and the ion drawout field are switched on and off out of phase, such that attachment processes take place only when the external field is not present. With this LPA method we have carried out electron attachment studies with an overall energy width of (0.1-0.2) meV, thereby improving the resolution in such experiments (Chutjian and Alajajian 1985) by about a factor of 50. Figure 12 (right part) shows the LPA cross section for SF_6^- formation, using a SF_6 gas target at T = 300 K (Klar et al. 1992a). At low energies the cross section approaches the behaviour for s-wave capture ($\sim E^{-1/2}$); the features I and III are associated with the ν_1- and ν_3-thresholds for vibrational excitation of SF_6 as indicated. The sharp peak II is due to the efficient attachment of slow electrons, associated with the (weak) process of $Ar^+(^2P_{1/2})$ formation in $Ar(2p_9)$ photoionization at wavelengths $\lambda_2 < 433$ nm. The width of peak II allows to estimate the experimental energy resolution, which for the shown data was about 0.2 meV (Klar et al. 1992a). In the future we plan to extend our LPA studies to molecular clusters.

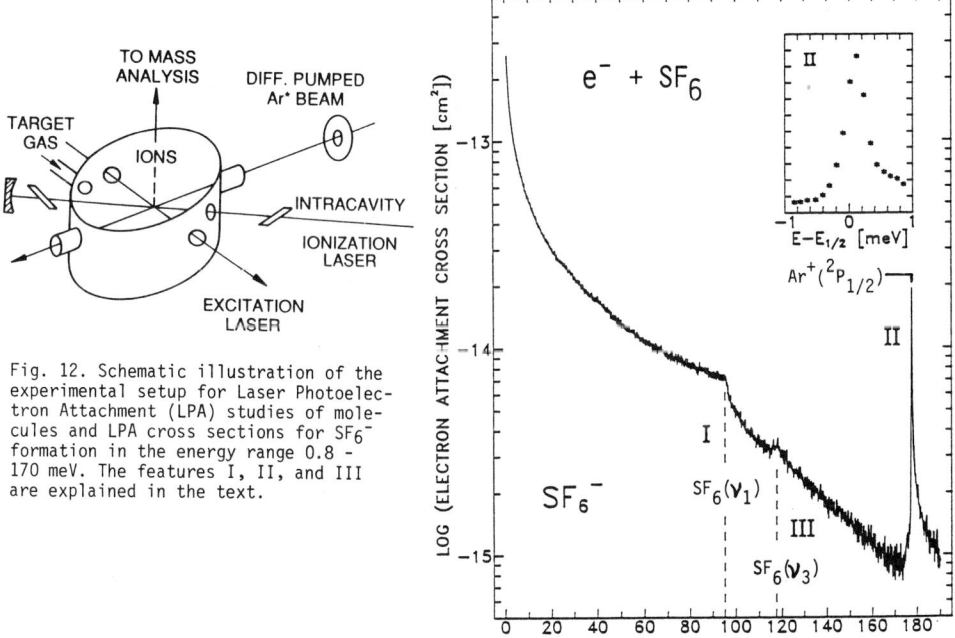

Fig. 12. Schematic illustration of the experimental setup for Laser Photoelectron Attachment (LPA) studies of molecules and LPA cross sections for SF_6^- formation in the energy range 0.8 - 170 meV. The features I, II, and III are explained in the text.

Our work has been supported by the Deutsche Forschungsgemeinschaft through Sonderforschungsbereich 91 and grant Ho 427/14-1. We gratefully acknowledge the contributions of J. Ganz, A. Siegel, K. Harth, M. Raab, T. Kraft, and M.-W. Ruf to the work presented in this paper, and we thank M.Ya. Amusia, K. Tsemekhman, and V. Tsemekhman for their cooperation and for discussions of their theoretical work on the Rg(ns', nd') resonances.

REFERENCES

Amusia M Ya, Tsemekhman V and Tsemekhman K 1992 Book of Abstracts, ECAMP 4, Riga, April 1992, p. 98

Bushaw B A, Cannon B D, Gerke G K and Whitaker T J 1987 Inst. Phys. Conf. Ser. **84** ed G S Hurst and C Grey Morgan (Bristol: Inst. Phys.) pp 103-8

Cannon B D and Janik G R 1989 Inst. Phys. Conf. Ser. **94** ed T B Lucatorto and J E Parks (Bristol: Inst. Phys.) pp. 217-20

Chang T N and Kim Y S 1982 Phys. Rev. A **26** 2728-32

Chutjian A and Alajajian S H 1985 Phys. Rev. A **31** 2885-92

Fano U 1961 Phys. Rev. **124** 1866-78

Ganz J, Lewandowski B, Siegel A, Bussert W, Waibel H, Ruf M-W and Hotop H 1982 J. Phys. B **15** L485-9

Ganz J, Siegel A, Bussert W, Harth K, Ruf M-W, Hotop H, Geiger J and Fink M 1983 J. Phys. B **16** L569-76

Ganz J, Raab M, Hotop H and Geiger J 1984 Phys. Rev. Lett **53** 1547-50

Hardis J E, Peifer W R, Cromer C L, Migdall A L and Parr A C 1989 Inst. Phys. Conf. Ser. **94** ed T B Lucatorto and J E Parks (Bristol: Inst. Phys.) pp. 237-40

Harth K, Ganz J, Raab M, Lu K T, Geiger J and Hotop H 1985 J. Phys. B **18** L825-32

Harth K, Raab M and Hotop H 1987 Z. Phys. D **7** 213-25

Hurst G S, Payne M G, Kramer S D and Young J P 1979 Rev. Mod. Phys. **51** 767-819

Hurst G S, Payne M G, Kramer S D, Chen C H, Phillips R C, Allman S L, Alton G D, Dabbs J W T, Willis R D and Lehmann B E 1985 Rep. Prog. Phys. **48** 1333-70

Johnson W R, Cheng K T, Huang K N and Le Dourneuf M 1980 Phys. Rev. A **22** 989-97

Johnson W R and Le Dourneuf M 1980 J. Phys. B **13** L13-7

Klar D, Harth K, Ganz J, Kraft T, Ruf M-W and Hotop H 1991 Proc. UK/USSR Seminar, Leningrad, 23-27 April 1990, ed M Ya Amusia and J B West (DL/SCI/R29, SERC Daresbury) pp 78-82

Klar D, Ruf M-W and Hotop H 1992a Chem. Phys. Lett. **189** 448-54

Klar D, Harth K, Ganz J, Kraft T, Ruf M-W, Hotop H, Tsemekhman V, Tsemekhman K and Amusia M Ya 1992b Z. Phys. D (in press)

Miller C M, Engleman R Jr and Keller R A 1985 J. Opt. Soc. Am. B **2** 1503-9

Schohl S, Klar D, Kraft T, Meijer H A J, Ruf M-W, Schmitz U, Smith S J and Hotop H 1991 Z. Phys. D **21** 25-39

Siegel A, Ganz J, Bussert W and Hotop H 1983 J. Phys. B **16** 2945-59

Stebbings R F, Dunning F B and Rundel R D 1975 Atomic Physics 4 ed G zu Putlitz, E W Weber, A Winnacker (New York: Plenum), pp. 713-30

Wang L-g and Knight R D 1986, Phys. Rev. A **34** 3902-7

Whitaker T A and Cannon B D 1989 Inst. Phys. Conf. Ser. **94**, ed T B Lucatorto and J E Parks (Bristol: Inst. Phys.) pp. 225-8

Inst. Phys. Conf. Ser. No 128: Section 2
Paper presented at RIS 92, Santa Fe, NM, USA, 24–29 May 1992

Non-ionization of Rydberg atoms by intense ps pulses

T. F. Gallagher, L. D. Noordam, H. Stapelfeldt,
D. G. Papaioannou, and D. I. Duncan

Department of Physics, University of Virginia, Charlottesville, VA 22901, USA

ABSTRACT: Exposure of Rydberg atoms to intense ps optical pulses does not ensure that they will be photoionized. If the pulses are of duration short compared to the classical orbit or round trip time of the Rydberg electron it is unlikely that photoionization will occur. Rather, a hole is burned in the Rydberg electron's wavefunction, and only those electrons at the ionic core when the ps pulse passes are ionized.

1. Introduction

It seems intuitively obvious that a loosely bound electron should be easily removed from an atom by an intense radiation pulse. However, if the frequency of the radiation is high compared to the binding energy of the electron, it is not likely that ionization will occur. The reason stems from the fact that a loosely bound electron is almost free, and a free electron can not absorb radiation. Similarly, when a Rydberg electron is near the outer classical turning point of its orbit it is unlikely to absorb visible radiation. Only when it is close to the ionic core is it likely to absorb the radiation. The near transparency of the Rydberg electron to visible light is the basis of the isolated core excitation method, which has been used to study the autoionizing states of alkaline earth atoms for some time (Cooke et al 1978). The basic idea is to carry out optical excitation of doubly excited autoionizing states starting from the bound Rydberg states. In essence, the outer Rydberg electron is a spectator while the inner electron absorbs the photons. Both one and two photon transitions of the inner electron have been used in a wide variety of spectroscopic experiments.

Here we describe the extension of the isolated core excitation experiments to higher intensities and shorter pulse lengths. As we shall see, with higher intensities photoionization of the Rydberg electron becomes possible, but the process terminates if the pulse length of the exciting laser is shorter than the round trip time of the Rydberg electron.

2. Experimental Approach

We have carried out several experiments with the Ba atom, chosen because it is a two valence electron atom with relatively low ionization potentials. The ionozation limit of Ba lies 42035 cm^{-1} above the Ba ground state, and the ionization limit of Ba$^+$ lies 80686 cm^{-1} above its ground state.

The Ba atoms are in a thermal beam which effuses from a resistively heated oven in a vacuum chamber. The atoms are excited with two ns dye lasers to a bound 6snd Rydberg state and then exposed to an intense ps laser pulse. Subsequent to the ps pulse the products are analyzed by applying a field pulse to the atoms. The Rydberg states of Ba and Ba$^+$ are analyzed by selective field ionization, and the charge states of the resulting ions are analyzed by their flight times to the detector.

3. Inner Electron Ionization

When the bound 6snd Rydberg atoms are exposed to a 5 ps pulse which is tuned to a Ba^+ 6s-nℓ multiphoton resonance, two different behaviors, shown in Figure 1, are observed, depending on whether or not the pulse duration, τ_{pulse} is longer or shorter than the round trip time, τ_{round}, of the Rydberg electron (Stapelfeldt et al 1991, Jones and Bucksbaum 1991). If $\tau_{pulse} < \tau_{round}$ the outer electron is not always ionized. Instead, the inner electron is ionized and the outer electron is projected from the initial Ba Rydberg state onto Ba^+ Rydberg states of the same physical size.   On the other hand, if $\tau_{pulse} > \tau_{round}$ the outer electron is simply photoionized.

Figure 1. Schematic representation of the interaction between the ps laser pulse and the Ba 6snd atom for (a) high n state and (b) a low n state.

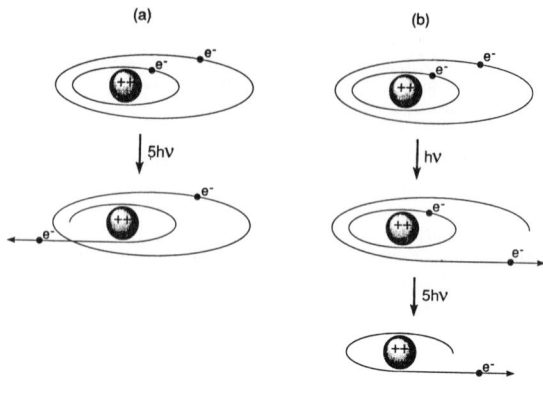

The experiment which led to the conclusion above was carried out in the following way. The atoms were excited to the bound 6snd Rydberg state via the route 6s6s-6s6p-6snd, using two ns dye lasers.   The atoms were exposed to the 5 ps laser pulse, and the atoms were subsequently exposed to a field pulse to field ionize Rydberg states and collect the Ba^+ and Ba^{++} ions. The wavelength of the second ns laser, which excited the atoms from the 6s6p state to the 6snd state, was scanned as the Ba^{++} signal was recorded.   In Figure 2 we show the resulting spectrum when the atoms were exposed to a 3.6 kV/cm field pulse and the ps laser was tuned to the Ba^+ $6s_{1/2}$-$7p_{3/2}$ transition.   As shown by the upper scale of Figure 2, for wavelengths to the blue of 417 nm the two ns lasers produce Ba^+, and it is subsequently ionized by the ps laser to form Ba^{++}, as expected.

Figure 2 The Ba^{++} signal recorded as a function of the wavelength of the ns laser from above the 6s limit to n=22

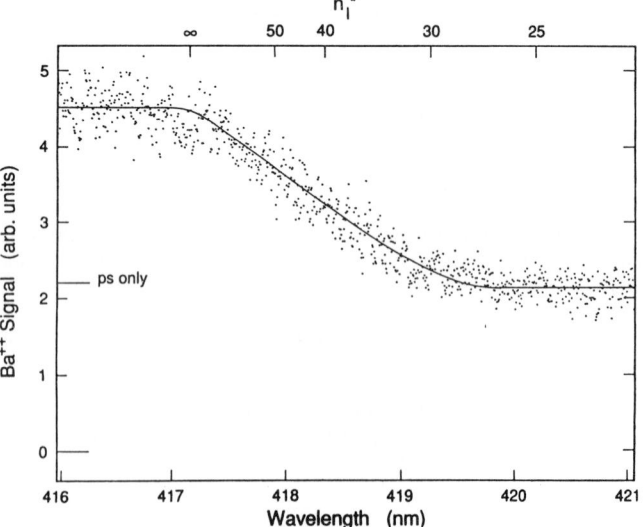

When the wavelength is increased, the Ba^{++} production does not drop at the ionization limit but decreases smoothly, reaching a plateau at a Ba effective quantum number, n_I^*, of 30. Although it is not apparent in Figure 2, the Ba^{++} signal exhibits the same Rydberg structure as the Ba^+ signal. Equally as interesting as the fact that the Ba^{++} signal does not disappear when the ns laser is scanned across the limit is where it does disappear. As shown by Figure 2, it disappears at n_I =30. The round trip time, in atomic units, is given by

$$\tau_{round} = 2\pi \, (n_I^*)^3$$

(1)

Evaluating Eq. (1) in conventional units, we find that τ_{round} = 5ps for n_I^* =32, which corresponds to the point at which the Ba^{++} signal begins to increase in Figure 2.

Figure 2 demonstrates that the increase in Ba^{++} only occurs for n_I^* in excess of 32. Examining the dependence of the signal on the strength of the field pulse demonstrates that the way in which Ba^{++} is actually formed is by field ionization of Ba^+ Rydberg states. When the field pulse is reduced from 3.6 kV/cm to about 700 V/cm, the signal is unchanged from Figure 2. However, for field pulses less than 600 V/cm the signal of Figure 2 no longer extends down to n_I^* =32, but is truncated abruptly at a higher value of n_I^*, as shown in Figure 3. The dependence on the field pulse demonstrates that the observed Ba^{++} comes from the field ionization of Ba^+ Rydberg states. We can relate the effective quantum numbers of the Ba and Ba^+ Rydberg states in the following way. The Ba effective quantum number, n_I^*, is known from the wavelength of the ns laser. The effective quantum number of the Ba^+ Rydberg state is determined from its ionization field. Explicitly, we assume that the ionization field is given by the classical field for ionization,

$$E = Z^3/16 \, (n_{II}^*)^4$$

(2)

Using Eq. (2) we can relate the field at which the Ba^{++} signal disappears to the effective quantum number of the Ba^+ Rydberg states produced by the ps laser. For field pulses beteween 82 and 590 V/cm we found that the ratio of n_{II}^*/n_I^* had an average value of 1.34. This value is very close to $\sqrt{2}$, the ratio for Ba^+ and Ba states in which the expectation values of the orbital radii are the same. Numerical calculations led to the ratio 1.40. In sum, it appears that the picture of Figure 1 is accurate.

Figure 3 Ba^{++} signal as in Figure 2 but for three smaller extraction fields. The arrows mark the positions of the thresholds. The curve drawn through the experimental points serves to guide the eye.

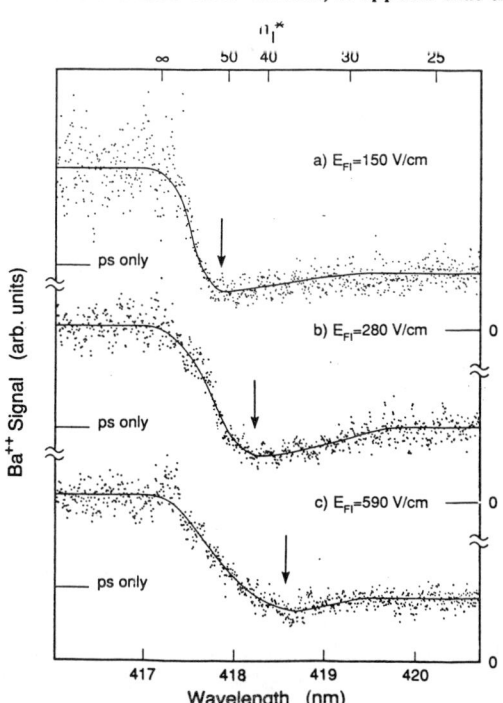

4. Redistribution of the Rydberg States

It is interesting to return to the question of why the time duration of the pulse is so critical. When $\tau_{pulse} < \tau_{round}$ only atoms in which the Rydberg electrons are at the core are ionized by the ps pulse. For all other atoms the Rydberg electron is too far from the core to be affected by the ps pulse, and all that happens to the atom is that the inner electron is removed. For the moment, let us ignore the inner electron and concentrate on the outer electron. A simple picture of what has happened is that the ps pulse has burned a hole in the Rydberg electron's wavefunction. In other words, we have created what might be called an anti wave packet. Instead of having the presence of a spatially localized wavefunction, we have its absence. The hole in the wave function propagates in time, just as a normal wave packet, and it periodically returns to the core. When the hole in the wavefunction is at the core, no ionization of the Rydberg electron can occur, so the presence of the hole affords the atom a transient and recurring stabilization against ionization (Fedorov and Movesian 1988, Parker and Stroud 1990, Burnett et al 1991).

A quantum mechanical picture of a wavepacket or anti wavepacket requires that we produce a coherent superposition of states. In our case the antiwavepacket is produced by a resonant Raman process via the continuum, as shown by Figure 4. The atoms are driven to the continuum and back down to a different 6snd Rydberg state by the ps pulse. The Raman process is resonant since the bandwidth of the ps laser must be broad enough to cover several Rydberg states.

The experiment we carried out **was performed in the same general way as the** inner electron ionization experiment(Noordam et al).

Figure 4 Schematic energy level diagram of the two ways of populating neighboring states starting from one Rydberg state. The Λ path corresponds to coupling via the continuum while the V path represents coupling via a virtual bound state (---). As shown, the initial Rydberg state is created by two step resonant ns laser excitation from the ground $6s^2$ state, via the 5d6p state, prior to exposure to the intense ps pulse.

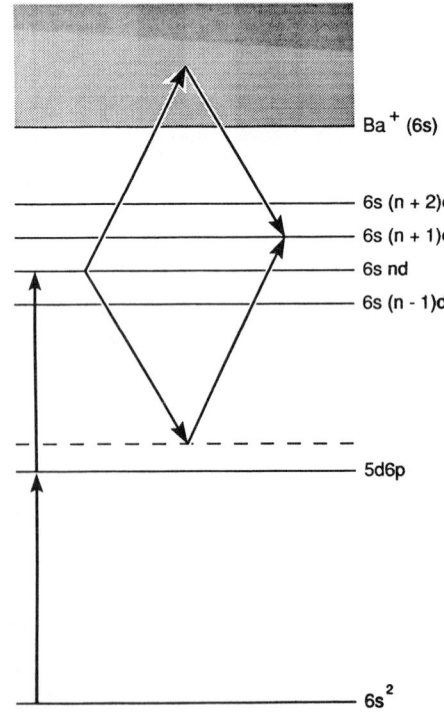

et al 1992). Ba atoms were excited to **the mixed Rydberg valence state 5d7d with two ns lasers** by the route 6s6s–5d6p–5d7d. The 5d7d **state is imbedded in the 6snd series at n≈26, and has 30%** 5d7d valence character. We chose to use the 5d7d state because it has a large photoionization cross section due to its valence character. The bound Rydberg atoms were exposd to a 1.5 ps pulse, after which the atoms were field ionized to analyze the Rydberg state distribution subsequent to the ps pulse. Typical observations are shown in Figure 4.

Figure 5 Field ionization spectrum of Ba Rydberg states when the initial state is the 5d7d state. The 5d7d and 6s26d states ionize at the same field and are indistinguishable in this spectrum. Trace (a) is with the ps laser while trace (b) is without it. The peaks to the left of the saturated 5d7d signal in trace (a) correspond to the excitation of the 27d-31d states by the ps laser.

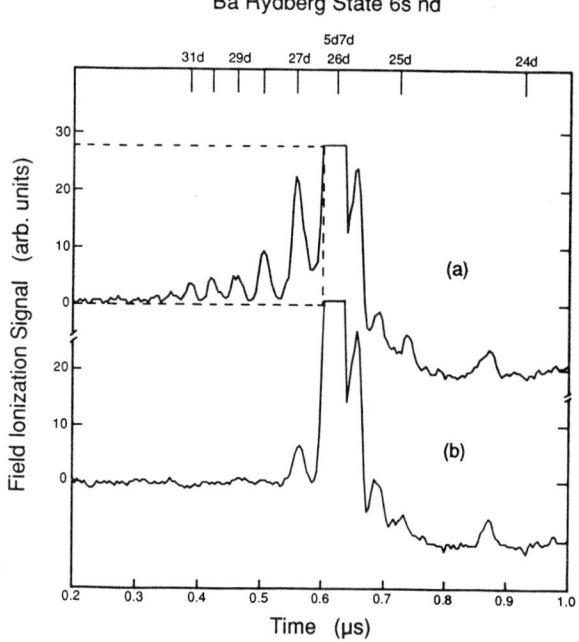

Without the ps laser, the field ionization spectrum consists of one strong, instrumentally truncated peak corresponding to the 5d7d state, and a subsidiary peak corresponding to the adjacent Rydberg state, which is not fully resolved by the ns dye laser. The ps laser evidently redistributes population to about five higher Rydberg states, up to n=31. This range of redistribution, over five states, is what we observe if we populate the Rydberg states from the 5d6p state using the ps laser. An important point is that the redistribution shown in Figure 5 does not occur by means of the V transition to a virtual state near the 5d6p state as shown in Figure 4. Tuning the ps laser across the 5d7d-5d6p resonance has no effect on the redistribution. Thus we can say with some assurance that the redistribution appears to occur by the continuum Λ Raman process of Figure 4, not via a bound state process.

5. Conclusion

Intense optical pulses alone are not sufficient to ensure ionization of a Rydberg atom. The pulse duration must exceed the orbit time to allow the electrons in all the Rydberg atoms to experience the pulse while they are near the ion core. If the pulse is short, a coherent superpeosition of Rydberg states is created and ionization of the Rydberg electrons ceases.

6. Acknowledgements

This work has been supported by the National Science Foundation.

7. References

Burnett K, Knight P L, Piraux B R M and Reed V C 1991 *Phys. Rev. Lett.* **66** 301
Cooke W E, Gallagher T F, Edelstein S A and Hill R M 1978 *Phys. Rev. Lett* **40** 178
Fedorov M V and Movesian A N 1988 *J. Opt. Soc. Am. B* **5** 850
Jones R R and Bucksbaum 1991 *Phys. Rev Lett.* **67** 3215
Noordam L D, Stapelfeldt H, Duncan D I and Gallagher T F 1992 *Phys. Rev. Lett.* **68** 1496
Parker J and Stroud C R Jr 1990 *Phys. Rev. A* **41** 1602
Stapelfeldt H, Papaioannou D G, Noordam L D and Gallagher T F 1991 *Phys. Rev. Lett.* **67** 3223

Inst. Phys. Conf. Ser. No 128: Section 2
Paper presented at RIS 92, Santa Fe, NM, USA, 24–29 May 1992

RIS measurement of AC Stark shifts and photoionization cross sections in calcium

J.B. Kim[*], X. Xiong[§], T.R. O'Brian, T.J. McIlrath[†] and T.B. Lucatorto

National Institute of Standards and Technology, Gaithersburg, MD 20899, USA

[*]*Atomic Spectroscopy Dept., Korea Atomic Energy Res. Inst., P.O.Box 7, Taejon, Korea*
[†]*Also with IPST, University of Maryland, College Park, MD 20742, USA*
[§]*Eastern Analytical Inc., College Park, MD 20742, USA*

The ac Stark shifts of four high-lying Ca levels in intense laser fields (10^8 to 10^{10} W/cm^2) are studied using two-photon resonant, three-photon ionization spectroscopy. The two-photon Rabi rates and photoionization cross sections for the excited states are also measured, and compared to values obtained from detailed modeling of the RIS profiles. Although three Rydberg levels studied display the expected ponderomotive energy shifts, a fourth doubly-excited level deviates significantly from ponderomotive behavior.

AC Stark shifts of highly-excited atomic levels in intense laser fields influence the photoelectron energy spectrum in above threshold ionization (ATI) experiments and provide insight into the interactions of atoms with strong fields [1]. Although second order modelling of Rydberg atoms in strong laser fields predicts ponderomotive ac Stark shifts (equal to the mean kinetic energy imparted to a free electron by the oscillating applied field), a number of experiments demonstrate significant departure from this predicted behavior [2]. In this paper we describe a resonance ionization experiment measuring the ac Stark shifts in four high-lying levels of Ca, using two-photon excitation of the upper levels to probe the energy shifts induced by a strong laser field (up to 10^{10} W/cm^2) of a different frequency.

A selected upper level, perturbed by the intense laser field, is populated by the relatively weak probe laser tuned to the two-photon resonance. The upper level population is interrogated (as a function of probe laser frequency) through ionization of the excited level by a single photon from the strong laser field. A related experiment measures the two-photon Rabi rates (due to the probe laser) and photoionization cross-sections (due to the strong laser) for the excited states, permitting detailed modeling of the RIS line profiles obtained. Three of the closely-spaced four levels studied exhibit the expected ponderomotive behavior. The fourth level departs significantly from ponderomotive behavior, probably due to strong coupling with continuum states.

The $4s9s$ $^1S_0$, $4s8d$ $^1D_2$, $4s10s$ $^1S_0$, and $3d5s$ $^1D_2$ levels of Ca, with energies near 47,000 cm^{-1} (compared with the ionization potential of about 49,300 cm^{-1}), are investigated. Ca atoms are produced in a vacuum chamber by heating the pure metal. To measure the energy shift of a level in an intense laser field, the energy of the two-photon resonance (relative to the ground state) is measured by scanning the probe laser (operating near 420 nm) through the resonance as the atoms are subjected to the intense field of a second laser (at 1064 nm). The excited atoms are ionized primarily by the strong laser and the ions detected with a simple time-of-flight mass spectrometer with single ion sensitivity.

Figure 1 schematically depicts the relevant energy levels and processes.

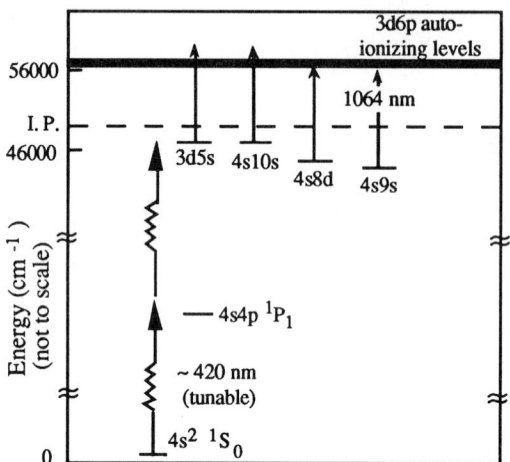
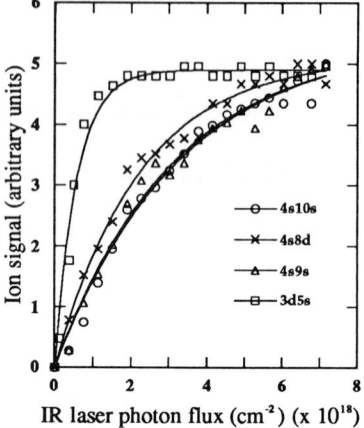

Figure 1. Relevant energy levels in Ca for measurement of two-color three photon RIS profiles.

Figure 2. Determination of ir photo-ionization cross sections as explained

The intense field is produced by a frequency stabilized injection-seeded pulsed Nd:YAG laser (1064 nm) operating in a single longitudinal mode, eliminating the need to consider complications from a stochastic, multimode perturbing field [3]. Part of the Nd:YAG radiation is frequency doubled to pump a tunable dye laser operating near 840 nm, and the dye laser radiation is frequency doubled to about 420 nm for two-photon excitation of a selected Ca upper level, exploiting near resonance with the intermediate $4s4p$ $^1P_1$ level. The ir and probe (blue) laser beams are carefully overlapped spatially and temporally and focused to the same position in the vacuum chamber, after the sizes of the laser beams are adjusted to compensate for lens dispersion and to ensure the blue beam waist is smaller than, and completely contained within, the ir beam waist. The ir and blue pulse durations are about 18 and 4 ns respectively. The polarization of each laser is linear and mutually parallel. Further experimental details are published elsewhere [4]. The probe (blue) laser intensity is kept sufficiently small that no ac Stark shift is detected from the blue laser alone and that the rate of photoionization due to the blue laser is negligible compared to the ir photoionization rate. The 1064 nm radiation couples the excited states to autoionizing levels in the $3d6p$ configuration enhancing the ionization rate. The ir laser intensity is varied from zero to about 10^{10} W/cm^2.

To interpret the resulting resonance ionization spectra, the two-color, three-photon resonant ionization process is modeled using the theory of L'Huillier et al. [5]. This model requires values for the photoionization cross sections (due to the strong ir laser) and the two-photon Rabi rates (for the weak probe laser) for the excited levels, and these data are estimated from related experiments. To measure the ir photoionization cross section of a level, the blue laser is tuned to the zero-field two-photon resonance while the ir laser is delayed by about 15 ns from the peak of the blue laser pulse. The blue laser intensity is held constant and small enough to produce negligible ionization, and the number of photoions is measured as the ir photon flux is varied. The lifetime of each of the excited levels is substantially longer than 15 ns, so the rate of ion production during the ir laser pulse can be approximated by the simple rate equation

$$dN_{ions}/dt = N_e\ \sigma_{ir}\ I_{ir}$$

with solution

$$N_{ions} = N_e(\tau)[1 - \exp(-\sigma_{ir}\ I_{ir}\tau)]$$

where N_{ions} is the number of ions from the ir photoionization, N_e is the number of excited state atoms, $N_e(\tau)$ is the number of excited state atoms at the end of the blue laser pulse, σ_{ir} is the ir photoionization cross section, I_{ir} is the ir laser photon flux density, and τ is the duration of the ir laser pulse. Figure 2 displays the results of the measurements along with best fits to the data, giving photoionization cross sections (in Mb): 0.32 (*4s9s*) , 0.45 (*4s8d*), 0.31 (*4s10s*), and 1.8 (*3d5s*). Note that the cross section for the *3d5s* state is significantly larger than the others.

The two-photon Rabi rates for the upper levels are estimated in a similar manner. The blue laser intensity is kept at the same low intensity to excite each of the four levels and a delayed ir pulse of sufficient intensity is used to ionize essentially all of the excited state population. During the blue laser excitation of the upper level (with no ir field) the ground and excited state populations are determined by the rate equations

$$d\,N_g/d\,t = - (N_g - N_e\,)\Omega$$
$$d\,N_e\,/d\,t = (N_g - N_e\,)\Omega$$

where N_g and N_e are the ground state and the excited state populations and Ω is the generalized Rabi rate, $\Omega = \Omega_{ge}^2/4\gamma_L$. Ω_{ge} is the two-photon Rabi rate between the ground and excited states, and γ_L is the blue laser linewidth (about 0.3 cm^{-1}). Solution of the equations above gives

$$N_{ions} = 1/2\ N_0[1- \exp(-2\Omega\tau\,)]$$

with N_0 the initial number of atoms subject to the blue laser pulse. Since N_0 cannot be accurately determined in this experiment, only relative values of Ω_{ge} can be measured. These relative measurements are normalized to a simple calculation of Ω_{ge}(*4s10s*) assuming a single intermediate state using the tabulated value for the *4s^2* $^1S_0 \Rightarrow 4s4p$ $^1P_1$ oscillator strength (*f* =1.75) [6] and a calculated value for the *4s4p* $^1P_1 \Rightarrow 4s10s$ $^1S_0$ oscillator strength (*f* = 3.6 x 10^{-4}) [7]. Based on this calculation of Ω_{ge}(*4s10s*) and the measured ratios, we obtain the following results for the Rabi frequencies (with probe laser intensity I_{420} in W/cm^2) : Ω_{ge}(*4s9s*) = 140 I_{420} rad/sec, Ω_{ge}(*4s8d*) = 46 I_{420} rad/sec, Ω_{ge}(*4s10s*) = 210 I_{420} rad/sec, and Ω_{ge}(*3d5s*) = 390 I_{420} rad/sec.

Using the measured photoionization cross sections and the estimated two-photon Rabi rates, the model of L'Huillier et al [5] is applied to the resonant ionization spectra of the levels investigated in this experiment. Figure 3 shows the measured (dotted) and calculated (solid) ionization signal as a function of probe (blue) laser frequency as the frequency is swept across the *4s10s* $^1S_0$ and *3d5s* $^1D_2$ two-photon resonances for three different ir laser intensities. Agreement between theory and experiment is good for the *4s10s* $^1S_0$ level. The *4s9s* $^1S_0$ *4s8d* $^1D_2$ data (not shown) and predictions show similar agreement. However, no choice of

σ_{ir} and Ω_{ge} gives ponderomotive RIS profiles that reasonably correspond to the data for the *3d5s* $^1D_2$ level. Figure 3b shows the best correspondence between the *3d5s* data and theory, achieved by using an energy shift equal to half the magnitude of the ponderomotive potential and values of σ_{ir} and Ω_{ge} significantly different from the measured values. The discrepancy

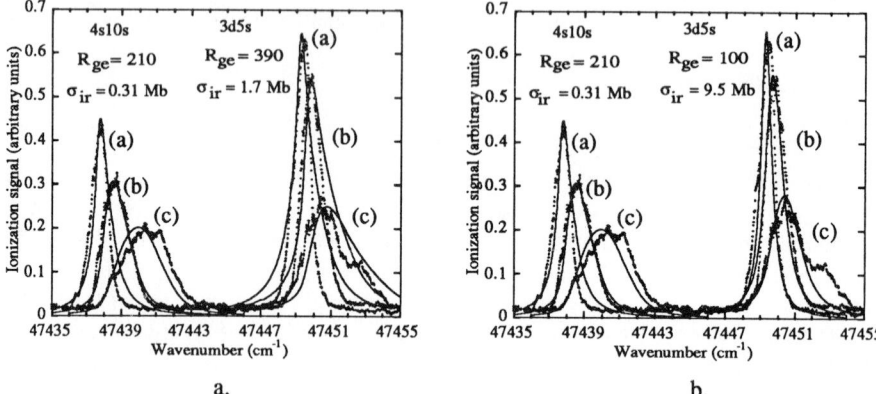

a.   b.

Figure 3. Measured (dotted) and caluculated (solid) RIS profiles for the *4s10s* $^1S_0$ and *3d5s* $^1D_2$ levels of Ca. $\Omega_{ge} = R_{ge}I_{420}$ a. Ponderomotive potential and experimental Rabi rates and photoionizaiton cross sections for both levels. Although the *3d5s* model reasonably predicts the peak energies, the predicted linewidths and shapes differ significantly from measured values. b. Half-ponderomotive potential for *3d5s* with "best fit" values of Ω_{ge} and σ_{ir}. IR laser intensities (x 10^9 W/cm^2): 0.66 (a), 2.0 (b) and 3.9 (c).

between the ac Stark shifts and RIS profiles of the *3d5s* $^1D_2$ compared to the other (Rydberg) levels may be due to strong coupling between the *3d5s* configuration and continuum levels[4], but further work is needed to resolve questions concerning resonance ionization phenomena in strong fields.

References

1.   T.J. McIlrath, P.H. Bucksbaum, R.R. Freeman, and M. Bashansky, Phys. Rev. A **35**, 4611 (1987); and P. Kruit, J. Kimman, H.G. Muller, and M.J. van der Weil, Phys. Rev. A **28**, 248 (1983).
2.   D. Normand, L.-A. Lompre, A. L'Huillier, J. Morellec, M. Ferray, J. Lavancier, G. Mainfray and C. Manus, J. Opt. Soc. Am. B **6**, 1513 (1989) and references therein.
3.   L.-A. Lompre, G. Mainfray, C. Manus and J.P. Marinier, J. Phys. B **14**, 4307 (1981), and P. Zoller, J. Phys. B **15**, 2911 (1982).
4.   Q. Li, J.B. Kim, X. Xiong, T.R. O'Brian, T.J. McIlrath, and T.B. Lucatorto, submitted to Opt. Lett.
5.   A. L'Huillier, L.-A. Lompre, D. Normand, X. Tang and P. Lambropoulos, J. Opt. Soc. Am. B **6**, 1790 (1989).
6.   W.L. Wiese, M.W. Smith, and B.M. Miles, <u>Atomic Transition Probabilities</u>, Vol. II (NSRDS-NBS 22, U.S. Gov. Printing Office).
7.   Q. Li, T.J. McIlrath, E.B. Saloman and T.B. Lucatorto, <u>Resonance Ionization Spectroscopy 1990</u>, J.E. Parks and N. Omenetto, eds. (The Institute of Physics, Bristol, 1991), p. 55.

Inst. Phys. Conf. Ser. No 128: Section 2
Paper presented at RIS 92, Santa Fe, NM, USA, 24–29 May 1992

Resonance ionization spectroscopy of zirconium atoms

Ralph H. Page, Stephen C. Dropinski, Earl F. Worden Jr., and J. A. D. Stockdale

AVLIS Program, Lawrence Livermore National Laboratory, P. O. Box 808
Livermore, California 94550 USA

We have examined the stepwise-resonant three-photon-ionization spectrum of neutral Zirconium atoms using three separately-tunable pulsed visible dye lasers. Lifetimes of even-parity levels (measured with the delayed-photoionization technique) range from 10 to 100 nsec. Direct ionization cross sections appear to be less than 10^{-17} cm^2; newly-detected autoionizing levels give peak ionization cross sections (inferred from saturation fluences) up to 10^{-15} cm^2. Members of Rydberg series converging to the 315 and 1323 cm^{-1} levels of Zr^+ were identified. "Clumps" of autoionizing levels are thought to be due to Rydberg-valence mixing.

Zirconium metal is widely used in nuclear reactors because it is tough and corrosion-resistant. Most importantly, it has a low thermal-neutron absorption cross section dominated by the contribution from 11%-abundant ^{91}Zr. Thus the neutronic properties are improved if the 91 isotope is selectively removed. This has been known for some time, but since 91 is an "interior" isotope (the others being 90, 92, 94, and 96,) purely mass-dependent separation techniques (e.g. diffusion) are much less attractive than optical methods (i.e. AVLIS—Atomic Vapor Laser Isotope Separation).

Hackett *et al.* (1988) have demonstrated two-color, mass-selective ionization of Zr atoms in a beam; the 53506-cm^{-1} ionization potential (Hackett 1986) can be reached with 3 photons from a copper-vapor-laser-pumped dye laser system. Since Doppler widths in a practical separator exceed the ~100 MHz isotope shifts of levels accessible with the first (λ_1) photon, isotopic selectivity was derived from the larger magnetic-dipole hyperfine shift (present only in ^{91}Zr). Three levels were identified as potential first steps (λ_1 settings) in photoionization pathways for the lowest thermally-populated levels, on the basis of favorably-grouped hyperfine transitions offset from the even-isotope envelope. Fixing λ_1 and scanning the second (λ_2) laser wavelength uncovered resonances (E_2 levels) around 36,000 cm^{-1}, resonantly enhancing two of the three photon-absorption transitions.

The design of an efficient photoionization scheme requires (at least) better knowledge of the photoionization (λ_3) spectrum, transition cross sections, and level lifetimes. Following the example of Carlson *et al.* (1976), we used a set of three Nd:YAG (532 nm)-pumped pulsed tunable rhodamine-class dye lasers, the third with variable timing, to measure these properties. With λ_1 and λ_2 set to pump an E_2 level, λ_3 was scanned to find a strong autoionizing resonance that gave a steady ion signal. Then, by delaying the λ_3 pulse's delay and fluence and noting the signal variations, we derived the E_2 level's lifetime and estimated its ionization cross section.

Beforehand, we confirmed the λ_2 searches (with an arbitrary λ_3 setting); in general, they extended to the 542-nm limit of the lasers' tunability.

Laser pulsewidths were about 5 nsec, fluences were typically 0.1 - 10 mj/cm², and linewidths were about 0.3 cm⁻¹; optogalvanic wavelength calibration with Ne lines gave level positions accurate to within 1 cm⁻¹. Our atomic beam source was a Zr-charged tungsten crucible surrounded by a slotted, resistively-heated cylindrical filament formed from a rolled sheet of tungsten. A slit collimated the effusive beam, which intersected the overlapped laser beams between a biased plate and grid that directed ions into the snout of a Galileo model 4716 channeltron. The ~100 V/cm ion-extraction field caused ion signals to be observed several tens of cm⁻¹ below the nominal ionization potential.

The λ_1 transitions (Fig. 1) terminate on levels with lifetimes of a few hundred nsec (Biemont 1981, Hannaford 1983), eminently suited to stepwise ionization with ~50 nsec copper vapor laser system pulses. Literature values (Biemont 1981, Corliss 1962) of the three largest oscillator strengths are 0.001 - 0.01; the approximate validity of the LS-coupling description of low-lying levels of Zr implies that some transitions (e.g. 0 - 18244) are especially weak because they violate the $\Delta J = \Delta L$ propensity rule.

Figure 2 (modeled after Fig. 4 of Hackett 1988) summarizes the salient results (for resonance ionization purposes) of our λ_2 and λ_3 scans. Level positions agree with those of Hackett *et al.* Three levels have been tabulated by Moore (1952): one at 35476 cm⁻¹ (not shown on the figure since no λ_3 scans were done); the 35210 level, and the 35860 level. All have 10 nsec lifetimes and have been attributed to a ⁵F term of the *5s6s* configuration. Their short lifetimes imply large

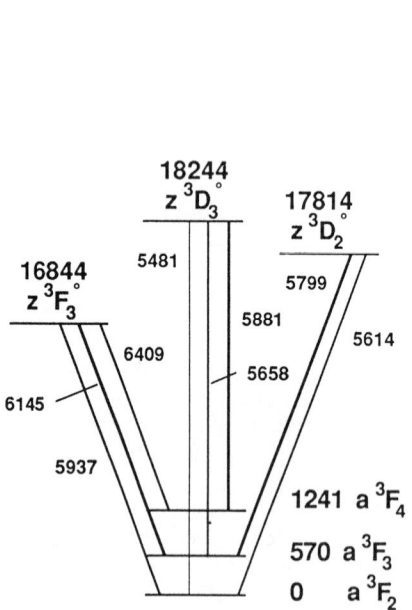

Fig. 1. λ_1 transitions with "good" hyperfine structure (lines tightly grouped and shifted from the even isotopes); oscillator strength increases with line weight.

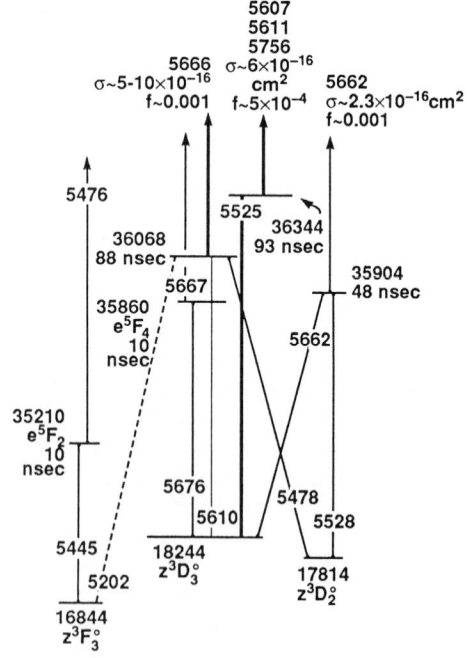

Fig. 2. E_2 levels and lifetimes, λ_2 wavelengths, and strongest λ_3 transition wavelengths and cross sections.

absorption and emission oscillator strengths, enhancing the signal strength and ease of detection when excited in a discharge. All other levels, for which no formal assignments exist, have longer lifetimes. Lines of varying weight subjectively depict relative transition strengths; we observed saturation broadening of the 5525 Å 18244 - 36344 transition, and believe its oscillator strength exceeds 0.001. This new lifetime data will facilitate the analysis of the emission spectrum recorded at the Kitt Peak Fourier transform spectrometer (Conway 1992.)

Figure 2 includes vacuum wavelengths of the strongest autoionizing transitions for each level. For the longer-lived levels, a cross section (saturation fluence) was deduced from the "knee" location in the ion signal-*vs*-fluence curve. Lifetimes not much longer than the laser pulse give gentler saturation curves, hindering the measurement. In particular, the 35210 level's λ_3 spectrum (which was short because of limited laser tunability) did not include any easily-saturated autoionizing transitions. Although the homogeneous linewidths of the narrowest peaks are still in question, the strongest signals we observed in all spectra were due to lines noticeably broader than the laser linewidth, and the integrated cross sections imply oscillator strengths around 0.001.

From a practical point of view, it is not yet obvious how to apportion a fixed amount of laser power (between three different wavelengths) to ionize the greatest number of zirconium atoms. However, we did note a potentially-useful coincidence between the 5662 Å λ_2 setting for the 18244 - 35904 transition, and the 5662 Å λ_3 peak ionization wavelength for the 35904 level. This would facilitate simple one-color detection of atoms in the 18244 level, or two-color ionization of zirconium.

While a detailed assignment of the features in the ionization spectra is not to be expected, we noticed some similarities in the spectra of different E_2 levels. The direct ionization cross section is seen to be quite small, with appreciable stretches of flat baseline between peaks. There are discernable "clumps" around 20 cm^{-1} wide (e.g. at 53710 cm^{-1}; Fig. 3) at several energies. Their detailed structures vary from level to level, but it seems reasonable to interpret them in terms of Rydberg-valence mixing. They undoubtedly contain spectral structure that could be better-resolved with narrow-band lasers and field-free ionization, possibly increasing the peak ionization cross sections dramatically. Figure 4 is a map of the clumps; it may aid in characterizing the E_2 levels and the valence (E_3) levels that appear to be the source of the oscillator strength.

Knowledge of the Zr I ionization potential and Zr II metastable level energies allowed us to identify portions of autoionizing Rydberg series converging to the ion's first (315) and third (1323 cm^{-1}) excited states. A few members ($n \sim 20$, with quantum defect near 0.95) of the series converging to the 315 cm^{-1} level are evident in the 35210 cm^{-1} spectrum. Their lines are not especially broad, in contrast with

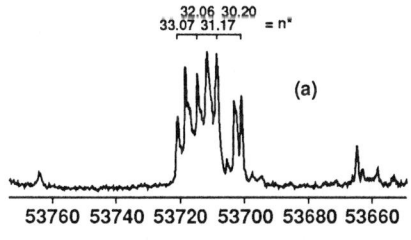

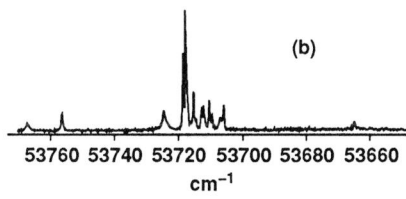

Fig. 3. The "clump" at 53710 cm^{-1} as observed from (a) the 35904 cm^{-1} level and (b) the 36344 cm^{-1} level. Quantum numbers refer to a Rydberg series converging to the Zr$^+$ 315 cm^{-1} level.

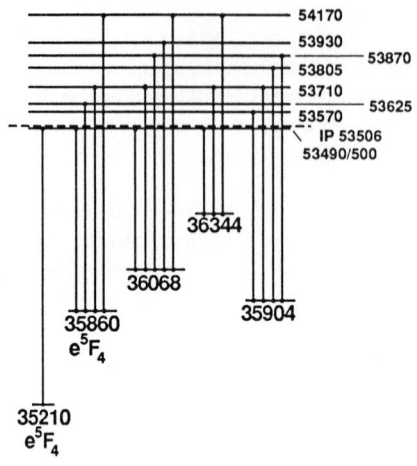

Fig. 4. Autoionizing level structure observed in transitions from more than one E_2 level.

a series observed from the 36068 cm^{-1} level, converging to the Zr II 1323 cm^{-1} level. In that case (again, the n ~ 20 region,) the peaks are somewhat asymmetric, the linewidths are about 1/3 the level spacings, and there is a modulation of the quantum defect and peak intensity as one passes through the maximum at 53950 cm^{-1} total energy. The clump in Fig. 3 appears to contain part of (at least one) series that converges to the 315 cm^{-1} level. Its highly-localized, steep-sided intensity distribution has yet to be explained.

We have provided a first glimpse into the Zr autoionizing level structure accessible with three visible photons. The derived lifetimes and transition strengths should suggest further studies and influence considerations of the feasibility of Zr AVLIS technology.

References

Biemont E, Grevesse N, Hannaford P and Lowe R M 1981 *Ap. J.* **248** 867

Carlson L R, Paisner J A, Worden E F, Johnson S A, May C A and Solarz R W 1976 *J. Opt. Soc. Am.* **66** 846

Conway J and Worden E 1992 work in progress (private communication)

Corliss C and Bozman W R 1962 *Experimental Transition Probabilities for Spectral Lines of Seventy Elements,* Natl. Bur. Stand. Monograph No. 53 (Washington, D.C.: U. S. Govt. Printing Office)

Hackett P A, Morrison H D, Bourne O L, Simard B and Rayner D M 1988 *J. Opt. Soc. Am. B* **5** 2409

Hackett P A, Humphries M R, Mitchell S A and Rayner D M 1986 *J. Chem. Phys.* **85** 3194

Hannaford P and Lowe R M 1983 *Opt. Eng.* **22** 532

Moore C E 1952 Atomic Energy Levels as Derived from the Analyses of Optical Spectra, vol. II, NSRDS-NBS 35 (Washington, D. C.: U. S. Govt. Printing Office)

Inst. Phys. Conf. Ser. No 128: Section 2
Paper presented at RIS 92, Santa Fe, NM, USA, 24–29 May 1992

Status report on the National Institute of Standards and Technology Resonance Ionization Spectroscopy/Resonance Ionization Mass Spectrometry Data Service

E. B. Saloman

Electron and Optical Physics Division, National Institute of Standards and Technology, Gaithersburg, Maryland 20899, USA

ABSTRACT: A resonance ionization spectroscopy data service is established at the National Institute of Standards and Technology. It provides atomic structure and laser information required for RIS/RIMS and publishes formatted data sheets. Three groups of data sheets have been published. They cover Al, As, Au, B, Bi, C, Ca, Cd, Co, Cr, Cs, Cu, Fe, Ge, Hg, Kr, Mg, Na, Ni, P, Pb, Sb, Si, Sn, and Zn. Preliminary versions of several more have been completed.

1. INTRODUCTION

The techniques of Resonance Ionization Spectroscopy (RIS) and RIS followed by Mass Spectrometry (RIMS) have been demonstrated in several state-of-the-art laser laboratories as analytical techniques of high elemental sensitivity. For example, Bekov (1989) and co-workers have demonstrated detection limits as low as .001 ppt (one part in 10^{15}) for Pt and Rh in sea water. Hurst and co-workers (1977) have demonstrated single atom detection and also the counting of 1000 atoms of one Kr isotope which was mixed with 2×10^5 atoms of other Kr isotopes and 1×10^{10} atoms of He (Hurst *et al* 1985).

These demonstrations present the RIS/RIMS technique as a potentially very powerful tool for monitoring environmental pollutants, performing trace impurity analysis in semiconductors, alloys, and other materials, measuring radioactive contamination, studying biological processes, as well as many other areas. However, to apply these techniques one must know the atomic structure of the element or elements to be analyzed and the structure of the other elements in the sample. One must know the transition energies connecting the different levels of the atoms and the associated oscillator strengths to choose an efficient ionization pathway (RIS scheme). One must also know the photoionization cross sections going to the continuum and the possibility of autoionizing resonances accessible with available lasers.

Often this information is not available and requires an expert in atomic structure calculations to provide reliable data. In addition to our service of providing data sheets, we have begun a new interactive service for the RIS/RIMS community in which we will provide calculations to estimate the ionization rates for the final steps of the RIS process as well as information about RIS schemes for particular elements to be detected under the specific experimental conditions of interest.

2. PHOTOIONIZATION CALCULATIONS

We make estimates of the excited state photoionization cross sections by using the Hartree-Fock codes with approximate relativistic corrections of Cowan (1981). These calculations generally involve configuration interactions, i.e. the inclusion of several configurations in the wavefunction of each level. Often the photoionization cross section is very dependent on the degree of configuration mixing. In fact, RIS measurements of these cross sections indicated that it was necessary to adjust the Hartree-Fock energies in those special cases where they deviated significantly from observed values (so as to obtain the correct configuration mixing) in order to achieve a reasonable approximation for the cross section. The estimated uncertainty of the calculated photoionization cross sections is a factor of two or three. Extra efforts can be made to locate autoionizing resonances in the range of interest where important measurements are involved.

3. DATA SHEETS

Calculations and data collection have been carried out and the first three groups of data sheets (for the elements Al, As, Au, B, Bi, C, Ca, Cd, Co, Cr, Cs, Cu, Fe, Ge, Hg, Kr, Mg, Na, Ni, P, Pb, Sb, Si, Sn, and Zn) have been published (Saloman 1990) (Saloman 1991) (Saloman 1992) in Spectrochimica Acta B. Preliminary data sheets for additional elements have been produced and work is continuing on producing more. Updates on existing sheets will be published when RIS/RIMS results indicate the need for revision of information in existing sheets. As in the past, we continue to solicit suggestions on how to improve our data service. We thank the many researchers who have advised us in the past.

The sheets list the element, its ionization energy, an introduction to the RIS/RIMS studies carried out on it, its stable isotopes, isotope shifts and hyperfine structure, RIS schemes, atomic energy levels, lifetimes, oscillator strengths, laser schemes, atom sources, and references for both the atomic data and the RIS/RIMS schemes. Also included are the results of calculations of excited state photoionization cross sections by Hartree-Fock techniques. Suggestions as well as reprints, preprints, and reports of RIS/RIMS work are solicited to allow us to include the information in future data sheets.

The table summarizes the schemes included in the data sheets completed so far (preference is given to experimentally proven schemes). For each element completed, the table lists the reference containing that data sheet. Then the first ionization energy of that element is given. Next the table specifies the types of RIS schemes for which data are provided for that element. The notation of the RIS schemes is explained by a series of examples which may be generalized to obtain cases not specifically stated: $\omega_1+\omega_1$ is a one-color two-photon RIS process in which the first photon excites the atom to a resonance level which is then photoionized by a second photon of the same color; $\omega_1+\omega_2$ is a two-color two-photon process in which the photoionizing photon has a different color than the excitation photon; $\omega_1+\omega_2+\omega_3$ is a three-color process involving two resonance steps and concluding with photoionization; $\omega_1+\omega_2^{AI}$ is a two-color process in which the first photon excites the atom to a resonance level which is then excited to an autoionizing state by the second photon; $\omega_1 + \omega_2^R$ is a two-color process in which the first photon excites the atom to a resonance level and the second photon excites the atom from this level to a high-lying Rydberg state which is subsequently ionized by collisions or by an electric-field pulse; and $2\omega_1 + \omega_1$ is a one-color process in which the first step is a two-photon transition to a resonance level followed by photoionization by a third photon of the same wavelength.

Summary of Completed Data Sheets for the RIS/RIMS Data Service

Symbol of Element	Ref. *	Ionization Energy (cm^{-1})	Types of RIS Schemes covered in the data sheet for this element		
Al	2	48278	$\omega_1 + \omega_1$ $\omega_1 + \omega_2 + \omega_3$	$\omega_1 + \omega_2$ $\omega_1 + \omega_2 + \omega_3^{AI}$	$\omega_1 + \omega_2^R$
As	1	78950	$2\omega_1 + \omega_1$	$2\omega_1 + \omega_2^{AI}$	
Au	1	74409	$\omega_1 + \omega_1$ $\omega_1 + \omega_2 + \omega_3$	$\omega_1 + \omega_2$ $\omega_1 + \omega_2 + \omega_3^{AI}$	$2\omega_1 + \omega_1$
B	1	66928	$\omega_1 + \omega_2 + \omega_3$		
Bi	3	58762	$\omega_1 + \omega_1$	$\omega_1 + \omega_2 + \omega_3$	$2\omega_1 + \omega_1$
C	1	90820	$2\omega_1 + \omega_1$		
Ca	2	49306	$\omega_1 + \omega_2 + \omega_1$ $\omega_1 + \omega_2 + \omega_3^{AI}$	$\omega_1 + \omega_2 + \omega_2$ $2\omega_1 + \omega_1$	$\omega_1 + \omega_2 + \omega_3$ $2\omega_1 + \omega_2^{AI}$
Cd	1	72540	$\omega_1 + \omega_1$ $\omega_1 + \omega_2 + \omega_1$ $2\omega_1 + \omega_2$	$\omega_1 + \omega_2$ $\omega_1 + \omega_2 + \omega_2$	$\omega_1 + \omega_2 + \omega_3$ $2\omega_1 + \omega_1$
Co	2	63564	$\omega_1 + \omega_1$	$\omega_1 + \omega_2 + \omega_3$	
Cr	2	54576	$\omega_1 + \omega_1$	$\omega_1 + \omega_2$	$\omega_1 + \omega_2^{AI}$
Cs	2	31406	$\omega_1 + \omega_1$	$\omega_1 + \omega_2 + \omega_3$	$2\omega_1 + \omega_2$
Cu	2	62317	$\omega_1 + \omega_1$ $2\omega_1 + \omega_1$	$\omega_1 + \omega_2$	$\omega_1 + \omega_2^{AI}$
Fe	1	63737	$\omega_1 + \omega_1$	$\omega_1 + \omega_2$	$2\omega_1 + \omega_1$
Ge	1	63713	$\omega_1 + \omega_1$ $\omega_1 + \omega_2 + \omega_3$	$\omega_1 + \omega_2 + \omega_1$	$\omega_1 + \omega_2 + \omega_2$
Hg	2	84184	$2\omega_1 + \omega_1$ $\omega_1 + \omega_2 + \omega_3$	$\omega_1 + \omega_2^{AI}$ $\omega_1 + \omega_2 + \omega_3^{AI}$	$\omega_1 + \omega_2 + \omega_2$

Kr	2	112914	$\omega_1+\omega_2$	$\omega_1+\omega_2+\omega_3$	$2\omega_1+\omega_1$
Mg	2	61671	$\omega_1+\omega_1$ $2\omega_1+\omega_1$	$\omega_1+\omega_2^{AI}$	$\omega_1+\omega_2+\omega_3$
Na	3	41449	$\omega_1+\omega_1$ $\omega_1+\omega_2^{R}+\omega_3$ $\omega_1+\omega_2+\omega_1$	$\omega_1+\omega_2$ $\omega_1+\omega_2+\omega_3$ $2\omega_1+\omega_1$	$\omega_1+\omega_2^{R}$ $\omega_1+\omega_2+\omega_2$
Ni	2	61619	$\omega_1+\omega_1$	$\omega_1+\omega_2+\omega_3$	
P	3	84581	$2\omega_1+\omega_1$		
Pb	1	59820	$\omega_1+\omega_1$	$2\omega_1+\omega_1$	$\omega_1+\omega_2+\omega_3$
Sb	3	69431	$\omega_1+\omega_1$ $\omega_1+\omega_2+\omega_3$	$\omega_1+\omega_2+\omega_1$	$\omega_1+\omega_2+\omega_2$
Si	1	65748	$\omega_1+\omega_1$	$\omega_1+\omega_2+\omega_3$	
Sn	3	59233	$\omega_1+\omega_1$	$\omega_1+\omega_2+\omega_2$	$2\omega_1+\omega_1$
Zn	1	75769	$\omega_1+\omega_1$ $\omega_1+\omega_2+\omega_3$	$\omega_1+\omega_2+\omega_1$ $2\omega_1+\omega_1$	$\omega_1+\omega_2+\omega_2$ $2\omega_1+\omega_2$

* References: [1] (Saloman 1990); [2] (Saloman 1991); [3] (Saloman 1992).

4. ACKNOWLEDGEMENTS

I wish to thank G. S. Hurst, J. E. Parks, J. C. Travis, and T. B. Lucatorto for their encouragement of this work. This project is supported in part by the U. S. Department of Energy Office of Health and Environmental Research under contract DE-AI05-86ER-60447.

5. REFERENCES

Bekov G I 1989 *Lasers in Atomic, Molecular, and Optical Physics* ed V S Letokhov (Singapore, World Scientific) pp 339-48
Cowan R D 1981 *The Theory of Atomic Structure and Spectra* (Berkeley, University of California)
Hurst G S, Nayfeh M H and Young J P 1977 *Appl. Phys. Lett.* **30** 229
Hurst G S *et al.* 1985 *Rep. Prog. Phys.* **48** 1333
Saloman E B 1990 *Spectrochim. Acta* **45B** 37
Saloman E B 1991 *Spectrochim. Acta* **46B** 319
Saloman E B 1992 *Spectrochim. Acta* **47B** 517

Inst. Phys. Conf. Ser. No 128: Section 2
Paper presented at RIS 92, Santa Fe, NM, USA, 24–29 May 1992

71

RIMS studies of high Rydberg and autoionizing states of the rare-earth element Dy

X Y Xu, H J Zhou, W Huang and D Y Chen

Department of Modern Applied Physics, Tsinghua University, Beijing, P.R.China

ABSTRACT: Two series of high Rydberg states $(n > 40)$, $4f^{10}(\,^5I_8)6snd$ and $4f^{10}(\,^5I_8)6sns$ of Dy were obtained. A new ionization potential of Dy, i.e. $IP = 47901.7 \pm 0.6 cm^{-1}$ was drawn. This is about one order more accurate than the results given in literature. Near the first ionzation limit of Dy, 235 new autoionizing states were found. To our knowledge, this is the first time to report the study of the autoionizing states of Dy.

1. INTRODUCTION

In the past the studies of atomic high–excited states were focused more on alkali and alkali–earth metals, but much less on rare–earth elements because the electronic structures of rare–earth elements are complicated; in which the mixed electron configurations and the perturbing states have a serious influence on their electronic energy levels, especially the high–lying excited states. With the development of laser resonance ionization spectroscopy(LRIS), the studies of high Rydberg states have been significantly improved through selecting proper experimental channels. The autoionizing states of rare–earth elements can also be found by LRIS, whose positions could be determined according to a certain relevant theory, so that we may study the mechanism of the autoionizing states from the profiles and intensities of the autoionizing spectra.

2. EXPERIMENTAL METHOD AND MEASURING RESULTS

The experimental setup (Xu 1990a) consists of a resonance ionization time–of–flight spectrometer, a laser system and a data aquisition system. A 5mg Dy sample was placed in an oven made of molybdenum, which was resistively heated to about 900℃. After ion–cutting with a DC electric field, a beam of neutral Dy atoms was produced and entered the intersective area. Two dye lasers pumped by an excimer laser were used to excite the atoms to Rydberg states and autoionizing states. In the measurement of Rydberg states, a high voltage pulse electric field was used to ionize the atoms after two laser resonance excitations and was applied to the electrodes 100ns later than the laser pulses. The ionized atoms were subsequently extracted and accelerated. Then they were allowed to travel in a one–meter drift tube. At the end they were collected by a pair of electron mutichannel plates. The signals were amplified and then entered a QDC (charge–to–digit converter). They were selected and sent to a computer system.

2.1. Measurement of Rydberg states

The ground state $4f^{10}6s^2(\ ^5I_8)$ of Dy can be excited to high–lying Rydberg states by two–step resonance excitation and ionized by a high–voltage pulse eletric field:

$$4f^{10}6s^2(\ ^5I_8)\ \xrightarrow{\ \lambda_1\ }\ 4f^{10}(\ ^5I_8)6s6p(\ ^3P^0_2)\ ^5I^0_8\ \xrightarrow{\ \lambda_2\ } \begin{cases} 4f^{10}(\ ^5I_8)6snd \\ 4f^{10}(\ ^5I_8)6sns \end{cases}$$

$$\underline{\text{electric field}}\ \ Dy^+ + e$$

where $\lambda_1 = 5547.25\,\text{Å}$ (in air). λ_2 was scanned from 3346Å to 3390Å. In the experiment the laser pulse energy of λ_1 was about 100μJ, whose linewidth was about 0.2cm^{-1} and the laser pulse energy of λ_2 was about 300μJ with linewidth 0.04cm^{-1}. A negative electric field pulse of 2000V was used for ionization and the Rydberg states (n$^* > 20$) could be detected experimentally. We obtained the calibrated wavelengths λ_2 as listed in the second column in Table 1 and Table 2 respectively.

Table 1 High–lying Rydberg series of $4f^{10}(\ ^5I_8)$6sns of Dy atom

1	$\lambda_2(\text{Å})$	$E_{expt}(\text{cm}^{-1})$	n*
1	3362.37	47762.62	28.15
2	3361.37	47771.49	29.10
3	3360.36	47780.41	30.15
4	3359.48	47788.20	31.18
5	3358.68	47795.29	32.20
6	3357.96	47801.66	33.22
7	3357.30	47807.53	34.25
8	3356.73	47812.57	35.21
9	3356.22	47817.14	36.15
10	3355.74	47821.43	37.11
11	3355.27	47825.54	38.11
12	3354.86	47829.24	39.08

Table 2 High–lying Rydberg series of $4f^{10}(\ ^5I_8)$6snd of Dy atom

1	$\lambda_2(\text{Å})$	$E_{expt}(\text{cm}^{-1})$	n*
1	3362.08	47765.21	28.30
2	3361.05	47774.33	29.29
3	3360.05	47783.13	30.36
4	3359.19	47790.76	31.38
5	3358.40	47797.75	32.41
6	3357.70	47804.05	33.44
7	3357.07	47809.59	34.42
8	3356.49	47814.71	35.42
9	3356.00	47819.12	36.34
10	3355.50	47823.53	37.35
11	3355.07	47827.39	38.27

2.2. Measurement of autoionizing states

The Dy atoms in the ground state were excited to autoionizing states in two steps. We selected four different intermediate states as exciting channels:

(1) $\quad 4f^{10}6s^2 \; {}^5I_8 \xrightarrow{\lambda_{11}} 4f^{10}({}^5I_8)6s6p({}^3P_2^0)^3I_7^0 \xrightarrow{\lambda_{12}}$ autoionizing states

(2) $\quad 4f^{10}6s^2 \; {}^5I_8 \xrightarrow{\lambda_{21}} 4f^9({}^6H^0)5d^2({}^3F)({}^8G^0)6s \; {}^9G_7^0 \xrightarrow{\lambda_{22}}$ autoionizing states

(3) $\quad 4f^{10}6s^2 \; {}^5I_8 \xrightarrow{\lambda_{31}} 4f^9({}^6F^0)5d6s^2 \; {}^7H_7^0 \xrightarrow{\lambda_{32}}$ autoionizing states

(4) $\quad 4f^{10}6s^2 \; {}^5I_8 \xrightarrow{\lambda_{41}} 4f^9 5d6s^2 \; {}^7K_7^0 \xrightarrow{\lambda_{42}}$ autoionizing states

where λ_{11}, λ_{21}, λ_{31} and λ_{41} were 5423.32 Å, 5395.57 Å, 5301.58 Å and 5451.31 Å (in air) respectively. λ_{12}, λ_{22}, λ_{32} and λ_{42} were scanned from 3330 Å to 3460 Å. In order to ensure both the saturated excitation of the first–step and a tolerable background, the pulse energy of the laser λ_{k1} (k = 1,2,3,4) were adjusted to about 1mJ and that of λ_{k2} (k = 1,2,3,4) about 2–3mJ. A partial spectrum of autoionizing state measured is shown in Fig.1

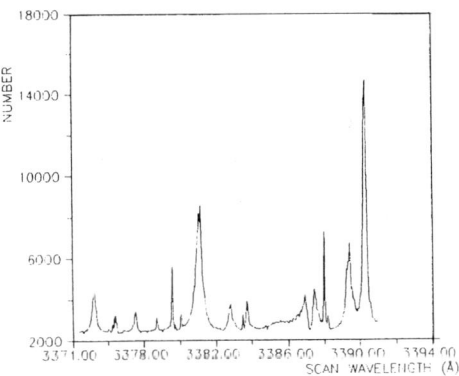

Fig. 1 Autoionizing spectrum of Dy by the excitation of $4f^{10}({}^5I_8)6s6p({}^3P_2^0)^3I_7^0$ state.

3. DATA ANALYSIS AND DISCUSSION

3.1. Determination of the first ionizing limit of Dy atom

So far as we know the first ionizing limit of Dy atom is $47900 \pm 5\,\text{cm}^{-1}$ (Martin 1978). The series of Rydberg states which is not influenced by perturbing state could be calculated in the frame work of single channel quantum defect theory. When n is large, the quantum defect of this series can be irrelevant to n (Xu 1990b):

$$E_n = IP - R/(n-\mu)^2 = IP - R/n^{*2}$$

The whole range of the original first ionizing limit (47895–47905 cm^{-1}) is divided into 100 steps. For each step, we can calculate the relevant n* values by using the non–perturbing Rydberg states measured in the experiment. Because the n^* values themselves are successively arranged, the quantum defects of this series differ only in decimal parts. By testing the consistency of the decimals, the most accurate IP can be obtained. Finally the first ionizing limit is taken as: $IP = 47901.7 \pm 0.6 cm^{-1}$.

3.2.   Determination of the positions and widths of autoionizing states

The total crosssection from an initial state to an autoionizing state is given by (Zhao 1991)

$$\sigma_{tot} = [B(\Gamma / 2) + A(E - E_0)] / [(\Gamma / 2)^2 + (E - E_0)^2] + \sigma_{nr}$$

The autoionizing states measured were non–linearly fitted to the above formula. According to the fitting results and the first ionizing limit obtained above, we got 62 autoionizing states in channel 1, 51 in channel 2, 97 in channel 3 and 25 in channel 4 respectively. Most of the linewidths of autoionizing states are great, in general, $\Gamma >$ $0.5 cm^{-1}$. The resonances of some states are very strong so that they can be used in LRIS hypersensitive analysis of Dy to improve the detecting sensitivity.

REFERENCES

Martin W C, Zalabas R and Hagan L 1978 Atmic Energy Levels —The Rare–Earth Element (Washington: U.S.Government Printing office)

Xu X Y, Tang J Z et al 1990a J. Phys. B **23** 3315

Xu X Y, Tang J Z et al 1990b Proc. 5th Int. Symp. on Resonance Ionization Spectroscopy and Its Applications (Bristol: IOP) pp247

Zhao W Z, Xu X Y et al 1991 Appl. Phys. B **52** 299

Inst. Phys. Conf. Ser. No 128: Section 2
Paper presented at RIS 92, Santa Fe, NM, USA, 24–29 May 1992

RIS studies of Rydberg structures of the lead atom

W.Y.Ma, Q.Hui, L.Q.Li, W.Z.Zhao, K.L.Wen and D.Y.Chen

Department of Modern Applied Physics, Tsinghua University, Beijing 100084, PRC

ABSTRACT: This paper describes the first work in measuring the even–parity 6pnp Rydberg series of the lead atom by means of the RIS–TOF mass spectrometer with a electrothermal atomization source. The Rydberg energy levels for the $6p(^2P^0_{1/2})np$ ($n = 37$–62) series were measured. We found the np series have fourfold structure. The energy levels and relevant quantum defects as well as the first ionization potential of Pb are obtained.

1. INTRODUCTION

Investigations of the high excited states of atoms are not only of significance in atomic physics, but also of great practical value in ultrasensitive analysis. Our goal of studying the Rydberg states of the lead atom is to find efficient ionization pathways applicable to ultrasensitive analysis in the environment science, such as the detection of lead concentration in ice and snow samples of the Antarctic.

The even–parity 6pnp Rydberg series ($J = 0$, $n = 10$–27; $J = 2$, $n = 10$–41) of the Pb atom have been measured with two–photon excitation from the ground state (Ding et al 1989). By exciting the intermediate states $6p(^2P^0_{1/2})7s\ ^3P^0_1$ or $6p(^2P^0_{3/2})7s\ ^3P^0_2$, the 6pnp Rydberg states of the Pb atom have been studied too (Young et al 1980, Martin et al 1982). Buch et al (1988) have investigated the odd–parity Rydberg states of lead with laser spectroscopic methods.But the data for Pb Rydberg states ($n > 40$) are much less complete so far. These data are important in studying the structure and selecting the optimal ionization schemes of Pb atoms. In this paper we report the measured results of the even–parity 6pnp Rydberg series of lead atoms by the RIS–TOF mass spectrometer.

2. EXPERIMENTAL METHOD

The RIS–TOF mass spectrometer includes several parts: electrothermal atomization, laser resonance excitation, electric field ionization and ion signal collection. The experimental layout is shown in Fig.1. The metallic lead sample was put into a graphite crucible which was heated by an alternating current to a temperature about 1000℃. After suppression of the thermal ions the collimated atomic beam of Pb was formed in a 5×10^{-4}Pa vacuum chamber. The two laser beams,which produced by two FL3002E dye lasers pumped by an excimer laser (EMG202), were directed so as to intersect the atomic beam perpendicularly between two electrodes, and the Pb atoms were stepwise excited and finally ionized by a electric field pulse, which was added to the electrode U_2. A DC 3.2kV voltage was also applied to the electrodes U_1 and U_2. The Pb ions were subsequently accelerated under the DC electric field. After travelling in a field–free drift tube of the TOF mass spectrometer they were detected by the microchannel plates (MCP). The current pulse signals from the MCP were amplified and then fed into the QDC (charge-to -digital converter). Finally the data can be obtained automatically by a microcomputer.

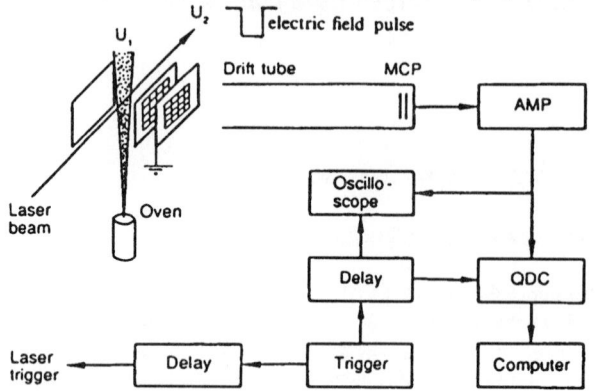

Fig.1   Schematic diagram of the RIS-TOF mass spectrometer

3. EXPERIMENTAL RESULTS AND DISCUSSION

The Pb atoms in ground state were excited to Rydberg states in two steps:

$$6p^2 \; ^3P_0 \; \xrightarrow[283.389nm]{\lambda_1} \; 6p(^2P^0_{1/2})7s \; ^3P^0_1 \; \xrightarrow[409.5-408nm]{\lambda_2} \; 6p(^2P^0_{1/2})np$$

The wavelength of λ_1 was 283.389nm. λ_2 was tunable in the range from 409.5nm to 408.0nm. The pulse energy fluxes of λ_1 and λ_2 were $14\mu J/cm^2$ and $2mJ/cm^2$ respectively. After excitation, a electric field pulse of $-330V/cm$ was applied and then the Rydberg atoms were ionized. The electric field pulse was delayed about 200ns relative to the laser pulse in order to reduce the Stark splitting. In the experiment, the resonance line corresponding to each Rydberg level of Pb can be obtained by fixing λ_1 and tuning λ_2 continuously. The experimental spectrum of $6p(^2P^0_{1/2})np$ ($n=37-62$) is obtained (shown in Fig.2) at the conditions that the linewidths of the two laser beams were $0.2cm^{-1}$ and the scanning step of λ_2 was 0.001nm. It is obvious that the measured 6pnp Rydberg states have fine structure. After narrowing the linewidth of λ_2 with the F-P etalon to $0.04cm^{-1}$ and decreasing the scanning step to 0.0003nm in the tunable range from 408.7nm to 408.2nm, we observed the fourfold structure of the np ($n=46$ -51) series. The experimental spectrum is shown in Fig.3. When n is greater than 51 in the np series, the fourfold structure can not be seen in this experiment becauseof the limit of energy resolution and detection sensitivity.

The energy level of $6p7s \; ^3P^0_1$ of the Pb atom is $35287.24cm^{-1}$ (Moore 1971). We chose the line (406.625nm) of Au atom which is produced by the transition from $6p \; ^2P_{1/2}$ to $6d \; ^2D_{3/2}$ as the calibration wavelength of λ_2. After correction of the refractive index of air, the experimental values of the Rydberg levels of the 6pnp series of the Pb atom were obtained.

The intermediate state of $6p \, (^2P^0_{1/2})7s \; ^3P^0_1$ can be represented by $(\frac{1}{2},\frac{1}{2})_1$ in jj notation. For the np series, each Rydberg state has four possible spectral terms, that is $(\frac{1}{2},\frac{1}{2})_{1,0}$ and $(\frac{1}{2},\frac{3}{2})_{2,1}$.

They correspond to the transitions of $(\frac{1}{2}, \frac{1}{2})_1 \rightarrow (\frac{1}{2}, \frac{1}{2})_{0,1}$ and $(\frac{1}{2}, \frac{1}{2})_1 \rightarrow (\frac{1}{2}, \frac{3}{2})_{1,2}$ according to the jj coupling selection rules.

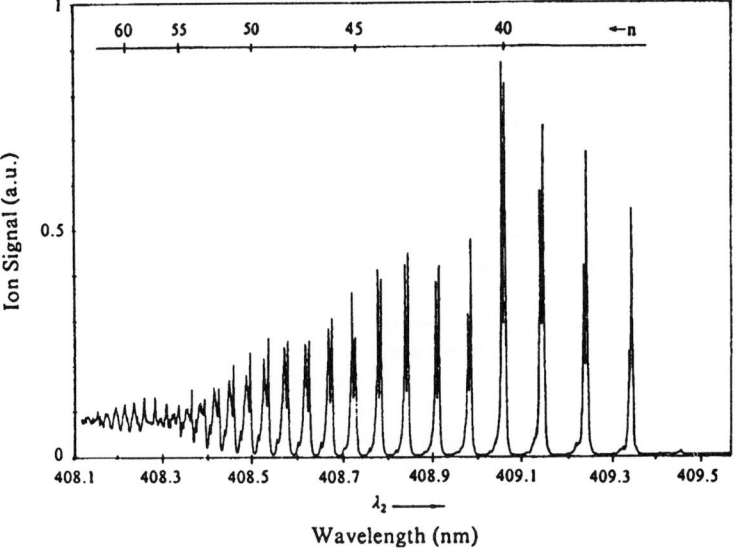

Fig.2　The $6p(^2P^0_{1/2})np$ Rydberg Spectrum of the Pb atom by two-step laser resonance excitation -electric field ionization. The abscissa is the scanning wavelength of λ_2.

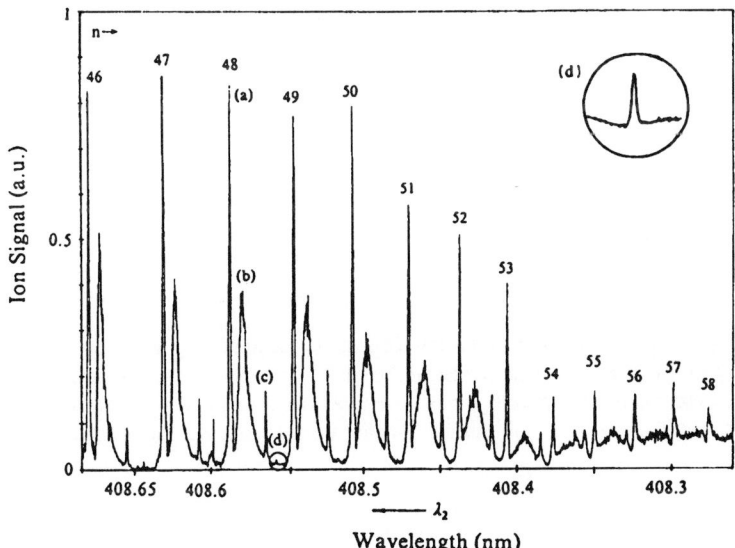

Fig.3　The fourflod structure of the $6p(^2P^0_{1/2})np$ series of the Pb atom

We evaluated the relative line strength of the four transitions. The evaluated results show that the line strength of transition corresponding to $(\frac{1}{2}, \frac{3}{2})_2$ state is the strongest and $(\frac{1}{2}, \frac{1}{2})_1$ state is the weakest one. The line strength of $(\frac{1}{2}, \frac{1}{2})_1 \rightarrow (\frac{1}{2}, \frac{3}{2})_1$ is stronger than that of $(\frac{1}{2}, \frac{1}{2})_1 \rightarrow (\frac{1}{2}, \frac{1}{2})_0$.

Compared with the experimental results, the specral terms of the fourfold structure are determined to be $(\frac{1}{2}, \frac{3}{2})_2$, $(\frac{1}{2}, \frac{3}{2})_1$, $(\frac{1}{2}, \frac{1}{2})_0$, $(\frac{1}{2}, \frac{1}{2})_1$, corresponding to the four resonance peaks (a), (b), (c) and (d) in Fig.3 respectively.

It is obvious that the line strength is changed greatly near n = 40. It is possible that the 6pnp series is perturbed by other states. The 6p40p state can be applied in sensitive detection, because its line strength is greater than others. The linewidth of the $(\frac{1}{2}, \frac{3}{2})_1$ series becomes more and more broader as the principal quantum number increasing, and finally it disappears when n > 56. Maybe it is caused by Stark braodening effect owing to the residual external electric field.

The limits and relevant quantum defects of the fourfold structure obtained by fitting the experimental data are given in table 1. On the average, the first ionization potential of the Pb atom is $59819.5 \pm 0.5 \mathrm{cm}^{-1}$.

Table 1   The ionization limits and quantum defects
of the fourfold structure of the $6p(^2P^0_{1/2})np$ series

Rydberg series	ionization limit (I) (cm^{-1})	quantum defect (μ)
$(\frac{1}{2}, \frac{3}{2})_2$	59819.0(3)	4.23(8)
$(\frac{1}{2}, \frac{3}{2})_1$	59819.8(0)	4.32(3)
$(\frac{1}{2}, \frac{1}{2})_0$	59819.6(9)	3.95(3)
$(\frac{1}{2}, \frac{1}{2})_1$	59819.5(1)	3.68(9)

4. CONCLUSIONS

We have shown that the resonance ionization TOF mass spectroscopy is a very effective tool to study Rydberg states. The fourfold structure of the $6p(^2P^0_{1/2})np$ series of the Pb atom has been observed and the corresponding spectral terms have been determined by comparing the evaluated branching ratio with the experimental results. Further work to analysis the measured spectra is in progress.

REFERENCES

Buch P, Nellessen J and Ertmer W 1988 *physica Scripta* **38** pp664–9
Ding Dajun, Jin Mingxing, Liu Hang and Liu Xuewen 1989 *J. Phys. B: At. Mol. Optics phys.* **22** pp1979–91
Martin T P 1982 *J. Chem. Phys.* **77**(7) 3815
Moore C E 1971 *Atomic Energy Levels* NSRDS–NBS 35, Vol. 3, pp209–210
Young W A, Mirza M Y and Duley W W 1980 *J. Phys. B: At. Mol. phys.* **13** pp3175–88

Inst. Phys. Conf. Ser. No 128: Section 2
Paper presented at RIS 92, Santa Fe, NM, USA, 24–29 May 1992

MBPT approach to the calculation of quantum defects of *nd'* (*J* = 1, 2) resonances in the rare gas atoms

V.Tsemekhman, K.Tsemekhman
A.F.Ioffe Physical-Technical Institute, 194021,
St-Petersburg,Russia

M.Amusia
Istitut fur Theoretische Physik, Universität
Frankfurt,D-6000 Frankfurt,FRG

Abstract

The problem of complex energies of highly excited Rydberg atoms is solved in the frames of MBPT. Formulas derived are used to provide quantum defects of $nd'(J = 1, 2)$-resonances in rare gases. An important role of correlation effects, dominantly of the self-energy part of Rydberg electron, is demonstrated. Most of the results are in good agreement with experimental data.

Complex energies of atomic Rydberg states in the MBPT

The energies of excited states, particularly, of highly excited Rydberg states, display the separate problem both for experimental and theoretical works. Very few *ab initio* calculations of these energies exist, however, so far (Johnson *et al*,1980). In the most of works the potential seen by highly excited electron is chosen to adjust the experimental values of energy (Aymar,1990). The main difficulty in the description of Rydberg electron is accurate account for its correlations with the motion of core electrons, especially, those connected with the polarization of the core. The effect of polarization of the ionic core cannot be described or even estimated by usual polarization potential $-\alpha/2r^4$ (α being static dipole polarizability of the core) since the latter is only an asymptotic expression for polarization interaction. Using the Many-Body Perturbation Theory (MBPT), we consistently investigate the energies of $nd'(J = 1, 2)$ states in the rare gases (Ar, Kr and Xe) described in the previous paper (Tsemekhman *et al*,1992, below referred as $TTA1$). The energy of atomic Rydberg state is given by:

$$E_n = -\frac{1}{2n^{*2}}, \tag{1}$$

n^* being effective principal quantum number,

$$n^* = n - \mu_l. \tag{2}$$

μ_l is the quantum defect (QD) which is independent of n for isolated resonances with rather high principal quantum number. Therefore, the problem of energies of Rydberg states can be reduced to calculation of the only parameter for the whole series, namely of the quantum defect. QD consists of its integral and fractional parts that have different origin.

Integral part of nl-state QD is determined by the number N_ℓ of occupied core states with the same angular momentum ℓ (to be published in details elsewhere). As it follows from the variational principle and from the orthogonality of nl-state wave function to the wave functions of the core occupied states with angular momentum ℓ, an integral part of nl-state QD $[\mu_l]$:

$$\mu_\ell = \begin{cases} N_0 - 1, & l = 0 \\ N_\ell, & l > 0 \end{cases} \tag{3}$$

Fractional parts of QD reflect the deviation of the core potential from the Coulomb one and the penetration of Rydberg electron into the core. To obtain them one can apply the MBPT methods. However, most of different MBPT methods share difficulties mentioned in the introduction to $(TTA1)$ and denoted there as **i** and **ii**. Shortly we discuss here how these problems are solved in the frames of perturbation theory (PT).

i Let ε_m and ϕ_m be the eigenvalues and eigenstates of Hamiltonian with some zero order many-electron potential with Coulomb asymptotics (say, Hartree-Fock Hamiltonian), and $\hat{V}$ be a short-range perturbation. The first terms of PT series on $\hat{V}$ can be written as:

$$\tilde{V}_{m\,k} = \tilde{V}_{m\,k}(\varepsilon) = V_{m\,k} + \sum_\ell{}' \frac{V_{m\,\ell}V_{\ell\,k}}{\varepsilon - \varepsilon_\ell} + \int_0^\infty \frac{V_{m\,\varepsilon'}V_{\varepsilon'\,k}}{\varepsilon - \varepsilon' + \imath\delta}d\varepsilon' + ..., \tag{4}$$

where $V_{m\,k} =< \phi_k|\hat{V}|\phi_m >$. In (5) ε is the energy of system that can be both positive and negative. Prime at the sum symbol denotes that the term with $\varepsilon_\ell = \varepsilon$ should be omitted. ε'- and ℓ-dependence of the short-range potential matrix elements for $\varepsilon' \ll 1$ and $m \gg 1$ reflects the simple energy dependence of the low energy Coulomb wave functions at small distances (see $TTA1$): $V_{\varepsilon'm}$ is independent of ε' and is $\sim 1/m^{3/2}$. Therefore, the contribution of small energies into integral in (5) grows logarithmically to infinity when $\varepsilon \to 0$. This growth escapes if one takes into account the second term in (5) corresponding to discrete spectrum intermediate states. To demonstrate this compensation we suggest the following procedure. If E is a certain magnitude so that $E \ll 1$ and $E < \varepsilon$, we can first include the contribution of $\varepsilon' > E$ into $\tilde{V}_{m\,k}$ according to:

$$\bar{V}_{m\,k} = V_{m\,k} + \int_E^\infty \frac{V_{m\,\varepsilon'}V_{\varepsilon'\,k}}{\varepsilon - \varepsilon' + \imath\delta}d\varepsilon' + \sum_{\ell=N_\ell+1}^{N} \frac{V_{m\,\ell}V_{\ell\,k}}{\varepsilon - \varepsilon_\ell} \tag{5}$$

Here, moreover, the lowest N discrete excitations whose energies are not small enough and, therefore, transition matrix elements do not show up simple energy dependence, are included, N_ℓ being defined in comments to (3). This is a free of peculiarities integral equation to be solved numerically. The remaining terms of PT with low energy continuum ($\varepsilon' < E$) and Rydberg discrete spectra intermediate states can be taken into account analyticaly for $|\varepsilon| \ll 1$ leading to:

$$\tilde{V} = \bar{V}/(1 - \bar{V} \cdot (2C - 2 \cdot \sum_{\ell=N_\ell+1}^{N} \ell^{-1} - \ln(2E - 2\varepsilon))), \tag{6}$$

C being Euler constant, $\tilde{V} = m^{3/2} \cdot \tilde{V}_{\varepsilon\,m}; \bar{V} = m^{3/2} \cdot \bar{V}_{\varepsilon\,m}$

ii Now we will use (4) - (6) looking for connection between the perturbation theory matrix elements and the complex energy of Rydberg state. For this purpose we employ Brillouin-Wigner perturbation theory (Thoules,1970):

$$E_n = \varepsilon_n + \sum_\ell{}' \frac{\bar{V}_{n\,\ell}\bar{V}_{\ell\,n}}{E_n - \varepsilon_\ell} + \int_0^E \frac{\bar{V}_{n\,\varepsilon'}\bar{V}_{\varepsilon'\,n}}{E_n - \varepsilon' + \imath\delta}d\varepsilon' + ... \tag{7}$$

where (5) is the definition of $\bar{V}$. The only but very important difference between the conventional PT (4) (usualy applied for continuum spectrum) and the one used in (7) is connected with their denominators: (7) requires that energy denominators contain yet unknown exact energy value E_n, $E_n = -1/2(n - \mu')^2$ (to compare with (1)). Equation (7) for E_n, and, therefore, for μ' can be solved analyticaly in the case of the potential with Coulomb asymptotics and results for $n \gg 1$ in:

$$\triangle \mu = \mu' - \mu = -\frac{1}{\pi} \cdot \text{arctg} \pi \tilde{V}. \tag{8}$$

This is the final form of relationship between PT matrix element $\tilde{V}$, connecting continuum low energy states and the quantum defect. If in (5) $\varepsilon > 0$, $\tilde{V}$ becomes complex. It means that there is a possibility for discrete state to decay. Hence, μ becomes also complex, its imaginary part being one half of the reduced width of the state (in a.u.) $(TTA1)$.

Summarizing the results we suggest a sort of rule how to calculate QD and widths entirely in MBPT.

i It is necessary to solve numerically integral equation (5) for PT matrix element $\tilde{V}$, including only "larger energies" intermediate states contribution.

ii To take into account the contributions of lower energy continuum and discrete spectrum intermediate states according to (6).

iii Using (8), one can receive the fractional part of quantum defect containing information on the position and on the width of considered resonance.

Calculation and discussion

The scheme proposed was employed to calculate the QD of nd' states in rare gases. Hartree-Fock approximation was used as zero order one. Energy values obtained in it already include Hartree-Fock QD. As it should be according to (3), their integral parts coincide with corresponding experimental values (Tables 3 and 4). On contrary, fractional parts of Hartree-Fock QD are strongly (by factors $1.5 \div 3$) different from experimental data. This fact proves the significant role of correlations to be taken into consideration. The main corrections to the fractional parts of QD in the language of MBPT are given by the RPAE matrix elements and by the self-energy part (SEP, Σ) of Rydberg electron. Consistent way to include into calculation these two perturbations requires that they be taken into account simultaneously. However we will first consider RPAE and SEP corrections separately to demonstrate their relative contributions into QD. Calculation of RPAE matrix elements is described in $(TTA1)$. For the correction

to QD the diagonal matrix element (Fig.1) is responsible. In it the thick line means the effective Coulomb interaction as explained in ($TTA1$). QD corrections due to RPAE correlations are denoted in Tables 3 and 4 as μ_{RPAE}. For $J = 2$ these values turn out to be the only that differ the energies of two

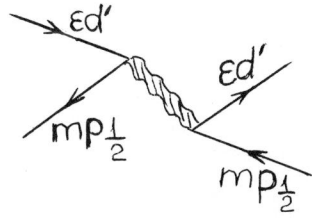

nd' resonances ($K = 3/2, K = 5/2$). The SEP of Rydberg electron describes the polarization of the core electrons by the field of far Rydberg electron. In the language of diagrams of MBPT,matrix elements of (Fig.2) appeared to be important in the first order of PT on Σ. The wave functions for all intermediate states have been calculated in the frozen core Hartree-Fock

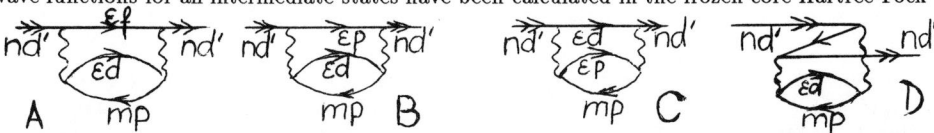

approximation. Contributions of diagrams $A - D$ into QD are presented in Tables 1 and 2. Treating $\Sigma = \Sigma_A + \Sigma_B + \Sigma_C + \Sigma_D$ as a perturbation, we received the matrix element $\tilde{\Sigma}$ after solving the Dyson integral equation (Fig.3) that is analogous to (4). Corrections to QD given

Table 1. J=1

	A	B	C	D	Total
Ar	0.0469	0.0154	0.0066	-0.0127	0.056
Kr	0.0665	0.0186	0.0062	-0.0163	0.075
Xe	0.1173	0.0220	0.0123	-0.0297	0.122

Table 2. J=2

	A	B	C	D	Total
Ar	0.0989	0.0159	0.0312	-0.0488	0.097
Kr	0.1144	0.0174	0.0206	-0.0436	0.109
Xe	0.1310	0.0139	0.0220	-0.0460	0.121

Table 3. Quantum defects, J=1.

	Hartree-Fock	$\Delta\mu_{RPAE}$	$\Delta\mu_\Sigma$	$\Delta\mu_{MBPT}$	μ_{THEORY}	μ_{EXP}
Ar	0.0737	0.0406	0.0589	0.127	0.200	$0.180 \div 0.200^a$
Kr	1.0703	0.0423	0.0838	0.144	1.214	1.223^b
Xe	2.1301	0.0342	0.1463	0.187	2.317	$2.324^c, 2.33^a$

Table 4. Quantum defects, J=2.

		Hartree-Fock	$\Delta\mu_{RPAE}$	$\Delta\mu_\Sigma$	$\Delta\mu_{MBPT}$	μ_{THEORY}	μ_{EXP}
Ar	$K=3/2$	0.1995	0.0548	0.1110	0.118	0.318	0.355^d
	$K=5/2$		0.0362		0.129	0.329	0.345^d
Kr	$K=3/2$	0.2328	0.0075	0.1245	0.104	1.337	1.342^d
	$K=5/2$		0.0251		0.130	1.363	1.361^d
Xe	$K=3/2$	2.3700	-0.0208	0.1355	0.090	2.460	–
	$K=5/2$		0.0113		0.113	2.483	2.475^d

a: Yoshino 1970; b: Ueda et al 1989; c: Maeda et al 1991; d: Klar 1987; e: Wang and Knight 1986.

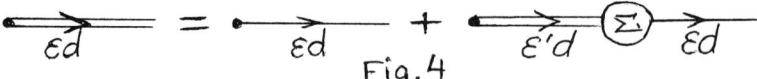

by Σ are denoted in Tables 3 and 4 as μ_Σ . Finally, we considered $\hat{V} = \hat{V}_{RPAE} + \hat{\Sigma}$ as a perturbation. In the diagram technique it means that SEP is included according to (Fig.4) into all intermediate, initial and final states of RPAE equation ($TTA1$, Fig.2). The final values for

Fig. 4

MBPT corrections to QD $\Delta\mu_{MBPT}$ are obtained by substituting of $\tilde{V}$ into (8). The agreement between theoretical and experimental quantum defects is good in the most of cases.

In these two works we presented the results of calculations of "internal" parameters of nd'-resonances: their positions and their widths. These parameters are adequate for description, for example, of resonant low energy scattering of electrons by positive ions (the work is now under consideration). For photoionization profile (Fano profile) at least one more parameter - profile index q - is required. It is different for every new initial state and can be easily calculated in the scheme presented in this work (to be published elsewhere). The authors are grateful to the Deutsche Forschungsgemeinschaft for financial support of their stays in Kaiserslautern, during which part of this research was carried out, and to Prof.H.Hotop and D.Klar for the helpful discussions and initiating of this work. We also thank Dr.G.Gribakin and Dr.M.Kuchiev for their interest in the subject.

Aymar M 1990, J Phys B **23** 2697-2716

Johnson W R, Cheng K T, Huang K-N and LeDourneuf M 1980 Phys.Rev.A **22** 989-988

Klar D 1987 Diplomarbeit, Universitat Kaiserslautern (unpublished)

Maeda K, Ueda K, Namioka T, Ito K 1991 Phys.Rev.A

Thoules D J 1970 Quantum Theory of the Many-Particle Systems, Academic Press

Tsemekhman V, Tsemekhman K, Amusia M Ya 1992 Preceding paper in this book

Ueda K, Maeda K, Ito K, Namioka T 1989 J Phys B **22** L481

Wang L g and Knight R D 1986 Phys.Rev.A **34** 3902

Yoshino K 1970 J.Opt.Soc.Am. **60** 1220

Inst. Phys. Conf. Ser. No 128: Section 2
Paper presented at RIS 92, Santa Fe, NM, USA, 24–29 May 1992

83

The widths of *nd'* (*J* = 1, 2) Rydberg autoionization resonances in the rare gas atoms. MBPT calculations

V.Tsemekhman, K.Tsemekhman
A.F.Ioffe Physical-Technical Institute, 194021,
St-Petersburg,Russia

M.Amusia
Istitut fur Theoretische Physik, Universität
Frankfurt,D-6000 Frankfurt,FRG

Abstract

Elaborate *ab initio* MBPT calculations are performed to provide reduced autoionisation widths of $Rg(nd', J = 1, 2)$ resonances for Ar, Kr, and Xe. The problems of highly excited states in MBPT are discussed and the solution found is applied. RPAE and self-energy correlations are taken into account. The results obtained are compared with experimental and theoretical data.

The photoionisation cross section of the heavier rare gases Rg = Ar, Kr, and Xe in the energy range between the $Rg^+(mp^5\ P_{1/2})$ and $Rg^+(mp^5\ P_{3/2})$ thresholds is dominated by the presence of $Rg(mp^5_{1/2}\ nl')$ photoionisation resonances, which have attracted the attention of both experimentalists and theoreticians since the pioneering work of Beutler. The widths of $Rg(nd', J = 1)$ have been measured in the single photon absorption experiments. All possible nd' resonances have been investigated by two photon excitation from the metastable state through the intermediate levels (Klar 1991,and references therein). Results of several semiempirical MQDT analyses have been reported. However, only two *ab initio* calculations (Johnson *et al* 1980, Taylor 1981), both dealing with $Rg(nd', J = 1)$, have been carried out.

The problems considered in this and in the following paper, denoted further as**TTA2**, are closely connected. Therefore the discussion below concerns both of our works.

It is used to consider that any Many Body Perturbation Theory (MBPT) method meets some strong difficulties when applied to the description of excited and autoionising states (Lee and Johnson 1980). Among them we would stress the two following.

i.　The problem of small denominators in the perturbation theory that arises due to a very small energy value of initial (particularly, autoionising Rydberg) state makes the direct numerical calculations almost impossible. One is urged to account accurately for the contribution of the terms with small denominators in some indirect way.

ii.　The problem of evaluating parameters characterizing both the excited state (its energy and width) and the process of ionisation (profile index q) from the calculated MBPT matrix elements becomes nontrivial due to the peculiarities of the perturbation theory in discrete spectrum.

To avoid these problems some methods combining MBPT with other theories (R-matrix and MQDT) have been introduced. In these two papers we suggest the direct MBPT calculation of energies and widths of autoionising states. For this purpose we derive analytical expressions that solve both problems in frames of MBPT. The method suggested is applied to the study of nd' resonances in heavier rare gas atoms Ar, Kr, and Xe. In this work we present the results of calculations of the widths of $Rg(nd', J = 1, 2)$ in these atoms.

The width of the autoionisation state Γ being proportional to the decay rate can be expressed as the imaginary part of the state's energy:

$$E = E_0 - i\frac{\Gamma}{2} .\tag{1}$$

If Γ is small compared to the energy spacing between two adjacent levels ΔE, the unitary relation yields another way for calculating the width:

$$\Gamma = 2\pi \sum_i \mid A_i \mid^2 , \qquad (2)$$

A_i being the amplitude of decay into a certain channel i. Difference between the two definitions turns out to be significant when $\Gamma \sim \Delta E$ and there is a noticeable uncertainty ($\sim \Gamma$) in the energy of initial state. Therefore, the final electron energy has uncertainty of the order of Γ, and the decay amplitudes vary with energy inside the width. It is not clear then how (2) can be applied. This situation takes place for almost all nd' resonances as they are rather broad: their widths are compared to $1/n^3$,(n is a principal quantum number), that is close to the spacing between two Rydberg levels in atomic units. In this case (1) remains the only precise definition of the width. In the last section we present results obtained both by (1) and by (2), where the final energy is equal to the energy of the resonance maximum.

The widths Γ_n of the Rg($mp^5_{1/2} nd'$) resonances exhibit a simple dependence on the effective principal quantum number n^*, namely: $\Gamma_n = \tilde{\Gamma}/n^{*3}$, where $\tilde{\Gamma}$ is the reduced resonance width which is independent of n^*. This result reflects the fact that the decay is caused by excited electron - ionic core Coulomb interaction at small distances $r \ll n^{*2}$, where the energy of highly excited state $-1/2n^{*2}$ and very low positive energy ε of the outgoing electron are negligible compared to the Coulomb potential energy.

Intermediate jl-coupling should be applied for description of the states under discussion . The whole set of quantum numbers characterizing Rg($mp^5 nd'(\varepsilon d)$) states consists of LSjlKJ. Here L,S,j are orbital angular momentum, spin, and total angular momentum of the mp^5 -core, respectively; K=j+l, l being angular momentum of the excited (or free) electron; **J=K+s**, $J = K + 1/2$, s being spin of the outer electron.

In terms of these quantum numbers Rg(mp^5nd')initial states and corresponding channels of decay are presented in Table 1.

All wave functions were calculated in the non-relativistic Hartree-Fock (HF) approximation; $mp_{1/2}$ and $mp_{3/2}$ holes are assumed to have the same radial wave functions; their energies were taken from the experimental data. According to what was explained above, small deviations of $mp_{1/2} - mp_{3/2}$ splitting from the exact value do not change the result of calculations. The electron radial wave functions of low positive energy used for description of both initial and final states were calculated in the frozen core approximation. The whole wave function of the system consisted of electron and frozen ionic core can be represented as a product of radial and angular parts. The latter one has to satisfy the symmetry corresponding to the appropriate set of jlKJ quantum numbers. For this purpose it is reasonable to expand it over angular parts of the wave functions of the states described in LS-coupling.The radial wave function is then calculated in the self-consistent field of required symmetry. It is worth to notice here that the radial wave functions of both $J = 2$ resonances appeared to satisfy the same HF equation. Therefore, the HF energies of these two resonances are equal;it means that they strongly overlap, and, possibly, interact. Later we explain why these states decay into different continua (see Table 1), and, so, do not interact in fact.

Table 1. Initial and Final States for the Decay of Rg $(nd', J = 1, 2)$.

	Initial states	Final states
		$mp^5_{3/2}\varepsilon d, K = 3/2$
J=1	$mp^5_{1/2}nd', K = 3/2$	$mp^5_{3/2}\varepsilon d, K = 1/2$
		$mp^5_{3/2}\varepsilon s, K = 3/2$
		$mp^5_{3/2}\varepsilon d, K = 3/2$
	$mp^5_{1/2}nd', K = 3/2$	
J=2		$mp^5_{1/2}\varepsilon s, K = 3/2$
	$mp^5_{1/2}nd', K = 5/2$	$mp^5_{3/2}\varepsilon d, K = 5/2$

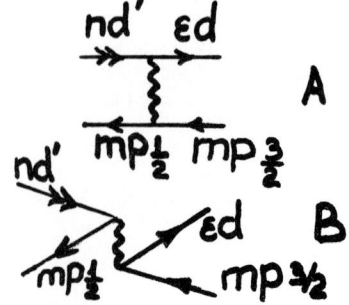

Table 2. The Autoionization Resonance Widths of Rg($nd', J = 1$)

	First order			RPAE			Im$\tilde{V}$	MBPT width	Other results
	$K_f = 3/2$	$K_f = 1/2$	Total	$K_f = 3/2$	$K_f = 1/2$	Total			
Ar	800	14100	14900	4100	17950	22050	22150	20400	30600[a] 21260[b]
Kr	2900	27250	30150	3850	17750	21600	24300	20400	21600[c] 33400[d] 35000[f]
Xe	9300	58000	67300	12100	29600	41700	46800	34750	35700[e] 41360[d]

For the calculation of resonance widths we employed the diagram techniques of MBPT. As noticed above, only (1) can yield the exact solution of the problem. It corresponds to the imaginary part of the diagrams presented below and in TTA2. Here, however, we show the diagrams describing the first order autoionisation amplitudes in order to simplify the qualitative explanation. In the first order of MBPT two diagrams, shown on Fig.1, are generally responsible for the decay. Single (double) arrow towards the right represents a continuum (bound) electron, a line with a single arrow to the left denotes a hole, and the wavy lines show the Coulomb interaction. Direct and exchange matrix elements with the same final state, corresponding to the diagrams A and B, respectively, usually have opposite sign and strongly compensate each other. Therefore, one can expect the autoionisation amplitude to be large only if one of the two diagrams is equal to zero, i.e. either direct or exchange transition is forbidden. The first case is realized if initial and final states have different quantum numbers K: since $\mathbf{K=J\text{-}s}$, and J is always good quantum number, it is necessary to change the direction of spin in order to change K value. Only exchange interaction affects the spin variables, while direct one does not. Due to this the final channel $K = 1/2$ dominates in the decay of $J = 1$ resonance. Moreover, the behavior of the $J = 1$ resonance width from Ar through Kr to Xe reproduces the behavior of the quantum defect: exchange interaction, being defined by overlapping of the outer electron's and mp^b - electrons' wave functions is the important in the decay of this state. However, the same penetration of the excited electron into the mp-shell defines the behavior of the quantum defect. For both $J = 2$ resonances, differing by K value, exchange

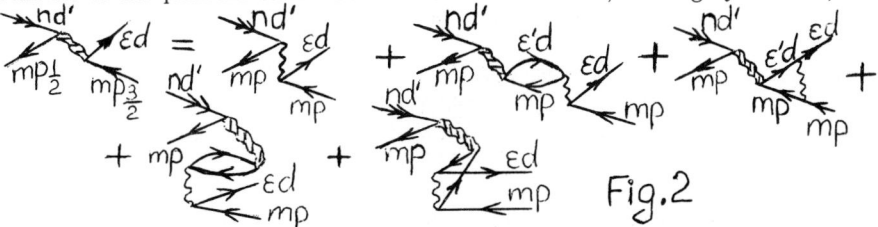

Fig.2

transition appears to be impossible. Therefore, K is conserved during the decay and two resonances have absolutely different channels of decay and do not interact. It is evident from Tables 2 and 3, that first order calculations well reproduce the behavior of the width while the values themselves, denoted as First order strongly differ from the experimental ones. So correlations have been taken into account, namely those corresponding to the RPAE (random phase approximation with exchange)(Amusia and Cherepkov 1975). The conventional RPAE has been modified to apply for jl- coupling. The graphic equation for the RPAE method is presented on Fig.2. Intermediate states, summation over which is supposed, were calculated in the jl-coupling. Thus, we solved 4-channel problem for $J = 1$ resonance, 3- and 2- channel problems for $J = 2, K = 3/2$ and $J = 2, K = 5/2$ resonances, respectively. Problem 1,

Table 3. The Autoionization Resonance Widths of Rg(nd', $J = 2$)

	$K = 3/2$					$K = 5/2$				
	First order	RPAE	Im V	MBPT width	Other results	First order	RPAE	ImV	MBPT width	Other results
Ar	23900	28000	37700	33950	26000[c]	5250	6700	10400	8800	8500[g] 5000[h]
Kr	21200	21800	22800	20650	14100[i]	5550	6600	7650	6450	4860[j] 3860[k]
Xe	16750	14600	9600	8700	12000±30%[f]	5070	5750	3350	2850	2250[l] 4000±30%[f]

mentioned in the Introduction, has to be solved here. The procedure of analytical account for the contribution of low energy intermediate states is described in TTA2. The results of the calculations accounting for RPAE are denoted in Tables 2 and 3 as RPAE.

In TTA2 we show that the problem of the resonance complex energy (1) requires for simultaneous calculation of both real and imaginary parts (see expression (8) there).It turned out that zero order Hartree-Fock approximation extended by the RPAE correlations being adequate for description the widths of nd' resonances, fails to predict their positions. So we introduced also the self-energy corrections into the complex energy calculations. The detailed analysis of the self-energy correlations is presented in TTA2. Here we would mention that, opposite to their role in the quantum defect, their contribution into the amplitudes of decay is rather small: they appear for the first time only in the third order of the perturbation theory. The width calculated by (2),including RPAE and self-energy corrections, is denoted in Tables 2 and 3 as ImV. The self-energy corrections become important after (8) from TTA2 is applied.

Final results are presented in Tables 2 and 3 as MBPT width. The comparison with the experimental values yields rather good agreement for almost all investigated states. The direct comparison with the other theoretical results is rather complicated: the widths of $J = 1$ resonances were not reported in the work of Johnson et al(1980) but obtained by Ueda et al(1987) by fitting the *ab initio* profiles of the former work. However, it is evident from Johnson *et al* (1980) that it is worth to take into account the RPAE correlations with inner shells. This problem is under the development now. The difference between our and other theoretical results can arise also due to different coupling applied: *jl-* coupling seems to be the most appropriate, at least for discrete states.

We express our acknowledgements in TTA2.

Amusia M and Cherepkov N 1975, Case Studies in At.Phys.5,47

Johnson W R,Cheng K T,Huang K-N,Le Dourneuf M 1980,Phys.Rev.A **22**,989

Lee C M and Johnson W R 1980,Phys.Rev.A22, 979/

Klar D 1987,Diplomarbeit,Universitat Kaiserslautern(unpublished)

Klar D,Harth K,Ganz J,Kraft T,Ruf M-W,Hotop H 1991, Proc.Seminar "Today and Tomorrow of Photoionization",DL/SCI/R29,SERC Daresbury

Radler K and Berkowitz J 1979a,J.Chem.Phys.**70**,216; 1979b J.Chem.Phys.**70**,221

Taylor K 1981,J.Phys.B **14**,L237

Tsemekhman K, Tsemekhman V, Amusia M 1992, Next paper in this book

Ueda K 1987,J.Opt.Soc.Am.B **4**,424

Wang L-g and Knight R D 1986,Phys.Rev A **34**,3902

Inst. Phys. Conf. Ser. No 128: Section 2
Paper presented at RIS 92, Santa Fe, NM, USA, 24–29 May 1992

87

Determination of term energy, hyperfine structure and life time of strontium Rydberg levels by resonance ionization spectroscopy in collinear geometry

K Wendt, G Herrmann[*], R Hohmann, H-J Kluge, S Kunze, J Lantzsch, L Monz, E W Otten, G Passler, J Stenner, K Stratmann, N Trautmann[*], K Walter[*], K Zimmer

Institut für Physik and [*]Institut für Kernchemie,
Johannes Gutenberg-Universität, D-6500 Mainz, Fed. Rep. Germany

ABSTRACT: The combination of collinear fast-beam laser excitation with particle detection via resonance ionization offers rather unique possibilities for the study of high lying atomic Rydberg states and provide high spectral resolution. Precise results for term energies, fine and hyperfine structures as well as configuration interaction parameters can be obtained. In addition life time measurements in the μsec range can be carried out by pulsed laser excitation and time resolved detection. Results on a number of $5sns\ ^3S_1$ and $5snd\ ^3D_3$ Rydberg states of Sr I in the range $17 < n < 35$ are reported; they form the experimental basis for ultra-sensitive trace analysis.

1. INTRODUCTION

In the last few decades the technique of collinear fast-beam laser spectroscopy has proven to be extremely successful for high-resolution studies of isotopic shift and hyperfine structure (Kaufman 1976). One of the advantages of this experimental approach is the high sensitivity, which results from compressing the full Doppler profile of an ensemble of particles into a very narrow and intense excitation profile in the collinear regime. The strong reduction of the Doppler width obtained by illuminating a fast atomic beam of about 50 keV energy with narrow-band laser light in collinear geometry results in narrow spectral line widths and permits high-resolution measurements mainly limited by the natural line width. So far the collinear excitation has been combined with a number of different detection techniques, starting from the simplest recording of the fluorescence light by photon counting up to highly sophisticated particle detection systems (Neugart 1987).

A second type of application of this laser spectroscopic technique has recently been developed in the field of ultra-sensitive trace analysis. The combination of the large Doppler shift of optical resonance frequencies with the narrow experimental line widths results in a suppression of neighbouring isotopes up to 10^9 per excitation step. One detection system well suited for trace analysis applies resonance ionization detection. Resonant (and highly selective) excitation into high-lying atomic Rydberg states and subsequent field ionization in an electric field are used to minimize the background, which is produced mainly from ionization of atoms in the fast beam by collisions with the residual gas molecules. The application of this technique to trace analysis of radioactive strontium isotopes is discussed in a separate contribution to this issue, which also gives a brief description of the experimental setup (Monz 1992).

2. EXPERIMENTS AND RESULTS

In this paper we focus on the possibilities offered by the RISICO (Resonance IoniSation In COllinear geometry) technique for high-resolution atomic physics studies. For quantitative understanding of the trace analysis capabilities of this method the atomic physics quantities, e.g., term energies, hyperfine structures and life times, have been studied with high precision. Measurements were made in the triplet system of strontium. Laser excitation can be carried out from the metastable 5s5p $^3P_2$ state, which is populated in the charge exchange process using cesium vapour with about 10% probability. Using an intra-cavity doubled single-frequency dye laser, direct single-step excitation with laser light of $\lambda \approx$ 325 nm into high lying 5sns $^3S_1$ and 5snd $^3D_3$ states is carried out. Most interesting for the subsequent field ionization and ion detection is the range of 17<n<35. States with these principal quantum numbers can be ionized with critical field strengths between 12 kV/cm (for n$\approx$17) and 0.5 kV/cm (for n$\approx$30).

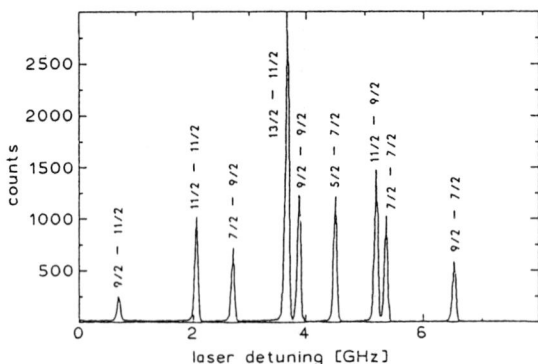

Fig. 1: Hyperfine structure of ^{87}Sr in the 5s5p $^3P_2$ -> 5s21s $^3S_1$ transition

A typical spectrum for the excitation into the 5s21s $^3S_1$ state of Sr I is given in Fig. 1. It shows the resolved hyperfine structure of ^{87}Sr (I =9/2). The absolute term energies can easily be evaluated via application of the Doppler formula from the simultaneously recorded iodine absorption spectrum. The results for a number of levels as measured so far are listed in Table 1. The precision of these data is increased up to a factor 50 over that of existing values (Beigang 82). In addition, the hyperfine structure has been extracted for a number of 5sns $^3S_1$ states. The hyperfine-induced mixing of these states to the corresponding 5sns $^1S_0$ states results in a shift of the hyperfine component with total angular momentum of F = I = 9/2. The absolute value of this effect can be evaluated theoretically with second-order perturbation theory from the known term energies and the hyperfine structure parameters. The measured shifts of the F = I HFS component increase from 31.6(5) MHZ for n=19 to 170.4(8) MHz for n=30 and are reproduced with a precision better than 5%.

Table 1: Term energy, quantum defect σ and life time τ of Sr Rydberg levels

Level	Term energy [cm⁻¹]	Quantum defect σ	Life time τ [μs]
5s17s $^3S_1$	45,341.24(4)	3.373	–
5s18s $^3S_1$	45,419.31(4)	3.373	–
5s19s $^3S_1$	45,482.876(2)	3.3722	1.83(20)
5s20s $^3S_1$	45,535.312(3)	3.3719	2.37(33)
5s21s $^3S_1$	45,579.068(3)	3.3718	2.73(49)
5s22s $^3S_1$	45,615.971(2)	3.3716	3.02(54)
5s23s $^3S_1$	45,647.384(3)	3.3711	3.36(71)
5s24s $^3S_1$	45,674.324(3)	3.3713	–
5s25s $^3S_1$	45,697.620(7)	3.3712	–
5s26s $^3S_1$	45,717.900(3)	3.3709	–
5s27s $^3S_1$	45,735.664(2)	3.3704	–
5s28s $^3S_1$	45,751.300(2)	3.3703	–
5s29s $^3S_1$	45,765.143(2)	3.3701	–
5s30s $^3S_1$	45,777.455(2)	3.3701	–
5s35s $^3S_1$	45,822.49(3)	3.656	7.50(4.40)
5s18d $^3D_3$	45,492.62(4)	2.200	0.63(3)
5s19d $^3D_3$	45,542.30(4)	2.224	0.75(3)
5s20d $^3D_3$	45,582.36(4)	2.289	0.48(3)
5s23d $^3D_3$	45,673.03(4)	2.423	0.77(3)
5s24d $^3D_3$	45,695.89(4)	2.450	0.94(5)
5s25d $^3D_3$	45,717.95(4)	2.368	1.13(8)
5s26d $^3D_3$	45,733.64(4)	2.491	1.33(10)
5s27d $^3D_3$	45,749.33(4)	2.503	1.58(14)
5s28d $^3D_3$	45,763.21(4)	2.517	1.77(19)
5s29d $^3D_3$	45,775.60(4)	2.529	–
5s30d $^3D_3$	45,786.73(4)	2.535	–

The HFS A factor of the 5sns states can be estimated theoretically via the semi-empirical Breit-Wills theory (see e.g. Kopfermann 1958) from the individual contributions of the two valence electrons. For high-lying Rydberg states the interaction between both electrons can be neclected and the contribution of the excited ns electron can directly be added to the dominant contribution of the low-lying 5s electron according to $A(5sns\ ^3S_1) = 1/2\ (a_{5s} + a_{ns})$. In very good approximation, a_{5s} is given by the A-factor of the ionic ground state $A(5s\ ^2S_{1/2}) = -1000.5(1.0)$ MHz (Buchinger 1990). A very small contribution of the excited electron of less than $a_{ns} \approx 0.5$ MHz for n > 18 is calculated by the semi-empirical Goudsmit Fermi Segrè formula. The estimate is in perfect agreement with the experimental A-factors of $A(5sns\ ^3S_1) = -500.3(4)$ MHz, which show no dependence on n within their errors. This agreement confirms the theoretial approach applied and the absence of any further configuration interactions.

With a modified experimental scheme life-time measurements of some strontium Rydberg levels have been carried out. We have used a pulsed amplification of cw laser light in a three-stage amplifier arrangement pumped by a high-repetition copper vapour laser. The atoms in the region between charge exchange cell and field ionization electrode are excited by the 20 nsec laser pulse. The length of this region is about 1.5 m and corresponds to a flight time of up to 5 μsec for a beam energy of 35 keV. During this time the ensemble of excited atoms passes through the ionization system and the ions are detected. Thus the particle signal can be recorded in a time-resolved mode. As an example Fig. 2 shows the exponential decay of the 5s19s $^3S_1$ level. In Table 1 the total life times of a number of levels are summarized. For further analysis these values can be separated into contributions from spontaneous decay into lower-lying levels and contributions caused by transitions between high-lying states induced by black-body radiation. This effect yields a considerable contribution of about 10% and lowers the total life time. It has been observed via the resolution of Rydberg states with different principal quantum number n. These states are ionized on different excess potentials in the field ionization region. Thus this detection mechanism allows for state selection. A typical analysis of the excess energy for excitation of the 5s19s $^3S_1$ state is shown in Fig. 3. While the dominant peak corresponds to the state which is directly excited by laser light, the smaller peaks represent the states with neighbouring principal quantum numbers populated via transitions induced by black-body radiation. This interpretation was confirmed by recording the time evolution of these side peaks which shows an exponential build-up as expected.

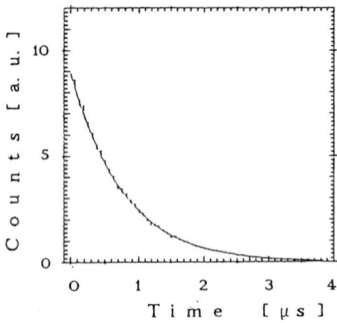 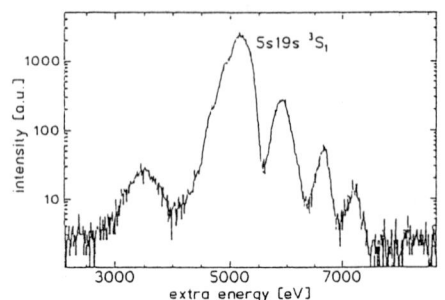

Fig. 2: Exponential decay of the 5s19s $^3S_1$ level

Fig. 3: Scan of the ion beam energy after field ionization, showing the 5s19s $^3S_1$ level which is directly excited together with states of neighbouring n. The latter are populated by transitions which are induced by black-body radiation

Conclusion and Outlook

Collinear fast-beam laser excitation in combination with detection via field ionization of high-lying Rydberg states can be used for a variety of studies of atomic physics quantities of highly excited states. By single-step or two-step narrow-band laser excitation starting from a metastable state, Rydberg levels are accessible for high-resolution studies. Results on the 5sns $^3S_1$ and 5snd $^3D_3$ levels were obtained. Further experiments to measure levels of the 5snp $^{1,3}P_J$, 5snd $^{1,3}D_J$ and 5snf $^{1,3}F_J$ series are planned. The pulsed excitation and time-resolved detection technique can be applied for life-time measurements in the time range between 50 nsec and 50 μsec. The energy selective detection scheme allows for the resolution of Rydberg states with different principal quantum number and thus the observation of effects induced by black-body radiation.

This work has been funded by the Bundesministerium für Umwelt, Naturschutz und Reaktorsicherheit under contract number StSch 4020.

REFERENCES

Beigang R et al 1982 Phys. Scr. **26** 183
Buchinger F et al 1990 Phys. Rev. **C 41** 2883
Kaufman S L 1976 Opt. Comm. **17** 309
Kopfermann H 1958 *Nuclear Moments* (New York: Academic)
Monz L et al 1992 contribution to this conference
Neugart R 1987 *Prog. in At. Spec.*, part D, (New York: Plenum) pp 75

Inst. Phys. Conf. Ser. No 128: Section 2
Paper presented at RIS 92, Santa Fe, NM, USA, 24–29 May 1992

91

*N*s*nl* (*l* > 6) double-Rydberg states in stray electric fields

P Camus, C R Mahon, P Pillet and L Pruvost

Laboratoire Aimé Cotton*, CNRS II, Bât. 505, Campus d'Orsay, 91405 Orsay Cédex, France.

* This laboratory is associated to the Université Paris-Sud

ABSTRACT: We report on the effect of stray electric fields on 7s and 10s$n\ell$ (ℓ=6 to n-1) doubly excited Rydberg spectra for n=13 and n=14. Here N and n are the principle quantum numbers of the inner and outer electrons respectively. The 7s$n\ell$ spectra show a single peak for each ℓ excited. We have measured their energy positions and deduced the 7s$n\ell$ quantum defects. The 10s$n\ell$ spectra unlike for the 7s$n\ell$, show a multiple peak structure common to spectra associated with different ℓ. This multiple peak structure is explained by taking into account the ℓ-mixing of the initial 6s$n\ell$ states from which the 10s$n\ell$ are excited.

INTRODUCTION

We have presented at the last RIS'90 conference (Camus *et al* 1991) in the study of doubly excited atoms that the use of high-ℓ angular momentum states produced by the Stark switching technique allows the observation of subtle correlation effects between the two excited electrons. This is made possible due to the much smaller autoionisation widths of the high-ℓ states. It was shown that this six peaks structure in the 7d$_{5/2}$nℓ spectra for n=11 ℓ=10 (Camus *et al* 1992) is due to the Coulomb electron-electron repulsion term, $1/r_{12}$, of the Hamiltonian and is well described in the $(j_1,\ell)K$ coupling scheme. Here j_1 is the total angular momentum of the inner 7d electron (j_1=5/2). The observed energy positions of the K components showed a strict dependence on the ℓ state excited.

OBSERVATION AND INTERPRETATION OF THE N$sn\ell$ SPECTRA

For the 7s$_{1/2}$nℓ states studied here, we are unable to resolve the two K=$\ell\pm$1/2 components in the $(j_1,\ell)K$ coupling scheme and we observe a single peak for each ℓ state excited. In Figure 1 we show spectra for 7s n=14 and ℓ=5 to 13. The red energy shift of these peaks relative to the parent ionic 6s$_{1/2}$-> 7s$_{1/2}$ transition energy is a direct measure of the quantum defect difference between the 6s$n\ell$ and the 7s$n\ell$ states. A general feature of these spectra is that the more penetrating low-ℓ states have the largest shift and therefore largest quantum defects due to the increased electron-electron interaction. For the same reason these low-ℓ states have the largest autoionisation linewidths, clearly aparent in the ℓ=5 spectra of Figure 1. The higher ℓ states have linewidths below our two-photon excitation laser resolution of 5 GHz.
We have measured the 7s$n\ell$ quantum defects and have fit them to the ℓ dependence equation

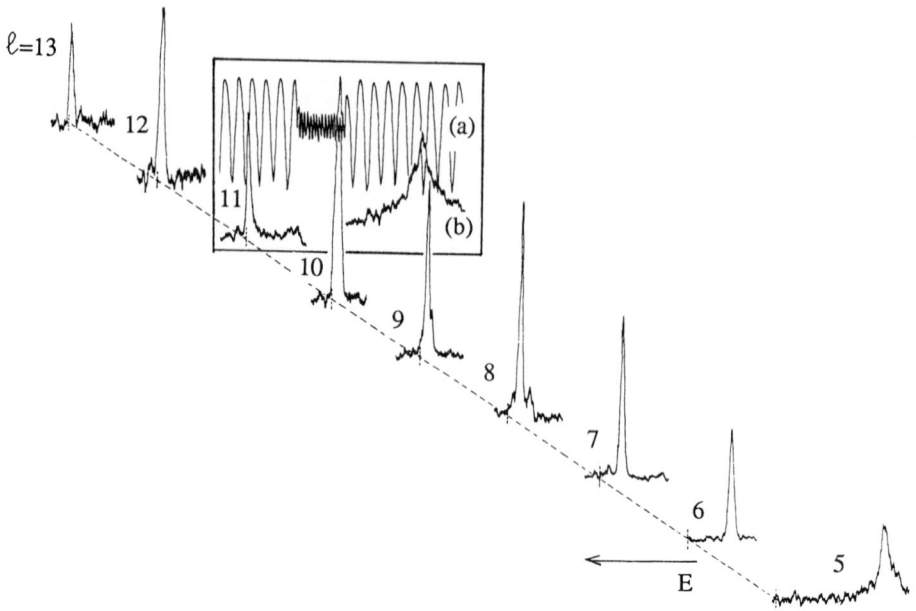

Fig.1 Two-photon excitation spectra $6s_{1/2}n\ell \rightarrow 7s_{1/2}n\ell$ for n=14 and ℓ=5 to 13. Spectrum in the window gives for ℓ=11 the 2 cm⁻¹ calibrating Fabry-Perot fringes (a) with respect to the position of the $6d5d\ ^1D_2$ - $6s7p\ ^1P_1$ transition of Ba I (b) used as a secondary standard reference. The other 7s n=14ℓ spectra are positionned on an absolute energy scale by referring the spectra to the calculated position of the parent ionic $6s_{1/2} \rightarrow 7s_{1/2}$ transition indicated by the dashed line.

$$C_{dip}\ \ell^{-5} + C_{quad}\ \ell^{-9} \quad (1)$$

where C_{dip} and C_{quad} are the effective dipole and quadrupole moments of the 7s inner electron. We obtain C_{dip}=634 and C_{quad}=19356 values. The data show that the electron-electron interaction is predominantly dipolar, with a maximun quadrupolar contribution of 6% for ℓ=4.

Unlike the 7snℓ spectra, the 10s n=14 ℓ spectra of Figure 2(a) show a multiple peak structure. In addition the spectral components are the same in spectra obtained for different "ℓ". This multipeak structure cannot be explained using the coupling scheme in the Coulomb electron-electron interaction model developed for the $7d_{5/2}n\ell$ spectra. In this coupling scheme only a maximum of two peaks is expected. The fact that the 10snℓ spectra are no longer ℓ characteristic implies that the initial 6sn"ℓ" states are ℓ-mixed. This is clearly shown in Figure 2(b) for which the same 10s n=14 ℓ=12 spectra is recorded in the presence of a 2 V/cm dc field. The same components are present in both spectra, though we note that the intensity distribution is different, reflecting the change in ℓ character of the initial 6snℓ Stark state. The spectra of Figure 2(a) is therefore taken, not in zero field, but in the presence of a stray electric field.

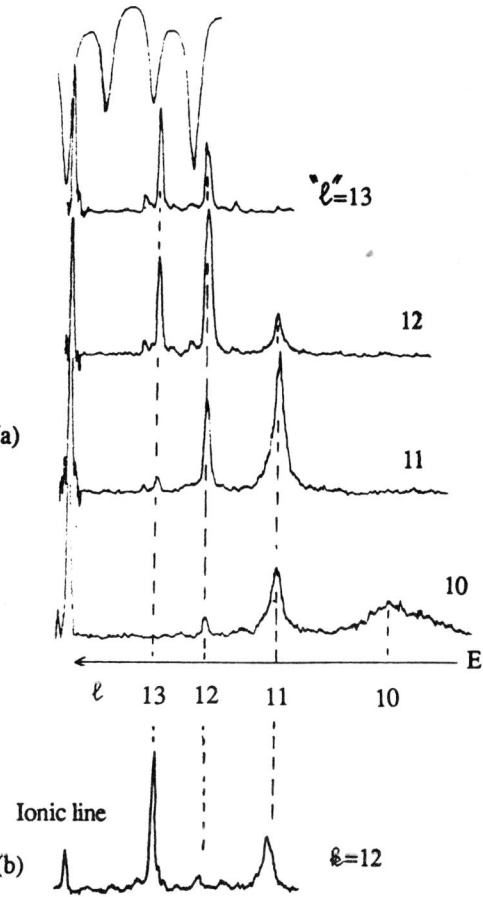

Fig.2 Two-photon excitation
spectra $6s_{1/2}n\ell \to 10s_{1/2}n\ell$. Spectra
are referred to the $6s_{1/2} \to 10s_{1/2}$
ionic transition and the Fabry-Perot
interval is 4 cm^{-1}.
(a) for n=14 and "ℓ" varying from
10 to 13 in zero static field applied.
(b) for n=14 ℓ=12 Stark state with
a 2 V/cm dc field applied.

We have studied this intensity variation as a function of the applied static field and have
compared it to the theoretical variation based on the calculated Stark mixing coefficients of the
$6sn\ell$ initial state. In Figure 3(a) we show the measured and calculated amplitude of the 10s
n=13 ℓ=12 state, excited from the 6s n=13 ℓ=11 state, as a function of the applied dc field.
The calculated field dependence in Figure 3(a) assumes no stray electric fields. The fact that at
zero applied dc field the measured signal does not go to zero as expected from the theory
implies again that there are stray fields present. If we introduce in our calculation a transverse
field of 1 V/cm added to the applied field, as well as adjusting by 20% the quantum defects of
the $6sn\ell$ states the agreement between theory and experiment is very good in Figure 3(b).
Adjustment of the $6sn\ell$ quantum defects is reasonable since they are know by a perturbative
calculation (Pruvost *et al* 1991) to only 20%. We have measured the quantum defect of $10sn\ell$
levels and have fitted them to the ℓ dependence of eqn (1). The resulting effective dipolar and
quadrupolar coefficients are C_{dip}=41421 and C_{quad}=-3366029. As for $7sn\ell$ states the dipolar

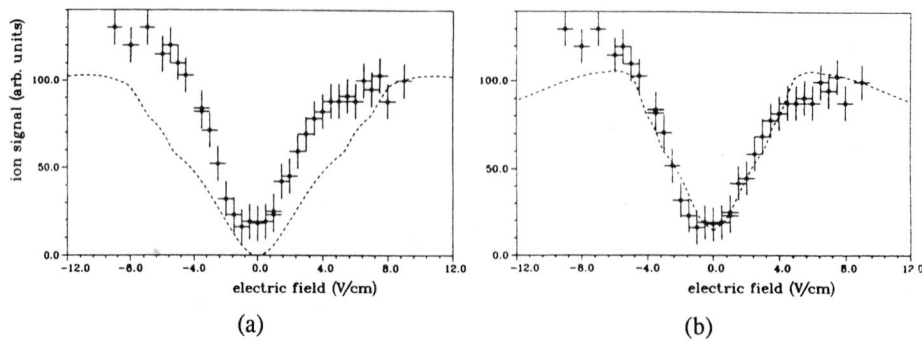

Fig.3 Comparison between theory and experiment for the 10s n=13 ℓ=12 peak amplitudes excited from the 6s n=13 ℓ=11 level vs static dc electric field.
(a) Theoretical dashed curve assumes no residual transverse stray field.
(b) Theoretical dashed curve assumes a 1 V/cm transverse stray field and a small adjustment of the 6snℓ quantum defects.

interaction dominates, with only a maximum quadrupolar contribution of 15% for ℓ=10. The 10snℓ states are unaffected by the presence of the stray electric field due to their much larger quantum defects. This implies that there is an effective stabilisation of the atom vis-à-vis electric fields upon exciting the core 6s electron to 10s.

CONCLUSION

We have shown that resonant excitation of high-ℓ double-Rydberg spectra is sensitive to very small stray electric fields. The effect of a stray electric field is to ℓ-mix the 6snℓ states used in laser excitation of double Rydberg states. We have demonstrated that the measured intensity variations in the 10snℓ spectra as a function of small applied electric field is an effective probe of the ℓ character of the 6snℓ initial Stark levels.

REFERENCES

Camus P, Lecomte J -M, Mahon C R, Pillet P and Pruvost L 1991 *Inst. Phys. Conf. Ser.* **114** 215
Pruvost L, Camus P, Lecomte J -M, Mahon C R, and Pillet P 1991 *J.Phys.B: At. Mol. Opt. Phys.* **24** 4723
Camus P, Lecomte J -M, Mahon C R, Pillet P and Pruvost L 1992 *J.Phys.II France* **2** 715

Inst. Phys. Conf. Ser. No 128: Section 2
Paper presented at RIS 92, Santa Fe, NM, USA, 24–29 May 1992

95

Saturation broadening effects in the resonance ionization spectroscopy of thorium

B M Tissue, C M Miller,[†] and B L Fearey

Isotope Sciences Group, INC-6, †Nuclear Chemistry and Analysis Group, INC-13
Los Alamos National Laboratory, Los Alamos, NM 87545

ABSTRACT: We describe the observation of a narrow peak superimposed on the saturation-broadened line profile in the Doppler-free spectrum of thorium. This peak was observed using counter-propagating laser beam resonant excitation of atomic thorium in cw resonance ionization mass spectrometry (RIMS). At low laser power a conventional Lamb dip was observed, however, as laser power was increased, the dip was replaced by a narrow peak. This narrow peak is attributed to a crossover peak arising from accidental degeneracies between different ac-Stark-split energy levels being simultaneously excited by the counter-propagating beam.

1. INTRODUCTION

A complete understanding of the effects of spectral saturation are necessary for our work using resonance ionization mass spectrometry (RIMS) methods for high-precision measurements of $^{230/232}$Th isotope ratios for geochronological applications (Fearey, *et al.* 1992). Saturation broadening can reduce the isotopic selectivity of the RIMS method and, in turn, limit the dynamic range of isotope ratio measurements by RIMS. A full understanding of the anomalous spectral features described in this paper is also necessary to develop a quantitative description of signal enhancement methods, such as retroreflecting the resonant and/or the ionizing laser beam, or alternatively, incorporating an external optical cavity around the ion source (Fearey, *et al.* 1991).

In the course of measuring isotope shifts of thorium using Doppler-free spectroscopy, we observed a narrow peak grow in on top of the usual saturation dip (Lamb dip) as the laser power was increased (see Figures 1 and 2). The anomalous peak was observed in RIMS measurements of ^{232}Th only when using counter-propagating laser beams. For our experimental conditions (laser power, lens focal length, sample temperature, *etc.*) this peak occurred only at the highest laser powers. At low laser powers a conventional Lamb dip was observed in the center of the Doppler-broadened line.

We have found no previous report of a narrow peak growing in with increasing laser power in Doppler-free spectroscopy. While the peak resembles an inverse Lamb dip (Siegman 1986), our experimental conditions can not produce such a phenomenon. We attribute the peak to a crossover peak arising from accidental resonances between ac-Stark-split energy levels in the ground and excited states (Demtröder 1982). This assignment is based on detailed calculations of the Einstein A coefficient and the Rabi splitting derived from the laser-power dependence of the saturation-broadened linewidths measured without the counter-propagating beam (*c.f.*, Bushaw 1989).

2. EXPERIMENTAL

A description of Doppler-free spectroscopy using cw RIMS can be found in Fearey, *et al.* (1990). Note that, since ^{232}Th has an even number of neutrons and zero spin there is no hyperfine structure. Ionization was achieved through a single-color, 1+1 process (photons to resonance + photons to ionization). The ultraviolet excitation was provided by a frequency-doubled, single-frequency Ti:sapphire laser pumped by an Ar$^+$ laser. This arrangement provided 15-30 mW of ultraviolet power in the 380-392 nm range with a linewidth of approximately 2 MHz. For power-dependence studies, the laser power was attenuated with calibrated neutral density filters. The laser beam was focused into the source region of the mass spectrometer with a 15-cm focal-length quartz lens and retroreflected with a 30 cm radius-of-curvature concave mirror. One μg of the thorium sample (ThO$_2$, derived from a natural thorium ore, in 1.5N HNO$_3$) and 1 μl of graphite slurry overcoat were deposited onto each of two rhenium side filaments of a conventional triple-filament thermal ionization assembly (Inghram and Chupka 1953). The pressure in the ion source region was typically $< 1 \times 10^{-7}$ torr. Ion detection was accomplished through a two-stage magnetic-sector mass spectrometer equipped with electron multiplier tube and pulse counter detection system. A PC computer collected the data from the pulse counter and simultaneously recorded frequency markers from a 300 MHz spectrum analyzer monitoring the Ti:sapphire-laser fundamental.

3. RESULTS AND DISCUSSION

Figures 1 and 2 show the cw-RIMS spectra at different laser powers exciting thorium at 26113.27 cm^{-1} in single-pass and in counter-propagating beam configurations, respectively. The laser powers

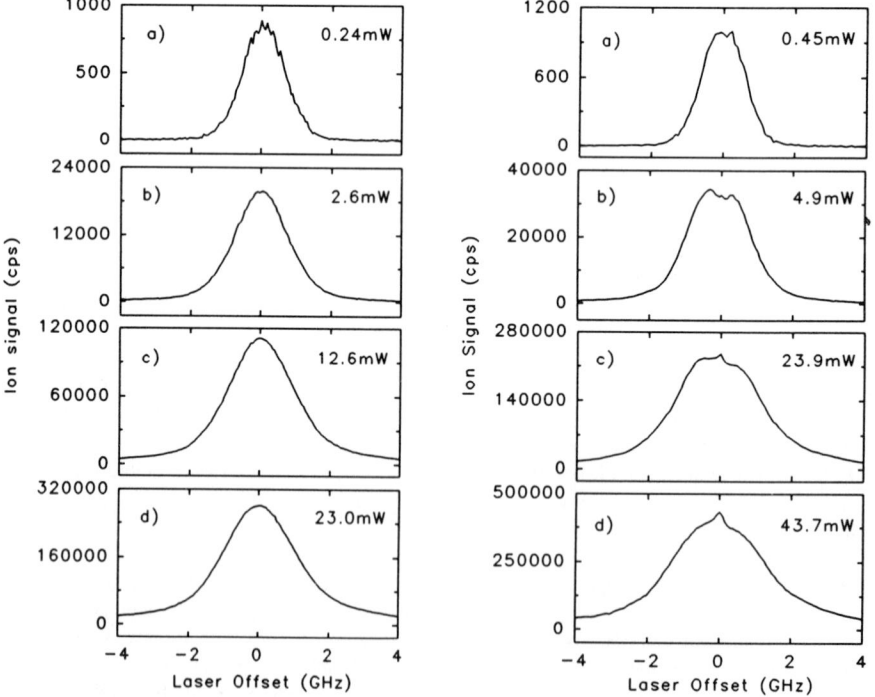

Figure 1 Single-beam RIMS spectra of Th.   **Figure 2** Doppler-free RIMS spectra of Th.

listed in the figures are for the single-pass and counter-propagating configurations. The total power with counter-propagating beams was approximately 1.9 times that of the single pass. In the Doppler-free spectra of Figure 2, a narrow peak is observed to grow in as the laser power increases, and the Lamb dip in the center of the saturation-broadened line becomes obscured.

The origin of the narrow peak in the Doppler-free spectra was investigated by analyzing the laser-power dependence of the saturation broadening in the single-pass spectra shown in Figure 1. The Lorentzian component of the lineshapes in Figure 1 were extracted by fitting the experimental lineshape to an approximation for the Voigt profile given by Whiting (1968). In general, the line profile was nearly a pure Gaussian at the lowest laser power and predominantly a Lorentzian at the highest power. The natural linewidth can then be determined by fitting the deconvoluted Lorentzian linewidths to the laser intensity using the equation given by Bushaw (1989):

$$\Gamma = \Gamma_0 \left(1 + 2\frac{I_l}{I_s}\right)^{1/2} \tag{1}$$

where I_l is the laser intensity, and Γ and Γ_0 are the measured and natural halfwidths (HWHM), respectively. The laser intensity was calculated using a measured effective beam radius of 20 μm. The saturation intensity, I_s, is given by:

$$I_s = 8\pi^2\Gamma_0 hc/3\lambda^3 \tag{2}$$

where h is Planck's constant, c is the speed of light, and λ is the wavelength of the transition. Figure 3 shows the fit of equation (1) to the power dependence data for the 26113.27 cm^{-1} transition. The natural halfwidth was determined to be 2.09 MHz, which is equivalent to a lifetime ($\tau = 1/4\pi\Gamma_0$) of 38.1 ns, and an Einstein A coefficient ($A = \beta/\tau$) of 2.35x10^7 s^{-1}. The Einstein A coefficient was calculated using branching ratios (β) determined from relative emission intensities taken from the thorium atlas (Palmer and Engleman 1983). The major source of error in these results is the uncertainty in the effective laser intensity due to uncertainties in the interaction length of the laser and atom cloud, the distribution of atoms in the beam path, and the extraction efficiency of ions into the mass spectrometer from different parts of the atom cloud (*i.e.*, the edges where the laser beam radius is larger). An effective beam radius of 20 μm was used which is the average of the radius at the beam waist and the radius at the edge of the atom cloud, estimated from the length of the heated filaments holding the sample. The uncertainty in the laser intensity leads to an overall estimated error in the calculated results of approximately fifty percent.

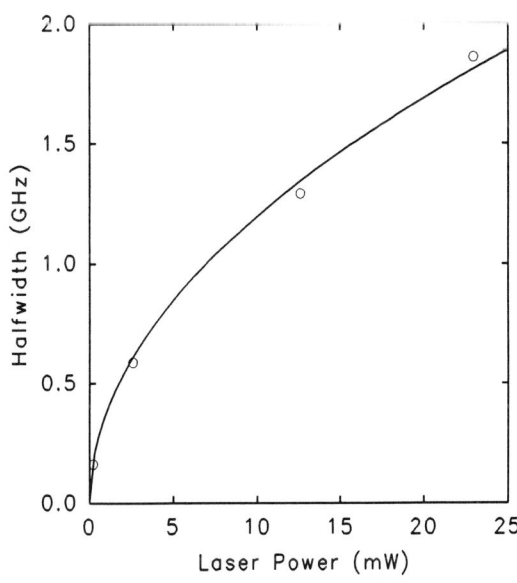

Figure 3 Fit of measured halfwidths to eqn. 1.

The Rabi splitting can be calculated from the Einstein coefficient using equation (3) (Bushaw 1989).

$$v_R = \left(\frac{A \; \lambda^3 \; I_l}{1.643 \times 10^{12}} \right)^{1/2}$$

(3)

The Rabi splitting calculated at the highest laser power (43.7 mW) is 1.67 GHz. This Rabi splitting is the splitting of the ground- and excited-state energy levels caused by coupling to the strong laser field (ac-Stark effect). The Doppler width of the transitions is about 1.3 GHz. These results suggest that when the Rabi splitting exceeds the Doppler width, the ac-Stark-split levels act as independent energy levels and produce a crossover peak. The crossover peak is the result of the laser frequency being simultaneously resonant with two different transitions for equivalent velocity groups, thereby increasing the relative ionization signal (Demtröder 1982, Siegman 1986, Fearey, *et al.* 1990). This crossover condition occurs halfway between the two ac-Stark-split transitions, which for thorium will be observed as a narrow Doppler-free crossover peak centered on top of the saturation-broadened absorption peak.

4. CONCLUSIONS

The anomalous narrow peak observed in Doppler-free spectra of thorium has been attributed to a crossover peak arising from splitting of the ground- and excited-state energy levels due to the ac-Stark effect. The concordance of the crossover peak arising most strongly when the Rabi splitting exceeds the Doppler width supports the explanation of the ac-Stark effect inducing the additional peak. Such anomalous spectral features were observed with readily achieved laser powers under certain experimental conditions, and must be taken into account for a quantitative description of RIMS.

5. REFERENCES

Bushaw B A 1989 *Prog. Analyt. Spectrosc.* **12** 247
Demtröder W 1982 *Laser Spectroscopy* (Berlin, Springer-Verlag) pp 489
Engleman Jr. R E and Palmer B A 1983 *J. Opt. Soc. Am.* **73** 694
Fearey B L, Parent D C, Keller R A and Miller C M 1990 *J. Opt. Soc. Am. B* **7** 3
Fearey B L, Johnson S G, and Miller C M 1991 in *Resonance Ionization Spectroscopy 1990 - Inst. Phys. Conf. Ser. 114* Parks J E and Omenetto N, eds.,(Bristol: Institute of Physics) pp 393
Fearey B L, Tissue B M, Olivares J A, Loge G W, Murrell M T and Miller C M 1992 *this volume*
Inghram and Chupka W A 1953 *Rev. Sci. Instrum.* **24** 518
Palmer B A and Engleman Jr. R E 1983 *Los Alamos National Laboratory Report*, LA-9615; see also Engleman Jr. R, *Gmelin Handbook, thorium suppl.* vol. A4
Siegman A E 1986 *Lasers* (Mill Valley, CA: University Science Books) pp 1199
Whiting E E 1968 *J. Quant. Spectrosc. Radiat. Transfer* **8** 1379

Inst. Phys. Conf. Ser. No 128: Section 2
Paper presented at RIS 92, Santa Fe, NM, USA, 24–29 May 1992

Two-photon resonance enhancement in calcium

A P Land [†], K W D Ledingham, R P Singhal and M Towrie[+]

Department of Physics and Astonomy, University of Glasgow, Glasgow G12 8QQ, UK
[†]*now at Genomyx Corp., South San Francisco CA 94080, USA*
[+]*now at Rutherford Appleton Laboratory, Didcot OX11 OQX, UK*

ABSTRACT: The resonance ionisation spectroscopy of calcium was studied over the wavelength range 413-437nm. The principal aim of this investigation was to examine the degree of enhancement through the close proximity of an intermediate state. In this range, two series of states, $4sns$ $^1S_0$ (n=8-13) and $4snd$ $^1D_2$ (n=7-11) could be reached from the ground state $4s^2$ $^1S_0$ by a two photon transition. Since only one intermediate state can be accessed from the ground state by a one photon transition in this wavelength region, the process should be relatively straightforward to analyse.

1. INTRODUCTION

When an atom or molecule is stimulated from one quantum state to another through the simultaneous absorption of two photons it is said to have undergone a two photon process. If this procedure is used in RIMS then it is called a 2+1 process (photons to resonance + photons to ionise). Most RIMS analytical work to date has utilised simple 1+1 schemes (Apel *et al* 1987), using either the fundamental or frequency doubled output from a single dye laser because it is perceived to be simple. On the other hand 2+1 schemes can use even less laser instrumentation and greatly extend the attainable ionisation potential and hence the range of elements accessible from a single dye laser. Furthermore through the use of counter-propogating beams a two-photon excitation can be made Doppler-free with consequent gains in both selectivity and sensitivity (Lucatorto *et al* 1984). The principal disadvantage to two-photon excitation is that the cross-sections are typically much lower than those encountered for one photon transitions. Many one-photon transitions can be saturated with laser fluences of less than $1\mu J/mm^2$ whereas a two-photon process typically requires $1J/mm^2$. The two-photon cross section can however be enhanced through the close coincidence of a suitable intermediate energy level with the energy of the exciting photons. In practise, this means that the upper level for the two-photon process is specified to be the one with energy closest to being double that of a level with a one-photon transition from the lower level.

It was decided to investigate the possibility of intermediate state enhancement for calcium in the wavelength range 413-437nm. In this range, two series of states, $4sns$ $^1S_0$ (n=8-13) and $4snd$ $^1D_2$ (n=7-11) can be accessed by a two photon process. In addition the possibility of enhancement is simple to identify, since only one state, $4s4p$ $^1P_1$ can be reached by a one-photon transition.

2. EXPERIMENTAL DETAILS

This has been described in detail elsewhere (Towrie *et al* 1990) and only the essential details will be descibed here. A schematic diagram of the Glasgow LARIS instrument is shown in fig.1. Calcium metal samples were fixed to stainless steel stubs and then mounted on a XYZΘ manipulator at the centre of a spherical sample chamber. The system was maintained at a pressure of 10^{-9} torr by turbomolecular and diffusion pumps. Sample changeover takes about 10 mins using a rapid transfer probe. The sample was vaporised using a Quantel YG585 Nd:Yag operated in fundamental mode at fluences about $1mJ/mm^2$. The post ablation ionising laser system consists of a Spectron SL2Q+SL3A Nd:Yag laser powering a Spectrolase 4000 dye laser with a bandwidth of $0.1cm^{-1}$ and an output energy of about $1mJ$ per pulse. The reflectron time of flight mass spectrometer system has an overall length of 3m with a FWHM mass resolution of about 700 for ions of about 50amu. A thin wire, 0.005cm in diameter, follows the ion flight path through the flight tube providing an electrostatic guide for the ions considerably increasing the transmission of the mass spectrometer.

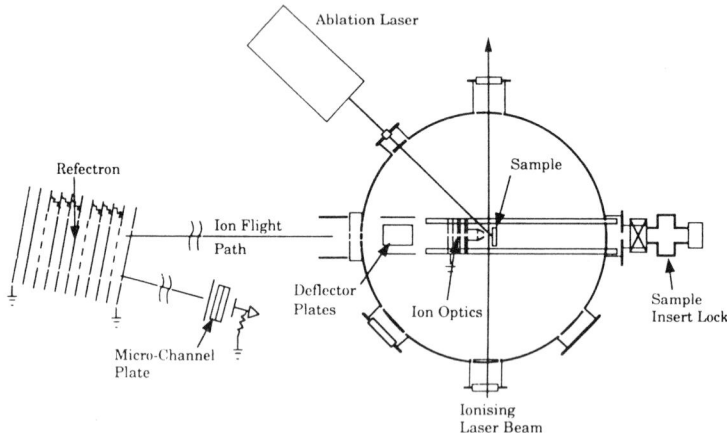

Fig. 1 The Glasgow LARIS Instrument

The data acquisition system records and stores mass spectra and laser pulse energies on a pulse-to-pulse basis. A Lecroy 2261 transient recorder linked to a COMPAQ 386/25 forms the basis of the system. Ion signals from the Galileo multichannel plate detector are digitised by the transient detector which provides 640 time channels (11 bit resolution) each of 10 - 100ns width. For wavelength scans, the ion signal integrated over a specified interval of the time spectrum is stored. For the spectra presented here, the time window was set for the m/z=40 peak and the time between the ablation and ionising lasers was chosen to be 2µs.

3. EXPERIMENTAL RESULTS AND DISCUSSION

Using a Coumarin 420 blue dye, it was possible to perform a constant laser power scan over the wavelength region 413-437nm with a fluence of about $500\mu J/mm^2$. This was achieved by a simple manual stabilization procedure using a Newport attenuator and was absolutely essential if accurate transition comparisons were to be made. The spectrum shown in fig. 2 was taken using linearly polarised light and transitions from the ground state $4s^2\ {}^1S_0$ to several $4sns\ {}^1S_0$ and $4snd\ {}^1D_2$ states are shown. Two-photon transitions from the ${}^1S_0$ ground state to $^1S_0$ excited states are forbidden for circular polarisation and fig.3 is a spectrum taken with circular polarisation showing almost complete suppression of the S-S transitions.

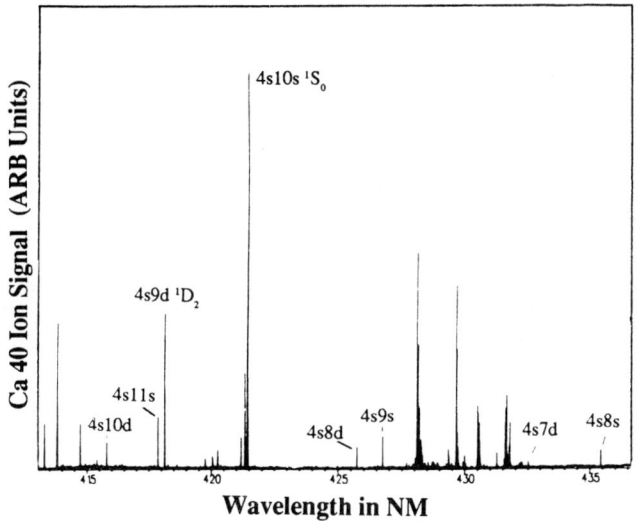

Fig. 2 ^{40}Ca spectrum taken   with linearly polarised light. The $^1S_0$ and $^1D_2$ series are clearly seen.

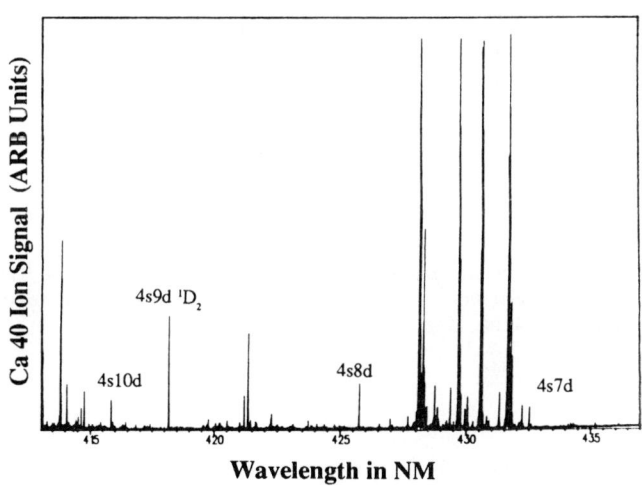

Fig. 3 ^{40}Ca spectrum taken with circularly polarised light. The $^1S_0$ series is completely suppressed.

At a basic qualitative level the date agrees well with the intermediate resonance enhancement model concept.   A clear resonance effect is observed for the 4sns $^1S_0$ series, with the strongest spectral feature being the 4s10s line which was expected to have the greatest enhancement. For the 4snd $^1D_2$ series, the 4s8d line is suppressed in line with the findings of Hansen (1983), with the 4s9d as the strongest member of the series again as might qualitatively have been expected.

If *if* are the initial and final energy levels and *r* is the intermediate state, the two photon cross section is given by:

$$\sigma_{if} = K \, f_{ir} f_{rf} / T \, E_{ir} \, E_{rf} \, (\lambda \Delta E)^2 \times \Sigma_m \mid C_{ir} C_{rf} \mid^2$$

where K is a known constant, f_{ij} is the oscillator strength for the *i-j* transition, T is the broader of the natural linewidth for the upper excited state and the laser linewidth, λ is the laser wavelength, ΔE is the energy mismatch with level r, C_{ij} is the Clebsch-Gordan coefficient for the transition *ij* and the summation is over all magnetic hyperfine states. The photoionisation cross-sections have been extrapolated from $\sigma_I = 5 \times 10^{-18}$ cm^2 (Apel *et al* 1987) for the 4s4d $^1D_2$ state and 6×10^{-19} cm^2 for the ground state 4s (Ditchburn and Hudson 1959) using the n^{*-3} rule given by Hurst and Payne (1988). The experimental results and the Glasgow population rate equation (PRE) model predictions are given in table 1. As can be seen, when using the calculated two-photon cross sections, the PRE predictions do not agree with experiment. Instead the PRE model suggests the two-photon steps are saturated, with the total ion yield following the unsaturated photoionisation cross sections. Most interestingly, there is reasonable correlation between the normalised two-photon cross sections and the experimental ion yields. It is possible that the theoretical calculated cross sections are wrong and further work, both experimetal and theoretical, is required to clarify this issue (Land 1990).

State	$\sigma_{2\omega}$	σ_I	Ion Yield (Theory)	Ion Yield (Expt.)
4s13s $^1S_0$	0.6	37	27	<1
4s12s	1.2	50	42	<1
4s11s	4.1	69	74	14
4s10s	100.0	100.0	100.0	100
4s9s	12.0	151	151	11
4s8s	2.4	247	223	4
4s11d $^1D_2$	14.0	56	56	11
4s10d	35.6	74	74	16
4s9d	100.0	100.0	100.0	100
4s8d	-	198	-	20
4s7d	25.0	300	308	6

Table 1   **Experimental and theoretical relative ion yields for 2 + 1 RIS scheme in calcium. S-state values have been normalised to the values for 4s10s, D-states to 4s9d**

REFERENCES

Apel E C, Anderson J E, Estler R C, Nogar N S and Miller C M 1987 *Appl. Opts.* **26,** 1045

Ditchburn R W and Hudson R D 1959 *Proc. Roy. Soc.*A256, 53

Hansen W 1983 *J. Phys. B : Atom. Mol. Phys.* **16,** 2309

Hurst G S and Payne M G, 1988, *Principles and Applications of Resonance Ionisation Spectroscopy* (Adam Hilger, Bristol, U.K)

Lucatorto T B, Clark C W and Moore L J, 1984 *Optics Comm.* **48(6)** 406

Land A P 1990, *Thesis,* University of Glasgow (unpublished)

Towrie M, Drysdale S L T, Jennings R, Land A P, Ledingham K W D, McCombes P T, Singhal R P, Smyth M H C and Mclean C J 1990 *Int. J. Mass Spectrom. Ion. Proc.* **96,** 309

Inst. Phys. Conf. Ser. No 128: Section 3
Paper presented at RIS 92, Santa Fe, NM, USA, 24–29 May 1992

The never-ending richness of resonance ionization

P. Lambropoulos

Department of Physics, University of Southern California, Los Angeles, CA 90089-0484
Foundation for Research and Technology-Hellas, Institute of Electronic Structure and Laser,
P.O. Box 1527, Heraklion 71110, Crete, Greece; and Department of Physics, University of
Crete, Crete, Greece

ABSTRACT: I review a class of new processes that involve coherent interactions within
the atomic continuum leading to laser-induced resonance structure which, in principle,
may be exploited for the realization of novel types of radiation sources.

I. INTRODUCTION

Resonance ionization, as practiced and discussed in this symposium, exploits the structure
of atoms and molecules in order to induce transitions discriminating one particular species,
hopefully, from everything else. Ideally, one would like to use as low an intensity as
possible so as to avoid distorting the atomic structure. Yet over the last few years, we
have seen an enormous variety of resonance phenomena[1]-[4] at quite high intensities,
demonstrating a surprising resilience of atomic resonances under such fields. The states
shift and broaden, but their signature appears quite strong either in scanning the ion signal
as a function of frequency, or even better, examining the energy and/or angle resolved
photoelectron spectrum. Some of the most surprising resonance effects in ionization have,
in fact, been observed under intensities such that the ionization threshold itself has shifted
upwards (due to the ponderomotive shift) by an amount exceeding the energy of one
photon. In all of these cases, it is the resonances with bound atomic states, shifted as they
may be, that are observed.

My purpose in this paper is to discuss resonances that are created artificially by the
imposition of an additional external laser beam. As a result, one can create resonances
where they do not exist in the bare atom, and in particular, in an atomic continuum which
without the extra laser would be smooth; in the sense that a photoionization cross section
in that energy range varies very slowly with energy. Such effects have recently been
receiving significant attention, both theoretically and experimentally, for reasons that range
from basic to the possibility of applications to new concepts of radiation sources. One
basic class of phenomena in this respect are referred to as LICS (Laser Induced Continuum
Structure). I review first the basic ideas behind these effects.

II. THE IDEA OF LICS.

In the traditional view of photoionization of a one-valence-electron atom, the absorption of the photon raises the electron from a bound state into the continuum instantly, so to speak, and irreversibly. There is no characteristic time (analogous to the spontaneous lifetime of discrete excited states) which we can associate with "how long" the electron "stays" in the continuum energy state to which it was raised. Another side of the same picture is that the dependence of the bound-free matrix element that determines the cross-section on the photon energy is smooth, exhibiting a slow variation originating from the oscillatory behavior of the wave functions. It shows no resonance-like structure. The situation changes significantly when the photoabsorption raises two electrons into the continuum. Then we encounter doubly excited discrete states embedded in (degenerate with) the single-electron continuum[5,6]. The process (at least as long as the field is not too strong) is still irreversible, but there is now a characteristic time which can be viewed as the lifetime of the discrete state that has been formed in the continuum. The dependence of the photoionization cross section on the photon energy is no longer smooth, but exhibits maxima and minima reflecting the interference between the amplitudes of the transition to the continuum and discrete parts of the wave function. By exciting the appropriate superposition of discrete and continuum wave function, we achieve a temporary localization and stabilization of an electron whose energy is above the ionization threshold. These so-called autoionizing states (or resonances) can have lifetimes ranging from less than a pico-second to microseconds, or in rare cases even be metastable against autoionization[7], depending on the atom and the configuration.

The above description tacitly assumed photoexcitation with traditional, weak, and incoherent radiation sources. The availability of coherent, more or less monochromatic, and strong sources (lasers) led to the idea of creating autoionizing-like features in the photoionization of even one-valence-electron atoms by using a second laser to couple an additional discrete (bound) state to the continuum. The essence of the idea in its simplest form, is depicted by the solid-line arrow in Fig. 1. The atom, initially, is in its ground state $|1>$ with two lasers of frequencies ω_a and ω_b present. If we have only ω_a, we will simply have photoionization with a smooth dependence on ω_a. If on the other hand, we also introduce ω_b, so that $E_2 + \hbar\omega_b$ lies above the ionization threshold and we scan ω_a around values that satisfy $E_1 + \hbar\omega_a \sim E_2 + \hbar\omega_b$, with ω_b fixed, we may expect to observe some resonance-like structure in the amount of ionization. One of the many ways of explaining this expectation qualitatively, is to view state $|2>$ as embedded in the continuum because of the presence of ω_b, which causes the discrete "atom + photon" state of energy $E_2 + \hbar\omega_b$ to be degenerate with the continuum at that energy. Then the absorption of ω_a encounters a superposition of discrete and continuum states (much like in autoionization), and depending on the details of the strength of the couplings, a more or less pronounced peak and/or dip ought to be observed in ionization as a function of ω_a. The first publication of this idea appeared in the papers of Heller and Popov[8]-[11] fifteen years ago. Due to the conceptual similarity with autoionization, and since in appearance it would suggest some structure in an otherwise smooth continuum, it goes by a number of names, such as "autoionizing-like resonance" or state, "pseudo-autoionizing resonance," "laser-induced continuum structure" (LICS), etc. In this paper, we shall use the term LICS for phenomena

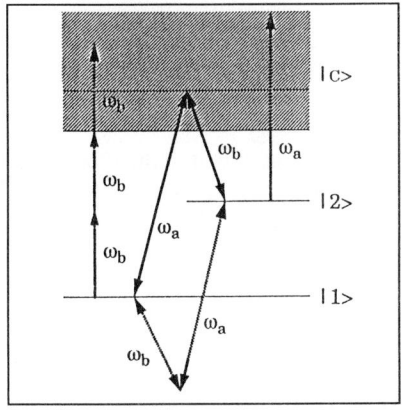

Figure 1: The simplest illustration for laser-induce continuum structure (LICS). $|1>$ and $|2>$ are two bound states of a one-valence electron atoms, $|c>$ represents the structureless continuum: ω_a, ω_b represent two different lasers.

involving the coupling of a discrete (bound) state to the respective continuum.

The solid-line arrows of Fig. 1 correspond to only part of the important interactions that come into play when both lasers are present. The rest are indicated, by dashed-line arrows and represent an additional (Raman-type) coupling of the states $|1>$ and $|2>$ (without a pole in the continuum), as well as ionization of state $|2>$ by photon ω_a, and (multiphoton) ionization of $|1>$ by ω_b. The latter two represent non-interfering decays of the atom into the continuum, a consequence of which can be the obliteration of the structure if ionization is monitored through the collection of ions or electrons, but without kinetic energy analysis. As has been shown by Dai and Lambropoulos[12], these additional couplings are as important as the basic solid-line paths, and cannot be ignored in any realistic assessment of the observability of the effect.

An obvious but significant generalization of the basic idea is to replace ω_a by a considerably smaller frequency, so that near resonance is achieved through the absorption of 3 photons, satisfying $E_1 + 3\hbar\omega_a \sim E_2 + \hbar\omega_b$. The LICS would now be expected either in the observation of 3-photon ionization as a function of ω_a (with ω_b fixed), or in the spectrum of third harmonic generation $3\omega_a$. The latter is particularly appealing if it can lead to the enhancement of the harmonic due to the oscillator strength that the embedding of state $|2>$ lends to the continuum. The number of additional (Raman) paths is now increased, but the fundamental idea remains the same.

In order to best pursue the analogy with autoionization, one would want to keep the intensity of ω_a weak compared to that of ω_b, so that the process, (whether it be ionization or harmonic generation) can be viewed as a "probe" of the structure created by ω_b. Recall that in autoionization of a two-valence-electron atom, the structure is created by the intraatomic (electron-electron) interaction, which comes with the atom and the state. Thus, photoabsorption around an autoionizing resonance represents a probe of the intraatomic structure. In LICS, we have more flexibility in that the value of ω_b and its intensity can be controlled externally, making the distinction between probe and coupling interaction only a matter of definition. There is, in fact, nothing that prevents us from keeping ω_a fixed and scanning ω_b. Insofar as harmonic generation is concerned, it should be noted that a third order process requires some intensity to be observable at all, with the consequence that, at least for observational purposes, the intensity of ω_a cannot be arbitrarily weak.

There is one more variety of the manifestation of LICS that has appeared in the literature. It involves the absorption of a weak (probe) beam from the ground state $6s(^2S_{1/2})$ of caesium into the $\varepsilon P_{1/2,3/2}$ continuum. The probe beam is linearly polarized. A right-circularly polarized coupling beam couples the excited state $8s(^2S_{1/2})$ to the same energy in the continuum, but due to the selection rules, the $8s$ is coupled only to the $\varepsilon P_{3/2}$ continuum. The probe beam, which due to its linear polarization is coupled to both continua, sees one continuum with and the other ($\varepsilon P_{1/2}$) without structure. Since the linearly polarized radiation can be decomposed into a superposition of left and right circular polarizations, its left component sees a smooth continuum while its right a structured one. The net result is that the two components are refracted unequally, leading thus to a rotation of the polarization of the probe beam as monitored after it has emerged from the interaction region. This effect has been predicted by Heller and Popov[11], and its observation reported by Heller *et al* [13]. It does, in fact, represent the first and, to this day, one of the few, documented and correctly interpreted observations of LICS in any context.

A rather voluminous theoretical literature[14] has accumulated since 1976, the time of appearance of the first papers by Heller and Popov. Most of such theoretical papers have dealt with idealized versions of the system involving two discrete levels and one continuum, and usually excluding the additional but inevitable couplings discussed above. In spite of the extensive body of theoretical work, only very few experimental attempts have proven fruitful. One was mentioned above in connection with the rotation of polarization of the probe beam. The second, in chronological order, has been published by Pavlov *et al* [15,16], who reported structure and enhancement of third harmonic generation (THG) in sodium vapor. In that experiment, the second harmonic (532 *nm*) of a *Nd: glass* laser was used to couple the excited state $5s$ to the continuum, and at the same time, to pump a tunable dye laser whose output was used to couple the ground state $3s$ to the continuum via a 3-photon process, which also produced third harmonic. The spectrum of the third harmonic, as a function of the frequency of the fundamental, showed a resonance and clear enhancement around that energy in the continuum at which the $5s$ was coupled by the 532 *nm* photon.

After a gap of a few years, Feldmann *et al*[17] reported observations showing some effects of LICS in 2-photon resonant 3-photon ionization of atomic sodium in the presence of a second laser. While some of the resonances observed can be attributed to LICS, as Felmann *et al*[17] recognized and Dai and Lambropoulos[12] later confirmed quantitatively, the Raman-type coupling dominated, leading to symmetric peaks, which is always the case when the couplings through the continuum are not dominant. Still, that was the first observation of the effect of LICS on the ionization signal. It is useful to note here that Feldmann *et al* reported observable structure only when the excited state $4d$ was coupled to the continuum, but not when $4d$ was replaced by either the $8s$ or $6d$. In all cases, the ground state was coupled to the continuum via 3-photon absorption which, as noted earlier and discussed by Dai and Lambropoulos[12] in detail, introduces a number of additional couplings. It is thus not too surprising that some excited states did not produce LICS. The lesson there is that the simple idea of Fig. 1 does not guarantee observability without a quantitative assessment that includes accurate atomic parameters.

The only other report of LICS in ionization was published by Hutchinson and Ness[18], who observed a peak in 2-photon resonant 3-photon ionization of *Xe* while a second laser coupled the excited state 10*p* to the appropriate energy in the continuum. Unfortunately, the reported structure is in serious disagreement with the theoretical calculation by Tang *et al*[19]. Until another experiment is performed, the result of Hutchinson and Ness cannot be counted among the documented and properly interpreted observations of LICS.

The first direct observation of asymmetric line shapes in ionization was reported most recently by two groups[20,21] working independently on the same atom (*Na*) but with different excited states. The critical development that made possible these observations was the design of the experiment so as to collect ions only from the center of the interaction region where both laser beams are sufficiently intense for the structure to develop above the background of the ionization by each of the two beams separately. A detailed analysis of the effects of the spatio-temporal distribution of the radiation on all effects of LICS, but especially harmonic generation, has been presented recently by Zhang and Lambropoulos[22].

III. GENERALIZATION AND FUTURE OUTLOOK

One can now go beyond these experiments and consider generalizations thereof. A first generalization is to consider LICS as depicted in Fig. 1 in a 2-electron atom so that we can arrange for an autoionizing state to be present at the position in the continuum where the photons ω_a and ω_b coincide. In that case, the photon ω_b would couple the excited state |2> to the autoionizing state, creating thus further structure to a continuum that already has structure due to the autoionizing state. If ω_a were replaced by a photon with one-third of its energy, we could then consider 3rd harmonic generation through an autoionizing state which is modified by the coupling of the autoionizing state to an excited bound state. One of the relevant questions here is whether we could exploit such an arrangement to enhance harmonic generation. A general scheme for this type of process is shown in Fig. 2 in the context of atomic *Ca*, which has been examined in some detail recently by Lambropoulos *et al*[23]. At this point, the theoretical predictions appear to be quite promising but experimental investigations are necessary before one can feel confident that unforeseen side effects will not cause complications. The alkaline earth atoms, and especially *Mg* and *Ca*, seem good candidates for the exploration of these effects with existing laser sources.

The above phenomena are intimately connected with the idea of lasing without inversion (LWI) or at least amplification without inversion which has recently been receiving considerable attention, at least theoretically[24]-[26]. The basic idea in the context of autoionization was proposed by Arkhipkin and Heller[24] almost ten years ago, and was revived most recently by Harris[25]. In short, if one were able to pump the discrete part of

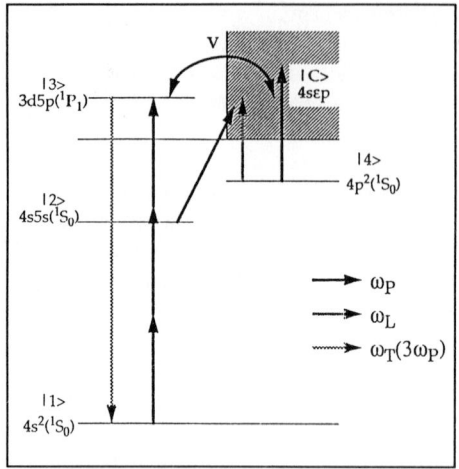

Figure 2: Schematic diagram of a proposed scheme for 3rd harmonic generation in *Ca* through an autoionizing resonance combined with LICS (ω_P = pump photon, ω_L = coupling photon for LICS).

an autoionizing state with a fairly asymmetric lineshape and probed with photons whose energy matched the minimum of the resonance with the ground state, it can be shown that the probe radiation can be amplified even if the number of atoms in the autoionizing state is smaller than those in the ground state. The physical origin of this effect can be traced to the asymmetry between absorption, which exhibits a minimum, and emission, which does not. Given the similarity between autoionization and LICS, one can envision variations on that scheme. In fact, any scheme, either natural or induced, where the absorption of photons exhibits a minimum due to interference, while the emission does not, can, in principle, be employed as a medium for LWI. As usual, the merits of such schemes for a realistic application to this process can be proven only after experimental investigation. Hopefully, during the next two years, we will see some of these schemes leading to the fruitful realization of amplification without inversion.

(This work was supported by the U.S. Department of Energy under Grant No. DE-FG03-87ER60504)

REFERENCES

[1] M.D. Perry and O.L. Landen, Phys. Rev. A$\underline{38}$, 2815 (1988).
[2] R.R. Freeman, P.H. Bucksbaum, H. Milchberg, S. Darack, D. Schumacher and M.E. Geusic, Phys. Rev. Lett. $\underline{59}$, 1092 (1987).
[3] H.G. Muller, H.B. van Linden van den Heuvell, P. Agostini, G. Petite, A. Antonetti, M. Franco and A. Migus, Phys. Rev. Lett. $\underline{60}$, 565 (1988).
[4] M.D. Perry, A. Szoke and K.C. Kulander, Phys. Rev. Lett. $\underline{63}$, 1058 (1989).
[5] U. Fano, Phys. Rev. $\underline{124}$, 1866 (1961).
[6] W.E. Cooke, T.F. Gallagher, S.A. Edelstein, and R.M. Hill, Phys Rev. Lett. $\underline{40}$, 178 (1978); W.E. Cooke and C.L. Cromer, Phys. Rev. A$\underline{32}$, 2725 (1985).
[7] C.A. Nicolaides and D.R. Beck, Phys. Rev. A$\underline{17}$, 2116 (1978).
[8] Yu.I. Heller and A.K. Popov, Sov.J. Quant. Electron. $\underline{6}$, 606, (1976).
[9] Yu.I. Heller and A.K. Popov, Op. Comm. $\underline{18}$, 8 (1976).
[10] Yu.I. Heller and A.K. Popov, Op. Comm. $\underline{18}$, 449 (1976).
[11] Yu.I. Heller and A.K. Popov, Zh. Eksp. Teor. Fiz. $\underline{78}$, 506 (1980), [Sov. Phys. JEPT $\underline{51}$, 255 (1980)].
[12] Bo-nian Dai and P. Lambropoulos, Phys. Rev. A$\underline{36}$, 5205 (1987).
[13] Yu.I. Heller, V.F. Lukinykh, A.K. Popov and V.V. Slabko, Phys. Lett. A $\underline{82}$, 4 (1981).
[14] For a review see, for example, P.L. Knight, M.A. Lauder, and B.J. Dalton, Phys. Rep. **190**, 1, (1990).
[15] L.I. Pavlov, S.S. Dimov, D.I. Metchkov, G.M. Mileva, and K.V. Stamenov, Phys. Lett. $\underline{89A}$, 441 (1982).
[16] S.S. Dimov, L.I. Pavlov, and K.V. Stamenov, Appl. Phys. B$\underline{30}$, 35 (1983).
[17] D. Feldmann, G. Otto, D. Petring and K.H. Welge, J. Phys. B: At. Mol. Phys. $\underline{19}$, 269 (1986).
[18] M.H.R. Hutchinson and K.M.M. Ness, Phys. Rev. Lett. $\underline{60}$, 105 (1988).
[19] X. Tang, Anne L'Huillier, and P. Lambropoulos, Phys. Rev. Lett. $\underline{62}$, 111 (1989).
[20] Y.L. Shao, D. Charalambidis, C. Fotakis, J. Zhang and P. Lambropoulos, Phys. Rev. Lett. $\underline{67}$, 3669 (1991).
[21] S. Cavalieri, Manlio Matera, and Francesco Pavone, Phys. Rev. Lett. $\underline{67}$, 3673 (1991).
[22] J. Zhang. P. Lambropoulos, Phys. Rev. A $\underline{45}$, 489 (1992).
[23] P. Lambropoulos, J. Zhang and X. Tang in Coherence Phenomena in Atoms and Molecules in Laser Fields, edited by A.D. Bandrauk and S.C. Wallace. Plemum Press, New York (1982).
[24] V.G. Arkhipkin and Yu.I. Heller, Phys. Lett. $\underline{98A}$, 12 (1983).
[25] S.E. Harris, Phys. Rev. Lett. $\underline{62}$, 1033 (1989).
[26] A. Lyras, Z. Tang, P. Lambropoulos, and J. Zhang, Phys. Rev. A$\underline{40}$, 4131 (1989).

Inst. Phys. Conf. Ser. No 128: Section 3
Paper presented at RIS 92, Santa Fe, NM, USA, 24–29 May 1992

111

Alterations of multiphoton-resonant processes through wave-mixing effects

W.R. Garrett

Oak Ridge National Laboratory, Oak Ridge, Tn 37830

ABSTRACT: A number of the manifestations of quantum interference effects on resonant multiphoton excitation processes are described. Examples of observions and predictions involving simple systems are summerized and discussed in the context of analytical applications.

1. INTRODUCTION

When a dipole-allowed atomic or molecular transition is resonantly driven by odd-multi-photon excitation or by stimulated hyper-Raman scattering, a nonlinear polarization of the resonant medium is generated at the transition frequency. This polarization serves as a source for generation of a multi-wave mixing field which can greatly alter the atomic response from that expected from the laser field acting alone. Oddly enough, the generated field may not produce observable photons (the medium can be quite opaque at the resonant frequency). Under circumstances where certain combination the molecular number density N, the oscillator strength $F_{0,n}$ between state $|n>$ and the ground state $|0>$, the path length, z, in the medium and the laser bandwidth Γ, satisfy the condition $(\pi Ne^2 F_{0,n}/mc)z/\Gamma >> 1$, a number of dramatic effects on odd-photon mediated processes can ensue as a result of interference between excitation pathways provided by the laser field and the internally generated wave-mixing fields. In some instances the effects occur even at quite low pressures (Payne and Garrett 1982) (down to values as low as 10^{-5} Torr with narrow-bandwidth laser pumping.)

The underlying basis for a number of effects associated with odd-photon pumping of optically allowed transitions is that for a given combination of resonant driving fields, the Rabi frequency term which mediates the transition also serves as a source term for generation of a nonlinear polarization at the sum or difference frequency. The field which is generated by this polarization has three characteristic features: it is strongly absorbed by the medium, since it is at the resonant frequency; it evolves 180° out of phase with the combined phase of the driving fields; it grows in magnitude until complete destructive interference is established between the laser and generated fields, at which point no additional excitations and no additional field strength are created.

We now have aquired a fairly global view of the influence of FWM fields on principal atomic and molecular transitions mediated by odd-photon resonant excitations. Contrary to the view held by all early investigators, including the present author, the internally generated field cannot be neglected in any beam geometry when the product mentioned above is large. The interplay of the externally supplied laser fields and the internally generated wave-mixing fields produce suppression and shifting of resonance lines which have a simple linear dependence on pressure, but a fairly complicated dependence on pump wavelengths, crossing angle between pump beams and mode of excitation (whether resonance occurs at a sum or difference frequency of the pump fields). Both the suppression and shifting processess are produced in a comprehensive treatment of the problem (Payne and Garrett 1990, Garrett et.al.1990). Excellent agreement between theory and experimental studies in noble gases and in metal vapors is illustrated for a number of predicted effects including complete suppression under unidirectional pumping and strong shifting under multidirectional pumping of resonance lines and hyper-Raman emissions (Garrett, Hart, Miller, Payne an Wray 1991) with single and multi-laser sources. Eleven predicted manifestations of the altered atomic or molecular response are experimentally verified, including dependencies of the interference-related effects on gas number density, oscillator strength, wavelength combination and relative propagation directions of pump laser beams. Confirmed also is the fact that, contrary to most other nonlinear effects, none of the suppression and shifting features (in the semi-classical regime) depends on the intensity of the driving laser field(s).

We give only a descriptive picture of the theoretical basis for the behavior of resonant excitations and stimulated hyper-Raman emissions under circumstances where wave mixing influences become germane. We concentrate instead on recent experimental results associated with these phenomena and their relevance to analytical applications.

2. THEORETICAL CONSIDERATIONS

The influence of internally generated fields on multiphoton excitation processes can be described by several alternative methods. Because of its capacity for including some interesting details, e.g., effects associated with laser bandwidths, etc., we have chosen to describe the atomic response with a two-state model in a density matrix approach. Though the discussion can be extended to include all odd-photon-mediated process in the category under consideration, we restrict ourselves to three-photon mediated transitions. The problem has been generalized to include two different modes for carrying out three-photon excitation by two independent laser beams. In the first mode two photons at frequency ω_{L1} are absorbed from laser 1 and one photon of frequency ω_{L2} is absorbed from laser 2. In the second mode two photons are absorbed from laser 1 and an emission is stimulated at ω_{L2} by laser 2. The unfocused laser beams with propagation vectors $\vec{k}_{L1}$ and $\vec{k}_{L2}$ are overlapped to form an excitation volumn where the beams cross at an arbitrary angle θ ($\theta = 0$ corresponds to parallel and $\theta = 180°$ to anti-parallel propagation). The system is described by two levels, $|0>$ as ground state, energy E_0, and $|1>$ as the excited state, with energy E_1. The two-state model leads to familiar equations for the elements ρ_{ij} of the density matrix, but the coupling, V_{01} between the two levels is composed of two components. The first component is direct three-photon pumping by two photons at ω_{L1} and one at ω_{L2}. The second part of V_{01} is the coupling produced by the internally generated field at the sum or difference frequency $\omega_m = 2\omega_{L2} \pm \omega_{L2}$, with local value $E_{\omega_m}^{\mathcal{L}}$ (to be calculated as part of the problem). In the rotating wave approximation the total interaction term is

$$V_{01}/\hbar = - \left(\Omega_{01}^{(3)} + \Omega_{10}^{(1)} e^{-\Delta k_r z} \right) e^{-ik_m z} e^{i\omega_m t}.$$

This is written here only show exlicitly how the interaction contains a term $\Omega_{01}^{(3)}$ due to pumping by three photons and a term $\Omega_{01}^{(1)} = \frac{1}{2\hbar} D_{01} E_0^{\mathcal{L}} e^{i(k_m + \Delta k_r)z}$ due to one-photon pumping by the four-wave mixing sum or difference-frequency field that is internally generated. Here D_{01} is the dipole matrix element between $|0>$ and $|1>$, $E_0^{\mathcal{L}}$ is the amplitude of the locally space-averaged FWM field at frequency ω_m, and Δk_r is the real part of the phase mismatch. At high concentration light generated at $\omega_m = 2\omega_{L1} \pm \omega_{L2} \cong (E_1 - E_0)/\hbar$ is resonantly absorbed in a distance less than a wavelength. Thus, the Lorentz approximation can be used to relate the local field at the location of the atom, $E_{\omega_m}^{\mathcal{L}}$ to the locally space averaged field, E_{ω_m}, which enters Maxwell's equations. The relation between the local and space averaged field is $E_{\omega_m}^{\mathcal{L}} = E_{\omega_m} + (4\pi/3)P_{\omega_m}$

where P_{ω_m} is the polarization of the medium at frequency ω_m. The space averaged field E_{ω_m} must be derived through solutions to Maxwell's equation, with the polarization P_{ω_m} as source term. The resonant part of the polarization can be expressed in terms of the density matrix elements as $P_{\omega_m} = NTr(\hat{\rho}\hat{D})$, where N is the number density and $\hat{D}$ is the electric dipole operator. Maxwell's equation can be solved for plane wave fields involving no incoming waves in terms of an integral of the time retarded polarization source term over the length of the medium. The local field can then be replaced by this space averaged field plus the Lorentz polarization to yield an expression for V_{01} which can be substituted into the equation fir the off-diagonal element of the density matrix to yield an integro-differential equation for ρ_{01}.

If it is assumed that ρ_{11} remains small ($\rho_{00} \sim 1$), then in presently applicable circumstances, where $\omega_m \Delta_0 L/c\Gamma_P \gg 1$, the equation can be solved exactly in the adiabatic limit. The solution can then be used to give an equation for the diagonal element ρ_{11}. A principal result is obtained thereby, namely the excitation rate, R, for producing $|1>$:

$$R = 2\Gamma_P \frac{|\Omega_{01}^{(3)}|^2}{\left(\delta_1 - \Delta_0\left(-1.11 + \frac{\lambda_{L1} \pm 2\lambda_{L2}}{[1-\cos(\theta)]\lambda_{mix}}\right)\right)^2 + \Gamma_P^2} = 2\Gamma_P \frac{|\Omega_{01}^{(3)}|^2}{\left(\delta_1 - \tilde{\Delta}_P\right)^2 + \Gamma_P^2}.$$

We defined $\lambda_{mix} = 2\pi c/(2\omega_{L1} + \omega_{L2})$ and $\delta_1 = 2\omega_{L_1} + \omega_{L_2} - (E_1 - E_0)/\hbar$. The excitation lineshape has the familiar form for a three-photon induced transition, but it now contains a total pressure dependent shift

$$\tilde{\Delta}_P = \Delta_0 \left[-1.11 + \frac{\lambda_{L1} \pm 2\lambda_{L2}}{[1 - \cos(\theta)]\lambda_{mix}}\right] = \Delta_P + \Delta_c$$

where $\Delta_P = \Delta_0(-1.11) = -\frac{4\pi}{3\hbar}N|D_{10}|^2 + \Delta_{bl}$ is the usual pressure induced shift and the last term is the cooperative shift

$$\Delta_c = \Delta_0 \frac{(\lambda_{L1} \pm 2\lambda_{L2})}{[1 - \cos(\theta)]\lambda_{mix}}.$$

This cooperative shift is indeed very novel in character. It is positive (violet) for the + mode and negative (red) for the − mode of excitation, it is linear in pressure (since Δ_0 is proportional to N) and is a function of the crossing angle θ. Additionally, we see that use of various combinations of λ_{L1} and λ_{L2} produce different shifts. These can be very large (many nanometers) if $\lambda_{L1} \gg \lambda_{L2}$ or if θ is small. Then there is an even more intriguing effect. If one of the beams, say λ_2, is retroreflected then a second coherent three-photon excitation path involving a new direction for L2, and thus a different Δ_c, is created. With this crossed and retroreflectd geometry a scan of ω_{L2} will produce two resonant peaks (see below). The width of the line (or lines) in terms of wave numbers and the shape remains the same and appears as a normal pressure broadened profile.

The total suppression of any excitation as $\theta \to 0$, which is the often-studied cancellation effect, can be obtained from a limit of the expression above. The usual result, that $R \simeq 0$ when $\delta_1 = 0$, is obvious. That is, no excitation at the position of the unshifted resonance. When θ becomes small, and Δ_c large, the excitation probability remains small everywhere. Thus for $\theta \to 0$ we have the result that ρ_{01} and ρ_{11}, go to zero and excitation is strongly suppressed for all detunings.

It is fairly easy to show (Garrett et.al. 1991) that the same considerations which produce a shift in the peak of the three-photon excitation profile also produces an identical shift in the peak of the gain profile for stimulated hyper-Raman scattering. This results in a pressure dependent shift in the SHR emission to shorter wavelengths. This prediction and an experimental confirmation in Xe is contained in a recent paper by Garrett et.al. (1991).

3. EXPERIMENTAL RESULTS

Since there are a large number of published studies involving suppression of three-photon (and five-photon) resonant excitations in atomic and molecular systems (with unidirectional beams), we forego introduction of any additional data on this well characterized feature of the present topic. Studies of the suppression of forward stimulated hyper-Raman emission associated with excitation of an optically allowed transition has so far only been reported for experiments in metal vapors (Garrett et. al. 1992), though the phenomenon should be universal. We report new results in connection with confirmation of the predicted SHR shift in Xe. (See also Garrett et.al. 1991).

We now present abbreviated data from a series of experiments involving two-color three-photon-resonant excitation (and subsequent ionization by an additional photon) in Xe. The experiments were intended to explore all of the predicted features of the suppression and collective lineshifting phenomena. Ten of these features were confirmed : 1) three-photon excitation is cancelled under unidirectional pumping (already discussed at length) ; 2) cooperative line shifts linear in number density of target gas are produced (for all geometries involving $\theta > \theta_{min}$); 3) two separate peaks are produced when two-color pumping in crossed geometry if one beam is retroflected to produce two complementary crossing angles; 4) different shifts are produced for a fixed geometry and pressure when different wavelength combinations are used for pumping a given state; 5) for different transitions, the slopes of shifts vs pressure are proportional to oscillator strength for coupling to the ground state; 6) the shifts obey the predicted relationship with respect to crossing angle; 7) forward Stimulated hyper-Raman emission is suppressed; 8) the collective shift for excitation in $2\omega_1 - \omega_2$ mode of excitation is opposite to that produced in the $2\omega_1 + \omega_2$ mode; 9) a collective shift, linear in pressure, is produced in stimulated hyper-Raman emission, and; 10) all of the effects are independent of the pump beam intensities.

With the exception of the last example below, all of the results described here were obtained with the use of a stainless steel ionization cell fitted with quartz windows. The cell contained a positively biased charge collection wire and it was traversed by beams from two dye lasers pumped by one Nd:YAG laser. Shown in Fig.1 are data for three-photon excitation of the $5d[3/2]_1^\circ$ level in Xe at pressures of 50, 200, 400, and 800 Torr. Laser 1 was held fixed and frequency doubled to 297.76 nm while laser 2 was scanned across three-photon resonance. The beams were overlapped at $\theta = \pi$ (i.e counterpropagated). Note the fairly large pressure dependent shift (which is actually a minimum as a function of θ at $\theta = \pi$). Shown in Fig.2 are data for excitation of the same $5d[3/2]_1^\circ$ state but for laser beams of a different wavelength combination arranged in a different geometry. Here the first laser was set at $\lambda_{L1}=291.9nm$ and the second was tuned across the 5d resonance, but at a crossing angle $\theta = 22.8°$. Xe pressure was 100 Torr in trace a) of the figure. In trace b) the pressure is unchanged, but laser 1 is retroflected with a dichroic mirror, thus providing an additional excitation pathway with $\theta = 157.2°$. This produces a second peak with only a small shift (position of the unshifted resonance is given by the vertical line). In trace c) the xenon pressure is 200 Torr and both peaks shift farther to the blue. Note that the shifts in Fig.2 that are produced at 22.8° are much larger than those in Fig.1 at corresponding pressure, as expected from the functional dependence of Δ_c on θ.

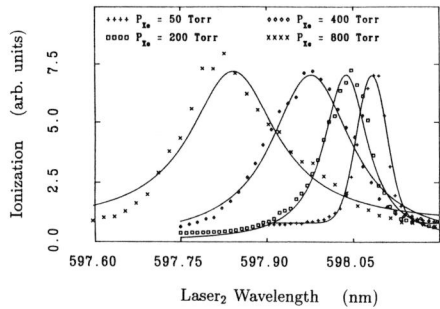 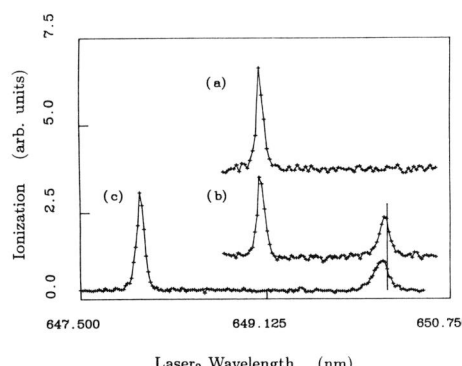

Fig.1 Resonant ionization signals for three photon excitation of the $5d[3/2]_1^0$ level in Xe at pressures indicated. Laser$_1$ held fixed, laser$_2$ scanned. Crossing angle $\theta = \pi$.

Fig.2 Resonant ionization profiles for $2\omega_{L_1} + \omega_{L2}$ excitation of the $5d[3/2]_1$ level in Xe at $\theta = 22.8°$. a) 100 Torr Xe, both beams single pass. b) 100 Torr, but with retroreflection of L$_2$. c) 200 Torr Xe. Vertical line at unshifted resonance position.

Now we turn to a consideration of measurements of stimulated hyper-Raman emission in Xe. An excimer pumped dye laser was frequency doubled to produce $\simeq 224.3$ nm photons which were focussed with a 10 cm f.l. lens. The focussed beam was in near two-photon resonance with the $6p'[3/2]_2$ level in Xe. Backward and forward stimulated hyper-Raman scattering associated with excitation of the $6s'[1/2]_1^0$ and production of $\simeq 834$ nm emission was measured. In Fig.3 we show example spectral traces of backward infrared emission in the region of $\simeq 834nm$ at a series of Xe pressures. The data are for Xe pressures of 200, 400, 800 and 1000 Torr, at a laser setting of 0.04nm from two-photon resonance with the $6p'$ level. We first note the dominant features of the spectra: for backward emission, two peaks are present; the separation between the peaks is pressure dependent; and that one peak occurs at a fixed wavelength while the other shifts with increasing Xe pressure; in the forward direction (not shown) only one peak appears, fixed in wavelength, at the position of the amplified spontaneous emission (ASE) peak on the right (though weaker and somewhat broader).

In the traces shown in Fig.3 the longer-wavelength peak, at fixed position, is ASE from the $6p'[3/2]_2$ level to the lower $6s'[1/2]_1^0$ state. The shorter-wavelength, shifting peak is SHR emission associated with excitation of the same $6s'[1/2]_1^0$ lower state through stimulated hyper-Raman scattering. With the uv pump intensities available in this study, SHR signals were observable only at very small detunings from two-photon resonance ($\delta_2 \le .05nm$). At elevated number densities two-photon excitation on the wing of the $6P$ level produces some ASE output even at detunings exceeding the laser bandwidth.

Thus in backward emission we expect to see ASE at 834.6nm, and a second SHR peak displaced from the ASE by an amount determined by the laser detuning and the predicted pressure dependent shift. With our laser power we were able to see SHR emission only at pressure-detuning combinations where some remnant of ASE was still observable.

Finally for a fixed laser setting, we predict that the SHR emission profile will show a large shift which varies linearly with Xenon pressure. Moreover, since the dipole matrix element (oscillator strength) which enters the expression for the collective shift is known for the $6S'$ to ground state transition, the shift can be accurately predicted. Taking 0.179 for this oscillator strength[21] the hyper-Raman shift is $\bar{\Delta}_P = -0.00039 P_{Xe}$ where P_{Xe} is in Torr and $\bar{\Delta}_P$ is in nm. The shift of the SHR-excited level is of opposite sign to that produced in three photon excitation of the $5d[3/2]_1$. That is, the excited level is shifted to lower energy, corresponding to a blue shift in the SHR emission with increasing P_{Xe}.

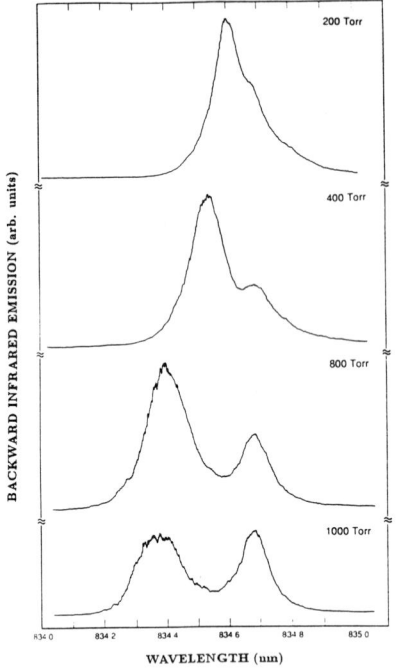

Fig.4 Stimulated hyper-Raman shift as a function of P_{Xe}. Data points are shown as squares, straight line is the theoretically predicted shift.

Fig.3   Spectral scans of backward infrared emission associated with excitation of the $6s'[1/2]_1$ level while tuning near two-photon resonance with the $6p'[3/2]_2$ state. Laser tuned 0.04nm from two-photon resonance. Xe pressure changed successively to the values indicated. Righthand peaks, at fixed wavelength, are ASE. Lefthand peaks are SHR emission.

Shown in Fig.4 is the magnitude of the shift, $\bar{\Delta}_P$, in the SHR profile as a function of P_{Xe}. The data are presented as squares. The predicted shift is given by the straight line. The data are in very good quantitative agreement with the predicted cooperative shift for $\Delta_P \leq .25nm$, i.e. at all but the highest pressures. Note that the lineshape at 1000 Torr also changes in appearence. This was typical of all the very high pressure results, including those described earlier. Presumably some other effect not included in the theory starts to become evident above a few hundred Torr. We conclude that in SHR excitation of a state which can also be excited from the ground-state by a single photon, the unavoidable generation of a FWM field at the resonant frequency $(E_1 - E_0)/\hbar$ not only suppresses SHR gain in the forward direction, but also produces a large pressure dependent shift in the SHR emission at nonzero angles.

4. CONCLUSIONS

The rather complicated suppression and shifting features associated with odd photon excitations of optically allowed transitions are rather counterintuitive in some instances and surprising in the efficacy of the observed effects over such wide ranges of experimental circumstances. The agreement between observations and experiments is uncommonly good. The independence of the effects on laser intensities, item (10) in the above list, is also found experimentally to hold over the entire range of available laser intensities. As predicted, the shifts and the suppression effects are, unlike most other nonlinear phenomena, not dependent on pump intensities.

Finally it is useful to note some of the implications of the phenomena under discussion in the context of analytical applications of RIS. The odd-photon interference effects discussed here are relevant only in cases involving multicolor excitation (e.g. where extreme selectivity might be sought through multiple resonant steps). The experiments described herein involved broadband lasers, thus in the chosen examples the effects are dramatic only for number densities corresponding to fractions of a mili-torr and above. Moreover very high pressure experiments were use to illustrate the very large magnitudes to which the modifications can lead. However with narrowband lasers interferences can occur in pulsed nozzle sources at 10^{-5} Torr (Gilligan and Eyler 1991). The modifications of atomic and molecular response are cause for caution in certain RIS schemes, but they also offer new opportunities for alternative analytical methods ranging from isotope separation schemes to remote sensing.

*Research sponsored by USDOE Office of Health and Environmental Research, under contract DE-AC05-84OR21400 with Martin Marietta Energy Systems, Inc.

5. REFERENCES

Garrett, W.R, Hart, R.C., Wray, J.C., Datskou, I., and Payne, M.G., Phys. Rev. Lett. **64**, 1717 (1990).

Garrett, W.R., Hart, R.C., Miller, J.C., Payne, M.G., and Wray, J.C. Optics Comm. **86**, 205 (1991).

Garrett, W.R., Moore, M.A., Hart, R.C., Payne, M.G., and Wunderlich, R.K., Phys. Rev.A **45**, 6687 (1992).

Gilligan, J.M., and Eyler, E.E., Phys Rev. **43**, 6406 (1991).

Miller, J.C., Compton, R.N., Payne, M.G., and Garrett, W.R., Phys. Rev. Lett. **45**, 114 (1980).

Moore, M.A., Garrett, W.R.,and Payne, M.G., Optics Commun. **68**, 310 (1988).

Payne, M.G., Garrett, W.R., and Baker, H.C., Chem. Phys. Lett. **75**, 468 (1980)

Payne, M.G., and Garrett, W.R., Phys. Rev. A **26**, 356 (1982).

Payne, M.G., and Garrett, W.R.,Phys. Rev. A **28**, 3409 (1983).

Payne, M.G., and Garrett, W.R., Phys. Rev. A **42**, 1434 (1990).

Payne, M.G., Garrett, W.R., Hart, R.C., and Datskou, I., Phys. Rev.A **42**, 2756 (1990).

Inst. Phys. Conf. Ser. No 128: Section 3
Paper presented at RIS 92, Santa Fe, NM, USA, 24–29 May 1992

119

Isotope biases in RIMS utilizing broad-band long-pulsed lasers

W. D. Brandon, S. L. Allman, W. R. Garrett, and C. H. Chen

Oak Ridge National Laboratory, P. O. Box 2008, Oak Ridge, TN 37831-6378 USA

M. G. Payne

Georgia Southern University, Landrum Box 8031, Statesboro, GA 30460

J. E. Parks

Institute of RIS, University of Tennessee, Suite 300, 10521 Research Drive, Knoxville, TN 37932

ABSTRACT: It has been shown (Fairbank *et al* 1990 and Lambropoulas *et al* 1990) that hyperfine structure plays a major role in producing "anomalous" odd to even isotope ratios in the stepwise excitation and ionization of an element with broad bandwidth lasers. This investigation is based on the suggestion (Payne *et al* 1991) that isotope biases may be avoided by detuning the laser from resonance by an amount greater than either the laser bandwidth or the hyperfine splitting, but less than the power broadened width. Our initial experiments have confirmed that detunings of 6 cm^{-1} are sufficient to eliminate these biases in the two-photon stepwise ionization by way of the $^3P_0 \rightarrow {}^3P_1$ resonance in Sn with a laser power density of 10^8 W/cm^2. Also, we have seen that a 1 + 2 resonantly enhanced ionization scheme faithfully reproduces odd to even ratios, which can be attributed to beam geometry.

1. INTRODUCTION

Those interested in a detailed theoretical treatment are referred to Payne *et al* (1991) in which two methods are described for eliminating isotope biases utilizing long-pulsed ($\tau_L >> \Delta_{hfs}^{-1}$) broad-band ($\gamma_L > \Delta_{hfs}$) LPBB lasers. For these parameters, the effects of hyperfine structure on ionization probabilities could be summarized as follows:

A) Selection rules
B) Dilution of oscillator strength
C) Beam geometry

A) For linearly polarized light the selection rules for the even and odd isotopes are $\Delta M_J = 0$ and $\Delta M_F = 0$ respectively. Clearly one must choose an ionization scheme in which J always increases with excitation energy since $F = I \otimes J$ provides a mechanism for breaking the M_J selection rule for the odd isotopes.

B) Even with a properly chosen ionization scheme (increasing J rule) hyperfine coupling leads to the dilution of the oscillator strength since the number of transitions is increased. Thus a larger percentage of the atoms exist in the excited state when the discrete-discrete transitions are all saturated. See Figure 1 for an illustration upon which our work is based.

C) Particularly large isotope biases occur in the low intensity regions at "the boundaries" of the laser beam even if the ionization probability is close to unity in the most intense regions. The hyperfine levels are excited coherently (since $\gamma_L > \Delta_{hfs}$) but have sufficient time to become dephased (since $\tau_L > \Delta_{hfs}^{-1}$) before ionization occurs leading to an enhanced isotope response. Thus, incorporation of a realistic beam geometry into the equations of motion is quite important in modeling this problem.

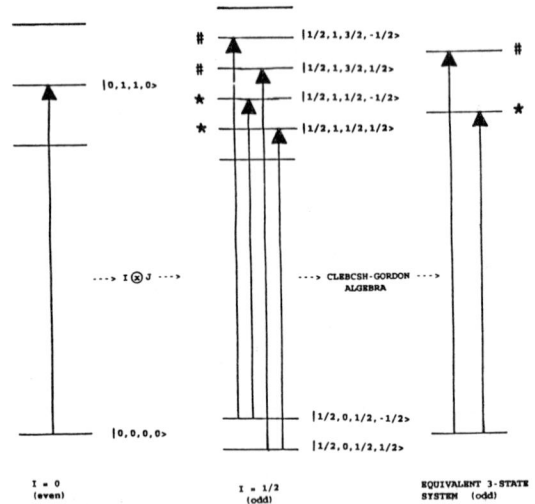

Figure 1 Energy level diagram in the $|IJFM_F>$ representation. Upon performing the C-G algebra, one finds the $\frac{1}{2},-\frac{1}{2}$ ground states behave identically. The oscillator strength splits such that $\frac{1}{3}$ is associated with $(\frac{1}{2} \to \frac{1}{2})^*$ transition and $\frac{2}{3}$ is associated with $(\frac{1}{2} \to \frac{3}{2})^\#$ transition. We note that $\Delta(E^\# - E^*) = \Delta_{hfs} = .25 \text{cm}^{-1}$ for the $^3P_1$ state in Sn.

Essentially, one may eliminate these biases utilizing "fast ionization". If ionization occurs before the hyperfine coupling acts to dephase the excited levels hfs effects are eliminated. In the case of LPBB lasers, the most experimentally simple of the two methods described (Payne *et al* 1991) for achieving fast ionization ($T_{ion} << \Delta_{hfs}^{-1}$) appeals to the uncertainty principle. We detune the laser from resonance by an amount greater than the laser bandwidth but less than the minimum power-broadened width in the hfs manifold: $\gamma_L < \delta < PBW_{min}$. For the ionized state to be created, the photons must be absorbed in a time $\tau_{abs} \sim \delta^{-1}$. Thus for sufficient δ: $T_{ion} \sim \delta^{-1} << \Delta_{hfs}^{-1}$; we expect no isotopic biases.

The ionization schemes used are shown in Figure 2. Scheme a, in which the $(1 + 1)$ competes with the more intense $(1 + 2)$ pathway yields no significant isotopic biases over a wide range of laser power densities. This can be attributed to an intermediate $|7p^1S>$ state and beam geometry. Since the doubling process is somewhat inefficient, any "edge effects" (large biases) caused by the $(1 + 1)$ ultraviolet absorption are compensated by the much higher power density of the fundamental visible. Fast ionization occurs over an overwhelming majority of the beam. However, this method is limited greatly by possible MPI of other background sources.

The "detuning method" is applied to scheme b of Figure 2. As stated earlier, we detune such that $\gamma_L < \delta < \eta\,\Omega_{min}$; η is a dimensionless function of δ and beam geometry and $\Omega_{min} = c_{min}\,I^{1/2}$ (Lambropoulas *et al* 1990), the minimum Rabi rate of the hfs manifold (see Figure 3). The PBW_{min} has been defined in this unconventional manner in hopes that some physical aspects might be clarified. To get an idea of the order of magnitude of the parameters involved, note that a laser intensity of 10^8 W/cm² produces a PBW_{min} of about 15 cm⁻¹. If one detunes by an amount $\delta = 5$ cm⁻¹, then $T_{ion} \sim \delta^{-1}$ or about 7 picoseconds. Comparing this to the hyperfine coupling time $\Delta_{hfs}^{-1} = 21$ ps, we expect this detuning to be sufficient to eliminate hfs effects.

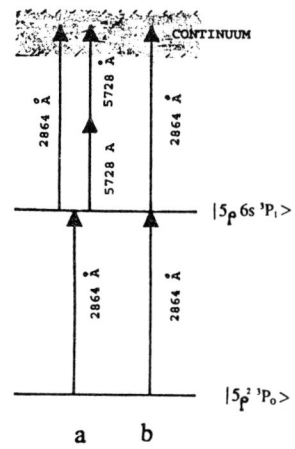

Figure 2 Schemes a and b.

2. EXPERIMENTAL CONDITIONS

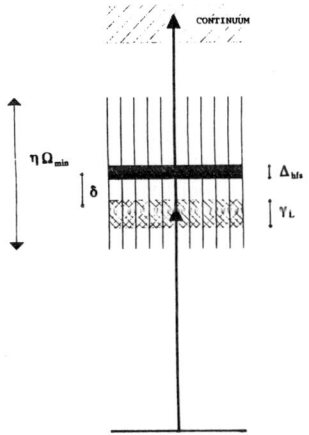

Figure 3 Fast ionization via detuning.

Experimental parameters were $I = 10^8$ W/cm², $\gamma_L = .6$ cm⁻¹, $\tau_L = 5$ns, laser rep rate = 10 Hz. One hundred twenty laser shots were averaged over each mass setting of a CVC quadrupole mass filter with ~ 30 mass settings per a.m.u. Mass settings were sampled randomly in order to reduce the effect of fluctuations in the atomic beam density which was created utilizing a ceramic "oven." Several spatial profiles of the laser beam taken with a linear diode array justify the incorporation of a chaotic field model with a Gaussian lineshape into the equations of motion (Payne *et al* 1991).

The data (Figure 4) clearly indicates a diminished anomaly with increasing δ. Experimental errors range from < 3% for $\delta < 1$ cm⁻¹; < 5% for 5 cm⁻¹ < δ < 10 cm⁻¹. We believe mode structure effects increase the experimental uncertainty in the 1 cm⁻¹ < δ < 3 cm⁻¹ range. However, the 5 cm⁻¹ < δ < 10 cm⁻¹ range offers a nice "δ operating region" somewhat analogous to the operating voltage region of a Geiger-Müeller counter. Fortunately, one has freedom in detuning to avoid MPI from other background sources. Even though the signal remains reasonably strong as long as $\delta < PBW_{min}$, loss of signal with increasing δ is a factor to be considered when utilizing this method. Somewhat amusing is the possibility of actually detuning the laser to a region in which $PBW_{odd} < \delta < PBW_{even}$. One might expect to find the odd response "suppressed" instead of "enhanced" in this region since $PBW_{odd} < PBW_{even}$.

3. CONCLUSION

The fast ionization via detuning method is promising for use as an analytical technique in measuring isotopic abundances. We also feel that a more intensive investigation into the relation between parameters such as ionization probabilities, power-broadened widths with various laser power densities, detunings, bandwidths, and pulse lengths coordinated with theoretical calucations is worth the effort involved.

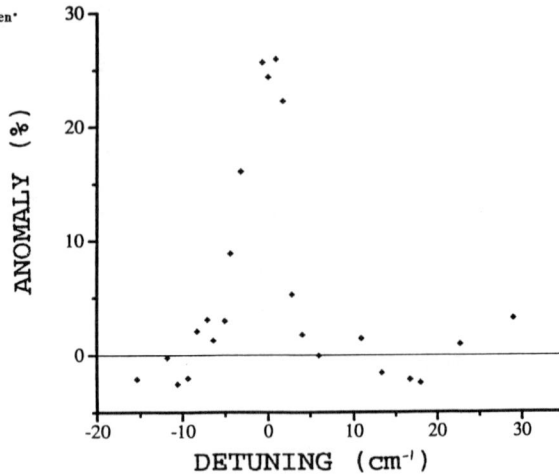

Figure 4 Anomaly vs detuning. Anomaly = ratio even/odd response (% difference from natural abundance).

4. ACKNOWLEDGEMENTS

Authors would like to thank D. Holt for preparation of manuscript, and R. C. Phillips, R. B. Jones, and J. J. DiCillo for valuable discussions. This research is sponsored by the Office of Health and Environmental Research, U.S. Department of Energy under contract DE-AC05-84OR21400 with Martin Marietta Energy Systems, Inc.

5. REFERENCES

Fairbank W M Jr, Sparr M T, Parks J E, and Hutchinson JMR, *Phys. Rev.* A40, 2195 (1990)
Lambropoulas P, Lyras A, *Phys. Rev.* A40, 2199 (1990)
Payne M G, Allman S L, Parks J E, *Spectrochimica Acta*, Vol. 46B, No. 11, pp 1439-1457 (1991)

Inst. Phys. Conf. Ser. No 128: Section 3
Paper presented at RIS 92, Santa Fe, NM, USA, 24–29 May 1992

Measurement of the odd–even effect in the resonance ionization of tin as a function of laser intensity

X. Xiong[†], J.M.R. Hutchinson and J. Fassett

National Institute of Standard and Technology, Gaithersburg, MD 20899

W. M. Fairbank, Jr.

Colorado State University, Fort Collins, CO 80523

[†]Also with Eastern Analytical, Inc., 335 Paint Branch Dr., College Park, MD 20742

ABSTRACT: The variation of the anomalous odd-even effect with laser intensity in single color 1+1 RIMS of tin at 286 nm is reported. The experimental results agree well with theoretical predictions from 10^5 to 3×10^9 W/cm^2, provided that the ionization cross sections are reduced by an order of magnitude in accordance with experimentally determined limits.

1. INTRODUCTION

The method of Resonant Ionization Mass Spectrometry (RIMS) plays an important role in isotope analysis, especially when the quantity of sample to be studied is very small. Combining the advantages of both lasers and mass spectrometers, RIMS has become a technique with high sensitivity and selectivity. However, some experiments in recent years have shown different responses between the even and odd isotopes, even with broadband laser radiation (e.g., Fairbank et al. 1989). Lambropoulos and Lyras (1989) have provided a theoretical explanation of the anomalous isotope ratios observed on one of the tin transitions (286.3 nm), and have predicted the dependence of this odd-even effect with laser intensity. In this paper, experimental results for resonant ionization of tin at 286.3 nm are presented for different laser intensities and compared to the theoretical predictions.

2. EXPERIMENTAL APPARATUS

The resonant ionization scheme used in this experiment is a simple 1+1 single-color resonant process using two ultraviolet (UV) photons at wavelength of 286.3 nm. The first UV photon makes the transition from Sn ground state $5p^2\ ^3P_0$ to the $5p6s\ ^3P_1$ excited state, which lies above half of the ionization potential of atomic tin. The second absorbed photon completes the transition into the ionization continuum.

A diagram of the experimental apparatus is shown in Figure 1. The UV radiation at 286.3 nm is generated by frequency doubling a tunable dye laser (Molectron DL18P), which is pumped by the second harmonic of a Nd:YAG laser (Molectron MY34-10) . The UV laser beam is sent, either unfocussed or focussed with a 25 cm focal length lens, into the source region of a magnetic sector mass spectrometer. There atomic tin is produced by thermal atomization of a metallic tin sample on a rhenium filament. The data acquisition system includes a boxcar integrator and a computer. Spatial profiles of the focussed and unfocussed beams in the ionization region taken with a CCD camera indicate that the laser beam is Gaussian with half

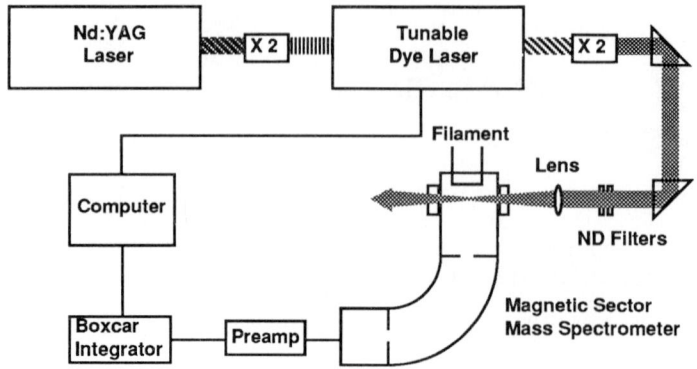

Figure 1. Experimental apparatus.

widths at half-maximum of 0.039 mm and 0.50 mm for the focussed and unfocussed laser beams, respectively. Neutral density filters are used to provide different laser intensities. The dye laser pulses have approximately 5 nsec FWHM duration and 0.3 cm^{-1} bandwidth.

3. RESULTS AND DISCUSSION

Sample results for one measurement of the RIMS response of all the stable tin isotopes with an unfocussed laser beam of 0.16 mJ (corresponding to intensity 3×10^6 W/cm^2 in the center of the beam) are presented in Table 1. These results represent an average over 100 laser shots at each mass. The errors are dominated by laser intensity fluctuations from shot to shot. The measured response, R, for each isotope is defined as the ratio of the RIMS and natural abundances. It is apparent that the odd mass isotopes all have an anomalously high response compared to the even mass isotopes and that any slope in the response within the even mass group is small. An overall measure of the odd-even anomaly is the parameter, $ß = 2(R_{odd} - R_{even})/(R_{odd} + R_{even})$, where R_{odd} and R_{even} are weighted averages over the odd and even isotopes, respectively. For this particular set of data, $ß = 0.280(24)$, where the error limits quoted are 1σ and are propagated in quadrature. This value is consistent with the measurement by Fairbank et al (1989) of $ß = 0.31(11)$ at 10^6 W/cm^2. For other intensities, measurements were taken only for the even and odd isotopes, 118 and 119. The measured odd-even effect at various intensities is presented in Figure 2.

Table 1. Abundances of tin isotopes measured by RIMS at intensity 3×10^6 W/cm^2.

Sn Isotopes	112	114	115	116	117	118	119	120	122	124
Natural Abund.	0.96	0.66	0.35	14.3	7.61	24.0	8.58	32.8	4.72	5.94
RIMS Abund.	0.929	0.640	0.492	14.92	10.29	26.69	11.22	32.93	4.55	5.57
Response R	0.968	0.970	1.406	1.043	1.352	1.112	1.308	1.004	0.965	0.938
Error	0.086	0.086	0.111	0.189	0.048	0.050	0.034	0.020	0.048	0.047

In order to compare theory directly to experiment, the predictions of Lambropoulos and Lyras (1989) must be averaged over the Gaussian spatial profile of the laser beam. The

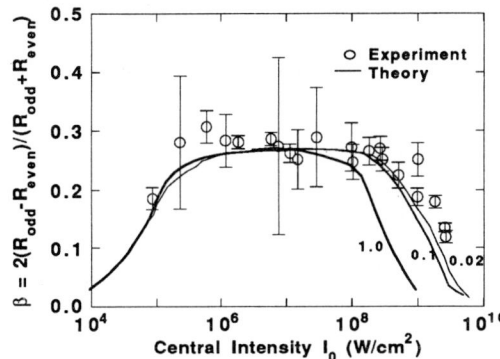

Figure 2. Intensity dependence of the odd-even effect. The solid lines represent theoretical calculations with all the ionization rates scales by the numbers next to the curves.

heavy line in Figure 2 is the result of this average. The predicted magnitude of the effect at intermediate intensity agrees well with the measurements, and the low intensity behavior is also well-represented. At high intensities, however, theory and experiment disagree on the intensity at which the odd-even effect decreases toward zero.

To understand this discrepancy, let us consider the dominant effects in the different intensity ranges which are responsible for presence or absence of an odd-even anomaly. At intermediate intensity, the resonant transition is completely saturated, but the ionization transition is not. For each group of interacting magnetic sublevels ($m_J = 0$ in the even isotopes and $m_F = \pm 1/2$ in the odd isotopes), the populations of the ground and resonant excited states become equalized early in the laser pulse by the competition of strong absorption and stimulated emission. In the even isotopes, this means that 50% of the atoms are available for ionization from the $m_J = 0$ level. On the other hand, in the odd isotopes there are two $m_F = 1/2$ excited states corresponding to F=1/2 and F=3/2. Thus when the populations of the three $m_F = 1/2$ states become equal, 2/3 of the atoms of the odd isotopes are available for ionization. While the predicted ionization rates (Lambropoulos and Lyras, 1989) for these two F levels are not equal (2.74 I and 3.08 I, respectively, where I is the intensity in W/cm^2), the average of the two ionization rates is close to that predicted for the even isotopes, 2.96 I. Thus, for this particular system, the odd-even effect at intermediate intensities can be understood as predominantly a population effect with 4/3 as much population available for ionization in the odd isotopes as in the even isotopes. In this approximation the value of the odd-even parameter is ß=2(4/3–1)/(4/3+1)=0.29, which reproduces well the flat region of the experimental and theoretical curves.

At low intensity, the resonance transition is not saturated, and the ground and excited state populations do not become equalized. The net result is approximately equal RIMS response for odd and even isotopes, as seen in Figure 2 for intensities below 10^4 W/cm^2. At high intensities, two effects occur which can cause the odd-even anomaly to disappear. First, when all isotopes in the laser beam are ionized, the RIMS response is the same for all isotopes, i.e., 100%. Second, the theoretical calculations indicate that above ~10^9 W/cm^2 the bound state populations are no longer able to become equal. Consequently, the odd-even effect is reduced, even when ionization is incomplete.

The observed disappearance of the odd-even anomaly at significantly higher intensities than predicted suggests that the ionization rates calculated by Lambropoulos and Lyras are too high. For the same transition, Saloman (1992) obtained a 6.5 times lower ionization cross section using a Hartree-Fock code with relativistic corrections.

The observed variation of the RIMS signal for an even isotope (mass 120) with laser intensity is shown in Figure 3a. The experimental conditions were approximately the same as in Figure 2, except that a 50 cm focal length lens was used. Three theoretical curves are shown which represent the results of numerically solving the density matrix equations for the system, as elucidated by Lambropoulos and Lyras (1989), with various multiples of their ionization cross section. Since there is little saturation evident in the experimental data, only

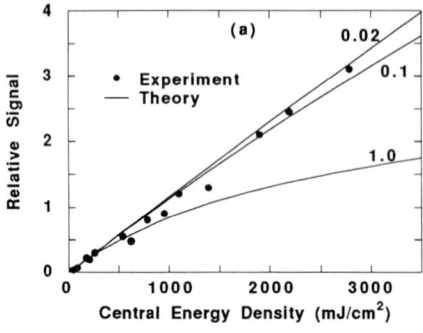
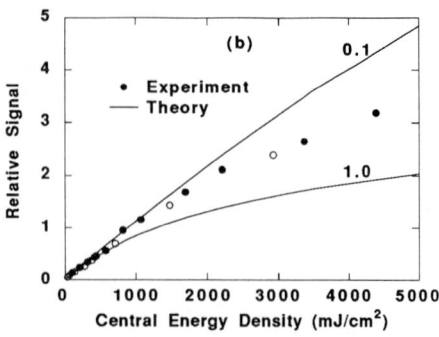

Figure 3. Experimental and theoretical saturation data for (a) single color [286 nm] and (b) two color [286nm, 295 nm] resonant ionization of tin. Theoretical curves for different scaling of all ionization rates are shown. Since the theoretical (3.5 nsec) and experimental (5 nsec) pulse widths differ, energy density E_0 is used instead of intensity ($I_0=E_0/\Delta t$).

an upper limit of 0.1 times the Lambropoulos value ($\Gamma<0.059$ I) can be extracted from the data. A second set of data shown in Figure 3b was taken with a constant, unfocussed beam ($\sim10^4$ W/cm^2) at 286.3 nm, which by itself produced negligible ionization, and a variable intensity beam at 295.2 nm, which was focussed with a 50 cm lens. This data shows a clearer saturation, corresponding to an ionization rate about 1/3 the Lambropoulos value.

The revised predictions of the Lambropoulos theory for the odd-even parameter ß as a function of laser intensity, with the all the ionization rates multiplied by factors of 0.1 and 0.02, are shown by the lighter lines in Figure 2. The agreement with experiment at high intensities is reasonably good in both cases.

4. CONCLUSION

The variation of the anomalous odd-even effect with laser intensity in the single color 1+1 scheme for RIMS of tin at 286 nm has been reported here. The experimental measurements agree well with theoretical calculations over a wide range of laser intensities, providing the ionization cross sections are reduced proportionally in accordance with experimentally determined limits.

5. ACKNOWLEDGEMENTS

The authors are grateful to Bill Bowman for his generous help and advice with the details of running the RIMS system. This work was supported by the U. S. Department of Energy under Contact No. DE-AI-05-86-ER-60447.

6. REFERENCES

Fairbank, W M, Jr., Spaar, M T, Parks, J E and Hutchinson, J M R 1989 *Phys. Rev. A* **40** 2195
Lambropoulos, P and Lyras, A. 1989 *Phys. Rev. A* **40** 2195
Saloman, E B "A Resonance Ionization Spectroscopy/ Resonance Ionization Mass Spectrometry Data Service III - Data Sheets for Sb, Bi, P, Na, and Sn (to be published).

Inst. Phys. Conf. Ser. No 128: Section 3
Paper presented at RIS 92, Santa Fe, NM, USA, 24–29 May 1992

Systematics of the odd–even effect in the resonance ionization of Os and Ti

R. K. Wunderlich*, G. J. Wasserburg, I. D. Hutcheon and G. A. Blake

Division of Geological and Planetary Sciences, California Institute of Technology, Pasadena, CA 91125, *Physikalisches Institut, Universität Augsburg, W-8900 Augsburg; FRG

ABSTRACT: Measurements of the odd-even effect in the mass spectrometric analysis of Ti and Os isotopes by resonance ionization mass spectrometry have been performed for $\Delta J = +1$, 0 and -1 transitions. Under saturating conditions of the ionization and for $\Delta J = +1$ transitions odd-even effects are reduced below the 0.5% level. Depending on the polarization state of the laser large odd isotope enrichments are observed for $\Delta J = 0$ and -1 transitions which can be reduced below the 0.5% level by depolarization of the laser field.

1. INTRODUCTION

The application of resonance ionization mass spectrometry (RIMS) to accurate isotopic analysis in geo- and cosmochemistry is complicated by the presence of laser induced shifts in measured isotopic ratios. Observed effects range from 3% (Walker and Fassett 1986) to over 40% (Fairbank et al 1989, Spiegel et al. 1992) with a predominant enrichment of the odd mass isotopes. Laser induced shifts can be broadly divided into wavelength tuning effects which originate from the finite overlap of the laser spectral distribution with the absorption maxima of the different isotopes (Wunderlich et. al. 1992 a and b) and the intrinsic odd-even effects (Fairbank et al 1989, Lambropoulos et al. 1989, Payne et al. 1991, Wunderlich et al. 1992 a and b). In practical applications of RIMS to isotopic analysis intrinsic odd-even effects can be compensated for by comparison of measured isotopic ratios with a standard of known isotopic composition (Spiegel et al. 1992). However, the size of the effects makes accurate determination of isotopic ratios difficult and precludes the ready identification of material of unusual isotopic composition. Furthermore, in some systems, like in the Re-Os chronometer, there is no standard for the radiogenic isotope available.

In this study we examined the odd-even effect in the resonance ionization of Ti and Os in order to devise methods to avoid or greatly reduce the size of these effects in isotopic analysis. Os and Ti were selected because of their importance in geo- and cosmochemistry.

2. ODD-EVEN EFFECTS AND THE MEASUREMENT OF ISOTOPIC RATIOS

Intrinsic odd-even effects can be divided into selection rule effects and dynamic effects. Selection rule effects can be present when the resonance transition involves angular momentum changes $\Delta J=0$ or -1. In this case a certain fraction of the ground state M_J and M_F components will not interact with a linearly or circularly polarized laser field. Because of the increased number of agular momentum states for isotopes with nuclear spin (odd isotopes) this fraction is smaller for the odd isotopes, resulting in odd isotope enrichment. Considering a $\Delta J=0$ transition between J=1 states, 1/3 of the ground state population does not interact with a linearly polarized laser field because of the forbidden $M_J = 0 \rightarrow M_J = 0$ transition. Upon addition of an half integer nuclear spin this selection rule is no longer in effect resulting in an enhancement of the odd isotopes. This situation is met in many resonance transitions in Ti and Os. In principle selection rule effects could be corrected by simple level counting. However, in many practical applications the laser radiation is not of a pure polarization state causing deviations of isotopic ratios obtained with this type of resonance transition from the expected values and a reduction of the odd-even effect (Wunderlich et al. 1992 a,b). Measured isotope ratios will then depend on the laser intensity, beam parameters and ion optical extraction conditions. The effects will be particularly pronounced in multistep ionization schemes with noncolinear laser beams because ionization can be produced from regions with widely different saturation of the different M_J ground state components. These considerations suggest the use of depolarized light for the excitation of the discrete resonances in RIMS. This also provides the advantage of not having to use all $\Delta J=+1$ resonance transitions.

Dynamic effects have been discussed in detail by Lambropoulos and Lyras (1989) and by Payne et al. (1991) including the effects of the laser spatial intensity distribution. Dynamic effects pertain to $\Delta J=+1$ transitions where from selection rule arguments no odd isotope enrichment should be present. A coherent and incoherent regime can be distinguished. In the first case with interaction times much smaller than the HFS coupling time a coherent superposition of HFS levels is excited. With regard to population balance ionization proceeds as if the HFS structure was absent and no odd-even effects should be observed. For interaction times much larger than the HFS coupling time and large laser bandwidhts excitation proceeds via independent HFS states. Due to the higher excited state degeneracy for isotopes with nuclear spin there will be on average more atoms in the resonance state for odd isotopes as compared to the even isotopes dending on the polarization state of the laser. This can result in a large odd isotope enrichment if the ionization is not saturated. Thus, saturation of the ionization is essential for obtaining a small odd-even effect. However, considering again the incoherent case depolarization of the resonance excitation laser should also result in an equal fraction of excited state atoms for the odd and even isotopes resulting in a large reduction of the odd-even effect.

3. EXPERIMENTS AND RESULTS

Details of the experimental set up and procedures have been described by Wunderlich et al. (1992 a). A frequency doubled pulsed laser system, thermal atomic beam and a modified quadrupole mass spectrometer were used for the RIMS experiments. Single photon 1 + 1 schemes with resonances near 40000. cm^{-1} were used for the resonance ionization of Ti and Os. The width of the HFS transition arrays for the odd Os and Ti isotpes were obtained from intracavity etalon scans. The dye laser fundamental could be operated with a bandwidth of 0.4 cm^{-1} to assure excitation of the whole HFS transition arrays.

The experiments on $\Delta J=+1$ transitions in Os and Ti were all performed in the incoherent regime, that is the pulse duration of 20 ns was much larger than the inverse HFS splitting. Results from a series of

Table 1. Odd-even effects in Os and Ti isotope ratios for $\Delta J=+1$ transitions.
Numbers in paranthesis indicate the 1σ error

Resonance State (cm^{-1})	$\delta^{47}Ti$ (⁰/oo)	$\delta^{49}Ti$ (⁰/oo)	$\delta^{189}Os$ (⁰/oo)	mode
39405.9			-3.4 (5)	sat
38544.4	-14 (12)	+2.8 (11)		sat
37851.5	+35 (11)	+4.6 (11)		sat
38159.5	+60 (15)	+144 (23)		nsat

measurements are given in Table 1. The isotopic ratios have been corrected for mass fractionation and are given in the per mil deviation from their standard values (δ-values) (Os ratios: Creaser et al. 1991, Ti ratios: Niederer et al. 1981). Reference isotopes were ^{46}Ti and ^{190}Os. Mode indicates saturating (sat) or nonsaturating (nsat) conditions of the ionization.

Under saturating conditions of the ionization no odd-even effect for Os is seen. This result differs from the one obtained by Walker and Fassett (1986) where a 3% enrichment in ^{189}Os was observed. With exception of the δ^{47}Ti = 35 value, the Ti results obtained under saturating conditions are compatible with the absence of odd-even effects. The values obtained under nonsaturating conditions show a considerable increase in the odd isotope ratios consistent with the arguments given above. The higher vavalue of δ^{49}Ti may be caused by the larger number of excited state components in ^{49}Ti compared to ^{47}Ti (I=5/2 for ^{47}Ti and I=7/2 for ^{49}Ti).

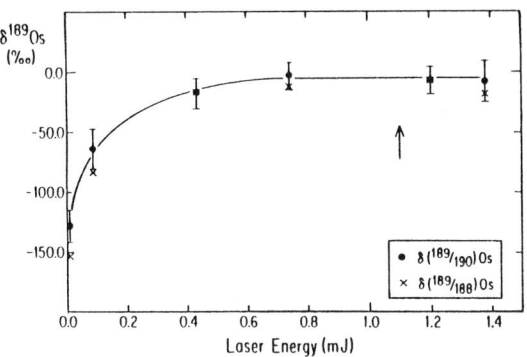

The power dependence of the ^{189}Os/^{190}Os and ^{189}Os/^{188}Os ratios measured for the ΔJ=+1 transition is shown in Fig.1. The arrow indicates the onset of ionization saturation. The ratios are constant for laser intensities larger than half the ionization saturation intensity and agree very well with the standard ratios after mass fractionation correction. The maximum laser intensity in the experiment was 5 x 10^7 Watt/cm^2 so that the constancy of the isotopic ratios can not be explained by laser bandwidth effects.

Fig.1. Power dependence of ^{189}Os/^{188}Os.

Selection rule and polarization effects were studied on ΔJ =0 and -1 transitions in Os and Ti. Some degree of depolarization was generated by passing the expanded laser beam through a quartz window exposed to localized stress creating a random amount of birefringence across the laser beam diameter. P_{ell} is the ratio of the maximum and minimum intensity measured in two orthogonal directions of a linear polarizer positioned behind the experiment chamber. In Table 2. odd-even effects obtained for different values of P_{ell} are listed in δ-values. With exception of the Os ΔJ=0 (40361.9 cm^{-1}) result odd-even effects are absent for a large degree of depolarization of the laser beam, P_{ell} =27. The Ti result is markedly different from the results obtained by Spiegel et al. (1992) in a multistep ionization scheme with a series of ΔJ=0 transitions where effects up to δ^{47}Ti = 350 were observed. Based on selection rules we calculate a maximum odd isotope effect of δ^{47}Ti = 225 for linearly polarized radiation for this type of transition.

Table 2.   Polarization dependence of the odd-even effect for ΔJ=0 and ΔJ=-1 transition in Os and Ti. Numbers in paranthesis indicate the 1σ error

P_{ell}	δ^{47}Ti (ΔJ=0)	δ^{189}Os (ΔJ=-1)	δ^{189}Os (ΔJ=0)
220	+137 (10)	+57 (8)	+89 (10)
27	+2.7 (10)	-1.3 (13)	+36 (8)

Thus we suggest that the size of the odd-even effect was most likely increased in the final autoionizing step with an additional ΔJ=-1 transition. The residual odd isotope enrichment of 3.6% for the Os ΔJ=0 as compared to the ΔJ=-1 resonance transition (39406.cm^{-1}) with P_{ell} =27 appears unusual considering the much larger oscillator strength of the ΔJ=0 transition (i.e. 0.3 compared to 0.003). As such a small admixture of elliptical polarization should saturate the $M_J=0$ to $M_J=0$ transition and no odd isotope enrichment should be observed.

The effect of the spatial intensity distribution in the laser beam on odd-even ratios is shown in Table 3. The $^{189}Os/^{190}Os$ ratio was measured for different focussing conditions for the $\Delta J=-1$ resonance transition and with $P_{ell} = 27$. The confocal parameter in all experiments was larger than the logitudinal extend of the ion collection region. It is seen that strong focussing results in a large reduction of the odd isotope enrichment as could be expected from the arguments given above. All experiments were carried out under saturating conditions of the ionization. The ionization rate R_I was smaller than the HFS splitting so that the ionization was generated from an incoherent superposition. In case the HFS splitting is much smaller than R_I it can be expected that odd-even effects will be absent for $\Delta J=0$ and -1 transitions.

Table 3. Dependence of $\delta^{189}Os$ on focussing conditions for $P_{ell} = 27$.

ΔJ	f = 100cm	f =35
-1	74 (11)	-1.3 (13)
0	57 (16)	36 (8)

4. SUMMARY AND CONCLUSIONS

Odd-even effects in the isotopic analysis of Os and Ti by RIMS were studied. It could be shown that under saturating conditions of the ionization odd-even effects were absent for $\Delta J=+1$ transitions or could be reduced below the 1% level. For $\Delta J=0$ and -1 transitions it was shown that the odd-even effects could be strongly reduced depending on the polarzation state and focussing conditions of the laser. In particular for one $\Delta J=0$ and -1 transition effects were absent for a large degree of depolarization and strong focussing conditions. The results given here show that with a careful choice of the experimental parameters it is possible to obtain an accuracy better 0.5% for isotopic analysis with RIMS.

ACKNOWLEDGEMENTS

This work was supported by DOE grant DE FG03-88ER-1351 and NSF grant EAR-8816936 to G. J. Wasserburg. The laser system was obtained through support from the Packard and Sloan Foundations and NASA grant NAGW -1955 to G. A. Blake. R. K. W. would like to acknowledge the support of Prof. H.-J. Fecht for this work.

REFERENCES

Creaser R A, Papanastassiou D A and Wasserburg G J 1991 *Cosmochim. Acta* **55** 597
Fairbank W M Jr., Spaar M T, Parks J E and Hutchinson J M R 1989 *Phys. Rev.* **A40** 2195
Lambropoulos P and Lyras A 1989 *Phys. Rev.* **A40** 2199
Niederer F R, Papanastassiou D A and Wasserburg G J 1981 *Geochim Acta* **45** 1017
Payne M G, Allman S L and Parks J E 1991 *Spectrochim. Acta* **46B** 1439
Spiegel D R, Pellin M J, Calaway W F, Burnett J W, Coon S R, Young C E, Gruen D M and Clayton R N 1992 *Anal. Chem.* **64** 469
Walker J R and Fassett J D 1986, *Anal. Chem.* **58** 2393
Wunderlich R K, Hutcheon I D, Wasserburg G J and Blake G A 1992 a *Int J. Mass Spectrom. and Ion Processes* in press
Wunderlich R K, Hutcheon I D, Wasserburg G J and Blake G A 1992 b submitted *Anal. Chem.*

Inst. Phys. Conf. Ser. No 128: Section 3
Paper presented at RIS 92, Santa Fe, NM, USA, 24–29 May 1992

Evidence on mode structure effects in the resonance ionization signal of strontium

H M Lauranto, T T Kajava, M I K Santala and R R E Salomaa

Department of Technical Physics, Helsinki University of Technology, Rakentajanaukio 2 C, SF-02150 Espoo, Finland

ABSTRACT: To investigate the causes of the fluctuations in RIS signals when using pulsed lasers a two-step excitation scheme of strontium was studied. Successive RIS signals, the corresponding pulse energies and the mode structure of the laser pulses of the resonant step were recorded. The fluctuations in the RIS signal were uncorrelated with pulse energy fluctuations of either step but correlated strongly with the intensity of the mode closest to the resonance line when small resonant field intensities were involved. With larger intensities the correlation diminished and completely disappeared in saturation.

1. INTRODUCTION

Tunable dye lasers are widely used in optical spectroscopy. In low intensity applications one can rely on continuos wave (CW) lasers where single mode operation can be routinely achieved. This results to a narrow laser bandwidth and consequently to a high spectral resolution. When more intense fields are required, as is, e.g., the case in non-linear atomic spectroscopy, one is compelled to use pulsed dye lasers capable of delivering high peak powers. The disadvantages of these lasers, compared to CW lasers, are their broader spectra, consisting usually of several longitudinal modes, and their variations in the pulse energy. Furthermore, also the mode structure exhibits strong pulse-to-pulse variations and frequency jitter, which may have unpredictable effects on the signal behaviour.

In this paper we show that an important source of fluctuations in the resonance ionization signal, when using pulsed dye lasers, is the pulse-to-pulse variations in the mode structure of the resonant laser field. We have chosen strontium ($Z = 38$) as a test case, because it is one of the few elements that can be excited from the ground state with a wavelength in the region 450–550 nm (a technical constraint, see sect. 2) and subsequently photoionized with current dye lasers without frequency doubling. Besides, strontium has a rich spectral structure with four naturally occurring isotopes: ^{84}Sr, ^{86}Sr, ^{87}Sr, and ^{88}Sr with abundancies 0.56, 9.86, 7.02, and 82.56 %, respectively, which offers a challenge for the achievable spectral selectivity.

2. EXPERIMENTAL

To reduce the number of effective fluctuating parameters we used a two-step excitation scheme with one resonant and one photoionizing step. We measured successive resonance ionization signals of strontium and simultaneously recorded the pulse energies of both lasers and the spectra of the laser pulses of the resonant step. A study of the pulse-to-pulse variations in the mode structures of our lasers has been reported previously (Kajava *et al* 1992).

Strontium atoms were excited from the ground state $5s^2$ $^1S_0$ to the lowest zero-spin, odd-parity state $5s5p$ $^1P_1^o$ (21698.482 cm^{-1}) and subsequently photoionized with a wavelength 412.4 nm, which excites the atoms just above the ionization limit (45925.6 cm^{-1}, Moore 1971; 45933.4 cm^{-1} according to our measurements). The resonant transition is very strong ($gA = 0.85 \cdot 10^8$ 1/s, Corliss and Bozman 1963) and could be saturated with a fluence of the order of 0.1 μJ/cm^2 while the photoionization transition was still unsaturated at the maximum fluences we were able to produce (0.5 mJ/cm^2).

In the transition $5s^2$ $^1S_0 - 5s5p$ $^1P_1^o$ the isotope shifts of the isotopes ^{84}Sr, ^{86}Sr, and ^{87}Sr relative to the most abundant isotope ^{88}Sr are 270.8, 124.8, and 46.3 MHz, respectively (Eliel *et al* 1983). The odd isotope ^{87}Sr ($I = 9/2$) has in addition a three component hyperfine splitting in the excited state; the width of the whole structure is 59 MHz (Kluge and Sauter 1974).

Our RIS apparatus consists of two dye lasers pumped with an excimer laser, a vacuum chamber and control and data collection electronics. The laser line widths were about 4.5 GHz and the longitudinal mode spacing in the laser pulse spectra were ca. 500 MHz. The diameters of the laser beams were adjusted to ca. 5 mm, and the wavelengths were calibrated to argon lines with an optogalvanic signal from a hollow cathode lamp. The atomic beam was created in a graphite furnace, crossed in the vacuum chamber with the laser beams between two electrode plates. The residual Doppler broadening of the transition line was approximately 150 MHz. After the simultaneous laser pulses a high voltage pulse was applied to sweep the photoionized atoms towards the ion detector. A more detailed description of the apparatus is given by Bekov *et al* (1987).

To record the successive single pulse laser spectra we have constructed a high resolution interferometric spectrum analyser. Its optical design is a modification of a Fabry–Perot interferometer. In this Fizeau configuration (Westling *et al* 1984) there is a small (≈ 25 μrad) tilt between the mirrors, which results to an interference pattern consisting of parallel, equally spaced fringes. The separation of the fringes can be easily controlled by adjusting the tilt angle between the mirrors. This makes the interferometer much easier to operate than a traditional Fabry–Perot interferometer. The resolution of the instrument depends on the reflectivities of the end mirrors, and only in the wavelength region 450–550 nm these

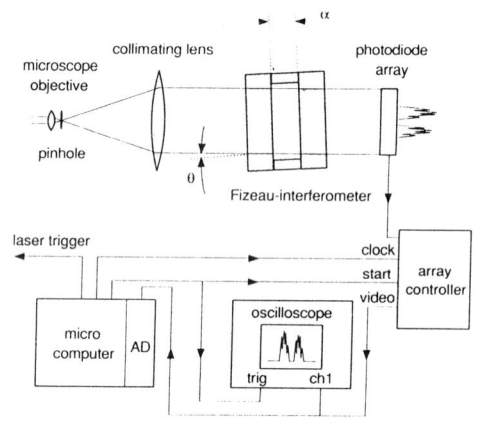

Figure 1. A sketch of the Fizeau spectrum analyser.

are such that the resolution is high enough to resolve the mode structure. The incident laser beam is spatially filtered and collimated with a microscope objective, a pinhole and an acromat lens. The interference pattern is recorded by a linear photodiode array (2048 diodes), which is controlled by a microcomputer. Figure 1 shows a schematic picture of the interferometer, which is described in detail in (Kajava *et al* 1992).

The measurements were performed at several pulse energies in the resonant step varying from rather small values to a region where the transition was clearly saturated. To reduce the amount of spectral data, we analysed the laser pulse spectra on an on-line basis and recorded the centre channels and the heights of the modes in each spectrum only. Each series of measurements consists typically of one hundred successive laser pulses. Before every succession of pulses we scanned over the transition line and tuned the wavelength to the signal maximum. In the spectra only the transition line of the most abundant isotope ^{88}Sr could be resolved, which is obvious when comparing the isotope shifts to the linewidth of the lasers.

3. SIGNAL DEPENDENCE ON EXCITATION FIELD

Our analysis of the measured data is based on linear regression models. The correlation coefficients measure how strong the linear relationship between two chosen variables is. In all the measurements, independently of the average pulse energies, the fluctuations in resonance ionization signal showed no correlation with the pulse-to-pulse fluctuations of the energies of either step. When using very small pulse energies in the resonant step, we discovered a strong dependence of the signal on a single mode intensity. This is manifested as a pronounced maximum in the signal–laser spectrum correlation curve in Figure 2. In the figure the abscissa is the ordinal number of the diode in the array, or in other words, it is a variable proportional to frequency. The ordinate of each point in the curve is the coefficient of correlation between the signal and the spectral intensity of the laser pulses at the abscissa channel. The intensities at each channel were calculated from the reduced mode structure data assuming a triangular mode shape for simplicity. The base width used for the triangle was 17 units, while the mode spacing was approximately 30 units in this scale. The curve has a clear maximum at the channel 644, which evidently means that the centre of the resonance line (of the most abundant isotope) lies close to that channel. It is worth noticing that the uncertainty in the location of the line is of the order of 15 channels (corresponding to ≈ 200 MHz $\approx$ Doppler broadened width of the transition), which is much better than is achievable by just scanning over the resonance line with a 4.5 GHz laser line. However, even this resolution is not sufficient to the resolve the less abundant isotopes.

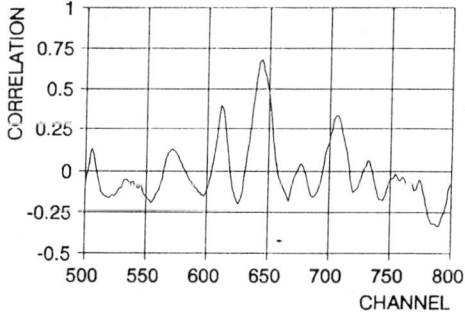 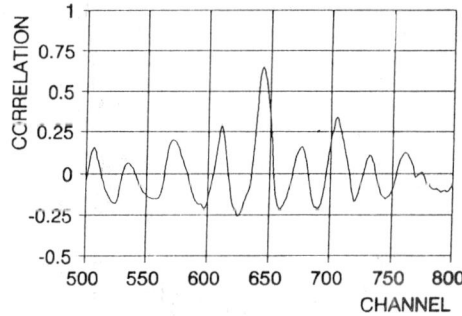

Figure 2. Signal–laser spectrum correlation curve in the weak field case. A pronounced maximum at the channel 644 indicates that the signal is mainly caused by the longitudinal mode centered closest to that channel.

Figure 3. Mode intensity correlation curve (scaled with a factor 0.65) calculated from the data of Figure 2. The residual structure in Figure 2 can be explained by the mode intensity correlations shown in this figure.

The residual structure of the curve in Figure 2 is explained by the mode intensity correlations in the laser pulses. Figure 3 shows the mode intensity correlation curve calculated from the same data as the curve in Figure 2. Here the ordinate of each point is the coefficient of correlation between the spectral intensity at the abscissa channel and the spectral intensity at the reference channel corresponding to the maximum of the signal–laser spectrum correlation curve. The agreement between Figure 3 and the residual residual structure in Figure 2 is good. In fact, if the linear theory were strictly valid, i.e., if the signal were caused only by the mode closest to the resonance line and were proportional to the overlap integral of the mode and the resonance line, the two curves should be identical, provided that the correlations would be calculated with the correct mode shape (convoluted with the resonance line shape).

Increasing the pulse energy of the resonant step resulted in a decrease of the correlation between the signal and a single mode intensity. The statistically significant correlation

Table 1. The maximum correlations found in the signal–laser spectrum correlation curves with different pulse energies in the resonant step. The transition started to saturate at the energies of the order of 0.030 μJ.

Energy (nJ)	16	20	65	100	220
Correlation	0.7	0.6	0.4	0.3	0.3

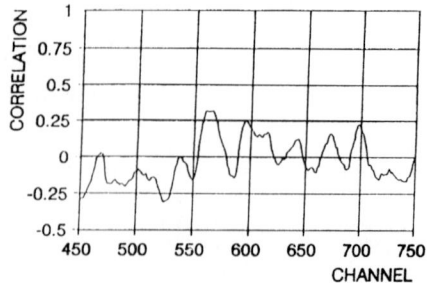

Figure 4. Signal–laser spectrum correlation curve in the strong field case, where the signal cannot be attributed to any particular single mode alone.

disappeared completely at the saturation. This is presented in Table 1. Figure 4 shows a signal–laser spectrum correlation curve in the strong field case, where the signal has no correlation with the intensity of any particular single mode.

4. CONCLUSIONS

In RIS based on pulsed dye lasers there are typically strong pulse-to-pulse fluctuations in the signal. To find out the reasons behind these fluctuations we studied a two-step excitation scheme of strontium and discovered that the fluctuations in resonance ionization signal do not correlate with pulse energy fluctuations of either the resonant or the phoionizing laser field. When the resonant field is weak, the fluctuations in the resonance ionization signal can be completely explained by the intensity variations of the longitudinal mode closest to the atomic resonance line, which means that the dynamics of resonance ionization can be described by a linear single mode theory. When the resonant field is increased the correlation between the fluctuations of the signal and a single mode intensity diminishes and disappears totally   at   the saturation value.   However, even in the saturated case the signal fluctuations had no correlation with the pulse energy fluctuations of the resonant field. This implies that also the strong field response is dependent on the mode structure of the exciting field, but in a more complex way than described by a linear theory. The observed signal–laser spectrum correlations can be utilized in noise reduction of the resonance ionization signal (Lauranto *et al* 1992).

REFERENCES:

Bekov G I, Kudryavtsev Yu A, Auterinen I and Likonen J 1987 *Resonance Ionization Spectroscopy 1986, Inst. Phys. Conf. Ser. No.* **84** eds G S Hurst and C G Morgan (Bristol: Institute of Physics) pp 97–100

Corliss C H and Bozman W R 1962 *Experimental Transition Probabilities for Spectral Lines of Seventy Elements, NBS Monograph* **53** (Washington: National Bureau of Standards) p 388

Eliel E R, Hogervort W, Olsson T and Pendrill L R 1983 *Z. Phys. A* **311** 1

Kajava T T, Lauranto H M and Salomaa R R E 1992 *Appl. Opt.* (in press)

Lauranto H M, Kajava T T, Santala M I K and Salomaa R R E "*Noise Reduction Techniques in RIS Utilizing Real-Time Laser Pulse Spectra*", these proceedings

Kluge H-J and Sauter H 1974 *Z. Phys.* **270** 295

Moore C E 1971 *Atomic Energy Levels, NSRDS-NBS* **35**/*Vol. II* (Washington, D.C.: National Bureau of Standards) p 189

Westling L A, Raymer M G and Snyder J J (1984) *J. Opt. Soc. Am. B* **1** 150

Inst. Phys. Conf. Ser. No 128: Section 3
Paper presented at RIS 92, Santa Fe, NM, USA, 24–29 May 1992

135

Noise reduction techniques in RIS utilizing real-time laser pulse spectra

H M Lauranto, T T Kajava, M I K Santala and R R E Salomaa

Helsinki University of Technology, Department of Technical Physics, Rakentajanaukio 2C, SF-02150 Espoo, Finland

ABSTRACT: Resonance ionization signal distributions and their dependence on laser mode variations are studied. Laser spectra are recorded for each signal value. Strong correlations between the signal and the closest mode intensity are found in the weak intensity range. By an active on-line rejection of unwanted mode patterns the RIS signal fluctuations can be reduced and an improved spectral selectivity achieved.

1. INTRODUCTION

The main sources of fluctuations in the ion signal in resonance ionization spectroscopy (RIS) are the lasers, atomization process and the detectors. In the present paper we concentrate on the effects arising from the lasers. The lasers used in RIS have usually a broad bandwidth involving several modes and their operation is pulsed. The laser spectrum varies from pulse to pulse. An obvious way to improve the system performance is to use intensity and frequency stabilized single-mode lasers. As this is not always practicable an alternative solution is to measure the laser spectra from each pulse and apply this information in the data processing of the RIS signal.

Large signal fluctuations in RIS are particularly harmful when one attempts to resolve a weak resonance heavily perturbed by the wing of a strong line, as is typically the case in isotope and isomer detection. Long averaging times are often excluded by the system drifts and the small samples available. One noise reduction technique could be the rejection of those data points which are obtained with undesirable mode patterns, e.g. those where the perturbing transition is resonantly excited. Another interesting idea is to study whether the signal statistics changes markedly with the detuning and saturation and whether these changes could be exploited in the noise reduction.

We describe typical signal distributions obtained from strontium which was chosen a test element because of its exceptionally simple RIS scheme and also because of its rich isotopic structure. We have performed similar, three-step ionization experiments also with niobium the ultimate goal in that case being the detection of the isomer ^{93m}Nb (Lauranto et al. 1991). The correlations between the strontium RIS signal and the characteristics of the laser spectra are studied. Supporting theoretical simulations have been done. Finally some noise reduction techniques will be discussed.

2. DATA ACQUISITION TECHNIQUE

The experimental setup used for measuring the strontium spectra is described in some detail by Lauranto et al. (1992). The transition $5s^2\,^1S_0 - 5s5p\,^1P_1^o$ at 460.7 nm was used as the first step and the second, ionizing step was excited by a 412.4 nm laser beam. Only the parameters of the laser coupling to the resonance transition were varied. The frequency and intensity fluctuations of the ionizing laser are expected to be of secondary importance because of the broad ionization cross-section.

The laser spectra were recorded with a Fizeau-type spectrum analyzer described by Westling et al. (1984) and Kajava et al. (1992). The individual laser pulse spectra show clear mode structure. From pulse to pulse the number and relative strength of the modes vary and, furthermore, there is a small jitter in the position of the mode-comb. In the measurements only the positions and peak heights of the modes were stored. Typically 100 pulses were used for each detuning and pulse energy level of the lasers. The pulse energy was measured for each shot. The effective value for the mode intensity $\tilde{x}_p(k)$ (p is the data point index and k the channel number, the diode array is 1×2048) is calculated from the raw data $\{x_p(l)\}$ by using a filter $g(k,l)$ (triangular, Gaussian, Voigt-profile): $\tilde{x}_p(k) = \sum_{l=1}^{2048} g(k,l)x_p(l)$. The filtering process reconstructs in an approximate way the original mode spectrum. An experimental deficiency in our spectrum analyzer is that the intensity to voltage ratio varies with the channel number, i.e., in $x_p(k) = \kappa(k)I_p(k)$ the proportionality coefficient $\kappa(k)$ is a function of k. The reason is the uneven illumination of the Fizeau interferometer. Note, however, that $\kappa(k)$ remains fixed as long as the interferometer is not readjusted. In the signal–laser spectrum or mode intensity correlation plots the proportionality coefficients $\kappa(k)$ cancel, provided that the width of the filter $g(k,l)$ is narrower than the inhomogeneity scale.

If the resonance line is narrow, the unsaturated signal is expected to depend mainly on the detuning and the intensity of the nearest mode. In the special cases, when two adjacent modes fall nearly symmetrically around the resonance line, both of the modes have to be accounted, but their cross-coupling is negligible to a first approximation. In the saturation region the contributions of individual modes diminish and multimode effects become increasingly important. The present study focusses on single (or a few) mode effects on an isolated resonance transition (degenerate or fully resolved); several previous papers have discussed the influence of the laser bandwidth on the RIS spectra in such cases where the signal arises from the interference between the incompletely resolved hyperfine components (see e.g. Capelle et al. 1987, Lyras et al. 1990).

3. SIGNAL FLUCTUATIONS, CORRELATIONS

The dependence of the signal fluctuations on the mode parameter variations is hard to predict theoretically and, therefore, we have relied in the data interpretation on statistical methods and simulations (Salomaa et al. 1991). Figure 1 (weak field, 1.5 nJ, spot size ca. 1 cm^2) and Fig. 2 (strong field, 15 nJ) display two samples of signal statistics at resonance. On the left hand side is the signal distribution which for weak fields is roughly an exponential whereas the strong field distribution resembles more a Gaussian. Similar shapes were observed in the niobium experiments. The right hand side plot is the signal versus the effective intensity of the mode that is located at the center of the resonance (determined from the position of the maximum correlation between the signal and the laser spectrum). A rectangular filter with a width of 31 channels (~200 MHz) was used. Note that the scales in Figs. 1 and 2 are relative and not the same.

In Fig. 1 the correlation between the signal and the intensity of the nearest mode is quite evident. The correlation coefficient equals 0.87, if the zero-intensity data points are ignored, and 0.70, if all the points are included (the two straight lines in the plots are the corresponding least square fits). In ordinary resonance ionization spectra of strontium, notable power broadening occurs at quite low pulse energies, of the order of 10 nJ. The nonlinearity of the atom–field interaction is manifested also in Fig. 2 where the single mode contribution to the signal is strongly decreased (the correlation coefficient is 0.52, if zero-intensity points are excluded, and 0.41 when all data points are retained). Multimode effects play an important role in the signal behavior in this region. It is, perhaps, worth pointing out that the relative variance of the pulse energy was less than 5 % which is far too small to explain the large signal fluctuations. In addition, the correlation between the signal and the pulse energy was statistically insignificant.

Even when the laser is detuned the signal–closest mode correlation prevails as long as the laser spectrum and the transition line overlap. For small pulse energies the detuning effects can be partly extracted from the resonance data using those mode patterns where the closest mode is slightly detuned but no other mode with considerable amplitude happens to lie in the vicinity. The triangular filtering distorts the single-mode spectral response curves and must be replaced by more accurate filter functions. If neighboring modes tend to coexist, then multiple regression analysis must be used. Once the atomic transition is outside the laser bandwidth the signal drops to uninterestingly small values, and furthermore the RIS apparatus is then anyway capable of resolving the lines. Cases where the overlap due to strong power broadening occurs require further studies.

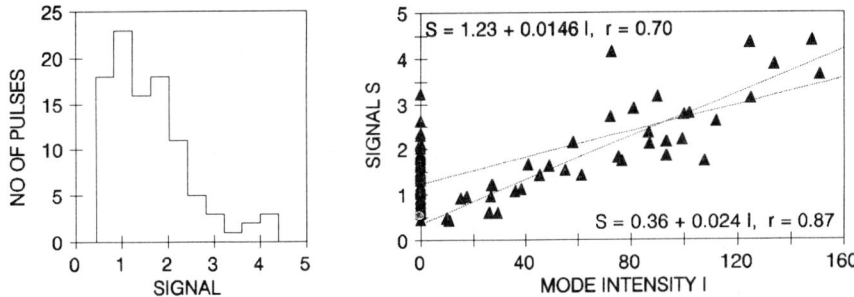

Fig. 1: The signal distribution at resonance at low laser intensities (left) and the dependence of the signal on the closest mode intensity (right).

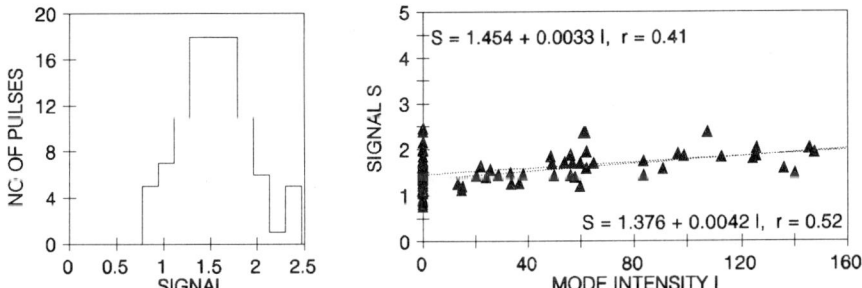

Fig. 2: Same as Fig. 1 but for strong laser pulses

4. NOISE REDUCTION

A straightforward way to increase the signal to noise ratio in the measurements is to increase the number of pulses used in the averaging. In the present strontium experiments typically 20–40 pulses per detuning step was needed to obtain decent statistics. In the weak field region much of the fluctuations can be removed by accepting only those signals where a mode hits a prefixed triggering window. If the mode intensity is simultaneously recorded, one is left with the residual scatter around the best-fit curve. Figure 3 displays the residual variance of the signal as a function of the width of the triangular window. Too narrow a window introduces poor statistics, since only a few pulses hit it, whereas in too broad windows the spectral selectivity is lost. As evidenced by Fig. 3 even in the region of medium saturation there still exists an optimal window width such that just the relevant modes are accepted. The scatter arises from the mode intensity fluctuations and could perhaps be partly reduced by using multivariable fits. In concentration measurements of strontium, the observed quantity at maximum signal (resonance) is proportional to $\langle S \rangle / \langle E \rangle$ where E is the pulse energy. The main uncertainty thus arises from the ion signal, because the fluctuations of E are rather small.

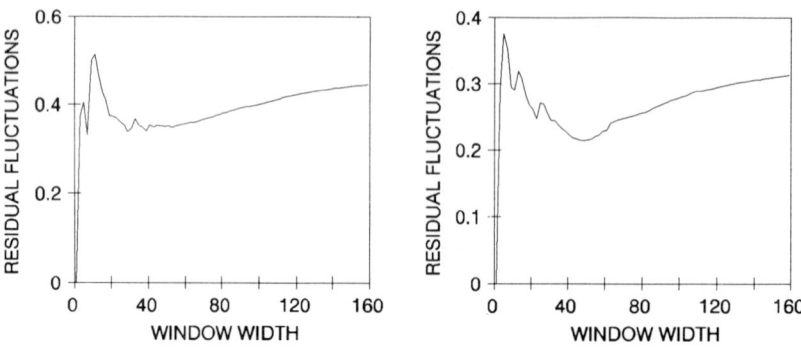

Fig. 3: The residual standard deviation (relative units) of the signal for various triggering window widths for weak (left) and strong fields (right). Mode spacing corresponds to ca. 30 channels.

A suggestion to resolve e.g. isotopic structure is to collect the data conditionally. One has to choose two channels one of which, say k_1, is close to the line center of the measured isotope and the second one k_2 close to the perturbing isotopic transition. Accepting only data points where there is a mode in k_1, but k_2 is empty, one is able to eliminate to some extent the disturbances. In our present configuration we have not yet achieved good enough statistics to be able to demonstrate this method in the rather difficult case of strontium.

When evaluating the feasibility of the present method one must recall that many of the multimode disturbances can be calibrated by using a reference beam. Fluctuations arising from atomization process or from electronic noise remain a problem. Here the application of the strong signal–laser spectrum correlation is demonstrated for weak fields which admittedly are not the goal in RIS. In a recent paper by Fee et al. (1992) a heterodyne-technique, closely related to the present ideas, was successfully applied to studies of frequency fluctuation effects in two-photon spectra where the resonant intermediate state is absent. Multiple regression analysis and more sophisticated signal triggering (involving several channels) may extend the applicability of the present method and certainly deserve further studies. Absolute wavelength calibration would also open new possibilities for the suggested technique.

REFERENCES:

Capelle G *et al.* 1987 J. Opt. Soc. Am. B **4** 445.
Fee M, Danzmann K, and Chu S 1992 Phys Rev A **45** 4911.
Kajava T, Lauranto H, and Salomaa R 1992 Appl. Opt. *Mode structure fluctuations in a pulsed dye laser*, (in press).
Lauranto H *et al.* 1991 *Resonance Ionization Spectroscopy 1990* eds J Parks and N Omenetto (Bristol: Institute of Physics) p 307.
Lauranto H, Kajava T, M Santala, and Salomaa R 1992 *Evidence of mode structure effects in resonance ionization spectroscopy*, these proceedings.
Lyras A, Zorman B, and Lambropoulos P 1990 Phys Rev A **42** 543.
Salomaa R *et al.* 1991 *Resonance Ionization Spectroscopy 1990* eds J Parks and N Omenetto (Bristol: Institute of Physics) p 223.
Westling L, Raymer M, and Snyder J 1984 J. Opt. Soc. Am. B **1** 150.

Inst. Phys. Conf. Ser. No 128: Section 3
Paper presented at RIS 92, Santa Fe, NM, USA, 24–29 May 1992

Simplified method for estimating multistep photoionization efficiency

M. Miyabe, I. Wakaida, K. Akaoka, M.Ohba and T. Arisawa

Japan Atomic Energy Research Institute, Tokai-mura, Naka-gun, Ibaraki-ken 319-11, JAPAN

ABSTRACT: A simplified method has been proposed to search for the multistep photoionization scheme with the highest quantum efficiency. The multistep scheme is equivalently converted to a single-step scheme so that the cross-sections for the equivalent schemes are compatible with each other. Measured two-step scheme cross-sections are in good agreement with the estimated values. The results show that the proposed method is useful for the development of energy-effective RIS schemes.

1. INTRODUCTION

Recently many investigations have been made to find efficient ionization schemes applicable to laser isotope separation and ultra-sensitive trace analysis (Bekov 1981, Fedoseev 1991, Herrmann 1991, Hui 1991). Many theories have been proposed thus far to estimate the overall multistep photoionization efficiency, but they have not dealt with optimization methods for searching for the most efficient scheme among many allowed multistep transition schemes. In this optimization, a scheme cross-section is defined as a function of photon flux, and this photon flux is optimized so that the photoionization efficiency is maximized for comparison with different multistep photoionization schemes. In most cases, the density matrix equations are used for the estimation of multistep photoionization efficiency. However, these equations are very complicated to be treated as an optimization method, which urges us to find a more simplified calculation procedure. In this report we present a simplified estimation method based on the rate equations.

2. CALCULATION OF MULTISTEP IONIZATION EFFICIENCY

Figure 1(a) shows the multistep resonance photoionization model, assuming that all the transitions occur simultaneously by the irradiation of multicolor laser beams, where N_m is the population of m-th level, n_m is the photon flux of the m-th transition, and the K-th level is an autoionizing level. For the photoabsorption transitions from the first through the (K-1)-th excitation step, stimulated emission processes are included, but in the transition involving the K-th state they are not included, because the autoionization rate is extremely large. Efficient excitation is realized when the transition for each step occurs so fast that the population of intermediate states could be regarded as constant in time. Then the multistep rate equations in Equation 1 are equivalently converted to the single-step rate equations in Equation 2 (Shore 1990). Using the relation of transition rate between the equivalent single-step system and the original multistep system, the cross-section σ_s for the equivalent single-step system which is called the scheme cross-section is calculated by Equation 3, where $\emptyset_m$ denotes the photon flux ratio defined by $\emptyset_m = n_m/n_1$.

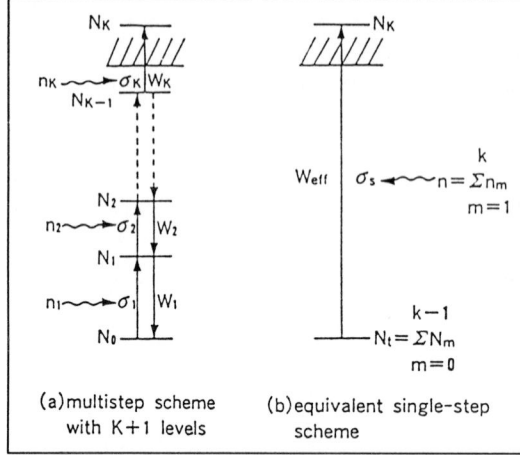

$$\begin{cases} \dot{N}_0 = -W_1 \ (N_0-N_1) \\ \dot{N}_1 = \ W_1 \ (N_0-N_1) \ -W_2 \ (N_1-N_2) \quad (1) \\ \cdots\cdots\cdots\cdots \\ \dot{N}_K = \ W_K \ N_{K-1} \end{cases}$$

$$\begin{cases} \dot{N}_t = -W_{eff} \cdot N_t \\ \dot{N}_K = \ W_{eff} \cdot N_t \end{cases} \qquad (2)$$

$$W_{eff} = \left(\sum_{m=1}^{k} \frac{m}{W_m} \right)^{-1}$$

$$\sigma_s = \frac{1}{\sum_{m=1}^{k} \phi_m} \ \frac{1}{\sum_{m=1}^{k} \frac{m}{\phi_m \sigma_m}} \qquad (3)$$

$$\sigma_{max} = \frac{1}{\left(\sum_{m=1}^{k} \sqrt{\frac{m}{\sigma_m}} \right)^2} \qquad (4)$$

$$\phi_m = \sqrt{m \frac{\sigma_1}{\sigma_m}}$$

Fig. 1. Photoionization scheme.

The maximum value of σ_s and the optimum laser photon flux ratio ϕ_m can be obtained by $\partial \sigma_s / \partial \phi_m = 0$ which results in Equation 4. Using this maximum value, we can compare the photoionization efficiency for each scheme.

3. VALIDITY OF SIMPLIFIED MODEL

3.1 Detailed Comparison of Simplified Scheme With Original One

In Equation 2 it is assumed that the population of each intermediate state is constant in time, but at the beginning of the actual transition with pulsed lasers, this population is continually increasing until it reaches steady state. Therefore, to verify the validity of Equation 2, the time evolution of the ionized population for the multistep scheme (Figure 1(a)) and the equivalent scheme (Figure 1(b)) is compared. Figure 2 shows examples of the time evolution of the ionized population. Curve 1 shows the sole solution for the single-step scheme, while curve 2 shows the solution for the multistep scheme with the specified set of photon flux ratios under which ionized populations calculated by the two

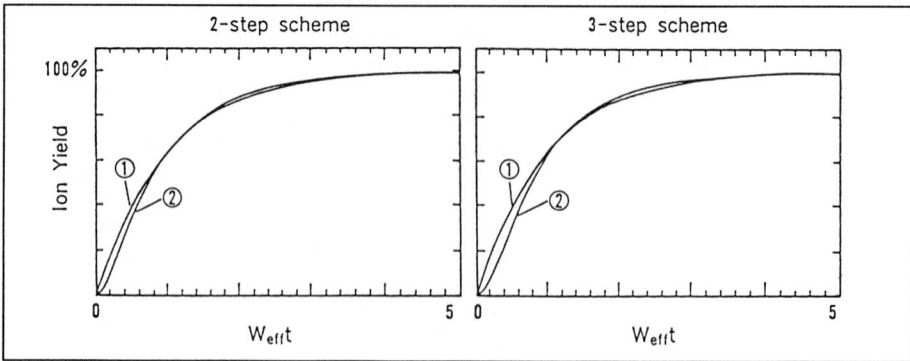

Fig. 2. Time evolution curves of multistep scheme and equivalent scheme.

different methods above become most deviated with each other. During the transient period ($W_{eff} = 0$) the deviation is large, but in the steady state region ($W_{eff} = 1$) which is expected to be close to the optimization condition, the deviation is only about 1% even in the worst case shown as curve 2.

3.2 Comparison of simplified scheme with experiment

(1) Experimental procedure

For demonstrating the applicability of the proposed method to the actual system, the two-step scheme is taken as a fundamental case, because the multistep scheme is in general considered to be composed of many two step systems. To measure the two-step scheme cross-section in the three level photoionization, the laser pulse timing was set as shown in Figure 3. The scheme cross-section was determined based on the photoion saturation characteristics. Since it depends on the photon flux ratio, the measurements of the saturation points were made by varying the total photon flux for various flux ratios.

(2) Experimental setup

Figure 4 shows the experimental setup. Three different colors were produced by three pulsed dye lasers (Lambda Physik FL3002E) pumped by two XeCl lasers (Lambda Physik EMG103 & 203). First of all, the second and the third colors were combined by a dichroic mirror and these combined beams were then recombined by the polarizing beam splitter. The atomic beam generated by the resistively heated ceramic crucible was irradiated by this collinear beam with three colors at right angles. Photoions produced by the three-step process were extracted by pulsed electric field and detected by a channeltron multiplier after traveling through a 1m drift tube of the TOF spectrometer. Ion signal was sent to a boxcar integrator and was processed by a microcomputer. For measuring the saturation characteristics, the cross sectional area of the laser beams was precisely adjusted using an aperture. The diameter of each laser beam in the interaction region was about 3mm. The photon flux was monitored by a pyroelectric probe (Laser Precision Corp. RJ-7620). The wavelengths of the lasers were monitored by a wavemeter. (Lasertechnics model 100F).

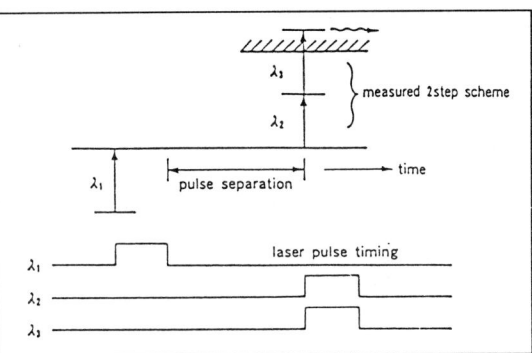

Fig. 3. Laser pulse timing.

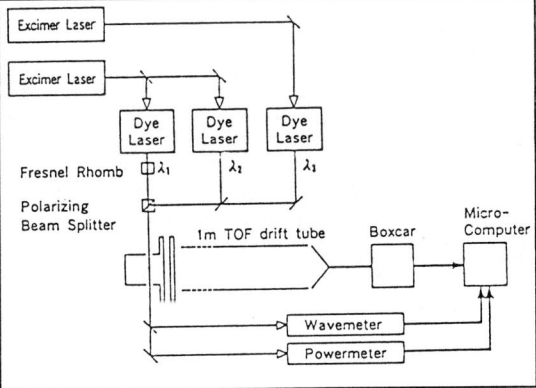

Fig. 4. Experimental setup.

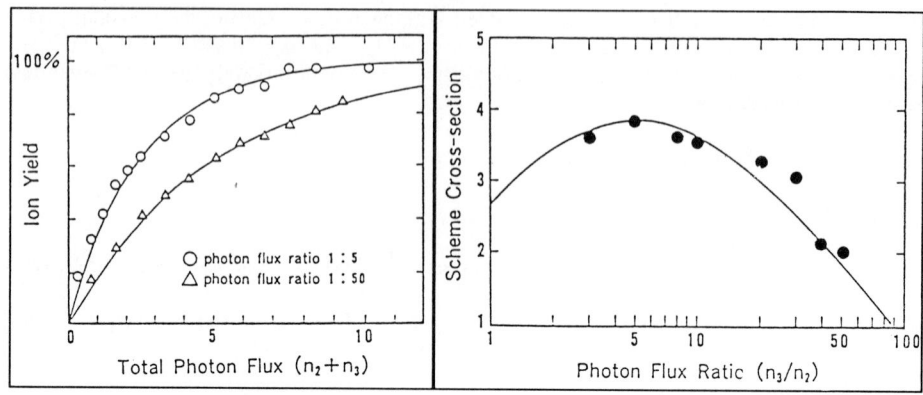

Fig. 5. Photoion saturation curves.   Fig. 6. Photoion saturation curves.

(3) Comparison

Figure 5 shows the typical saturation curves taking the photon flux ratio with $n_2:n_3=1:5$ (circle) and 1:50 (triangle) as examples. The total photon flux is expressed in units of 10^{23}/s/cm^2. The solid curve denotes the ionized population calculated by the simplified model. The agreements are good for both cases. Figure 6 shows the measured scheme cross-section (in 10^{-16} cm^2) as a function of the photon flux ratio. The optimum photon flux ratio giving the maximum of the scheme cross-section is about 1:5, which is in good agreement with the value calculated from Equation 5 with the measured ratio of the cross-section for the second and third steps $\sigma_2:\sigma_3=13:1$. These agreements suggest that proposed characterization method is applicable to the actual system.

This work was performed under contract with the Atomic Energy Bureau of the Science and Technology Agency of Japan.

REFERENCES
Bekov G I, and Vidolova-Angelova E P 1981 *Sov. J. Quantum. Electron.* **1** 1137
Fedoseev V N, Mishin V I, Vedeneev D S, and Zuzikov A D 1991 *J. Phys.* **B24** 1575
Herrmann G, Riegel J, Rimke H, Sattelberger P, Trautmann N, Urban F-J, Ames F, Otten E-W, Ruster W and Scheerer F 1991 *Resonance Ionization Spectroscopy 1990*, eds N Omenetto and J E Parks (Bristol:Inst. of Physics) pp 251–254
Hui Q, Chen D Y, Niu J G, Cheng Y, Xu X Y, and Zhao, W Z 1991, *Resonance Ionization Spectroscopy 1990*, eds N Omenetto and J E Parks (Bristol: Inst. of Physics) pp 297–300
Shore B W, 1990 *The Theory of Coherent Atomic Excitation*, vol. 2 (New York: John Wiley & Sons, Inc.) p 816

Inst. Phys. Conf. Ser. No 128: Section 3
Paper presented at RIS 92, Santa Fe, NM, USA, 24–29 May 1992

143

Results of kinetic modelling of a four-level atomic system for RIS

J A Strauss, H G C Human and P E Walters*

Atomic Energy Corporation, P O Box 582, Pretoria 0001, South Africa
*Merensky Institute of Physics, University of Stellenbosch, Stellenbosch 7600, South Africa

ABSTRACT: Numerical calculation has been employed to find the conditions necessary for achieving saturation in a three step atomic RIS system. The finite risetime and width of the pulsed excitation lasers as well as radiational decay of the excited levels were taken into account. The results show i.a. that a fast atomic system requires much higher laser energy than was previously accepted, and that with a slow atomic system a delay of the consecutive laser pulses w.r.t. each other is advantageous.

1. INTRODUCTION

In order to benefit from the excellent sensitivity that the RIS technique offers in principle, it is necessary to saturate every bound-bound transition in an atomic excitation scheme, including the final ionisation step. A numerical model, based on the rate equations approach, was developed to determine the influence of laser parameters on the saturation and relative ion yield of atomic systems. A three step excitation scheme was considered, which is necessary for a general situation in which the use of UV laser beams is avoided.

2. THE RATE EQUATIONS

The rate equations for a four-level atomic system are as follows:

$$\frac{dN_0}{dt} = - W_{01} N_0 + W_{10} N_1 + \frac{N_1}{\Gamma_{10}}$$

$$\frac{dN_1}{dt} = W_{01} N_0 - W_{10} N_1 - W_{12} N_1 + W_{21} N_2 - \frac{N_1}{\Gamma_{10}} + \frac{N_2}{\Gamma_{21}}$$

$$\frac{dN_2}{dt} = W_{12} N_1 - W_{21} N_2 - W_{23} N_2 - \frac{N_2}{\Gamma_{21}}$$

$$\frac{dN_3}{dt} = W_{23} N_2$$

where N_i is the population number density of level i, W_{ik} is the excitation rate from level i to level k, and Γ_{ki} the relaxation time from level k to level i. Because all ions are removed very fast from the radiation zone to the detectors, no relaxation mechanism has been provided for the third (ionised) level. Time dependence of both the excitation rates (W's) and populations (N's) are implied in the above expressions.

The analytical models described in the literature by Liu (1974), Letokhov

(1977) and Arisawa et al (1983) considered a three-level system in which the relaxation terms (the terms containing Γ_{ki} in the above expressions) were ignored, implying excitation rates so high that in comparison the relaxation rates can be neglected. Furthermore, the time dependence of the laser intensity was not observed, i.e. the laser pulse was taken as a step function. Due to these simplifications they were able to apply the Laplace transform and find an analytical solution for the relative population of the third level, which is the relative ion yield of their atomic system.

In the work presented here the time dependence of the laser pulse was taken into account by defining the excitation rates as

$$W_{ik}(t) = \frac{\sigma_{ik} \lambda_{ik} I_{ik}}{h \, c} \, p(t)$$

with σ_{ik} and λ_{ik} the absorption cross section and wavelength for the 3 transitions i $\rightarrow$ k, respectively, I_{ik} the laser peak intensity, h and c is Planck's constant and the velocity of light, respectively, and

$$p(t) = \exp \left[- \frac{4 \, (t - t_0)^2}{\Delta t_p^2} \right]$$

is a Gaussian time dependent function which is kept equal for all three excitation steps. With the introduction of this time dependent pulse and consideration of the relaxation rates, it was not possible to find an analytical solution for the four level system. Numerical methods have therefore been resorted to, with the accompanied limited generality of such solutions. However, in an endeavour to observe trends in the behaviour of such a system we calculated the interaction of a *short* (12 ns) and a *long* (55 ns) laser pulse with a *fast* as well as with a *slow* atomic system whose typical transition probabilities (gA) and J-values are given in Table 1. The wavelengths chosen are suitable for use of copper vapor laser pumped dye lasers in order to take advantage of their high repetition rate. The kinetic response of the level populations was calculeted under *weak* as well as *strong* saturation conditions (listed in Table 2), resulting in a 10% and 90% ionisation yield respectively.

Table 1. Parameters of the atomic systems considered.

Transition	Fast atomic system			Slow atomic system		
	λ(nm)	gA($10^8 \mathrm{s}^{-1}$)	J	λ(nm)	gA($10^8 \mathrm{s}^{-1}$)	J
0 $\rightarrow$ 1	560	1.0	0.5	610	0.01	5
1 $\rightarrow$ 2	605	10.0	1.5	540	0.02	6
2 $\rightarrow$ 3	820	1.0	2.5	650	0.01	7

Table 2. Energy values (μJ) for weak and strong saturation.

Saturation	Fast atomic system			Slow atomic system		
	E_1	E_2	E_3	E_1	E_2	E_3
weak	0.2	0.02	2	5	0.5	0.5
strong	2	0.2	20	50	5	5

3. RESULTS

The differences between the analytical and numerical model are clearly shown when considering the predictions of level populations as a function of time for the long laser pulse (55 ns) acting on the fast atomic system under weak saturation conditions, shown in Figure 1. While the analytical model predicts a very fast depopulation of the ground level until total depletion is reached, with a subsequent fast increase in ionisation up to 100%, the numerical model predicts a slower response and lower ionisation, as well as an increase of the ground state population after termination of the laser pulse due to radiational decay from upper levels.

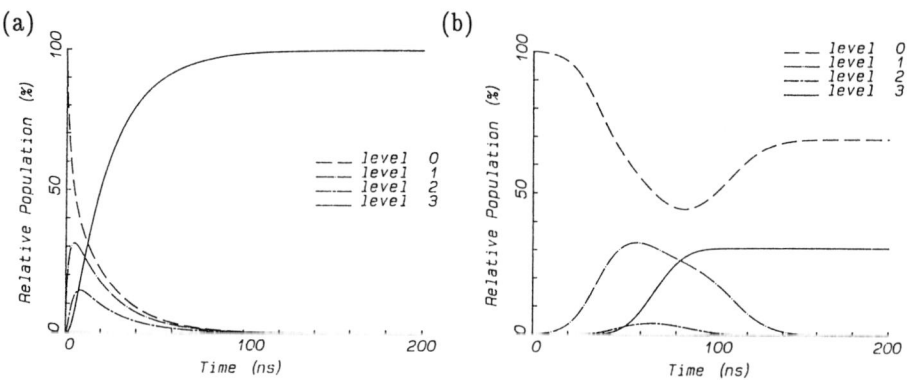

Fig. 1   Time response predicted by (a) the analytical and (b) the numerical model.

3.1   Pulse Duration

For a slow atomic system, there is no real difference in the population response for the short and long laser pulses, where an ion yield of 90% is obtained with strong saturation in both cases. A similar behaviour is observed with weak saturation. The fact that the same energy values were used (Table 2) for both the long and short pulse is an important result, showing that with a slow atomic system, where the influence of radiational decay is delayed, there is no need to increase the laser energy when a long pulse is used.

Far a fast atomic system the ionisation yield is always higher with a short laser pulse : to achieve 90% ionisation the laser using the long pulse requires 1.8 times higher energy than the short (12 ns) pulse.

3.2   Saturation

The importance of strong saturation of every excitation step is clearly shown by the calculations. For weak saturation of the 2nd and 3rd steps the maximum ion yield is ~ 15%, even for a highly saturated first step. Weak saturation in any one of the intermediate steps blocks the ionisation yield. Strong saturation of all three steps is required for > 90% ionisation.

3.3   Radiational Decay

Exclusion of the relaxation terms N_1/Γ_{10} and N_2/Γ_{21} from the rate equations does not have much effect on the ion yield of a slow atomic system under conditions of strong saturation. In fact, the energies required for strong saturation according to the numerical model compares very well with the calculated saturation intensities from the conventional equation

$$I_{sat} = \frac{h\ c}{\lambda\ \sigma\ \tau}\ ,\ \text{(where } \tau \text{ is the pulse duration)}$$

which does not take radiational decay into account.

However, there is a marked influence of the relaxation terms on the fast atomic system. When using the saturation intensities calculated with the above equation as input to the numerical model, an ion yield of only 35% is obtained. The energies had to be increased by more than ten times to obtain > 90% ionisation, showing that the inclusion of radiational decay in a kinetic model for fast atomic systems is important if the correct saturation intensities need to be known.

3.4   Time Delay

From the results thus far obtained one would expect to see some effect on the slow atomic system if the consecutive laser pulses were delayed in time. An increase in the ion yield from 8% to 14% was indeed observed when the second and third laser pulses of the short pulse system were delayed by 12 and 24 ns respectively, but only for a weakly saturated system. The benefit of pulse delay can only be realised if the laser pulse is much faster than the response of the atomic system, and if strong saturation of all transitions cannot be obtained.

4. CONCLUSIONS

The numerical model developed for calculating the kinetic response of a four-level atomic system to laser excitation, proved to be invaluable for detailed insight into the level populations at any time. A wealth of information is furnished by the numerous graphs which could not be shown here. In particular, the effects of laser pulse length and radiational decay of excited levels were shown to be marked for a fast atomic system.

REFERENCES

Arisawa T, Maruyama Y 1983 *J. Phys. D : Appl. Phys.* **16** 2415
Letokhov VS 1977 *Frontiers in Laser Spectroscopy* (Amsterdam, N-Holland)
    Ch.11
Liu YS 1974 *Appl. Opt.* **13** 2505

Inst. Phys. Conf. Ser. No 128: Section 3
Paper presented at RIS 92, Santa Fe, NM, USA, 24–29 May 1992

147

Collective effects in non-resonant multiphoton ionization: a theoretical and experimental analysis

C. Altucci, R. Bruzzese, C. de Lisio, S. Solimeno and V. Tosa[*]

Dipartimento di Scienze Fisiche, Pad. 20 Mostra d'Oltremare, 80125 Napoli (Italy)
[*]Institute of Isotopic and Molecular Technology, Cluj-Napoca, Romania

ABSTRACT: We present in this work a kinetic model which describes the evolution of charged particles produced by high-intensity, ultrashort laser pulses in nonresonant multiphoton ionisation (MPI). The numerically predicted waveforms of the ion detector signal are compared to those observed in nonresonant MPI of Xe with 30 psec, 10^{13} W/cm^2, 1.064 μm laser pulses. The comparison highlights the relevance of space charge effects in typical nonresonant MPI conditions, and shows the effectiveness of our model in accounting for them.

1. INTRODUCTION

In nonresonant MPI experiments ultrashort laser pulses are focused on a gas jet inside an evacuated scattering chamber. The ionised atoms produce a plasma of small dimensions and low density. A factor that can be of importance in determining the expansion of the ion cloud and, as a consequence, the final ion yield is the space charge field thus created. It follows that for obtaining a more realistic description of the detection process when a time-of-flight (TOF) mass spectrometer is used, it is necessary to analyse the ion cloud evolution during its flight from the production region towards the detector. To this end we have developed a kinetic model based on the Vlasov-Maxwell equation and on the approximation of a space charge field with a linear profile. The model has been used for predicting the waveforms of the ion detector signal in the case of 11-photon nonresonant ionisation of Xe atoms studied with our specific TOF apparatus. The numerical results have been compared to the experimental ion signals obtained by irradiating Xe atoms with intense (10^{13}-10^{14} W/cm^2), 30 psec laser pulses at 1.064 μm.

2. THE KINETIC MODEL

Our TOF spectrometer is schematically shown in Fig.1. Ions are separated by elecrons by means of an extracted (dc) field and reach their end by hitting a detector placed beside the ionisation chamber , after a flight through acceleration and drift regions. However, the separation of the plasma in two oppositely charged clouds does not occur instantaneously. Depending on whether or not the Debye length of the plasma is grater than the plasma effective size, the extraction potential can either penetrate into the whole cloud and separate immediately the ions from electrons, or it takes a finite time for the extraction potential to leak through the plasma cloud. When the laser intensity is so high to produce a wealth of energetic (>1 eV) above-threshold-ionisation (ATI) electrons, (see Bruzzese et al 1989) most of them escape almost immediately from the potential well created by the space charge. This has allowed us to ignore in our analysis the electron contribution and include only the effects of an external inhomogenous electric field. Moreover, given the negligible values of the particle collision frequency at our gas pressures, we have omited discrete particle interactions and assumed that collective effects dominate on the time and length scales of interest. Accordingly, we have described the ion distribution evolution by means of a Maxwell-Vlasov equation (Krall and Trivelpiece 1973) with electric fields consistent with the space charge densities. This equation has been integrated by approximating the space charge field with a linear profile. For this class of fields the particle distribution function is obtained from the initial one by applying a time-dependent linear transformation to the phase-space coordinates (see de Lisio et al 1992 for the details).

The initial ion distribution function has been assumed to be Gaussian in the phase-space coordinates. The spatial sizes of the initial ion packet have been chosen according to the experimental values of the Rayleigh length and focal waist of the laser beam, while the ion velocity distribution corresponds to the thermal energy of the atoms.

In our case the analysis is complicated by the presence of six more abundant isotopes of Xe gas. The kinetic model in its analytical form does not take into account the separation of different isotopes during the flight along the spectrometer. This difficulty has been circumvented in the computer program by neglecting the isotopic composition of the ion cloud as long as the centers of the relative distributions are separated by a distance smaller than the size of the ion cloud. Successively, different isotopes are treated separately as noninteracting clouds.

The full discussion of the integration of the Vlasov equation for space charge fields varying lineary with the distance and in the presence of a static external electric field, as well as the analysis of the transport of ionic species through an electrostatic TOF spectrometer can be found in de Lisio et al (1992).

3. EXPERIMENTAL APPARATUS

The experimental apparatus consists of a 30 cm TOF mass spectrometer attached to a scattering chamber (see Fig.1). The whole system is bakeable and can be evacuated to a background pressure of 4×10^{-9} Torr. Gas samples are introduced into the system through an effusive beam downstream a 100 μm diameter stainless steel capillary whose tip is positioned 2.5 mm from the focal region. The local density has been estimated to be about 900 times the background one. The TOF resolution is adequate to resolve the signals produced by the different Xe isotopes.

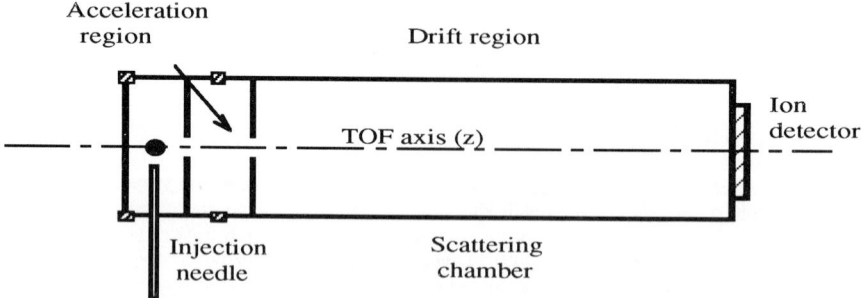

Fig. 1 Schematic view of the TOF spectrometer

In our experiments we have used a Nd:YAG laser system ($\lambda = 1.06$ μm) whose output consists of 30 psec, 30 mJ pulses (at 10 Hz) with an approximately Gaussian transverse profile. A biconvex lens having a focal length of 175 mm focuses the laser to a spot size of ~ 12± 1 μm, and the resulting intensity lies in the interval $10^{13} \div 10^{14}$ W/cm^2.

Details about the experimental apparatus and the procedures carried out to ensure an accurate evaluation of the number of ions produced per laser shot is given in deLisio et al (1992).

4. RESULTS AND CONCLUSIONS

According to the kinetic model briefly described in Sec.2, we have developed a computer program that is able to obtain ion waveforms which fit quite well the experimental ion signals for a wide range of laser energies and number of ions. We have analysed both conditions where the peaks relative to different isotopes can be clearly distinguished in the signal, and conditions leading to a single, broad (unresolved) ion peak. The fit of the measured signals withe the reconstructed (computed) waveforms appears to be satisfactory in both cases.

A first result of our computer simulation is that the final size of the ion pocket increases drastically (up to 20-30 times) by increasing the number of produced ions. As a consequence, when ions are detected as a function of the laser intensity, the fraction of charged particles missing the detector can increase with the laser intensity. This space charge effect might cause deviations of the power law dependence of the ion yield versus laser intensity from that predicted by the lowest order perturbation theory (11 in our case).

Typical comparisons between experimental and theoretical results are shown in Figs.2 and 3. In Fig.2a we report the measured ion signal for a 2.3 mJ laser pulse energy and a local gas pressure of 5×10^{-2} Torr. In these experimental conditions the estimated number of ions produced in the interaction zone is about 10^3. The corresponding theoretical signal, computed according to this value, is plotted in Fig.2b. The agreement between the two signals is good

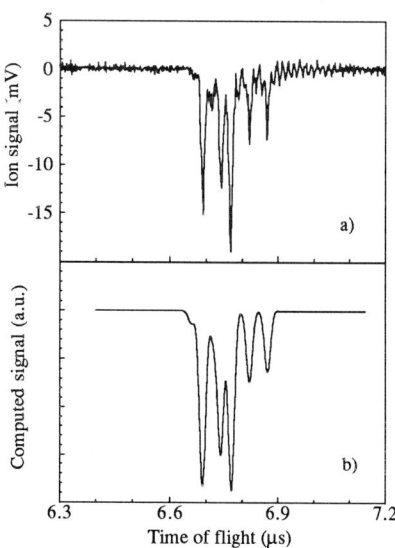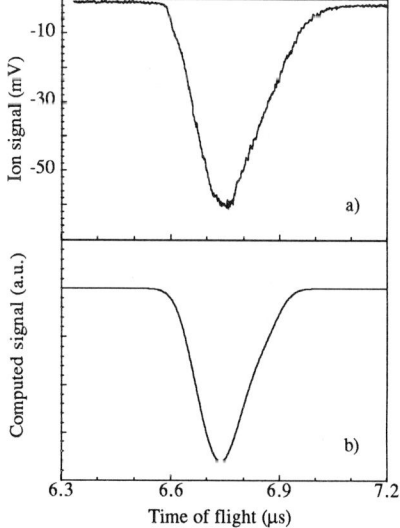

Fig. 2 (a) Experimental ion signal for 2.3 mJ laser energy, and 5×10^{-2} Torr gas pressure; (b) computed signal in the corresponding case of N=10^3 ions.

Fig. 3 (a) Same as in Fig.2a but for 11 mJ laser energy; (b) same as in Fig. 2b but for N=10^7 ions.

both with regard to the relative temporal delays between the five main peaks corresponding to the five more abundant Xe isotopes, which are clearly resolved in this case, and in the height ratios which reflect the isotope natural abundances.

In Fig. 3 we report a similar comparison between experimental (a) and numerical (b) results at the same gas pressure but with a laser energy of 11 mJ. This corresponds to a production of about 10^7 ions. The agreement between the overal shape and the FWHM of the two signals is again good. The isotopic structure of the ion signal has now completely disappeared. This is due to the spreading, and consequent overlapping, of the single isotopic ion packets which gives rise to a structureless signal.

In conclusion, we have presented a kinetic description of the evolution of the ionic species produced in nonresonant MPI of atoms with high intensity laser pulses. The kinetic model, based on the integration of the Vlasov-Maxwell equation with the assumption of a linear profile for the space charge field, has shown to be able to successfully predict the waveforms of the ion signal for our experimental set-up, and can be easily extended to other nonresonant MPI experiments.

The main result of our analysis consists in revealing the existence of a transition from a situation where the charge evolution can be described in terms of a single-particle formalism (see Fig.2), to a regime where space charge (collective) effects play an essential role in the ion packet evolution (Fig.3). The parameter characterising the evolution regime is the initial ion density. Although the transition towards the regime where collective effects have to be accounted for occurs in a continuous way, it is possible to single out an ion number, about 10^5, above which the space charge effects become dominant. In our experimental conditions, this corresponds to an ion density of about 10^{12} cm^{-3}.

References

Bruzzese R, Sasso A and Solimeno S 1989 *La Rivista del Nuovo Cimento* **12**(7), 1-105

Kroll N.A. and Trivelpiece A. 1973, *Principles of Plasma Physics* (McGraw Hill: New York)

de Lisio C, Di Palma T, Altucci C, Bruzzese R, Solimeno S and Tosa V 1992 *J Phys B: At Mol Phys* , in print

Inst. Phys. Conf. Ser. No 128: Section 4
Paper presented at RIS 92, Santa Fe, NM, USA, 24–29 May 1992

Interactions of laser induced fluorescence and ionization techniques in atomic and molecular spectroscopy

N. Omenetto

CEC, Joint Research Centre, Environment Institute, Ispra, Italy

ABSTRACT: The various combinations of the laser induced fluorescence and ionization techniques applied to atmospheric pressure, as well as to low pressure atomic and molecular reservoirs are illustrated. It is shown that, although each technique has in itself many attractive analytical and diagnostic capabilities, their interaction results not only in a better understanding of the dynamics of the excited states, but also provides new spectroscopic information and several fundamental parameters of the system investigated.

1. INTRODUCTION

Laser Induced Fluorescence (LIF), Laser Enhanced Ionization (LEI) and Resonance Ionization Spectroscopy (RIS) are nowadays well established analytical techniques in different types of atomic and molecular reservoirs which can operate at atmospheric pressure (flames, plasmas), at reduced pressure (glow discharges) and under vacuum (jet cooled molecules, atomic beams). During the past years, as testified by numerous publications, it became more and more evident that the simultaneous use of the fluorescence and ionization techniques could be greatly beneficial in analytical and spectroscopic studies performed on many different atoms and molecules. The aim of this paper is to focus on the interactions of LIF with LEI or RIS. As a result, the high number of investigations, whether of analytical or diagnostic character, carried out with both techniques **alone** will not be discussed here, **unless** the aim of an independent LIF measurement was to evaluate some ionization parameter, and viceversa.

An attempt is made at first to outline the motivations for undertaking simultaneous LIF-LEI or RIS measurements and their relevance to chemical analysis. Then, various types of fluorescence-based methods will be reviewed and it will be shown how fundamental parameters of the interaction and/or data relevant to the ion formation and decay processes can be derived from these experiments.

2. ANALYTICAL STUDIES

In all the studies reported below, excimer-pumped or Nd-YAG pumped dye lasers were used as excitation sources, with flames and graphite furnaces as atom reservoirs.

One of the first motivations to conduct ionization measurements in a flame, while at the same time monitoring the atomic fluorescence signal, is that an absolute measurement of the number of photons emitted in one particular transition, in optically saturated conditions, reflects directly the number density of the excited level and therefore the total number density. If this **absolute analysis** is attempted, then one needs to know the number of excited atoms lost in the ionization process. For an excitation pulse in the nanosecond range and for transitions whose excited state lies several eV below the ionization potential, LEI does not cause a substantial depletion of the total atomic population. On the other hand, when two lasers are used in the excitation process, as in double-resonance fluorescence experiments in a graphite furnace (Vera et al., 1989), a significant amount of excited atoms will be lost because of ionization: in this case, the accuracy of the result will be severely degraded if these losses are not accounted for. In addition, simultaneous ionization measurements in a graphite furnace will have another advantage, which should not be overlooked. In fact, if the ion signal is compared to the fluorescence signal **for different matrices**, the role played by essential analytical parameters such as the atomisation efficiency and the interferences in the vapour phase can be investigated.

LEI and RIS have been used to **detect fluorescence photons.** This can be accomplished with the so-called "resonance monochromator" approach (Matveev 1983, Omenetto et al., 1989b), in which fluorescence from the sample to be analyzed is excited in one analytical cell and then transferred into another cell, filled with pure analyte vapour, which constitutes the ionization detector. Calculations on the limiting noises and signal-to-noise ratios attainable in different atom reservoirs have led to the conclusion that spectacular sensitivities, close to ppt levels in concentration units and to sub-femtogram amounts in absolute units, should be reached. However, spurious scattering problems will be significant and need to be overcome, or else the attractiveness of this approach will be lost. No analytical applications to real samples have yet been reported with this technique.

An interesting observation on the possibility of detecting **anomalous contributions to the LEI signal** in a flame from atoms positioned outside the laser interaction volume was made (Axner and Sjöstrom, 1990). The experiment was performed on strontium atoms, with two spatially non-overlapping laser beams (two-step LEI) and it was concluded that "scattered" light, which might have included fluorescence, was responsible for the effect observed.

As a final example, pertinent to the field of combustion diagnostics, a technique, called Photoionization Controlled - Loss Spectroscopy (PICLS) (Lucht et al., 1983) was devised to obtain **quenching-free fluorescence measurements** in flames, and was applied to atomic hydrogen (Salmon and Laurendeau, 1986). In PICLS, after being excited to a given level, the atoms are photoionized at a rate which is made higher than the downward quenching rate: the limitations due to the difference in the quenching coefficient at different locations in the flame are therefore removed by making the photoionization process the dominant loss mechanism for the excited state atoms.

3. SPECTROSCOPIC STUDIES

As is often the case also for other laser-based techniques, the applications of LIF/LEI to fundamental diagnostic studies are much more abundant than those dealing with strictly analytical problems. In order to cover them adequately, and also to emphasize the similarity, and yet the versatility, of the different approaches (together with the original nomenclature proposed by the authors), several publications are collected in Table 1. Some comments are in order here for a correct interpretation of this Table: (i) a choice was made among many papers, and the list is certainly not exhaustive; (ii) mainly those papers describing the simultaneous use of the fluorescence and ionization techniques, or the use of LIF to get ionization parameters, were considered; and (iii) literature references are listed in chronological order but this does not necessarily imply a priority either in the inception of the idea or in its application.

Table I: Literature scan of selected fluorescence and emission studies relevant to ionization diagnostics in different atomic and molecular reservoirs.

Technique and associated name	Mechanism Studied - Information provided	Authors (year)
Modulated Fluorescence	Photoionization σ - Cs atomic beam	Gilbert et al. (1984)
Optical Detection of LEI	Decay kinetics of Sr^+ in flames and plasmas	Turk et al. (1986, 1987)
Exctinction Spectroscopy	Lifetimes of core-excited levels - HCD, Li heat pipe	Pedrotti (1986)
Laser Depletion Spectroscopy	Level positions, line-widths, f's, τ's of core excited levels - Rb heat pipe	Spong et al. (1987, 1988) Harris (1988)
Fluorescence Dip Spectroscopy	Photoionization σ - Mg atoms, argon plasma	Omenetto (1989a)
Two-Colour Dip Spectroscopy	Energy and dynamics of excited, jet-cooled molecules	Ito (1989)
Fluorescence Reduction Spectroscopy	Photoionization σ - Cs atoms, heated gas cell	Bonin et al. (1989)
Fluorescence Dip Spectroscopy	Simultaneous LIF/LEI - Theory in flames	Axner et al. (1989)
Fluorescence Dip Rydberg Spectroscopy	Ionization Potentials of Ti, V, Fe, Co, Ni - Low pressure sputtering chamber	Page and Gudeman (1990)
Fluorescence Dip and Ion Dip Spectroscopy	Spectroscopy of single rovibronic state - Benzene molecular beam	Weber et al. (1990)
Fluorescence Dip Spectroscopy	Autoionization σ. Ion yield, Ionization efficiency. Mg atoms	Petrucci et al. (1991) Omenetto et al. (1991)
Fluorescence Studies of Multiphoton Ionization (MPI)	MPI of Ca and Sr. Low pressure oven	Haugen and Stapelfeldt (1991)
Fluorescence Dip Spectroscopy	Charge Transfer in an argon plasma	Farnsworth et al. (1991)

It would clearly be too long to describe in detail all the experiments reported in the Table and therefore only some examples and considerations, mainly taken from our own work, are reported here. In the studies involving the decay of Sr ions, one laser was used to ionize the atoms and another laser, tuned to an ionic fluorescence transition, was delayed in time until the ions formed disappeared, thereby allowing the direct evaluation of the time decay of the signal back to the original (thermal) level and the influence of easily ionizable elements (e.g., K) on this decay (see Fig. 1).

An inspection of Table 1, however, reveals a common methodological ground of many experiments, i.e., the modulation of the fluorescence or emission signals caused by the onset of the ionization at a rate comparable to or higher than the radiative deactivation rate. It was shown (Alkemade, 1985; Omenetto et al., 1989c) that a careful study of the simultaneous dependence of the fluorescence and ionization signals upon the laser photon irradiance provides the key for distinguishing LEI from RIS in flames. As Figure 2 shows, at a given I, Q_F decreases. This modulation of the fluorescence is indeed extremely useful and has led to a variety of applications, from the determination of the ionization potentials of several atomic species to the assignment of level positions, autoionizing lifetimes, oscillator strength of the transitions and ionization cross sections. The **easier accessibility** of the intermediate (depleted) level as compared to the final level of a multistep excitation scheme is amply stressed in several of the papers listed in Table 1. Of particular relevance is also the technique of **ion dip** spectroscopy, which allows better spectral resolution than the complementary fluorescence dip technique and was largely applied to the study of ground state vibrational levels of jet-cooled large molecules (e.g., jet-cooled phenol).

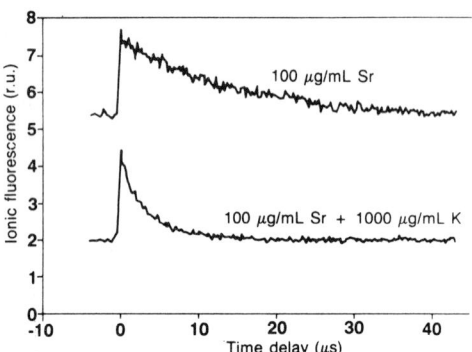

Fig. 1 Decay of strontium ions in an argon plasma. The addition of electrons (due to K) produces a significant shortening of the decay time.

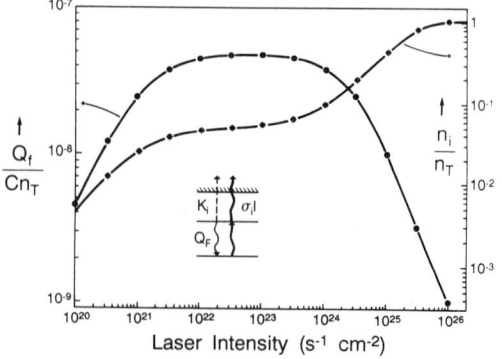

An important concept, which has already been stressed (Omenetto et al., 1991) is that of the usefulness of **time resolution within the laser pulse** (see Figure 3). In fact, in a few tens of ns, a metastable level and the ionization continuum (**m** and **i** in Fig. 3) will act as traps in which the atoms accumulate. The fluorescence will then decay during the excitation and the loss rates, R's, evaluated. As an example, Fig. 4 shows two waveforms obtained with a fast microchannel plate photomultiplier and a digitizing signal analyzer. The striking difference in the time behaviour observed at high laser irradiance could be due to direct photoionization of the 3 p level.

Fig. 2 Theoretical dependence of the integrated fluorescence (Q_F) and ion yield upon the laser photon irradiance. $k_i = 10^7 s^{-1}$; $\sigma_i = 10^{-17}$ cm^2; n_i and n_T are the ion and total number densities (cm^{-3}); C is a constant (J cm^3 s^{-1}).

DIRECT EVALUATION OF
COLLISIONAL RATE COEFFICIENTS

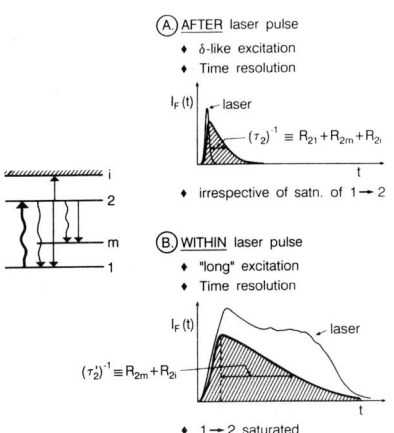

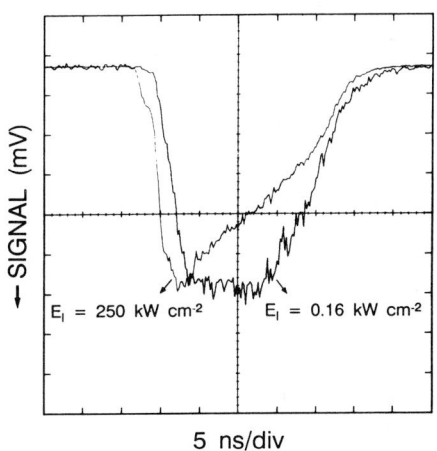

Fig. 3 Idealized scheme of different time-resolved fluorescence signals.

Fig. 4 Time-resolved resonance fluorescence waveforms for magnesium in an argon plasma at different laser irradiances.

The usefulness of time resolution extends also to cases in which a reaction such as the **charge transfer** between magnesium atoms and argon ions is investigated. In an argon plasma, charge exchange is believed to be responsible for the direct population and depopulation of the Mg II 4s ^{2}S and 3d ^{2}D levels, whose energy is "quasi-resonant" with that of the ground state argon ion. Two lasers were used to ionize the Mg neutrals and to enhance the population of the quasi-resonant levels: by monitoring the corresponding variations in the ion and atom emission signals, the above mechanism could be experimentally demonstrated (Farnsworth et al. 1991). Time resolution within the laser pulse would allow the direct evaluation of the rate constants of the charge transfer reaction.

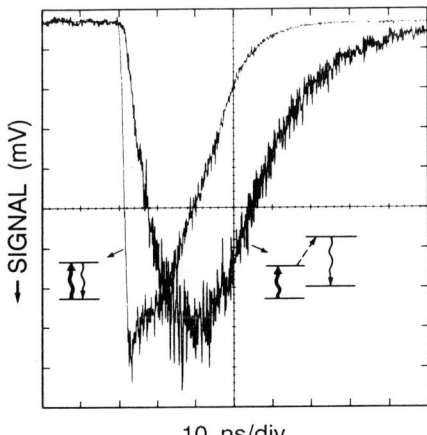

Fig. 5. Time-resolved resonance fluorescence and collision-induced fluorescence for thallium in an argon plasma.

A final example is shown in Fig. 5. Here, resonance fluorescence of thallium was measured at 377.572 nm, while collision-induced fluorescence, still excited at 377.572 nm, was recorded at 351. 924 nm. The energy difference between the levels shown is 1.2 eV. It can clearly be seen that a time lag of 20 ns is necessary for the maximum build-up of the population in the fluorescent level. In our plasma, this collisional transfer is essentially due to electrons and argon atoms and varies drastically with the discharge conditions. A systematic evaluation of the fluorescence waveforms originating from a sufficient number of levels located between that excited by the laser and the ionization continuum will then provide a direct insight on the magnitude of the collisional transfer rates involved.

Two outcomes of this study seem to be worth of stressing: (i) in the case of disconnected, two-step LEI, the correct time delay between the two pulses can be accurately established; and (ii) the ion **production** mechanism, whether due to direct photoionization (RIS) or collisionally-assisted ionization (LEI), might be better elucidated when, with a fast detection system, the true ionization pulse shape is investigated.

4. CONCLUSIONS

Only a few examples have been reported in this paper to illustrate the versatility and attractiveness of combining several ionization and fluorescence measurements in many analytical and physical studies on atoms and molecules present in largely different environments. The key point of most studies is that spectroscopic data relevant to an excited state can be directly inferred by monitoring the modulation in the population of an intermediate state connected to it in the excitation ladder. It is expected that the number of publications describing the usefulness of such approach will constantly grow in the future.

5. LITERATURE

Alkemade C Th J 1985 *Spectrochim. Acta* **40B** 1331.
Axner O, Norberg M and Rubinsztein-Dunlop H 1989 *Spectrochim. Acta* **44B** 693.
Axner O and Sjöström S 1990 *Appl. Spectrosc.* **44** 864.
Bonin K D, Gatzke M, Collins C L and Kadar-Kallen M A 1989 *Phys. Rev. A* **39** 5624.
Farnsworth P B, Smith B W and Omenetto N 1991 *Spectrochim. Acta* **46B** 843.
Gilbert S L, Noecker M C and Wieman C E 1984 *Phys. Rev. A* **29** 3150.
Harris S E 1988 *Optics News* **11**.
Haugen H K and Stapelfeldt H 1992 *Phys. Rev. A* **45** 1847.
Ito M 1989 *Int. Rev. Phys. Chem.* **8** 147.
Matveev O I 1983 *J. Anal. Chem. USSR* **38** 561.
Omenetto N 1989a *Proceedings of RIS-88, Inst. Phys. Conf. Ser. N. 94,* IOP Publishing Ltd., 141.
Omenetto N, Smith B W and Winefordner J D 1989b *Spectrochim. Acta, Special Supplement,* 91.
Omenetto N, Smith B W, Jones B T and Winefordner J D 1989c *Appl. Spectrosc.* **43** 595.
Omenetto N, Smith B W and Farnsworth P B 1991 *Proceedings of RIS-90, Inst. Phys. Conf. Ser. No. 144,* IOP Publishing Ltd, 369.
Page R H and Gudeman C S 1990. *J. Opt. Soc. Am. B* **7** 1761.
Pedrotti K D 1987 *Opt. Comm.* **62** 250.
Petrucci G A, Stevenson C L, Smith B W, Winefordner J D and Omenetto N 1991 *Spectrochim. Acta* **46B** 975.
Salmon J T and Laurendeau N M 1986 *Opt. Lett.* **11** 419.
Spong J K, Kmetec J D, Wallace S C, Young J F and Harris S E 1987 *Phys. Rev. Lett.* **58** 2631.
Spong J K, Imamoglu A, Buffa R and Harris S E 1988 *Phys. Rev. A* **38** 5617.
Turk G C and Omenetto N 1986 *Appl. Spectrosc.* **40** 1085
Turk G C, Axner O and Omenetto N 1987 *Spectrochim. Acta* **42B** 873.
Vera J A, Stevenson C L, Smith B W, Omenetto N and Winefordner J D 1989 *J. Anal. At. Spect.* **4** 619.
Weber Th., Riedle E and Neusser H J 1990 *J. Opt. Soc. Am. B* **7** 1875.

Inst. Phys. Conf. Ser. No 128: Section 4
Paper presented at RIS 92, Santa Fe, NM, USA, 24–29 May 1992

Spatially selective detection of anomalous contributions to laser-enhanced ionization in flames

Gregory C. Turk

National Institute of Standards and Technology, Gaithersburg, MD 20899 USA

ABSTRACT: By control of the voltage applied to the detection electrodes for laser-enhanced ionization in flames, some degree of spatial selectivity can be achieved. This spatially selective detection can be used to study anomalous contributions to flame LEI caused by indirect scattered laser radiation in regions of the flame which are not directly within the laser beam.

1. INTRODUCTION

Recent research by Axner *et al* (1990) has demonstrated that under certain experimental conditions a significant, or even dominant, fraction of the signal in laser-enhanced ionization (LEI) spectroscopy originates from the ionization of atoms which lie outside of the laser beam in a flame. This phenomenon is the result of light that can reach atomic vapor outside the laser beam by scattering or fluorescence from the laser-irradiated volume of the flame, or stray laser radiation from optical elements in the experiment. The implications of this anomaly can be important -- affecting the interpretation of important experimental observations such as line profiles and laser power dependence. Turk (1981) has noted that the accuracy of LEI for chemical analysis is dependent on the proper alignment of the laser beam with the ionization detection electrodes, and consequently any anomalous LEI from inappropriate regions of the flame will have deleterious effects. Thus, it is important to be able to locate the source of LEI within an atomic vapor cell. This paper describes a procedure for making spatially selective measurements of the active LEI volume for pulsed laser excitation in an air-acetylene flame by manipulation of the properties of conventional LEI detection electrodes.

Understanding the principles behind this procedure requires a knowledge of the details of signal detection in an LEI measurement, first described by Havrilla *et al* (1984) and expounded upon in the work by Travis *et al* (1984). In an experiment using a pulsed laser, the LEI signal is the result of the current induced in the detection circuit by the motion of the laser-created ions and electrons in response to an externally applied electric field. In the most common experimental arrangement, this field is applied using a water-cooled cathode at a negative high voltage inside the flame above the laser beam, and the metal burner head at ground potential as the anode. As a result of the large difference in the mobilities of positive ions and electrons in the flame, a negative space charge develops around the cathode, forming what is known as the ion sheath. A non-zero electric field exists only within this ion sheath, and thus for an LEI signal to be detected, the ionization must occur within the sheath.   The size of the sheath is determined by, among other things, the magnitude of the applied voltage.   To a first approximation, the sheath can be described by an electric field which is maximum at the cathode and decreases linearly in the direction of the anode to zero at the edge of the sheath, such that the entire applied voltage is dropped across the sheath.   The extent of the sheath increases as the applied voltage increases.   By control of the applied voltage, the size of the sheath can be controlled and thus the volume of flame from which ionization is detected.   It is this control of the sampled volume of the flame from which spatial information regarding the LEI active volume is obtained.

2. EXPERIMENTAL

All measurements were made in an air-acetylene flame using a 5-cm single slot burner head as the anode and a water-cooled 0.635-cm diameter stainless steel tube as the cathode.   The bottom edge of the cathode was 2.5 cm above the burner head.   Sodium was used as the analyte, aspirated into the flame in the conventional manner at a concentration of 1 mg/L. The laser was a Nd:YAG pumped dye laser operated at 10 Hz with a pulse energy of approximately 1 mJ at 589 nm.   The laser beam was apertured to a 0.10 cm high by 0.15 cm wide shape, and was aligned 1.1 cm below the cathode.   Excitation spectra were collected using the conventional LEI data collection system consisting of a current-to-voltage preamplifier, 10 kHz - 1 MHz filter, and computer-interfaced gated integrator with a 1-µs gate width.

3. RESULTS AND DISCUSSION

Figure 1(a) shows an approximation of the electric field strength within the flame as a function of position below the cathode with a potential of -300 V applied to the cathode. The position of the edge of the sheath, where the field strength drops to zero was determined empirically in a series of measurements in which the relationship between LEI signal and applied voltage was measured at a variety of alignments of the laser beam below the cathode, as described elsewhere by Turk (1992).   Under these conditions the region of non-zero electric field extends 0.5 cm below the cathode, and thus no electric field is present within the laser beam, which is 1.1 cm below the cathode.   When the negative potential applied to the cathode is increased to -800V, as seen in figure 1(b), the extent of the sheath increases to 1.2 cm below the cathode.   In this case the laser beam now experiences the electric field.   More detailed measurements of the electric field within the flame are presented in the work of Travis *et al* (1984).

Figures 2 (a) and (b) show LEI excitation spectra across the sodium D lines under the electric field conditions shown in parts (a) and (b) of Figure 1, respectively.   An LEI signal is clearly seen in part (a), despite the fact that no electric field is present to detect LEI from the laser beam.   The source of this signal is indirect laser excitation, from scattering or other sources, within the region 0.5 cm below the cathode.   The line profiles are sharp, without any indication of saturation broadening, as would be expected for the low power radiation being received indirectly in this region of the flame.

A much different excitation profile is seen in part (b) of Figure 2, where the higher applied potential now samples LEI signal from the laser beam, as well as from the region of the

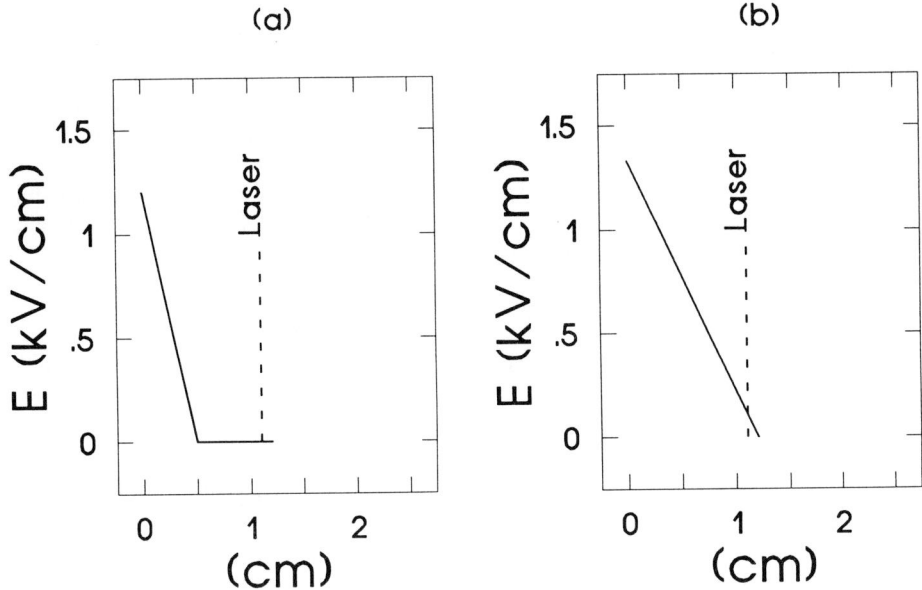

Fig. 1. The electric field within the flame as a function of distance below the cathode with 1 mg/L of sodium being aspirated and (a) -300 V applied to the cathode and (b) -800 V applied to the cathode.

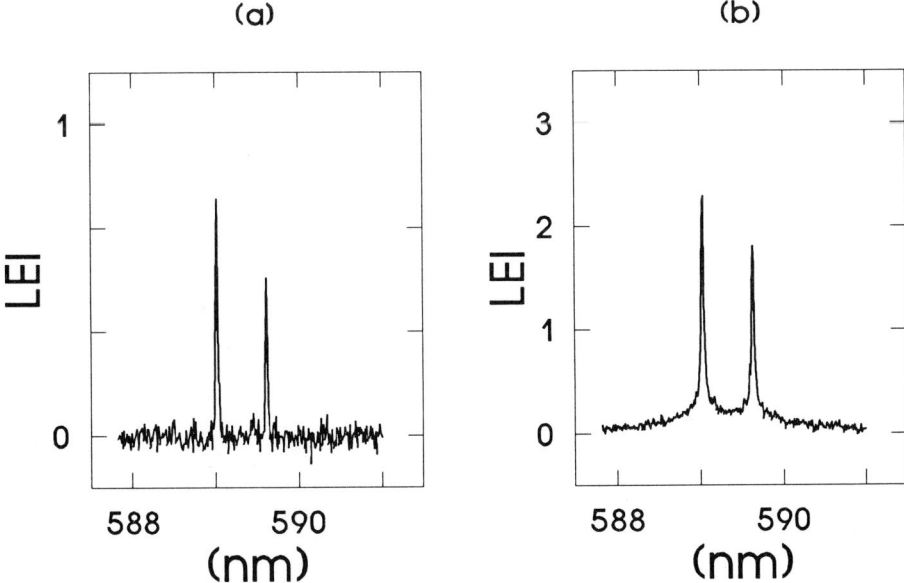

Fig. 2. LEI excitation profiles for the sodium D lines under the conditions described in Fig. 1 (a) and (b). The arbitrary LEI signal units are consistent for both parts of the figure.

flame between the cathode and the laser beam. What is seen is a combination of LEI from within the laser beam, under conditions of high laser irradiance, and indirect low power laser excitation in the zone between the cathode and the laser beam. The excitation profile reflects this combination of vastly differing excitation conditions, with a saturation broadened pedestal, from the LEI within the laser beam, with a sharp line profile superimposed on top, from the LEI outside the laser beam. Despite the much higher laser power inside the laser beam, the effect of saturation and the fact that the volume being indirectly excited is much greater than that within the laser beam results in a situation in which the dominant source of signal at the line peaks is the result of anomalous indirect laser excitation.

REFERENCES

Axner O, and Sjöström S 1990 *Appl. Spectrosc.* **44** 864

Turk G C 1981 *Anal. Chem.* **53** 1187

Havrilla G J, Schenck P K, Travis J C, and Turk G C 1984 *Anal. Chem.* **56** 186

Travis J C, Turk G C, DeVoe J R, Schenck P K, and Van Dijk C A 1984 *Prog. Analyt. Atom. Spectrosc.* **7** 199

Turk G C 1992 submitted to *Anal. Chem.*

Inst. Phys. Conf. Ser. No 128: Section 5
Paper presented at RIS 92, Santa Fe, NM, USA, 24–29 May 1992

Photoionization dynamics and cation spectroscopy with coherent VUV radiation

Ralph. T. Wiedmann, Russell G. Tonkyn, Edward R. Grant[†], Michael G. White

Chemistry Department, Brookhaven National Laboratory, Upton, NY 11973

[†]*Department of Chemistry, Purdue University, West Lafayette, IN 47907*

Abstract: Pulsed field ionization (PFI) has been used in conjunction with coherent VUV radiation to investigate the rotational state distributions of molecular cations following single photon ionization. For photoionization of the H_2X (X = O, S) molecules, transitions between asymmetric top levels involving the rotational angular momentum projections, K_a and K_c, permit resolution of the photoelectron continua according to symmetry. The observed spectra also clearly demonstrate the importance of the non-spherical nature of the molecular ion potential which leads to photoelectron final states which are unexpected from atomic-like analogies.

1. Introduction

Photoelectron spectroscopy provides information on the dynamics of the photoionization process by revealing the internal state distribution of the molecular fragment (molecular cation) through the measurement of the kinetic energy distribution of the "light" fragment (photoelectron). The dynamics are considerably different than for neutral fragmentation, yet both involve similar concepts, especially as regards the evolution of a photoexcited "complex" into the observed asymptotic channels and the molecular forces which influence their branching ratios. With the development of laser-based, zero kinetic energy (ZEKE) and pulsed field ionization (PFI) techniques it is has now become possible to extend photoelectron spectroscopy to measurements of cation *rotational* state distributions and hence, fully resolve quantum state distributions following photoionization (Müller-Dethlefs 1991, Grant 1991). We have used the PFI technique in conjunction with a high-resolution VUV laser source to investigate the spectroscopy and one-photon ionization dynamics of several linear molecules (O_2, N_2O, Xe_2, HCl and OH(OD)) in which we have characterized the angular momentum constraints in rotational photoionization transitions (Tonkyn 1989, Braunstein 1990, Wiedmann 1991, Tonkyn 1991a, Tonkyn 1992, Wiedmann 1992b). Our concentration on one-photon processes arises from the extensive body of information which exists for molecular VUV photoionization using laboratory or synchrotron radiation sources, but for which only H_2 has been rotationally resolved. In this paper, we present results of recent studies on the H_2X (X = O, S) molecules which explore the symmetry properties of allowed rotational photoionization transitions in non-linear molecules.

2. Experiment

Figure 1 is a schematic diagram of the photoelectron spectrometer and VUV source which are described in detail elsewhere (Tonkyn 1989b). Briefly, we use a small, separately pumped frequency tripling chamber attached to a photoelectron/photoion time-of–flight (TOF) spectrometer. Tunable VUV radiation in the range 150 nm to 90 nm is produced by non-resonant third harmonic generation ($\omega_{vuv} = 3\omega_{uv}$) as well as resonant and non-resonant four-wave sum/difference frequency mixing ($\omega_{vuv} = 2\omega_{uv} \pm \omega_{vis}$) in free jet expansions of krypton, xenon and molecular nitrogen (Page 1987). The collinear laser beams ($\omega_{uv}, \omega_{vis}$) are focused by a 100 mm focal length achromatic lens into the tripling chamber and very near the 1 mm dia. exit nozzle of a piezoelectric pulsed-valve (Proch 1989). The diverging fundamentals (visible and UV) and harmonically generated VUV radiation are captured by a pyrex capillary tube (30 cm long; 1 mm i.d.) and directed into the TOF spectrometer, where it passes between two parallel plates which define the extraction field for the TOF spectrometer. The capillary acts as an efficient differential pumping barrier separating the VUV generation chamber ($\sim 10^{-3}$ Torr) and the spectrometer chamber ($\leq 1 \times 10^{-7}$ Torr), while also acting as a light guide which provides a collimated VUV beam without refocussing. The VUV intensity at the spectrometer is estimated to be $10^9 - 10^{10}$ photons per pulse, with an energy bandwidth of 0.7 cm^{-1}.

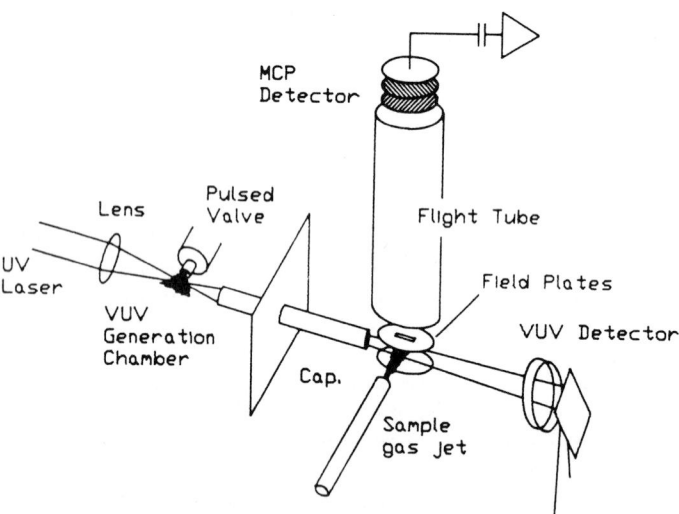

Figure 1. Schematic diagram of time-of-flight photoelectron spectrometer used for one-photon photoionization measurements.

High-resolution threshold photoelectron spectra are obtained by the delayed, pulsed field ionization (PFI) method as first demonstrated by Reiser, *et al* (1988). PFI takes advantage of the well known Stark shift of an ionization threshold in an external electric field (given by $\sim 6\sqrt{F}$ cm^{-1}, where F is the field strength in volts/cm). Bound Rydberg levels very near the ionization threshold which are stable in a field free environment become open ionization channels when a Stark shift larger than their binding energy is applied. For fields used in these studies (0.3-0.5 V/cm) only Rydberg levels with $n \geq 150$ can be field ionized. Rydberg states with such high principal quantum numbers are very long lived due to rapid l mixing induced by a small DC electric field (~ 0.05 V/cm). The small positive DC voltage on the repeller (lower) plate also sweeps out any slow or near ZEKE photoelectrons produced directly by the laser pulse. After a delay of 700 ns, a fast, negative pulse ($0.3 - .5$ V) is applied to the repeller plate which field ionizes the metastable, high-n Rydberg states. Electrons produced by this pulsed field are readily distinquished from photoelectrons arising from direct ionization as the arrival time of the former depend only on the pulse delay and voltage.

3. Threshold Photoionization of H_2X (X = O, S)

High resolution (~ 1 cm^{-1}) threshold photoelectron spectra for jet-cooled (15 K) samples of H_2O and H_2S were obtained for rotational transitions between the

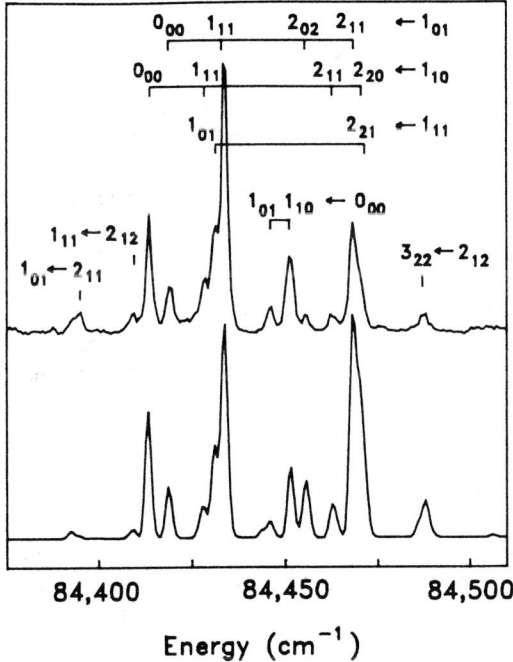

Figure 2. Upper trace: Rotationally resolved, pulsed field ionization spectrum for one-photon ionization of H_2S to the $\tilde{X}^2B_1$ (000) level of H_2S^+. Lower trace: Simulated photoionization spectrum of H_2S at 15 K including both type A ($\Delta K_c = \pm 1, \Delta K_a = 0$) and type C ($\Delta K_c = 0, \Delta K_a = \pm 1$) rotational transitions.

$X\,^1A_1$, (000) neutral ground state and the $\tilde{X}\,^2B_1$, $(\nu_1\nu_2\nu_3)$ vibrational levels of the ionic ground state (Tonkyn 1991b, Wiedmann 1992a). For H_2O, rotationally-resolved data were obtained for ionization into the (000), (100) and (010) vibrational levels, while for H_2S^+ only the (000) and (010) levels could be probed due to inefficient VUV production at wavelengths near the (100) symmetric stretch. The PFI spectrum for the $\tilde{X}\,^2B_1$, (000) level of H_2S^+ is shown in Figure 2. The H_2X^+ spectra could be readily assigned to two types of rotational photoionization transitions corresponding to specific changes in the asymmetric top angular momentum projection quantum numbers, K_a and K_c. Most of the stronger lines can be classified as type C rotational transitions $(\Delta K_c = 0, \Delta K_a = \pm 1)$ but type A $(\Delta K_c = \pm 1, \Delta K_a = 0)$ transitions are also clearly evident, particularly in the jet-cooled spectra. The utility of this classification stems from the fact that general symmetry arguments show that these rotational transitions are associated with only one photoelectron symmetry, i.e. ka_1 with type C and ka_2 with type A. The appearance of type A transitions with $\Delta K_a = 0$ is in variance with the predictions of a multichannel quantum defect theory (MQDT) analysis of photoionization of asymmetric top molecules by Child and Jungen (1990). These authors predict that photoionization will only involve a subset of type C transitions with $|\Delta K_a| = |K_a^+ - K_a''| = 1$ and $|\Delta N| = |N^+ - N''| \leq 1$, where N is the total angular momentum exclusive of spin. These limits on the changes of core angular momenta arise from the assumption that the $1b_1$ molecular orbital can be described exclusively in terms of an atomic p_z orbital. More recently, Gilbert and Child (1991) have presented a model based on field-induced autoionization in an effort to explain the appearance of type A transitions in the threshold spectra of H_2O. Although such rotational autoionization processes could give rise to the appearance of rotational final states which are nominally forbidden by optical selection rules, the specific application to H_2O also invokes an atomic-like description of photoexcitation from the $1b_1$ molecular orbital.

A more straightforward explanation for the appearance of type A transitions can be derived from a very recent *ab initio*, Schwinger variational photoionization calculation on H_2O (Lee 1992). In this calculation, no assumptions concerning the atomic character of the initial and final states are made and the continuum is calculated in the full anisotropic potential of the ion core. The calculations predict the appearance of type A photoionization transitions without recourse to final state interactions (field-induced autoionization) and the calculated H_2O^+ rotational spectrum is in nearly quantitative agreement with experiment. A comparison of the H_2O^+ $\tilde{X}\,^2B_1$, (000) PFI spectrum and the theoretical near threshold photoionization calculation is shown in Figure 3.

A partial wave analysis shows that type A transitions are accompanied by nearly pure p wave photoelectrons. Furthermore, the type C transitions are not exclusively d partial waves as expected from the atomic analogies, but include significant and for some lines (e.g. $0_{00} \leftarrow 1_{10}$) dominant s wave character. These calculations emphasize the importance of the non-spherical nature of the molecular ion core which can "scatter" or torque the escaping photoelectron into different partial waves. For H_2S, we expect that similar dynamics are applicable since the $1b_1$ molecular orbital is primarily localized on the sulfur atom. However, the participation of the nominally unoccupied $3d$ levels centered on the heavier sulfur atom could lead to additional high l components in the continuum and modify the rotational line strengths.

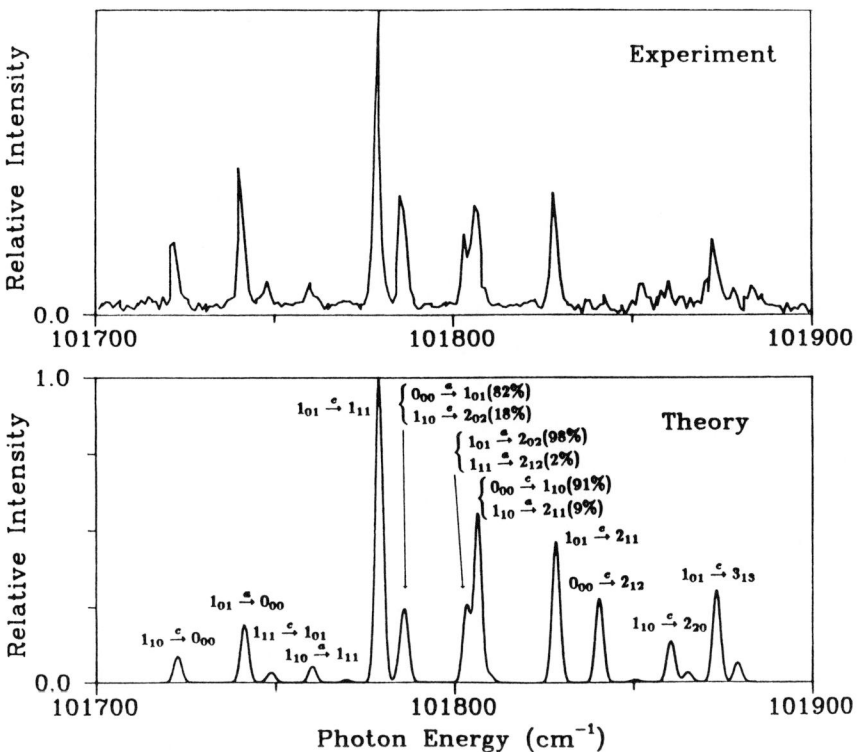

Figure 3. Experimental (top) and calculated (bottom) threshold photoelectron spectrum for photoionization from the $1b_1$ orbital of H_2O. The labels a and c refer to type A and type C rotational branch transitions.

4. Acknowledgements

This work was performed at Brookhaven National Laboratory and supported by the US Department of Energy, Office of Basic Energy Sciences under contract No. DE-AC02-76CH00016. E.R.G. acknowledges support from the National Science Foundation under Grant No. CHE-8920555.

5. References

Braunstein M, McKoy V, Dixit S N, Tonkyn R G and White M G 1990
   J. Chem. Phys. **93** 5345
Child M S and Jungen Ch 1990 *J. Chem. Phys.* **93** 7756
Gilbert R D and Child M S 1991 *Chem. Phys. Lett.* **187** 153
Grant E R and M G White 1991 *Nature* **354** 249
Lee M T, Wang K, McKoy V, Tonkyn R G, Wiedmann R T, Grant E R and
   White M G 1992 *J. Chem. Phys. in press*
Müller-Dethlefs K and Schlag E W 1991 *Annu. Rev. Phys. Chem.* **42** 109
Page R H, Larkin R J, Kung A H, Shen Y R and Lee Y T 1987 *Rev. Sci. Instrum.*
   58 1616
Proch D and Trickl T 1989 *Rev. Sci. Instrum.* **60** 1989
Reiser G, Habenicht W, Müller-Dethlefs K and Schlag E W 1988
   Chem. Phys. Lett. **152** 119
Tonkyn R G, Winniczek J W and White M G 1989a, *Chem. Phys. Lett.* **164** 137
Tonkyn R G and White M G 1989b *Rev. Sci. Instrum.* **60** 1616
Tonkyn R G and White M G 1991a *J. Chem. Phys.* **95** 5582
Tonkyn R G, Wiedmann R T, Grant E R and White M G 1991b *J. Chem. Phys.*
   95 7033
Tonkyn R G, Wiedmann R T and White M G 1992 *J. Chem. Phys.* **96** 3696
Wiedmann R T, Grant E R, Tonkyn R G and White M G 1991 *J. Chem. Phys.*
   95 746
Wiedmann R T and White M G 1992a *Proceedings of the SPIE: Optical Methods
   for State- and Time-Resolved Chemistry* **1638** 273
Wiedmann R T, Tonkyn R G, White M G, Wang K and McKoy 1992b
   J. Chem. Phys. in press

Inst. Phys. Conf. Ser. No 128: Section 5
Paper presented at RIS 92, Santa Fe, NM, USA, 24–29 May 1992

167

Molecular surface analysis by laser ionization of desorbed molecules

M. J. Pellin, K. R. Lykke, P. Wurz, and D. H. Parker

Materials Science/Chemistry Divisions,
Argonne National Laboratory.
Argonne, Illinois 60540

ABSTRACT:While elemental analysis of surfaces has progressed dramatically over the past ten years, quantitative molecular surface analysis remains difficult. This is particularly true in the analysis of complex materials such as polymers and rubbers which contain a wide compliment of additives and pigments to enhance their material characteristics. For mass spectrometric analysis the difficulty is two fold. First, desorption of surface molecules must be accomplished with minimal fragmentation and collateral surface damage. Second, the desorbed molecules must be ionized for subsequent mass analysis with high efficiency and without significant cracking. This paper focuses on the second of these problems.

---

*Work supported by the U.S. Department of Energy, BES-Materials Sciences, under Contract W-31-109-ENG-38.

1. INTRODUCTION

The two techniques most likely to be used for molecular surface analysis, particularly for complex surfaces, Secondary Ion Mass Spectrometry (SIMS) (see for instance Gardella 1990) and Laser Induced Mass Spectrometry (LIMS) (see for instance Asamoto 1990, Johlman 1990, Li 1990, or Huang 1988), attempt to accomplish molecular desorption and ionization in a single step. Unfortunately molecular analysis often requires careful optimization of both the desorption and the laser ionization step. For instance, the amount of fragmentation of molecular species during the laser desorption depends on many factors including the energy absorbed by the surface, the wavelength of the laser, the pulse length of the laser (Lazare 1989, Feldmann 1987, and Srinivasan 1989). Several authors have shown that separating the ionization and desorption steps can allow characterization of complex surfaces(Grotemeyer 1989, Becker 1990, Lubman 1990). Here we detail studies of complex surfaces where the desorption step and the ionization step are separately optimized. While many problems remain in understanding the desorption of molecules from surfaces, this paper will focus on the ionization step.

The use of laser post ionization has been found to have significant advantages in the analysis of complex materials. First, laser ionization can be efficient, discriminative, and relatively "gentle." (see for instance Hunt 1991, Lubman 1990 or Nogar 1985) Second, the spectral content of the postionization spectrum can provide valuable information to mass spectrum which are always crowded and complex (Hunt 1991, Lubman 1990 or Lustig 1991). Finally the technique can be coupled with a variety of desorption techniques including sputtering, electron stimulated desorption and laser ablation. For molecules of intermediate mass, laser postionization can provide a unique method in the analysis of complex samples.

For high mass molecules laser multiphoton ionization appears to be more difficult. Recent studies suggest, however, that the cross-section for multiphoton laser ionization decreases with increasing mass (Campbell 1991, Schlag 1992, Wurz 1991a,b,1992) . When "large" molecules are photoionized, the ionization process does not necessarily proceed by direct, prompt electronic excitation, but rather involves extensive, rapid internal conversion. The physical process of ejecting an electron from these molecules containing enormous vibrational degrees of freedom is then similar to thermionic emission from solids. In this view rapid excitation by many photons leads to molecular heating. The molecule then "cools" through the ejection of electrons and/or molecular fragments. Thus multiphoton laser postionization of large molecules may not be possible for labile molecules.

2. EXPERIMENTAL

Results from two experimental apparatuses will be discussed in what follows. Because each apparatus has been discussed in detail elsewhere, they will only be briefly described here. The details of the coherent light generation can also be found in these publications.

The first apparatus is a time of flight (TOF) mass spectrometer (see figure 1) which utilizes laser desorption from a sample surface to introduce surface molecules into the gas phase. Following postionization, the photoions are accelerated to 8 KV in one or two steps and then traverse a field free region striking an ion detector. Details of the apparatus may be found in several publications (Hunt 1991, Lykke 1991,1992)

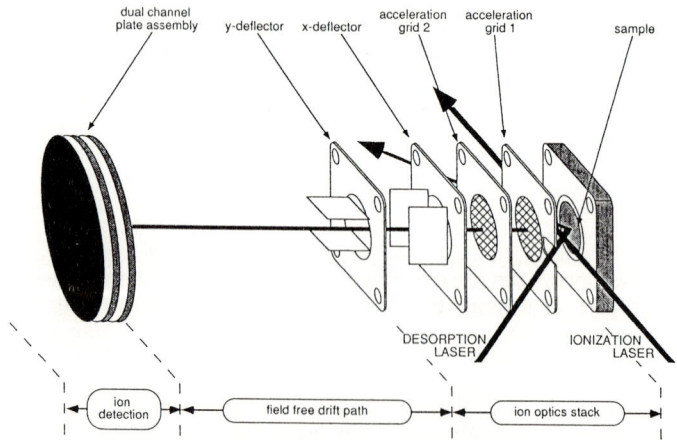

Figure 1. Schematic of the Laser Desorption Time of Flight apparatus. The system is in a ultra high vacuum apparatus which possesses a sample load lock and fused silica windows.

The second apparatus is a laser desorption fourier transform mass spectrometer (FTMS) consisting of a three region vacuum chamber with each chamber separated by differential pumping apertures. This system allows sample introduction through a vacuum interlock. The vacuum system is on a moveable cart allowing the sample and analyzer cell to be placed in the bore of a 7 T superconducting magnet. The FTMS experiments were performed with an Ionspec Omega data acquisition system. RF chirp excitation is used to accelerate photions into cyclotron orbits inside the cell. Detection of the image charge generated by this coherent motion on the cell walls can be converted to a mass scale. The long transients achievable in this system allows measurements with extremely high mass resolution. A complete description of this apparatus may be found in Parker et al 1992. A diagram of the appartus may be found in figure 2.

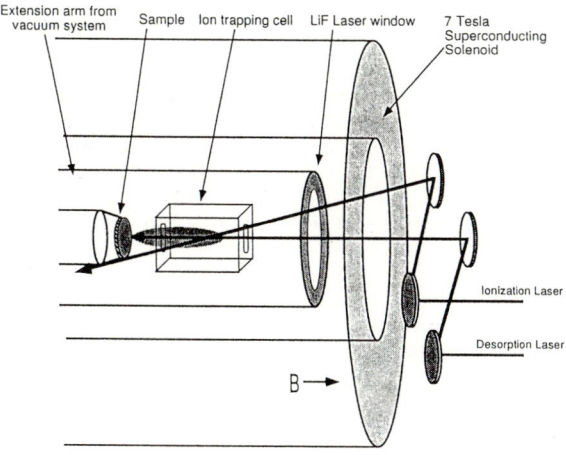

Figure 2. Schematic of the Laser Desorption FTMS apparatus showing the position of the analyzer cell in the 7 T superconducting magnet. The laser entrance port allows access both by a desorption laser and by a photoionization laser.

The principal advantage of FTMS is its ability to achieve extremely high mass resolutions (see for example Li 1990 and references therein). FTMS is a particularly attractive method when postionization is being considered because the mass resolution is not in first order affected by

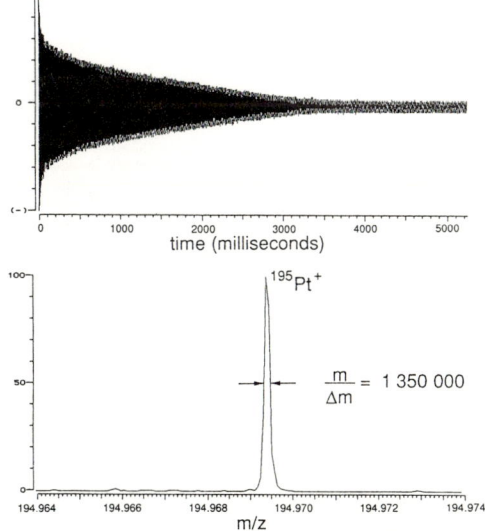

Figure 3. FTMS spectrum of a laser desorbed Pt isotope. The mass resolution displayed exceeds 1.3×10^6.

the formation position of the ions. Thus high mass resolution can be achieved even for large ionization volumes. Figure 3 displays the mass resolution achievable for a Pt isotope. The top panel of figure 3 shows the time dependent image charge signal. This coherent transient lasts nearly three seconds. Fourier transform of this signal yields the mass spectrum shown.

3. RESULTS

Commercially available rubbers are complex mixtures of polymer molecules and various additive molecules. These additive molecules impart certain desirable characteristics of the polymer (Latimer 1989,1988,1986). Detection of polymer additives has been difficult due to the complexity of the mass spectrum. Figures 4 and 5 display laser desorption followed by subsequent laser postionization TOF mass spectra for a vulcanizate rubber sample. While a more complete description of this work can be found elsewhere (Hunt 1991, Lykke 1992) the two mass spectra are illustrative of both the difficulties and the promise of this type of analysis.

Figure 4 shows a mass spectrum utilizing 308 nm desorption and 118 nm ionization. Since 118 nm light is sufficient to ionize in a one photon process all of the molecules in the desorbing flux, this spectrum is representative of the molecules present in the desorbing flux. This flux is dominated by low molecular weight fragments of the rubber polymer and demonstrates some of the difficulties in the desorption step.

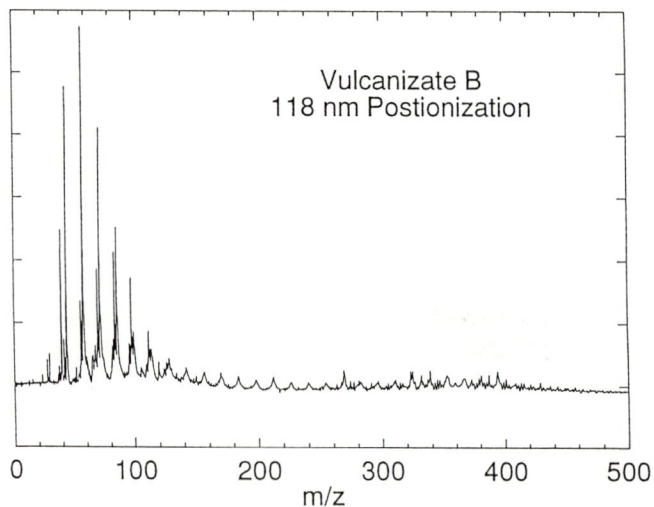

Figure 4. Postionization of a vulcanizate rubber using 118 nm (10.5 eV) radiation. The low mass region represents fragments of the rubber polymer backbone. These molecules appear to dominate the desorbing flux.

Figure 5 shows a wide variety of different ionization experiments following 308 nm desorption. Four different laser wavelengths have been used for postionization - 355 nm, 308 nm, 266 nm, and 212 nm. Even with such a crude spectral analysis it is possible to use

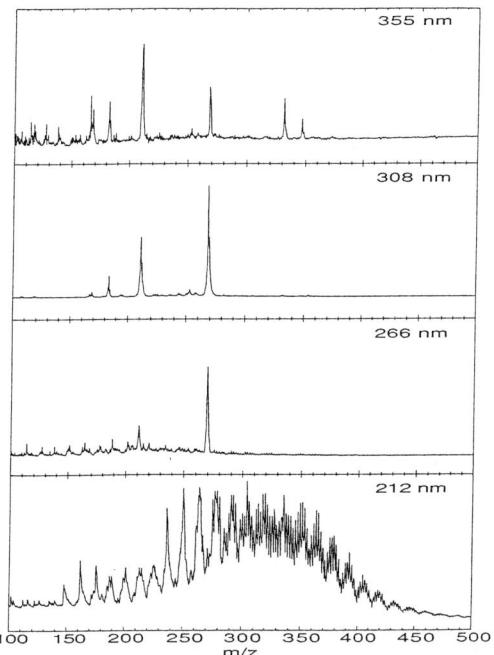

Figure 5. Postionization of a vulcanizate rubber using 118 nm (10.5 eV) radiation. The low mass region represents fragments of the rubber polymer backbone.

this laser wavelength information to gain insight into the molecular content of the rubber surface(Lubman 1990, Lustig 1991). Different laser wavelengths access different polymer additives. Radiation at 355 nm selectively ionizes additives that contain aromatic groups, while 212 nm light tends to ionize all but the small ablation fragments of the polymer.

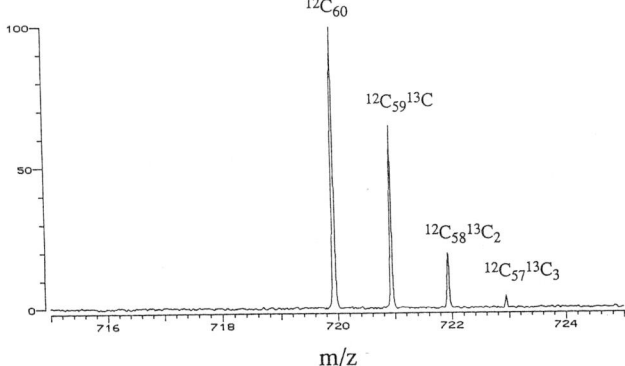

Figure 6. FTMS spectrum of a positive ions of C_{60} produced in the laser desorption process. Clearly evident is the isotopic content of the molecule.

For larger molecules intersystem crossing and subsequent fragmentation make laser multiphoton ionization difficult. In these cases, two schemes are viable. First as in figure 4 one photon ionization can be an effective tool. A second alternative is to use the ions produced in the desorption process itself. An example of the direct ionization of a large molecule is displayed in figure 6. The positive ion of C_{60} is cleanly displayed. Note the isotopic content of the molecule is in accordance with the natural abundance of C.

4. CONCLUSIONS

Laser ionization of desorbed molecules is a sensitive discriminative means for the analysis of molecules on surfaces. Significant work on the understanding of both the desorption and the ionization process of molecules remains, however, before this can be considered a viable method for quantitative surface analysis.

5. REFERENCES

Asamoto B, Young J R and Citerin R J 1990 *Anal. Chem.* **62**, 61-70.
Becker C H, Jusiniski L E, and Moro L 1990 *Int. J. Mass Spect. Ion Proc.* **95** R1-R4.
Campbell E E B, Ulmer G, and Hertel I V 1991 *Phys. Rev, Lett.* **67** 1986-1988.
Feldmann D, Kutzner J, Laukemper J, MacRobert S, and Welge K H 1987 *Appl Phys B* **44** 81-85.
Gardella J A, Pireaux J-J 1990 *Anal. Chem.* **62**, 645A-660A.
Grotomeyer J,Boesl U, Walter K, and Schlag E W, 1986 *J. Amer. Chem. Soc.* **108** 4233.
Huang L Q, Conzemius R J, Junk G A, and Houk R S 1988 *Anal. Chem.* **60**, 1490-1494.
Hunt J E, Lykke K R, and Pellin M J 1991 *Methods and Mechanisms for Producing Ions from Large Molecules* (Plenum Press, New York) pp. 309-314.
Johlman C L , Wilkins C L, Hogan J D, Donaovan T L, Laude D A, and Youssefi M-J 1990 *Anal. Chem.* **62**, 1167-1172.
Lattimer R P, Harris R E, and Rhee C K, 1986 *Anal. Chem.* **58**, 3188.
Lattimer R P 1988 *Rubber Chem Tech.* **62** 548.
Lattimer R P and Harris R E 1989 *Rubber Chem Tech.* **62** 548.
Lazare S, and Granier V 1989 *Laser Chem* **10** 24.
Li Y, McIver R T, and Hemminger J C 1990 *J. Chem. Phys.* **93** 4719.
Lubman D M, 1990, *Lasers and Mass Spectrometry* (Oxford University Press, New York) pp 1 - 545.
Lustig D A, and Lubman D M 1991 *Int. J. Mass Spect. Ion Proc.* **107** 265-280.
Lykke K R, Pellin M J, Wurz P, Gruen D M, Hunt J E, and Wasielewski 1991 *Mater. Res. Soc. Proc. Symp.* **206** 679.
Lykke K R, Parker D H, Wurz P, Hunt J E, Pellin M J, Gruen D M, Hemminger J C, and Lattimer R P 1992 submitted to Anal Chem.
Nogar N S, Estler R C and Miller C M *Anal. Chem.* **57**, 2441-2444.
Parker D H, Chatterjee K, Wurz P, Lykke K R, Pellin M J, Stock L M, and Hemminger J C 1992 accepted for publication in *Carbon*.
Srinivasan R, and Braren B 1989 *Chem. Rev.* **89** 1303.
Wurz P, Lykke K R, Pellin M J and Gruen D M 1991a *J. Appl. Phys.* **70** 6647-6652.
Wurz P, and Lykke K R 1991b *J. Chem. Phys.* **95** 7008-7010.
Wurz P, and Lykke K R 1992*J. Chem. Phys.* submitted.
Schlag E W, Grotomeyer E, and Levine R D 1992 *Chem. Phys. Lett.* **190** 521-527

Inst. Phys. Conf. Ser. No 128: Section 5
Paper presented at RIS 92, Santa Fe, NM, USA, 24–29 May 1992

173

Nitric oxide containing heteroclusters probed by two-photon ionization in a supersonic expansion

Sunil Desai,[*] C. S. Feigerle,[*] and John C. Miller

Oak Ridge National Laboratory, P.O. Box 2008, Oak Ridge, TN 37831-6125

ABSTRACT: Mixed clusters of the form $(NO)_m Ar_n (m \leq 8, n \leq 54)$ are produced in a supersonic expansion, photoionized by nonresonant absorption of 266 nm or 532 nm photons and detected by time-of-flight mass spectrometry. Anomalously large relative intensities are observed for the cluster ions $NO^+ Ar_n (n = 12,18,22,54)$ and $(NO)_2^+ Ar_n$ $(n = 17,21)$ and are attributed to extra stability of these ions. These "magic numbers" at $(m + n) = 13,19,23,55$ are compared to those observed in rare gas clusters and other $M^+ Ar_n$ heteroclusters and assigned to icosahedral structures.

Upon doping the NO/Ar expansion with small amounts of N_2O, NO_2, CO_2, and H_2O, other mixed clusters can be synthesized. Various cluster ions of the form $(NO)_m^+ (N_2O)_n$, $(NO)_m^+ (NO_2)_n$, $(NO)_m^+ (H_2O)_n$, $H_3O^+ (H_2O)_n$, $(NO)_m^+ (CO_2)_n$, and $(NO_2)_m^+ Ar_n$ have been observed. In $NO/N_2O/Ar$ expansions the cluster ion series $(NO)_3^+ (N_2O)_n$ $(n = 0$ -12$)$ is considerably more abundant than clusters of N_2O about nitric oxide monomers, dimers, or tetramers suggesting an unusually stable structure for the trimer ion cluster.

1. INTRODUCTION

With the marriage of the tunable laser and the pulsed supersonic nozzle, the spectroscopic study of weakly bound molecules and clusters has accelerated explosively during the last ten years. The understanding of cluster structure, bonding, and chemistry has many practical implications in such diverse fields as catalysis, semiconductor processing, and atmospheric modeling. In particular, related to the latter subject, ionic van der Waals molecules such as $(NO)_2^+$, $NO^+ \bullet N_2$, $NO^+ \bullet CO_2$, $NO^+ \bullet H_2O$, $(O_2)_2^+$, $O_2^+ \bullet H_2O$, etc. have been observed in the atmosphere in rocket-borne mass spectrometer studies. The nitric oxide-containing clusters have been particularly important in the so-called D-region where solar Lyman-α radiation creates NO^+ ions which convert into water cluster ions via a series of ligand exchange reactions (Ferguson *et al* 1979).

In the work described here, we will present experiments on laser ionization of clusters of nitric oxide with rare gases and with other species of atmospheric importance such as N_2O, NO_2, CO_2, and H_2O. Major emphasis will be placed on the cluster distribution and the identification of so-called "magic numbers." These are clusters of a certain magic size which seem to be especially stable due to structural, electronic, or dynamic factors.

2. EXPERIMENTAL

The present apparatus has evolved in several stages from that used in our earlier studies of ultracold molecules and clusters which employed a nanosecond tunable dye laser, free jet expansion, and low resolution time-of-flight (TOF) mass spectrometer (Miller and Compton 1986, Feigerle and Miller, 1989). At an intermediate stage, a new apparatus was constructed with a picosecond laser providing resonant and nonresonant ionization (Smith and Miller 1989, 1990). The present version (Desai *et al* 1992) employs a differentially pumped source and detection region, a skimmed supersonic jet, and an improved mass spectrometer (resolution ~ 300). Typically two- or three-photon ionization is effected with either 266 nm (~ 4 mj) or 532 nm (~ 30 mj) pulses of 30 ps duration. The pulsed valve and mass spectrometer are of conventional design, and details may be found in the papers cited above.

3. XENON CLUSTERS

Since the first observation of magic numbers in rare gas cluster distributions, researchers have been intrigued with the possible reasons for the extra stability of certain sized clusters (Echt 1981). A model has evolved which involves assigning the magic numbers to especially symmetric structures of the ionized cluster and recognizing that extensive fragmentation occurs during the ionization process (Haberland 1985, Klots 1991). Magic numbers have been identified in cluster distributions of many atomic and molecular species including the now famous Buckyball or C_{60}.

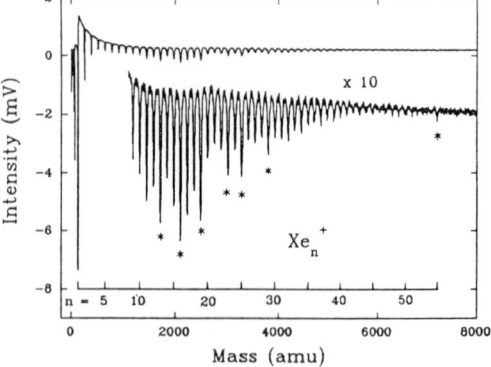

Fig. 1. Xe expansion; $\lambda = 266$ nm.

Figure 1 shows the distribution of Xe_n^+ clusters observed on the present apparatus following three-photon ionization. Magic numbers, marked with asterisks, are observed for clusters with n = 13, 16,19,23,25,29 and 55. The most prominent of these correspond to highly symmetric icosohedral structures. Calculations (Böhmer and Peyerimhoff 1987) indicate that Xe_{13}^+ and Xe_{19}^+ are especially stable when arranged as an icosahedron and double icosahedron respectively about a Xe^+ ion. Xe_{54}^+ arises after completing another solvation sphere. A similar spectrum, obtained with nanosecond laser pulses, has been published previously (Echt *et al* 1987).

4. NITRIC OXIDE - RARE GAS CLUSTERS

Small clusters composed of nitric oxide and various rare gas partners have been studied in our labs previously (Miller and Cheng 1985, Miller 1987, 1989, Smith and Miller 1989, 1990). Recently we have been able to synthesize very large clusters of the form $(NO)_m(Ar)_n$ (Desai *et al* 1992). The range of the indices m $(1-8)$ and n$(1-54)$ depend on the relative concentrations and expansion conditions.

Figure 2 shows a spectrum of NO^+ $Ar_n(n \leq 54)$ following 532 nm ionization. Since the ionization potential of NO is lower than that of Ar, one expects structures where the charge is localized on the NO^+ core, and the argon atoms arrange themselves in solvent spheres. The magic numbers for n = 12,18,22, and 54 indicate that icosohedral structures are especially stable, similar to rare gas clusters. In separate experiments, ionizing with 266 nm photons, the solvated nitric oxide dimer $[(NO)_2^+Ar_n]$ magic num-

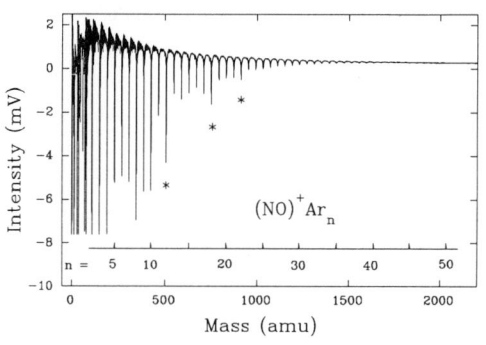

Fig. 2. NO/Ar expansion: $\lambda = 532$.

bers are only observed for n = 17 and 21. Apparently, the dimer is too big to be stabilized within a single icosahedron. A double icosahedron, however, has two central sites and can accommodate the dimer (Desai *et al* 1992).

5. MIXED MOLECULAR CLUSTERS

By expanding three-component gas mixtures, we have observed an array of mixed molecular cluster ions such as $(NO)_m^+(N_2O)_n$, $(NO)_m^+(NO_2)_n$, $(NO)_m^+(CO_2)_n$, and $(NO)_m^+(H_2O)_n$. These ions are important in stratospheric photochemical cycles as mentioned in the introduction.

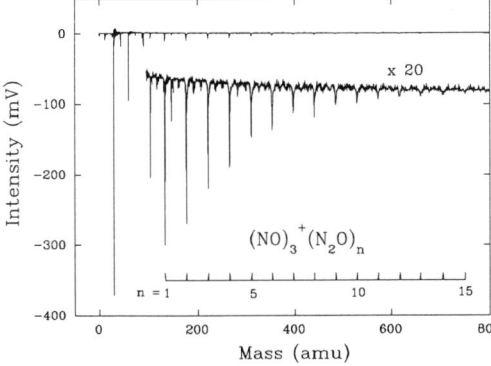

Fig. 3. NO/N₂O/Ar expansion; $\lambda = 266$ nm.

Figure 3 shows a particularly interesting distribution of $(NO)_m^+(N_2O)_n$ clusters. The clusters $(NO)_3^+(N_2O)_n$ form the prominent series and extend to about n = 15. Other series, such as $(NO)_1^+(N_2O)_n$ and $(NO)_2^+(N_2O)_n$, are observed but are considerably weaker and have a lesser extent. Apparently, the $(NO)_3$ or $(NO)_3^+$ has some special stability when bound to N_2O clusters. $(NO)_3^+(CO_2)_n$ and $(NO)_3^+(H_2O)_n$ clusters similarly appear to be more stable than clusters with other numbers of nitric oxide molecules. It should be noted that in $(NO)_m^+Ar_n$ clusters the trimer containing species are not especially prominent.

Again the magic-sized cluster may be unusually stable due to several factors. For example, since NO has an unpaired electron, all odd NO clusters have an unpaired electron available for chemical bonding. However, once ionized, the even NO clusters have an unpaired electron. Alternately, the nitric oxide trimer (or its ion) may be able to form some uniquely stable structure when bound to N_2O. For example, Jones (1992) has performed MNDO studies which indicate that an $(NO)_3$ species may be cyclic. Further experiments are planned to further illuminate the structure of these interesting species.

6. ACKNOWLEDGMENTS

This research was sponsored by the Office of Health and Environmental Research, U.S. Department of Energy under contract DE-AC05-84OR21400 with Martin Marietta Energy Systems, Inc. John C. Miller acknowledges use of a NATO International Collaboration Grant (0474/87) during the course of this work.

7. REFERENCES

Böhmer H U and Peyerimhoff S D 1988 *Z. Phys. D* **8** 91
Desai S, Feigerle C S and Miller JC 1992 *J. Chem. Phys.* **97** xxxx
Etch O, Cook M C and Castleman A W Jr 1987 *Chem. Phys. Lett.* **135** 229
Echt O, Settler K and Recknagel E 1981 *Phys. Rev. Lett.* **47** 1121
Feigerle C S and Miller J C 1989 *J. Chem. Phys.* **90** 2900
Ferguson E E, Fehserfeld F C and Albritton D L 1979 in *Gas Phase Ion Chemistry* ed
    M T Bowers (Academic Press, New York) vol I p 45
Haberland H 1985 *Surf. Sci.* **156** 305
Jones W H 1991 *J. Phys. Chem.* **95** 2588
Klots C E 1991 *Z. Phys. D.* **20** 1001
Miller J C 1987 *J. Chem. Phys.* **86** 3166
Miller J C 1989 *J. Chem. Phys.* **90** 4031
Miller J C and Cheng W C 1985 *J. Phys. Chem.* **89** 1647
Miller J C and Compton R N 1986 *J. Chem. Phys.* **84** 675
Smith D B and Miller J C 1989 *J. Chem. Phys.* **90** 5203
Smith D B and Miller J C 1990 *J. Chem. Soc. Faraday Trans.* **86** 2441

*The University of Tennessee, Knoxville, Tennessee 37996

Inst. Phys. Conf. Ser. No 128: Section 5
Paper presented at RIS 92, Santa Fe, NM, USA, 24–29 May 1992

A systematic study of the resonance ionisation spectra of Br_2, I_2, IBr and ICl

R J Donovan,   R Flood,   J Goode,   K P Lawley,   R Maier,
T Ridley and A J Yencha*

Department of Chemistry, University of Edinburgh, West
Mains Road, Edinburgh, EH9 3JJ, Scotland, U.K.

ABSTRACT: A systematic study of the Rydberg and ion-pair
states of Br_2, I_2, IBr and ICl, observed using resonance
ionisation techniques, is presented. We show that for
Br_2 and I_2 Rydberg states dominate spectra obtained by
coherent two-photon excitation, whilst ion-pair states
dominate optical double resonance spectra. In contrast
ion-pair states are observed both in the coherent
two-photon excitation and optical double resonance
spectra of IBr and ICl.

1. INTRODUCTION

We have recently carried out systematic studies of the
Rydberg states of Br_2 and I_2 using resonance enhanced (2+1)
multiphoton ionisation in the region 48,000-85,000 cm^{-1}.
This systematic study has allowed numerous Rydberg series
to be identified and some previously observed series to be
reassigned (Ridley *et al* 1990, Donovan *et al* 1992).
Surprisingly, transitions to ion-pair states are absent or
extremely weak.

In the work presented here, we show that the resonance
enhanced multiphoton ionisation spectra of IBr and ICl are
dominated by ion-pair transitions, with Rydberg transitions
being relatively weak. Vibrationally resolved spectra of
these ion-pair states have been obtained using
jet-cooling.

We also present some very recent results obtained using
optical double resonance with jet-cooling. This technique
strongly favours the observation of ion-pair states for
both homonuclear and heteronuclear halogen molecules.

*Permanent address: Department of Chemistry and Department
of Physics, State University of New York at Albany,
Albany, N.Y. 12222, U.S.A.

2. EXPERIMENTAL

The experimental arrangements for resonance enhanced multiphoton ionisation and optical double resonance have been described previously (Ridley *et al* 1990, Holmes *et al* 1991). In addition to these techniques we have combined optical double resonance with jet-cooling for some of the work presented here. A commercial pulsed valve (General Valve, nozzle diameter $250\mu m$) was directed into a vacuum chamber, pumped by a 1500 ls^{-1} oil diffusion pump fitted with a liquid nitrogen trap. A background pressure ⟨ 5×10^{-4} Torr, was obtained with the valve running. Typical backing pressures of 0.5 atm. were used with a gas mixture of *ca* 0.2-1% halogen in argon. Wavelength calibration was achieved using the I_2 atlas. Ionisation was detected using a pair of parallel plate electrodes positioned on either side of the jet and separated by 2 cm.

3. (2+1) RESONANCE IONISATION

Spectra of I_2 and Br_2 were obtained using (2+1) resonance enhanced multiphoton ionisation (REMPI) in a static gas cell at 300 K. Numerous strong Rydberg transitions were observed in the region between the energetic threshold for (2+1) ionisation and the molecular ionisation limit. The detailed results and an analysis of the Rydberg systems observed have been presented elsewhere (Ridley *et al* 1990, Donovan *et al* 1992). Here we emphasize that the spectra recorded at 300 K showed no clear evidence for the presence of ion-pair transitions. However, with jet-cooling a weak ion-pair system, adjacent to the $[^3/_2]5d$ Rydberg system, could be resolved. The spectrum is rather unusual for an ion-pair system, in being limited to a narrow wavelength region. The appearance of this system and its proximity to Rydberg transitions suggests that it has gained intensity from mixing with the close lying $[^3/_2]5d$ Rydberg state. The assignment of the ion-pair state and the details of the Rydberg/ion-pair mixing will be studied in detail in future work.

The analogous (2+1) REMPI spectrum of ICl, obtained using jet-cooling, is shown in Figure 1. The spectrum is dominated by an extended ion-pair system which is readily assigned to, $E0^+(^3P_2) \leftarrow X0^+$. Some narrow Rydberg transitions of similar intensity are also observed.

Similar results have also been obtained for IBr, $E0^+(^3P_2) \leftarrow X0^+$. Again, jet-cooling was essential in order to resolve vibrational structure. The reason that the $E0^+$ ion-pair states of ICl and IBr are observed, whilst the analogous states of I_2 and Br_2 are not observed with (2+1) REMPI, is straightforward. The $E0^+$ ion-pair state has the electronic configuration $\sigma 1_\pi 4_\pi *4_\sigma *1$. This configuration has *ungerade* symmetry for the homonuclear halogens and thus the analogous transitions are forbidden using (2+1) REMPI. However, the question remains as to why none of the other ion-pair states (there are 20 such states in total) are observed using (2+1) REMPI. We have tentatively suggested

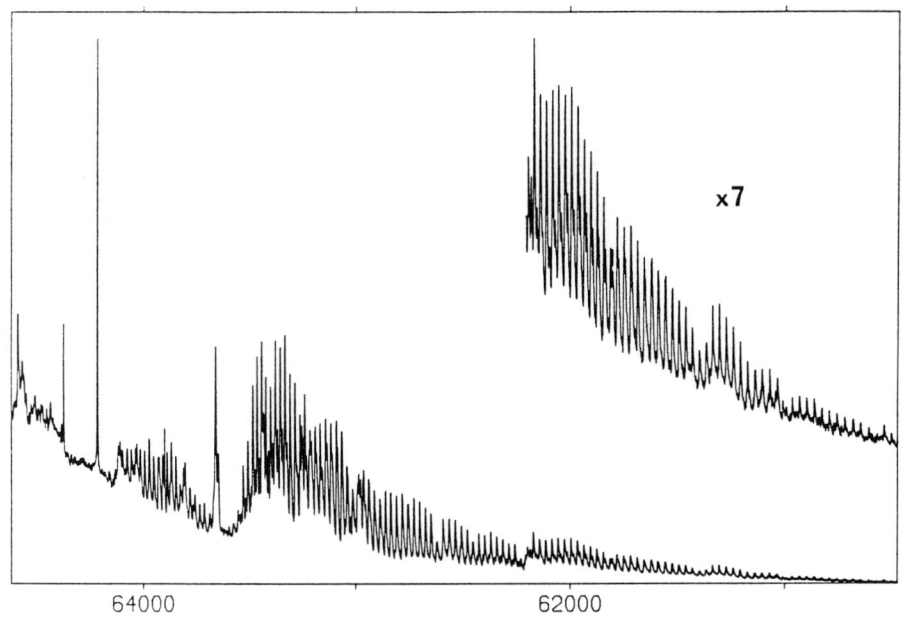

Figure 1. (2+1) REMPI spectrum of jet-cooled ICl.
A long progression in the $E0^+(^3P_2)$ ion-pair state
dominates the spectrum (v'=254-384)

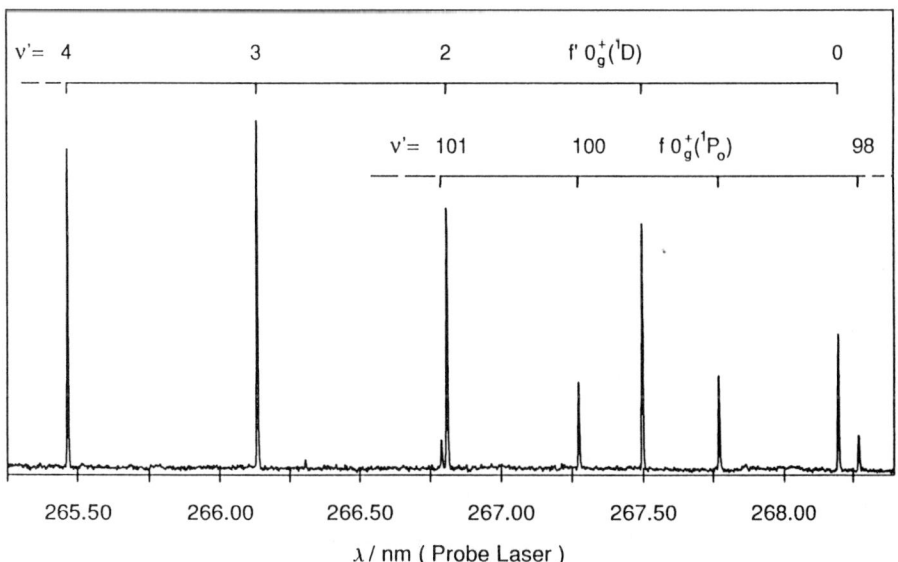

Figure 2. Jet-cooled spectrum of I_2 obtained using
optical double resonance with resonance ionisation

that this is due to massive predissociation in the Franck
Condon region, but this requires further confirmation.

4. OPTICAL DOUBLE RESONANCE

It is now well established that optical double resonance
techniques give ready access to the manifold of ion-pair
states (Holmes *et al* 1991, Ishiwata *et al* 1992).   Ion-pair
excitation spectra are normally monitored by observing
fluorescence but we have shown that resonance ionisation
techniques can also be used.   By using jet-cooling in
conjunction with optical double resonance and resonance
ionisation detection, we have been able to achieve a marked
increase in sensitivity and quality of the observed
spectra.   An example is shown   in Figure 2 where the
$f'0_g^+(^1D)$ ion-pair state of I_2 is observed.   The following
excitation scheme was used.

$$f'0_g^+(^1D), v'=0\text{-}18 \longleftarrow B0_u^+, v=14,22 \longleftarrow X0_g^+, v''=0$$

The observed spectra are very sharp (almost atom like) and
of high quality.   The assignment of the vibrational
progression   is straightforward and the origin is readily
identified from the dispersed fluorescence.   Only a brief
report on high vibrational levels of this state has been
given previously (Hoy and Taylor 1987).   Using the $B0_u^+$
state as intermediate we have also been able to observe the
$E0_g^+(^3P_2)$ and $f0_g^+(^3P_0)$ ion-pair states of I_2.

Similar spectra have also been observed for ICl and we have
identified the $\beta1(^3P_2)$, $G1(^3P_1)$, $H1(^1D)$ $E0^+(^3P_2)$, $f0^+(^3P_0)$
and $f'0^+(^1D)$ ion-pair states in this way.

5. CONCLUSION

Optical double resonance with jet-cooling and resonance
ionisation provides a powerful new approach to ion-pair
spectroscopy which will greatly facilitate future studies.
The question as to why so few ion-pair states are observed
using (2+1) REMPI spectroscopy remains open.

6. REFERENCES

Donovan R J, Flood R V, Lawley K P and Ridley T 1992
   Chem. Phys. (in press)

Holmes A J, Lawley K P, Ridley T, Donovan R J and
   Langridge-Smith P R R 1991 *J. Chem. Soc. Faraday Trans.*
   87 15

Hoy A R and Taylor A W 1987 *J. Mol. Spec.* 126 484

Ishiwata T, Osamu N and Obi K 1992 *J. Mol. Spec.* 151 513

Ridley T, Lawley K P, Donovan R J and Yencha A J 1990 *Chem.
   Phys.* 148 315

Inst. Phys. Conf. Ser. No 128: Section 5
Paper presented at RIS 92, Santa Fe, NM, USA, 24–29 May 1992

Laser induced fragmentation processes in nitrobenzene

A Marshall, A Clark, R Jennings, K W D Ledingham, J Sander and R P Singhal
Department of Physics & Astronomy, University of Glasgow, Glasgow G12 8QQ, Scotland

ABSTRACT: Laser induced ionisation and fragmentation has been observed in a number of nitroaromatic species. In an attempt to understand the processes taking place in the laser / molecule interaction, we have conducted a number of experiments on the nitrobenzene molecule. Time-of-flight (TOF) mass spectrometry and wavelength dependence studies of fragment ion yields (especially the NO^+ fragment) suggest an initial predissociation into neutral fragments followed by multiphoton ionisation and further fragmentation.

1 INTRODUCTION

The study of aromatic molecules by multiphoton absorption has been an area of continuing interest over many years (Johnson (1975), Zhu *et al* (1990)). In particular, a large degree of similarity has been observed in the wavelength dependence of both parent and fragment ions for several molecular species (Zandee and Bernstein (1979), Marshall *et al* (1991)). This suggests a fragment pathway which proceeds through the intermediate excited states of the parent molecule. It is known that in the case of nitroaromatic species, irradiation by ultra violet light leads to predissociation (Zhu *et al* (1990)), and thus any intermediate excited state features of the parent molecule would be absent from the predissociation products.

A program of wavelength dependence and TOF experiments has been carried out using nitrobenzene in order to try to understand more about the pathways followed from the parent molecule to the various fragments. Nitric oxide (NO) and nitrogen dioxide (NO_2) have also been investigated in an attempt to determine the origin of the NO^+ ion in the nitrobenzene mass spectrum.

2 EXPERIMENTAL

Figure 1 shows a schematic of the experimental arrangement. A Lumonics TE 860-M excimer laser was used to pump a Lumonics EPD330 dye laser system. With suitable dyes (coumarin 47, 102 and 307), and frequency-doubling using an Inrad auto-tracking system, tunable laser radiation in the wavelength range 224-250nm could be generated. The pulse energy could be attenuated using a Newport attenuator and the maximum pulse energy was ~1mJ for the coumarin 102 dye.

Sample vapour was admitted into the high vacuum chamber through a needle valve controlled capillary line. The base pressure was ~5×10^{-8}mbar and sample pressures of ~10^{-6}mbar and ~10^{-5}mbar for nitroaromatics and NO and NO_2, respectively, were required for clear signals. Samples were used as received from suppliers without further preparation.

Ions produced in the laser interaction were extracted and separated in a 1.2m TOF mass spectrometer with a resolution ~220 at m/z 77. Detection was via a Thorn EMI electron multiplier, and both ion and pulse energy signals were recorded through an 11-bit ADC on a CAMAC controlled LSI-11 data recording station.

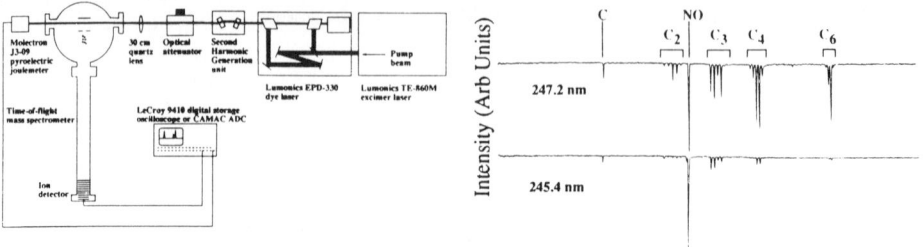

Fig 1 Experimental configuration

Fig 2 TOF spectra of nitrobenzene at 247.2 and 245.4nm

3 RESULTS AND DISCUSSION

Laser ionisation of nitrobenzene has already been carried out at the fixed wavelength of 248nm using a quadrupole mass spectrometer (Apel and Nogar (1988)). One of the advantages of a TOF system is that a complete spectrum of ions can be recorded for every laser pulse. Figure 2 shows the TOF spectrum recorded at two distinct wavelengths (247.2 and 245.4nm). At 245.4nm the m/z 30 fragment shows a marked enhancement relative to the other fragment intensities. This fragment is the nitric oxide cation (NO^+), and the other fragment groups correspond to hydrocarbon clusters. When the laser is scanned over the wavelength region 245-250nm and the wavelength dependent ionisation yield of various fragment ions is recorded (Figure 3), the difference between the NO^+ and hydrocarbon ions is obvious. The resonance features observed in the NO^+ spectrum are identified as originating from the rovibrational structure in the excited electronic state ($A^2\Sigma$ (v=0) ← $X^2\Pi_{1/2,3/2}$ (v=2)) of the neutral NO molecule. Enhancements observed in the hydrocarbon spectra are not evident in the NO^+ spectrum, which suggests that two separate pathways have arisen by this stage of the absorption mechanism: one involving the hydrocarbon ring, the other the nitro group.

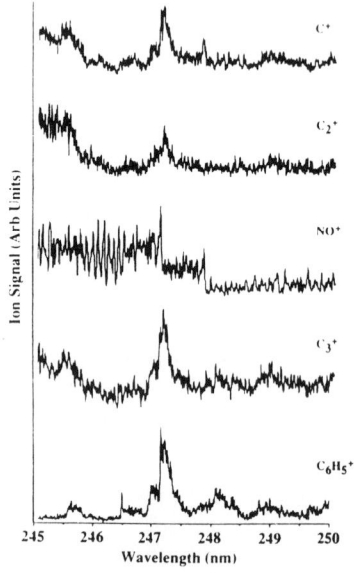

Fig 3 Wavelength dependence of some fragment ions of nitrobenzene

The hydrocarbon data suggests that, after the initial predissociation involving the nitro group has taken place, there is some quasistable structure left which can absorb additional photons and undergo further fragmentation. This is supported by the observation that the highest mass fragment (m/z 77) has the same enhancements as the lowest mass fragment (m/z 12).

In order to try to more fully understand the precise mechanisms taking place in the break -up of the nitrobenzene molecule, an extensive study of the nitrobenzene NO^+ ionisation yield was undertaken. This was compared with the NO^+ yield from NO and NO_2 bulk gases, since the two most probable routes to NO^+ formation in nitrobenzene are:

$$M + h\nu \rightarrow (M\text{-}NO) + NO \qquad \text{Option 1}$$

and $\qquad M + h\nu \rightarrow (M\text{-}NO_2) + NO_2 \qquad$ Option 2

(the rovibrational structure in the NO^+ spectrum indicates that the pathway must pass through the neutral NO ground state at some point, ruling out NO^+ formation directly from the parent molecule).

Figure 4 shows the wavelength dependence of the NO^+ yield from nitrobenzene, NO_2 and NO over the wavelength range 224-250nm. The ratio in the NO spectrum of the $0\leftarrow0$ to $0\leftarrow1$ intensities (0 and 1 denote the vibrational levels in the ground or excited states) is ~1000:1, and the $0\leftarrow0$ feature is on a different scale from the rest of the spectrum.

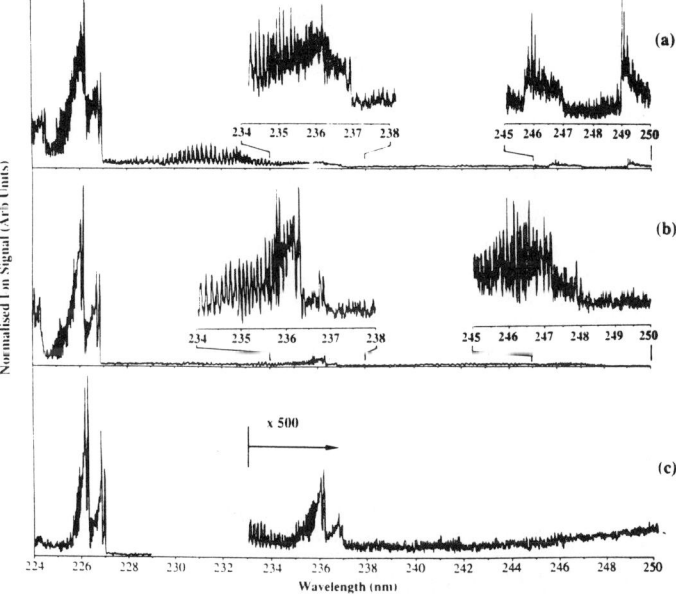

Fig 4

NO^+ ion wavelength dependence from samples of
(a) NO_2,
(b) nitrobenzene and
(c) NO,
in the wavelength range 224-250nm

Comparison of the nitrobenzene and NO_2 spectra show a number of similarities. In particular, the relative intensities of the $1\leftarrow1$, $0\leftarrow0$, $0\leftarrow1$ and $0\leftarrow2$ transitions compare well. In addition, the rotational spectra associated with these transitions show clear similarities, except for the case of the $0\leftarrow2$ transition in NO_2 where a rotational inversion seems to be taking place. On the other hand, the NO_2 band at 246nm has a much smaller intensity in nitrobenzene, while the band at 249nm does not appear at all in the nitrobenzene spectrum.

When the nitrobenzene data is compared with the NO spectrum, there does not seem to be a significant correlation. A Boltzmann distribution of population in the vibrational levels gives reduction of around three orders of magnitude in going to the next highest level. If it were the case that NO was dissociating from the nitrobenzene parent molecule, then a non-Boltzmann distribution might be expected, as there would be excess internal energy available to populate higher vibrational levels in NO.

The internal energy argument could also be used to explain the absence of the NO_2 249nm band from the nitrobenzene spectrum. This band is due to the B(2)←X(4) transition (Ledingham *et al* (1992)) following dissociation of NO_2 below 5.08eV (=244nm). The v=4 level is occupied because of a population inversion in the particular dissociation channel followed. NO_2 predissociating from the parent molecule could be expected to gain internal energy which would open an alternative dissociation channel above 5.08eV which does not display this inversion.

Neither argument accounts for the absence of the band at 228-234nm which is due to the A(1)←X(2) and A(2)←X(3) transitions in neutral NO.

The most illuminating data for determining the pathways would seem to be that of the TOF spectra. Figure 5 compares the TOF spectrum of NO_2 with that of nitrobenzene. The NO_2 signal shows a strong peak at the NO^+ position, accompanied by a much smaller peak at the NO_2^+ position. Close inspection of the nitrobenzene spectrum shows the presence of an NO_2^+ peak in approximately the same ratio to the NO^+ peak as in the case of NO_2 (Ledingham *et al* (1992)).

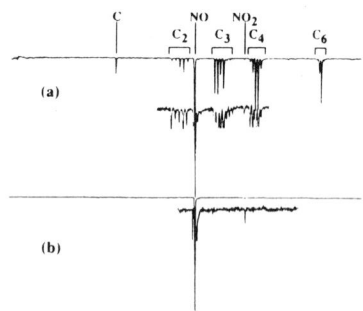

Fig 5 TOF spectra of nitrobenzene and $NO2$ showing NO^+ / $NO2^+$ ratios

4 CONCLUSION

On exposure to UV laser radiation, the nitrobenzene molecule predissociates to form a predominantly hydrocarbon structure plus a nitro derived group. The evidence of the NO^+ / NO_2^+ ratio plus the highest recorded mass being m/z 77 suggest that the NO_2 group predissociates from the parent molecule and then undergoes subsequent predissociation and ionisation to form NO^+, while the hydrocarbon structure absorbs further photons and then fragments to give a range of fragment ions.

5 ACKNOWLEDGEMENTS

This work has been carried out with the support of the Procurement executive, Ministry of Defence. AC and JS are indebted to the SERC for postgraduate funding.

6 REFERENCES

Apel E C and Nogar N S, Int. J Mass Spec. and Ion Proc. **70** (1986) 243
Johnson PM, J. Chem. Phys. **64** (1976) 4143
Ledingham K W D, Marshall A, Clark A, Jennings R, Sander J and Singhal R P, (1992)
        To be published
Marshall A, Clark A, Jennings R, Ledingham K W D and Singhal R P, Meas. Sci. and Tech.
        2 (1991) 1078
Zandee L and Bernstein R B, J. Chem. Phys. **71** (1979) 1359
Zhu J, Lustig D, Sofer I and Lubman D M, Anal. Chem. **62** (1990) 2225

Inst. Phys. Conf. Ser. No 128: Section 5
Paper presented at RIS 92, Santa Fe, NM, USA, 24–29 May 1992

Resonance ionization spectroscopy of the $E, F^1\Sigma_g^+$ state of H_2

Michael P. McCann, Chung-hsuan Chen, and Marvin G. Payne

Department of Chemistry, University of Central Florida, Orlando, FL 32816-0366

ABSTRACT: A (2+1) resonance ionization scheme was used to ionize H_2. A fixed VUV photon (118 nm) and tunable visible light excited ground state hydrogen $(E, F^1\Sigma_g^+ \leftarrow\leftarrow X^1\Sigma_g^+)$ and a residual UV photon (355 nm) ionized the molecule. From the ion signals and the laser pulse energies, the two-photon rate constants were calculated. Tunneling of a vibrational level in the outer well of the $E, F^1\Sigma_g^+$ potential into the inner well was observed.

1. INTRODUCTION

Resonance ionization spectroscopy (RIS) schemes and resonance-enhanced multi-photon ionization (REMPI) schemes frequently employ a step involving the absorption of two photons simultaneously. The use of lasers makes such nonlinear absorption schemes not only possible but the process can be easily saturated. There is a good deal of motivation for employing these two-photon steps. In atoms or molecules with a center of symmetry (such as H_2) one-photon and two-photon allowed transitions are mutually exclusive. The ground state to excited state transition $(E, F^1\Sigma_g^+ \leftarrow\leftarrow X^1\Sigma_g^+)$ examined in this study is not allowed by a one-photon transition. Thus two-photon absorptions open up excited states that were previously difficult to study. Another reason for employing various non-linear schemes is to study excited states at high energy. Molecular hydrogen is transparent in the infrared, visible, and ultraviolet region of the electromagnetic spectrum. Thus light sources in the above regions were useless for study of atoms or molecules with the lowest-lying electronic state at very high energies.

There are a few interesting aspects of this study. We have examined an excited state that has become quite popular because it is the lowest one-photon forbidden electronic state in the simplest neutral molecule (Senn and Dressler 1987),(Rinnen *et al* 1991). We employed a very simple scheme for exciting this state. The output of the dye laser was used directly rather than employing expensive doubling, mixing crystals and a tracking system to keep those crystals accurately tuned. Instead we used a coherent vacuum ultraviolet (VUV) beam of rather low intensity $(\sim 10^{10}$ photons per pulse) and which was divergent in the ionization region of the experimental chamber. Each VUV photon in the ionization region was effectively converted to an ion. Thus some variation on this theme could be used as a tunable VUV photon detector. Finally, we observed tunneling of the highest vibrational level in the outer well (which is not observed due to Franck-Condon factors) into the inner well.

2. EXPERIMENTAL DESIGN

The experimental chamber is shown in Figure 1. The third harmonic (355 nm) from a pulsed Nd:YAG laser system is split to a dye laser and a tripling cell. The tripling cell is filled with xenon and is mounted on one of the windows of /the chamber shown in Figure 1. The third harmonic (355 nm) of the Nd:YAG laser is tripled in xenon and yields a coherent light pulse at the fixed wavelength of 118 nm. Neat hydrogen from the pulsed valve is ejected into the chamber. Once the supersonic gas jet is in the ionization region, the jet is irradiated with coherent visible light from the dye laser through a quartz window and coherent VUV light (118 nm) from the tripling cell through a LiF window. UV light (355 nm) also passing through the tripling cell, ionizes the hydrogen molecule once it has been excited by the visible and VUV photons.

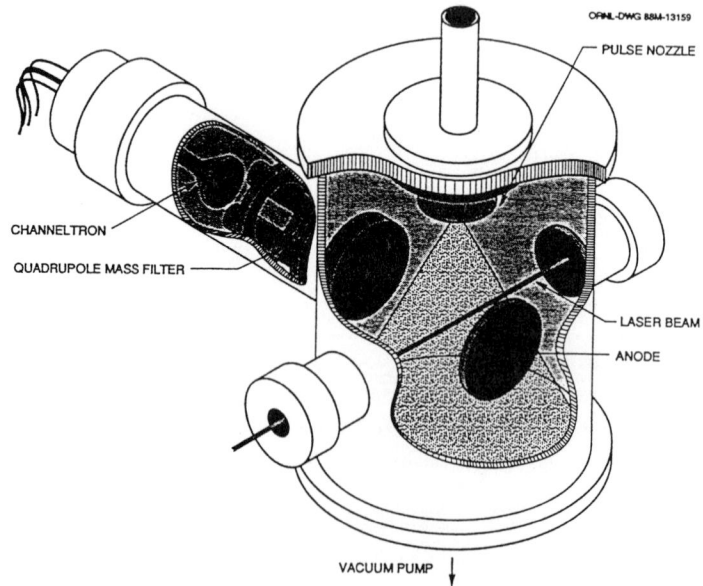

ORNL-DWG 88M-13159

PULSE NOZZLE

CHANNELTRON

QUADRUPOLE MASS FILTER

LASER BEAM

ANODE

VACUUM PUMP

Figure 1. Experimental Apparatus

3. RESULTS AND DISCUSSION

A hydrogen spectrum is shown in Figure 2. For $^1\Sigma_g^+ \leftarrow\leftarrow \ ^1\Sigma_g^+$ transitions, Q peaks are the dominant feature (Rudolph 1987). The two-photon rate constant

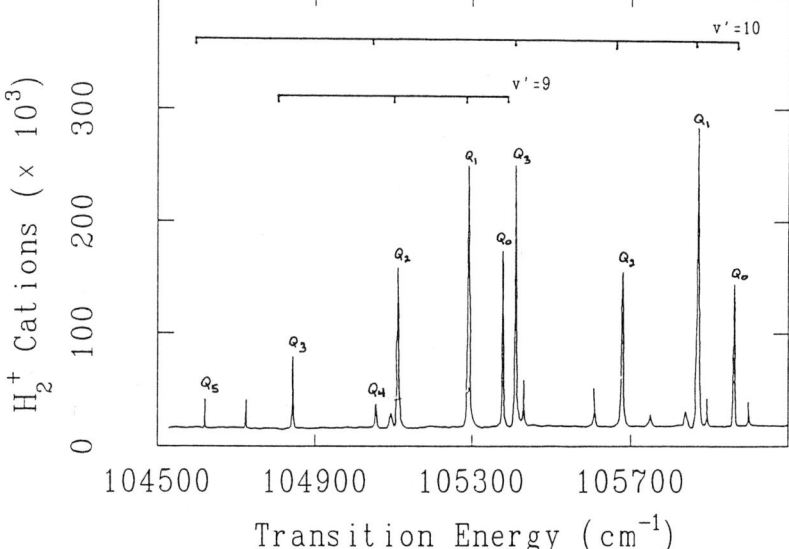

Figure 2. $E, F \, {}^1\Sigma_g^+(v' = 9, 10) \leftarrow\leftarrow X \, {}^1\Sigma_g^+(v'' = 0)$ spectrum

was calculated from the signal strength, and laser pulse energies (McCann *et al* 1988) and is tabulated in Table 1.

Though the two-photon rate constant or two-photon cross section is very useful for measuring and comparing a species ability to absorb two photons, this parameter is an awkward unit. The magnitude of this parameter is dependent on the bandwidth of the laser or lasers used to promote the two-photon transition. So in comparing two-photon rate constants from one experiment to another, some sort of correction must be made for different laser bandwidths. We propose the following parameter that will not contain any bandwidth dependence and which will be termed as the "two-photon constant" and is defined as follows:

$$T_2 \equiv K_2 \sqrt{\Delta_1^2 + \Delta_2^2}$$

where
    T_2 = two-photon constant
    K_2 = two-photon rate constant
    Δ_1 = bandwidth of Laser 1 at FWHM (rad/s)
    Δ_2 = bandwidth of Laser 2 at FWHM (rad/s).
The "two-photon constant" will have the units of cm^4. This convention will allow the direct comparison of two-photon experiments without further manipulation.

Table 1

Transition Energies and Two-Photon Rate Constants for H_2

Transition	Energy (cm^{-1})	K_2 (cm^4 s)	T_2 (cm^4)
$E,F\ ^1\Sigma_g^+, v'=3 \leftarrow X\ ^1\Sigma_g^+, v''=0$	101,486	5.0×10^{-50}	2.7×10^{-38}
$E,F\ ^1\Sigma_g^+, v'=6 \leftarrow X\ ^1\Sigma_g^+, v''=0$	103,554	1.9×10^{-49}	7.2×10^{-38}
$E,F\ ^1\Sigma_g^+, v'=7 \leftarrow X\ ^1\Sigma_g^+, v''=0$	103,833	5.0×10^{-50}	1.9×10^{-38}
$E,F\ ^1\Sigma_g^+, v'=8 \leftarrow X\ ^1\Sigma_g^+, v''=0$	104,724	3.2×10^{-49}	1.2×10^{-37}
$E,F\ ^1\Sigma_g^+, v'=9 \leftarrow X\ ^1\Sigma_g^+, v''=0$	105,377	3.2×10^{-49}	1.2×10^{-37}
$E,F\ ^1\Sigma_g^+, v'=10 \leftarrow X\ ^1\Sigma_g^+, v''=0$	105,960	3.0×10^{-49}	1.1×10^{-37}
$E,F\ ^1\Sigma_g^+, v'=11 \leftarrow X\ ^1\Sigma_g^+, v''=0$	106,709	3.8×10^{-49}	1.0×10^{-37}
$E,F\ ^1\Sigma_g^+, v'=12 \leftarrow X\ ^1\Sigma_g^+, v''=0$	107,421	4.1×10^{-49}	1.1×10^{-37}
$E,F\ ^1\Sigma_g^+, v'=13 \leftarrow X\ ^1\Sigma_g^+, v''=0$	108,096	2.9×10^{-49}	7.8×10^{-38}
$E,F\ ^1\Sigma_g^+, v'=14 \leftarrow X\ ^1\Sigma_g^+, v''=0$	108,790	2.1×10^{-49}	5.7×10^{-38}

Quantum mechanical tunneling of the $v' = 7$ vibrational level of the outer or so called "F-well" into the inner well is observed as can be seen in Figure 3.

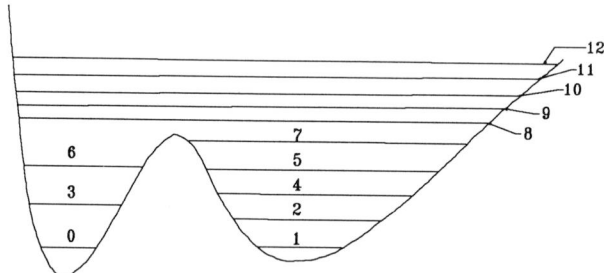

Figure 3. $E, F\ ^1\Sigma_g^+$ Potential Energy Curve

Research sponsored by the Office of Health and Environmental Research, U.S. Department of Energy under contract DE-AC05-84OR21400 with Martin Marietta Energy Systems, Inc. C. H. Chen is head of the Photophysics Group at Oak Ridge National Laboratory, Oak Ridge, TN 37831-6378. M. G. Payne is a professor in the Department of Physics, Georgia Southern University, Statesboro, GA 30460.

REFERENCES

McCann M P, Chen C H, and Payne M G 1988 *J. Chem Phys.* **89** 5429

Rinnen Klaus-Dieter, Buntine Mark A, Kliner Dahv A V , and Zare Richard N 1991 *J. Chem. Phys.* **95** 214

Rudolph H, Lynch D L, Dixit S N, McKoy V, and Huo Winifred M 1987 *J. Chem. Phys.* **86** 1748

Senn P, and Dressler K, 1987 *J. Chem. Phys.* **87** 1205

Inst. Phys. Conf. Ser. No 128: Section 5
Paper presented at RIS 92, Santa Fe, NM, USA, 24–29 May 1992

189

Atomic pathways in the fragmentation of nitroaromatic molecules

A Clark, KWD Ledingham, R Jennings, A Marshall, J Sander and RP Singhal

Department of Physics and Astronomy, University of Glasgow, Glasgow G12 8QQ, Scotland

ABSTRACT: During the course of studies on resonance enhanced multiphoton ionisation (REMPI) spectra of nitro-aromatic molecules, a number of sharp atomic transitions were observed in the wavelength region 243-260nm. These have been unambiguously identified as transitions in atomic carbon and hydrogen. In the case of carbon, transitions originating from both ground and metastable excited states have been observed. Laser fluence measurements on these transitions, both on and off-resonance, indicate that at least three photons are required to liberate a neutral carbon atom from the parent molecule together with a further three photons needed for ionisation.

1. INTRODUCTION

Interaction of intense UV light with nitro-aromatic vapours results in the production of a number of $C_nH_m^+$ type fragment ions as well as carbon ions (C^+) (Marshall *et al*, 1991).

The wavelength dependence of the $C_nH_m^+$ fragment ion yield generally exhibits broad resonances. These features are also present in the carbon ion spectrum but additionally a number of sharp transitions, particularly at higher laser fluences, were observed. The sharp transitions will be shown to arise by the resonant ionisation of atomic carbon, both in the ^3P ground state and the ^1S and ^1D metastable excited states. Similar sharp structure attributed to atomic carbon resonances has been observed in the ionisation spectra of several aromatics in the range 279.4-320.4nm (Whetten *et al*, 1983) and in the ionisation spectra of CCl_4 in the range 286.7-287.4nm (Pratt *et al*, 1985). Resonant three-photon ionisation of atomic hydrogen at 243.15nm was also observed in the time of flight mass spectrometer (TOF). In particular, the hydrogen ion signal is extremely strong and shows the production of a very large number of neutral hydrogen atoms compared with hydrogen ions in the molecular fragmentation process. In this paper, observations of the ion yields from the atomic fragments in nitrobenzene (NB) and nitrotoluene (NT) are discussed in the wavelength range 243-260nm.

2. EXPERIMENTAL DETAILS

UV laser radiation of pulse length 6ns and pulse energy of maximally 1mJ was generated by frequency doubling the output of a Lumonics EPD-330 dye laser which was pumped by a Lumonics TE-860M XeCl excimer laser. A BBO crystal was mounted in an Inrad 510 autotracking unit which permitted continuous wavelength tuning in the range 220- 260nm. The laser fluence could be varied using a Newport attenuator. The gaseous samples were introduced to a vacuum chamber at a base pressure of typically 2×10^{-8} mbar. via a needle valve controlled stainless steel capillary tube. The sample pressure was nominally 10^{-6} mbar. Laser produced ions were extracted and detected in a 1.2m linear Time-of-Flight mass spectrometer of mass resolution 220 at m/z = 70.

3. RESULTS AND DISCUSSION

Figure 1 shows a laser-induced mass spectrum of nitrobenzene vapour recorded at 247.2nm. As the laser fluence was increased, there is more fragmentation and a marked enhancement in the carbon ion signal. In Figure 2, the wavelength dependence of the carbon ion yield is presented in the range 245-260nm for nitrotoluene. Nitrotoluene provides a much smoother molecular background relative to which the carbon ion signal is measured. Table 1 indicates the initial and final atomic states involved in the ionisation process of the neutral carbon atoms released. The energies of the atomic states have been taken from Moore (1948). The peak at 246nm corresponds to the $^3P \rightarrow {}^3P$ transition and is shown in much greater detail in Figure 3.

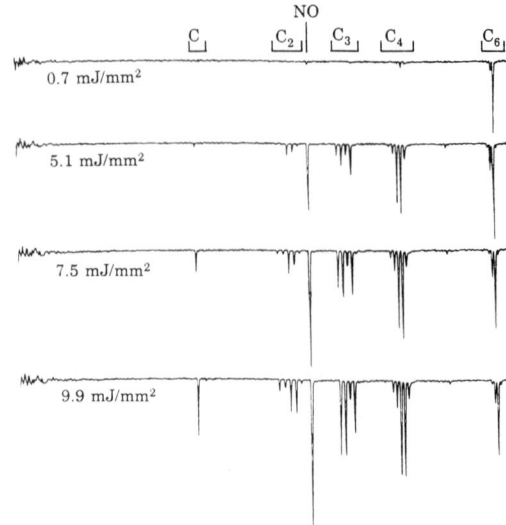

Figure 1: Mass spectra of nitrobenzene at various fluences.

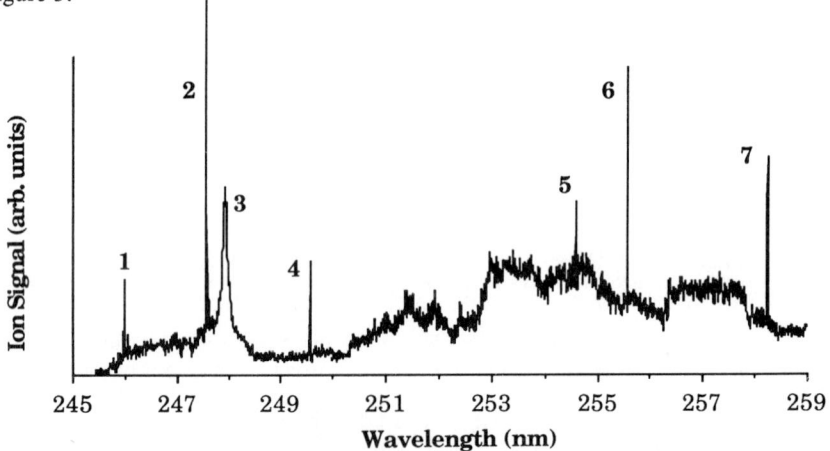

Figure 2: Ionisation spectrum of carbon (m/z=12) produced from p-nitrotoluene.

The carbon atomic transitions identified in the present experiment are illustrated in the partial Grotrian diagram shown in Figure 4, where the transitions are labelled for clarity. Many of the transitions originate from the metastable ^1D and ^1S levels which suggests that the fragmentation process results in the production of neutral carbon atoms in ground and excited states. The multiplet at 246nm (Figure 3) is due to transitions between triplet even parity states (Table 1). Two-photon selection rules restrict the number of allowed transitions in the multiplet to seven since $J=0 \leftrightarrow J=1$ are forbidden (Melikechi *et al*, 1985). All seven transitions were observed with widths (FWHM) of 0.005nm and their wavelengths agree with the expected values to 4 parts in 10^6.

Peak Number	2ω Wavelength (nm)	State Initial	State Final	J value Initial	J value Final	State energy (cm⁻¹) Initial	State energy (cm⁻¹) Final	Excitation photons	Ionisation photon	Parity change
	245.867	$2p^2$ ^3P	$2p^2$ ^3P	0	2	0	81344.48	$2\omega + 2\omega$	1ω	No
	245.918	"	"	1	2	16.4	81344.48	$2\omega + 2\omega$	1ω	"
	245.968	"	"	0	0	0	81311.52	$2\omega + 2\omega$	1ω	"
1	245.972	"	"	1	1	16.4	81326.33	$2\omega + 2\omega$	1ω	"
	246.000	"	"	2	2	43.5	81344.48	$2\omega + 2\omega$	1ω	"
	246.054	"	"	2	1	43.5	81326.33	$2\omega + 2\omega$	1ω	"
	246.100	"	"	2	0	43.5	81311.52	$2\omega + 2\omega$	1ω	"
2	247.51	$2p^2$ ^1S	$4p$ ^1S	0	0	21648.4	82252.31	$2\omega + \omega$	ω	No
3	247.93	$2p^2$ ^1S	$3s$ $^1P^o$	0	1	21648.4	61982.2	2ω	2ω	Yes
4	249.49	$2p^2$ ^1S	$4p$ ^1D	0	2	21648.4	81770.36	$2\omega + \omega$	ω	No
5	254.60	$2p^2$ ^1S	$4p$ ^1P	0	1	21648.4	80563.57	$2\omega + \omega$	ω	No
6	255.69	$2p^2$ ^1D	$3p$ ^1P	2	1	10193.7	68858.0	$2\omega + \omega$	ω	No
7	258.27	$2p^2$ ^1D	$6d$ $^1D^o$	2	2	10193.7	87632.0	$2\omega+\omega+\omega$	ω	Yes

Table 1: Carbon resonance wavelengths with initial and final states identified.

Transitions from the ground state multiplet to the nearby ^3D and ^3S levels have energies within the scanned wavelength range but were not observed. The reason for this is uncertain. The fluence dependence of the carbon ion signal was measured both on and 0.05nm off-resonance for the 247.51 and 249.49nm transitions. Off-resonance, both exhibit third order power dependences, whereas sixth order dependences are obtained on resonance. The molecular fragment ions also show a third order power dependence. A credible model will be to surmise that both carbon ions and atoms are produced in the dissociation process which also results in the production of hydrocarbon fragment ions and are characterised by the cubic dependence. The carbon atoms require a further three photons for ionisation at 247.51 and 249.49nm (Table 1) and show a sixth order power dependence. The on-resonance power dependences were measured after subtracting the non-resonant dissociative component.

In the dissociation of nitro-aromatic molecules, production of atomic fragments other than carbon is also expected. In the mass spectrum signals for hydrogen, nitrogen and oxygen were not observed in the wavelength range 245 to 260nm. However in this wavelength range, these atoms have no resonant transitions. Fig.5 shows the hydrogen ionic signal at m/z = 1 when the laser (fluence equal to 15mJ/mm^2) is tuned to 243.15nm. The ionisation of H atoms proceeds through a 2+1 process also shown in Figure 5. The large on-resonance hydrogen signal corresponds to a very significant yield of neutral hydrogen atoms in the fragmentation of nitroaromatic molecules.

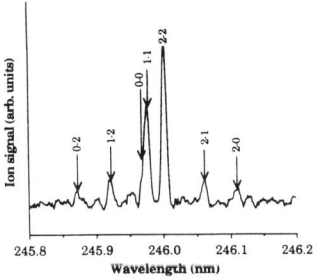

Figure 3: $^3P \to {}^3P$ multiplet in greater detail.

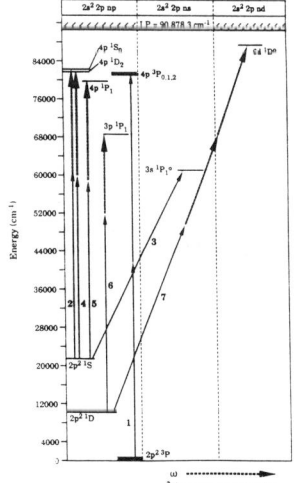

Figure 4: Partial Grotrian diagram of carbon.

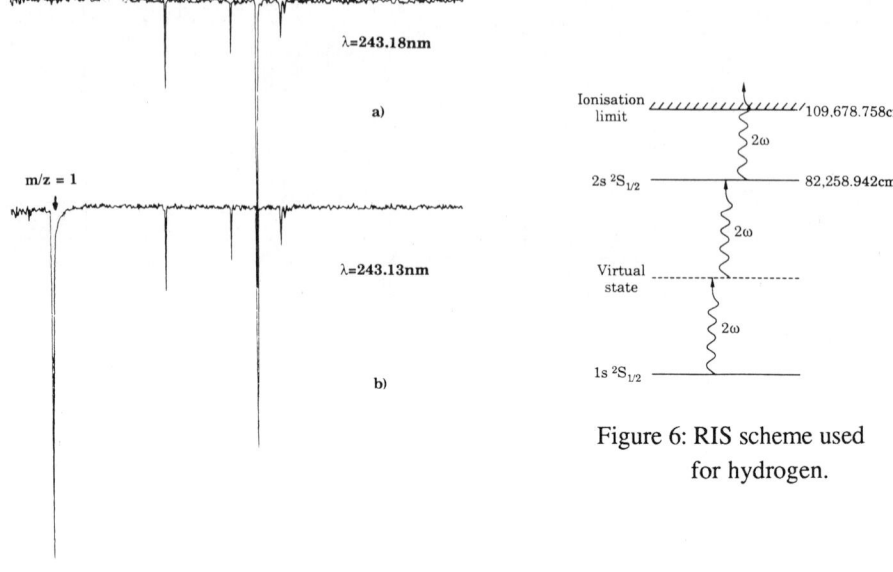

Figure 6: RIS scheme used
for hydrogen.

Figure 5: a) Mass spectrum of nitrobenzene at 243.18nm
     b) Mass spectrum of nitrobenzene recorded
        at the hydrogen resonance, 243.13nm.

4. CONCLUSIONS

The UV multiphoton dissociation of nitroaromatic molecules has been shown to be a source of efficient production of both carbon and hydrogen atoms. The 13 carbon resonances may serve as an excellent means of wavelength calibrations during REMPI experiments.

5. REFERENCES

Marshall *et al* (1992), Int. J. Mass Spectrom. and Ion Proc. 112, 273
Melikechi *et al* (1985),J. Opt. Soc. Am. B3, 41
Moore C E (1971) Atomic energy levels NSRDS-NBS 35, US Govt. Printing Office,
  Washington DC
Pratt *et al* (1985), J. Chem. Phys., 82(2), 676
Whetten *et al* (1983), J. Phys. Chem., 87(9), 1484

Inst. Phys. Conf. Ser. No 128: Section 6
Paper presented at RIS 92, Santa Fe, NM, USA, 24–29 May 1992

193

Limitations of extreme trace analysis from the standpoint of analytical chemistry

G Tölg

Institut für Spektrochemie und angewandte Spektroskopie (ISAS) und
Laboratorium für Reinststoffanalytik (LRA) des
Max-Planck-Institutes für Metallforschung, Stuttgart,
Bunsen-Kirchhoffstraße 11, W-4600 Dortmund 1, Germany

ABSTRACT: In extreme trace analysis chemists and physicists follow different ways in order to optimize detection power and reliability of their methods. Here we discuss the possible strategies leading to optimal results in the case of real complex samples.

1. INTRODUCTION

The continuously growing challenge to find the optimal balance between our technological progress and the risks that unavoidably go with it, implies a more and more efficient and problem-oriented analysis. The analysts permanently try to improve the most important quality criteria of analytical methods: detection power, reliability and economical aspects (Tölg and Garten 1985, Tölg 1988). To reach these goals physicists and chemists committed to analysis follow different motivations and strategies.

The physical methods of analysis are based on the determination of analytical signals directly dependent on the analyte concentration in a sample. This approach works the best if the analyte signals are not perturbed by neighboring (overlapping) signals, and if the signal intensity is not affected by some remaining components in the sample. In that case even single atoms are detectable. However, real samples are mostly complex multicomponent systems, and the results of such conceived determination are applicable only to a very limited number of real cases since practically all classical methods of atomic spectroscopy are strongly influenced by matrix effects. These effects allow the determination of accurate direct dependence between the element concentration and the corresponding signal intensity (calibration curve) in very few cases. The compensation of these cross-interferences is possible using calibration standards of very similar composition as the sample which is to be analyzed. On the other hand, the separation and determination procedures based on chemical and electrochemical reactions hardly permit specific multielement analysis, these operations are lengthy and relatively insensitive. The attainable separation factors have an order of magnitude 10^4 in the best case. Nevertheless, chemical single element determination procedures possess very good reproducibility and relative standard deviations of 1% to 2 %. Consequently, in modern trace analysis the optimal solution of an analytical problem can be expected only if the chemical and the physical points of view coincide, and if physical and chemical determination methods are linked together.

2. MAIN PROBLEMS IN EXTREME TRACE ANALYSIS

2.1 Systematic errors

Some 25 years ago at the Max-Planck-Institut of Metal Research in Stuttgart we started the research on the analytical characterization of high-purity metals. The impurity concentrations of interest were in the ppb range, in those days a special challenge for an analyst (Tölg 1981). Very reproducible results have been obtained, but differences between results of different methods where bigger than one order of magnitude. This was the reason to start the first interlaboratory comparison analysis in Germany, analyzing parts of the same sample in different analytical laboratories. We chose a relatively simple example: the determination of Hg in milk powder. The final results are shown in Fig. 1. It was not the average value that came close to the real one, but the lowest one which would otherwise have been eliminated by statistical error analysis (Tölg 1976). We found that the systematic errors were the cause for strong deviations in our results, and their influence was by far exceeding the statistical error of individual measurements. Generally, they represent the main difficulty in extreme trace analysis (Tölg 1972, 1979, and Tschöpel et al 1982). They grow up rapidly, if lower and lower contents have to be determined (Tölg et al 1987). Deviations of orders of magnitude from the real value are a very often case.

In order to minimize the influence of systematic errors in extreme trace analysis no generalizations or extrapolations concerning the systematic errors are allowed. Their origins vary from case to case, and their identification and minimization must be done with a lot of experience and criticism. The accuracy of the results is guaranteed only when several mutually independent methods lead to the same value.

Laboratory No#	Mean value of the laboratory No# (ppb)	Standard deviation (ppb)	Coefficient of variation (%)
1	10		
2	1		
3	100		
4	8		
5	*0.5*	*This value is close to the true one*	
6	10		
7	44		
8	1.4		
9	50		
10	11		
11	136		
12	54		
Total:	**35.5**	44	124

Fig. 1. Interlaboratory comparison analysis of Hg in milk powder.

In this context it is easy to realize that a doubtful or even wrong result of an inadequate analytical procedure can be a lot more expensive than the application of a more expensive but adequate combined multistep procedure. Therefore, the experience and the criticism of analysts influence the economical aspect of the methods of analysis very substantially.

2.2 Matrix effects

Physical methods are direct and relative instrumental methods, subjected to matrix dependent spectral and non-spectral interferences. These methods can be accurately calibrated using standard reference samples compensating in this way systematic errors. In cases where there are still no standards available it is possible to obtain optimal detection power and reliability only if the trace element quantity which is to be determined, exists well isolated from the other components in a very small volume or area, enabling only excitation of the analyte. Therefore, we are often dependent on combined multistep procedures where chemical methods of decomposition and separation (preconcentration steps) must precede a disturbed physical method of determination. Of course, such multistep procedures are not free of systematic errors, but the errors and their sources are better controllable. In this sense, the minimum difficulties and uncertainty cause instrumental neutron activation analysis (INAA), sputtered neutral mass spectrometry (SNMS), and X-ray fluorescence spectrometry (XRFA), if very thin samples are used. Consequently, in order to minimize matrix effects, extreme trace analysis assumes a tight symbiosis of the chemical separation and the physical determination methods.

2.3 Contamination

It is particularly difficult to keep low and constant contamination caused by the most abundant elements (Fig. 2) which may be already present in or introduced into the analytical system.
Detection limits of analytical methods described in the literature are mainly obtained by extrapolation, but real detection limits in many cases are determined by blank values. For specially abundant elements (Si, Al, Fe etc.) they can be many times larger than the extrapolated, "theoretical" detection limits. Therefore, the minimization of blanks caused by the contamination of the equipment, the reagents and the laboratory air plays a decisive role in extreme trace analysis. The intrinsic contamination by the materials used for the tools and the equipment presents a serious problem. Suitable materials of high chemical stability and purity must be chosen. Usually, quartz glass is the most appropriate option. Furthermore, the equipment surfaces must be thoroughly cleaned by special cleaning procedures and reagencies. The contamination by air dust can be avoided working in clean room atmosphere.

2.4 Detection power

In natural matrices the lowest elemental concentrations to be determined correspond to the omnipresent concentrations of the elements which lie in the range > 1 ng/g for the most of the elements. However, there are two exceptions. First, in the case of the platinum group metals concentrations of omnipresence are in the pg/g range. Second, in the case of special

× 1		× 10^{-2}		× 10^{-3}		× 10^{-4}		× 10^{-5} %	
O	46.40	Ti	57	Zn	7	Sc	22	Tb	9.1
Si	28.15	H	14	Ce	6	B	10	J	5
Al	8.23	Mn	10	Y	3.3	Pr	8.2	Tl	4.5
Fe	5.63	P	10	La	3	Sm	6.0	Cd	2
Ca	4.15	F	6.3	Nd	2.8	Gd	5.4	Bi	1.7
Na	2.36	Ba	4.3	Co	2.5	Ge	5.4	In	1.0
K	2.09	Sr	3.7	Li	2	Hf	3	Hg	1.8
Mg	2.33	S	2.6	N	2	Dy	3	Ag	0.7
	99.34	C	2.0	Nb	2	Yb	3	Se	0.5
		Zr	1.6	Ga	1.5	Er	2.8	Pd	0.1
		Ci	1.3	Pb	1.2	Be	2.8	Pt	0.05
		V	1.3	Th	0.9	U	2.7	Re	0.05
		Cr	1.0	Sn	0.2	Ta	2	Au	0.04
		Rb	0.9	W	0.1	As	1.8	Te	0.01
		Ni	0.7		34.5	Mo	1.5	Ir	0.01
		Cu	0.5			Ho	1.2	Rh	0.01
			117.2			Eu	1.2		26.57
						Cs	1.0		
						Sb	0.2		
							83.2		

Fig. 2. Abundance of the elements in eruptive rocks (after Taylor 1964).

natural matrices like Arctic- or Antarctic ice concentrations of the contaminants of interest are even smaller and lie in the lower pg/g range. For comparison, the typical detection power of contemporary trace analysis methods is between ng/g and the high pg/g range. Therefore, an improvement of detection power in bulk analysis is generally requested in the analysis of high-purity materials of interest in different areas of technology. It also must be urged that bulk analysis offers integral evidences, represented by a larger sample mass. Modern trace analysis is more interested to obtain detailed informations about the distribution of elements in microareas, or on grain boundaries and surfaces as well as about (their) chemical binding forms. Here, the limited sample mass imply determination methods with absolute detection abilities as high as possible. If, for example, the concentration of about 1 ng/g Al (the order of magnitude of Al concentration in normal blood serum) in a biological cell cluster with the mass of about 1 μg has to be determined, the absolute detection power of about 1 fg is necessary. Today, we are still far away from this goal. Nowadays, for the determination of such a low Al concentration, a sample of at least 100 mg is needed. The demand for orders of magnitude better absolute detection limits raises the question where the limits of applied elemental analysis really are. The answer is given by the root-N-rule of statistical analysis. For example, if the statistical error in a quantitative determination of Zn has to be less then 1 % at least 10^4 Zn atoms are needed. This corresponds to a mass of about 1 attogram Zn (10^{-18} g Zn, Broekaert et al 1987).

3. THE MOST IMPORTANT DETERMINATION METHODS IN EXTREME TRACE ANALYSIS: A CRITICAL COMPILATION

The features of the most important determination methods used today in extreme trace analysis are compared in Fig. 3 (Broekaert et al 1987, Tölg 1975). The classical chemical methods - spectrophotometry, fluorometry, chelate gas chromatography, chelate HPLC, ion chromatography, and voltametry have their safe place in the toolboxes of analysts. They possess good detection abilities (some of them in the pg range), require relatively simple devices, and can easily be calibrated. However, all these methods demand a large special experience, they are lengthy and not easily automatized. As a rule, they are suitable only for a single-element determination or for a simultaneous determination of only several elements. The limits of the majority of these methods are well known. Their detection capabilities are generally limited by blanks and other sources of systematic errors (adsorption, desorption, volatilization, etc.).

The possibilities and limitations of radiochemical methods are well known (Parry 1991). It must certainly be emphasized that activation analysis, the application of radiotracers and other radiochemical methods in innovative extreme trace analysis are indispensable, first of all in the preparation of standard reference samples.

When methods of atomic spectrometry are concerned recent developments in atomic absorption (AAS), optical emission (OES), fluorescence spectrometry (OFS), X-ray fluorescence analysis (XRFA), and mass spectrometry (MS) can be discussed.

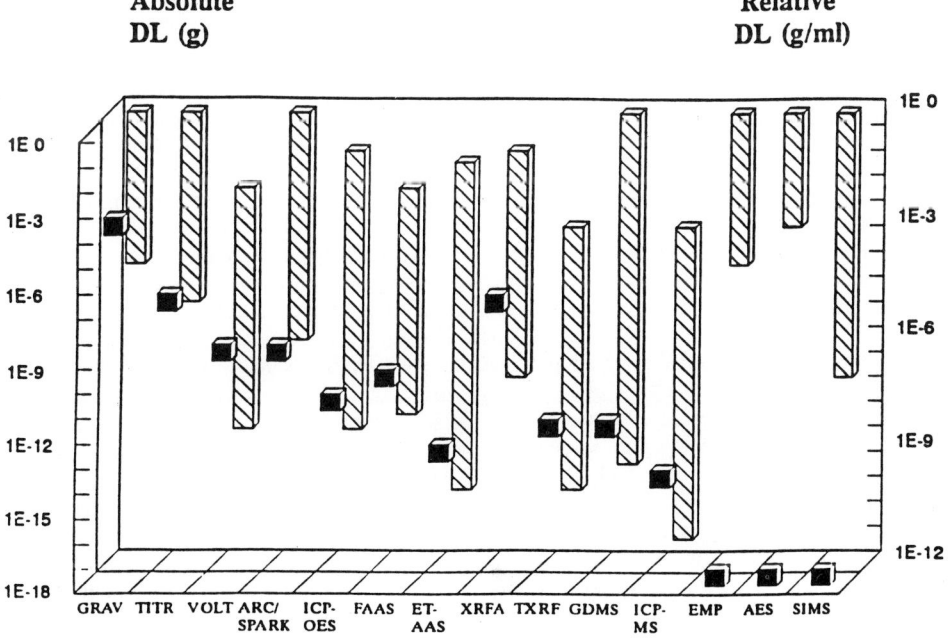

Fig. 3. The detection power and the range of application of the most important determination methods in the elemental trace analysis.

The AAS will still be the choice for direct and indirect determination of the majority elemental traces in different matrices (Sturgeon 1990, Slavin 1991). Its still existing limitation in the determination of individual elements is compensated by its high detection ability. By electrothermal atomization in furnaces relatively reliable detection limits in the pg range can be obtained for many elements. The absolute detection power of AAS can be improved for about one order of magnitude through electronic signal processing. However, there are additional very efficient procedures based on chemical and physical preconcentration methods, which can improve detection power (down to the fg range) and reliability. Among these are the gas-phase separation and the enrichment of the content of the element which has to be determined. The examples are the separation of Hg as metal vapor, of As, Se, Te, Sb and Sn as hydrides and of Ni as Ni-tetracarbonyl. The decisive step preceding the actual determination is the condensation of the volatile component in an adsorber system, followed by the pulsed release of the condensate by suddenly heating the adsorber (trapping technique). It allows, for example, an accurate determination of Se (Alt et al 1987) and Ni (Drews et al 1990) in the pg range.

In the OES we are faced with many critical comparative investigations of different new excitation principles with high frequency (ICP) or microwave plasmas (MIP), cathodic sputtering, and laser ablation (Tölg 1975). These methods open new opportunities for the problem-oriented multi-element determination, however, without substantial improvement of the detection power.

In analytical practice, the ICP-OES (Thompson et al 1989, Hieftje 1989 and Olesik 1991) and MIP-OES (Sanz-Medel 1991, Broekaert et al 1985, Heltai et al 1990 and Richts et al 1991) are today the most important methods. Their detection limits lie between the flame-AAS and furnace-AAS, and can be improved for at least one order of magnitude combining new sample introducing techniques (e.g. a high pressure injection of the analyte solutions) and chemical preconcentration methods.

The greatest difficulty for the improvement of detection power of all optical (atomic) spectroscopy methods is the compromise (existing from the time of Bunsen and Kirchhoff) that sample volatilization, atomization and excitation are made by the one and the same source. While the atomization needs high energy in order to break chemical bonds, the excitation should be done with as low as possible energy in order to keep an optimal signal-to-noise ratio. Recently, it has been possible to optimize different steps separately using lasers for ablation (atomization) in the first step (Moenke-Blankenburg 1989). The possibility to vaporize very small parts of a sample by a focussed laser beam - also in the case of electrically non-conductive samples - opens new ways for in situ micro-distribution analysis of elements possessing a lateral resolution of about 1 μm. According to the results of Niemax (1989 and 1990) the earlier problem of very poor reproducibility of the laser ablation technique at the standard atmospheric pressure or in high vacuum is no more relevant if the vaporization is performed by laser pulses of low energy and in the presence of about 100 Torr of a noble gas, responsible for the energy transfer (Leis et al 1989, Niemax et al 1990, Sdorra 1988 and Quentmeier 1990).

Lasers have certainly a bright future in applied analysis and the most important innovations in methods of optical atomic spectroscopy can be expected in laser spectroscopy (Niemax 1990, 1991 and 1992). This conclusion is probably valid for all types of excitation leading to analytical signals obtained by absorption, fluorescence or emission after the irradiation of an atomic cloud by a laser tuned to a resonant frequency.

For example, the laser-AAS (LAAS) opens ways to multielement determination with a great detection ability and dynamic range if simple laser systems with semiconductor diode lasers can be used (Niemax 1991, Hergenroeder et al 1989).

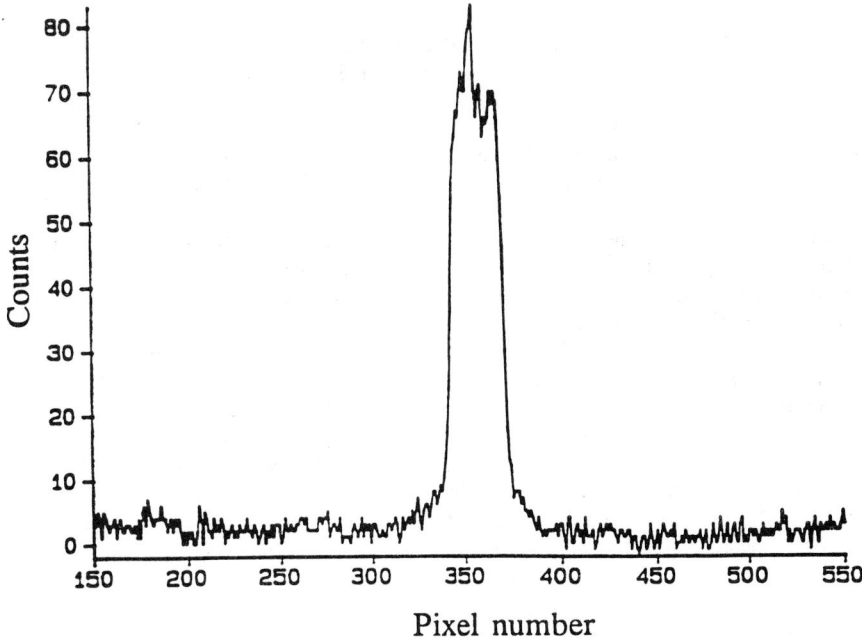

Fig. 4. LIF spectrum of the magnesium resonance line recorded with one laser shot. Argon buffer gas pressure was 140 hPa (After Niemax and Sdorra 1990).

On the other hand, the laser induced fluorescence (LIF) already allows the determination of a few elements down to the fg range (Phab 1991, Bolshov et al 1991, Niemax and Sdorra 1990). For example, Fig. 4 shows the signal of about 45 fg magnesium in a steel sample after laser ablation and determination through LIF. The evaporated amount of the sample is about 30 ng and the Mg concentration is about 1.5 ppm. Consequently, this gives an absolute detection limit of 5 fg Mg (Niemax and Sdorra 1990). Fig. 5 shows the distribution of Si (100 μg/g) in the case of homogeneous and inhomogeneous distribution in a steel sample (Sdorra 1988).

The coupling of mass spectrometry, e.g. RIMS (Rinke et al 1989, Saloman 1991, Barthe et al 1991 and Nogar et al 1992), and the Doppler-free methods using CW lasers allows a very good resolution and even isotope selective spectrometry can be carried on. In this way the problem of calibration can be solved using the isotope dilution technique. Monitoring products of ionization process after resonant laser excitation the detection limits in the attogram range can be obtained (Niemax 1991).

The final judgements of attainable detection limits of the laser spectroscopy methods are still premature, because all principles regarding their analytical usefulness are still under development. Temporary, the biggest difficulty for a quick introduction of laser spectroscopy in analytical praxis is still a very high price of commercial laser systems. However, they can very soon be replaced by substantially cheaper semiconductor diode lasers, if they can reach shorter wavelength range.

The complete separation of the atomization and excitation steps in spectroscopic analysis makes the foundation for better detection power and reliability. Therefore, the combination of particle sputtering techniques, for example laser ablation, and laser spectroscopy is a very promising concept in optical atomic spectroscopy. It offers a high absolute detection ability (attogram range) for almost all elements, an in situ micro-analysis of atomic distribution by laser ablation, a lateral resolution of about 1 μm, and the calibration through internal standardization.

The possibilities and the limitations of the conventional wavelength- and energy-dispersive X-ray fluorescence spectrometry (XRFA) are well known (Janssens et al 1989). The method has optimal detection limits of about 0.1 μg for elements with an atomic number about 30. The most important factors which limit its detection power are Compton- and Rayleigh scattering caused by primary radiation. The high matrix dependence requires a calibration with standard samples. Better detection limits and higher precision can be reached only if the signal-to-background ratio is improved and the matrix effects reduced. It can be accomplished by using total reflection X-ray fluorescence analysis (TXRFA, Klockenkämper 1991, Knöchel 1990), particle induced X-ray emission spectrometry (PIXE, Klockenkämper 1991, Knöchel 1990, Johansson et al 1988, Vis 1990) or synchrotron X-ray fluorescence spectrometry (SXRFA, Knöchel 1990). In TXRFA the excitation beam impinges on the sample holder (made of quartz, glassy carbon or acryl glass) under a very small angle of

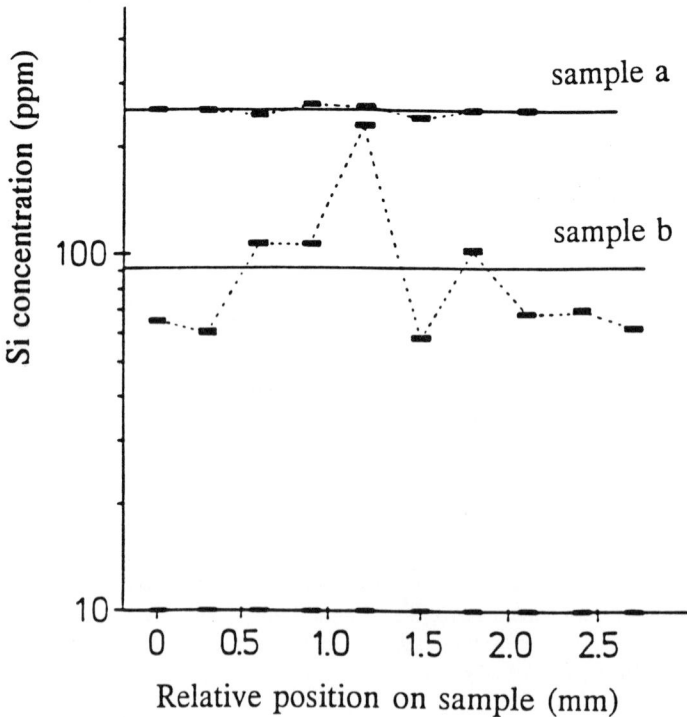

Fig. 5. Measurements of the position dependent density of Si for two different steel samples showing homogeneous (sample a) and inhomogeneous (sample b) spatial distribution of Si.

incidence, and it is totally reflected. Only very thin samples on the target can be transmitted by primary beam and secondary radiation excited. Owing to lower scattered radiation background the detection limit is improved by a factor of 10^3 in comparison to standard techniques, and the pg range is accessible. Regarding the detection abilities TRXFA is superior to PIXE. However, the PIXE compared to the conventional XRFA reaches higher absolute detection limits. The advantages of the substantially more expensive PIXE which needs a particle, i.e. proton accelerator, are in the possibility to focus the particle beam to a diameter of about 0.1 μm allowing simultaneous multi-element microdistribution analysis in the μg/g range. For example, simultaneous multi-element microdistribution analysis can be performed over the cross-section of a human hair. Even more expensive is the SXRFA, assuming a synchrotron radiation source, so that its advantages in regard to detection limits and good lateral resolution can be used only in special cases. All three new excitation methods in XRFA represent a big step ahead for micro-distribution analysis of elements. However, the TRFA is the most suitable because of its convenient expenses. Many other interesting biological applications described in the literature are often noncritically overestimated because of the favourable cost-to-benefit ratio and the reliability of the results.

Inorganic mass spectrometry (MS) is at this moment the most universal and the most sensitive instrumental method for simultaneous determination of practically all elements (Adams et al 1988). In comparison to OES the MS-spectra show only a few lines which can be easily interpreted. Very good accuracy can be obtained with analytes in solution using the isotope dilution techniques (except for one-isotope elements).

The conventional solid state mass spectrometry (SSMS) is still without competition for the simultaneous characterization of impurities in high purity materials. However, it is usable as a quantitative method only after calibration by standard samples.

Furthermore, thermal ionization in combination with the isotope dilution method is not surpassed in the case of single element and multielement determination owing to its detection abilities and reliability.

The most advanced development in applied micro- and trace analysis can be found in inductively coupled plasma mass spectrometry (ICP-MS), glow discharge mass spectrometry (GD-MS) and laser-mass spectroscopy. In ICP-MS (Broeckaert 1990) the detection limits for majority of elements are in the range between 0.001 to 1 ng/ml. That is, in average, several orders of magnitude better than in ICP-OES where detection limits for different elements vary over 5 orders of magnitude. Further improvements of detection limits in ICP-MS can be realized through systematic optimization of ionization parameters (flow rate of aerosols, electric power and the observation height of the plasma) and the already mentioned high pressure injection technique of the analyte solution into the plasma.

The GD-MS (Stüwer 1990a and 1990b) is until now a multiply promising tool for direct bulk-analysis of conducting samples and the analysis of thin layers of solid samples. It can be applied likewise for the microanalysis of solutions. As shown by our first results in the determination of platinum, a significant improvement of absolute detection limits down to the fg range could be obtained (Jakubowski 1991). Consequently, the detection limits of GD-MS are very promising for direct analysis and may even surpass ICP-MS.

If isotope dilution techniques are applied very good precision and accuracy can be obtained not only for thermionic techniques (Tancer et al 1989, Herzner et al 1991) but also for ICP-MS and GD-MS.

The SNMS (sputtered neutrals mass spectrometry) opens new ways for the further matrix independent characterization of bulk samples and surfaces of solid samples in the μg/g range (Reuter 1989, Husinsky et al 1991, Wilhortitz et al 1991, Schmidt et al 1991 and Jenett et al 1992). The recently introduced new generation of laser-MS and TOF-SIMS in combination with successive ionization of neutral particles sputtered from the sample leads to new, very sensitive and reliable micro-probe techniques having also a very good lateral resolution (6th ASA Proceedings 1991).

4. CONCLUSION

A general statement, which analytical method for trace element analysis should be recommended for the characterization of real complex material samples, is not possible. Very often we incline to over-estimate a specific method what may hinder an innovative analysis. The best method or the most suitable combination of methods can be discussed only in regard to the actual analytical problem. The optimal solution of an analytical problem is less a question of tools or methods than a question of the right analytical strategy. Consequently, the advanced analysis is enormously accelerated through a narrow symbiosis of chemical and physical methods. Finally, only with efficient analysis it will be possible to balance correctly the positive and the negative effects of todays technological advancement.

REFERENCES

Adams F, Gijbels R and van Grieken R 1988 *Inorganic mass spectrometry-Chemical Analysis Vol 95* eds. J D Winefordner and I M Kolthoff (New York: Wiley)

Alt F, Messerschmidt J and Tölg G 1987 *Fresenius Z. Anal. Chem.* **327** 233

Barthe M E, Debrun J L, Gibert T and Dubreuil B 1991 *J. Trace and Microprobe Techn.* **9** 1

Bolshov M A, Boutron C F, Ducroz F M, Görlach U, Kompanetz O N, Rudniev S N and Hütsch S 1991 *Anal. Chim. Acta* **251** 169

Broekaert J A C and Leis F 1985 *Mikrochim. Acta* 261

Broekaert J A C and Tölg G 1987 *Fresenius Z. Anal. Chem.* **326** 495

Broekaert J A C 1990 *Analytiker Taschenbuch Vol. 9* (Berlin: Springer) pp 127-163

Drews W, Weber G and Tölg G 1990 *Anal.Chim.Acta* **231** 265

Heltai G, Broekaert J A C, Burba P, Leis F, Tschöpel P and Tölg G 1990 *Spectrochim. Acta* **45B** 857

Hergenroeder R and Niemax K 1989 *Trends Anal. Chem.* **8** 333

Herzner P and Heumann K G 1992 *Mikrochim. Acta* **106** 127

Hieftje G M 1989 *J. Anal. At. Spectrom.* **4** 117

Husinsky W, Wurz P, Traumfellner A, Betz G 1991 *Fresenius J. Anal. Chem.* **341** 12

Jakubowski N, Stüwer D and Tölg G 1991 *Spectrochim. Acta* **46B** 155

Janssens K H and Adams F C 1989 *J. Anal. At. Spectrom.* **4** 123

Jenett H, Kikuta Y 1992 *Spectrochim. Acta* **47B** 143

Johansson S A E and Campbell J L 1988 (London: Wiley)

Klockenkämper R 1991 *Analytiker Taschenbuch Vol. 10* (Berlin: Springer) pp 111-152

Knöchel A 1990 *Fresenius J. Anal. Chem.* **337** 614

Leis F, Sdorra W, Ko J B and Niemax K 1989 *Mikrochim. Acta* 185

Moenke-Blankenburg L 1989 *Laser Micro Analysis-Chemical Analysis Vol. 105* ed J D Winefordner (New York: Wiley)

Niemax K and Sdorra W 1990 *Appl. Optics* **29** 5000

Niemax K 1990 *Fresenius J. Anal. Chem.* **337** 551

Niemax K 1991 *Analytiker Taschenbuch Vol. 10* (Berlin: Springer) pp 1-28
Niemax K 1992 *Naturwiss.* (in press)
Nogar N S and Estler R C 1992 *Anal.Chem.* **64** 465
Olesik J W 1991 *Anal. Chem.* **63** 12A
Parry S J 1991 *Activation Spectrometry in Chemical Analysis Vol. 119* (New York: Wiley)
Phab J 1991 *Anal. Proc.* **28** 415
Proceedings of the 6th Working Conference on Applied Surface Analysis Kaiserslautern 1991 *Fresenius J. Anal. Chem.* **341** 1
Quentmeier A, Sdorra W and Niemax K 1990 *Spectrochim. Acta* **45B** 537
Reuter W 1989 *Trends Anal. Chem.* **8** 203
Richts U, Broekaert J A C, Tschöpel P and Tölg G 1991 *Talanta* **38** 863
Rinke H, Herrmann G, Mang M, Mühleck C, Riegel J, Sattelberger P, Trautmann N, Anes F and Kluge H-J, Otten E-W, Rehklau D, Ruster W and Scheerer F 1989 *Mikrochim. Acta* 223
Saloman E B 1991 *Spectrochim. Acta* **46B** 319
Sanz-Medel A 1991 *Mikrochim. Acta* 265
Schmidt U C, Fichtner M, Goschnick J, Lipp M and Ache H J 1991 *Fresenius J. Anal. Chem.* **341** 260
Sdorra W 1988 *Diplomarbeit* (Dortmund: University)
Slavin W 1991 *Anal. Chem.* **63** 1033A
Sturgeon R E 1990 *Fresenis J. Anal. Chem.* **337** 538
Stüwer D 1990*a* *Fresenius J. Anal. Chem.* **337** 737
Stüwer D 1990*b* *Analytiker Taschenbuch Vol. 9* (Berlin: Springer) pp 165-189
Tancer D and Heumann K G 1991 *Anal. Chem.* **63** 1984
Thompson M and Walsh J N 1989 *Handbook of Inductively Coupled Plasma Spectrometry* (New York: Chapman and Hall)
Tschöpel P and Tölg G 1982 *J. Trace and Microprobe Techniques* **1** 1
Tölg G 1972 *Talanta* **19** 1489
Tölg G 1975 *Wilson and Wilsons's Comprehensive Analytical Chemistry Vol III* ed G Svehla (Amsterdam: Elsevier) pp 1-184
Tölg G 1976 *Naturwiss.* **63** 99
Tölg G 1979 *Fresenius Z. Anal. Chem.* **294** 1
Tölg G 1981 *Aim and Methods of Microchemistry* eds H Malissa, M Grasserbauer, R Belcher (Wien: Springer) pp 203-230
Tölg G and Garten R P H 1985 *Angew. Chemie* **97** 439
Tölg G and Tschöpel P 1987 *Anal. Sci.* **3** 199
Tölg G 1988 *Fresenius Z. Anal. Chem.* **331** 226
Vis R D 1990 *Fresenius J. Anal. Chem.* **337** 622
Wilhartitz P and Ortner H M 1991 *Fresenius J. Anal. Chem.* **341** 125

Inst. Phys. Conf. Ser. No 128: Section 6
Paper presented at RIS 92, Santa Fe, NM, USA, 24–29 May 1992

TARIS analysis in biomedical, environmental and geochemical researches

F Asaro*, GI Bekov, RS Belkin, VM Gulevich, NG Khomyakov, AN Kursky** and
D Yu Pakhomov**

Institute of Spectroscopy Russian Academy of Sciences, Troitsk, Moscow Region, 142092
RUSSIA; *Lawrence Berkeley Laboratory University of California at Berkeley, Berkeley,
CA 94720, USA; **Central Research Institute of Prospecting for Base and Noble Metals,
Moscow, RUSSIA

ABSTRACT: The method of Thermal Atomization of a substance in vacuum combined
with Resonance Ionization Spectroscopy (TARIS) has been used for the detection of Pt
traces in biological samples and for the measurement of Ru and Rh concentrations in
ancient geological deposits and rock samples.

1. INTRODUCTION

RIS experiments carried out during the last several years in different laboratories have shown
that in spite of the high sensitivity and selectivity of RIMS, its application to the analysis
of complex samples is faced with many difficulties. These difficulties are mainly caused by
incomplete atomization of the substance under analysis. As a result, for the calibration of
analytical signals obtained, one needs a number of standard samples, the creation of which
is itself a great analytical problem. Also, multiphoton nonresonant ionization of nonatomized
matrix particles produces a wide mass spectrum of background ions which can interfere with
the element ions. The moderate resolution of mass filters usually used in RIMS do not allow
discrimination between element and molecular ions with the same masses. One of the ways
to solve the problem is to use sophisticated mass filters with good transmission and
resolution of the order of 10^4. But the problem of quantitative analysis will still exist with
this approach. Another more radical way is the development of atomization methods to
realize close to unity efficiency of sample atomization, and as a result, the suppression of
molecular component in the atomic vapor by several orders of magnitude.

The thermal atomization of a substance in the vacuum, as has been shown in a number of
our experiments, (Bekov et al. 1988a) makes it possible in many cases to suppress
substantially matrix effects, and to carry out with RIS direct quantitative elemental analysis.
However, the thermal atomization in its simple version used is not completely free of matrix
effects when directly analyzing biological or geological samples. We have suggested a two-
stage thermal atomizer (Bekov et al. 1991) where the vaporization and the atomization of
a sample is separated in time and space. The experiments carried out with this atomizer
have demonstrated the possibility of direct, standardless analysis of biological and geological
samples. This paper presents results on the determination by the TARIS technique of Pt
traces in human blood, the concentration of Ru and Rh in ancient deposits for studying their
origin and history, and in rocks for forecasting their diamond content.

2. EXPERIMENTAL AND RESULTS

The laser analytical spectrometer used in the experiments comprises of a pumping excimer laser (308 nm), an optical block with three dye lasers, and an analytical part consisting of a vacuum chamber with electro-thermal atomizer and time-of-flight mass filter. The multistep excitation of a selected Rydberg state followed by electric field pulse ionization has been used for effective and low background ionization of atoms of the element under study. The spectrometer and some analytical procedures have been described in detail elsewhere (Bekov et al. 1989, 1991) Table 1 presents the atomic data and laser wavelengths used for Rydberg state excitation of Pt, Rh and Ru atoms.

Table 1. Atomic states and laser wavelengths in TARIS experiments

Element	$\mid n_0 \rangle$	λ_1(nm)	$\mid n_1 \rangle$	λ_2(nm)	$\mid n_3 \rangle$	λ_3(nm)	$\mid n_R \rangle$
Pt	$5d^9 6s$ $^3D_3$	306.47	$5d^9 6p$ $^3D^0_2$	505.95	$5d^9 7s$ $^3D_3$	510.05	$5d^9 17*$ p $(^3D_2^0)$
Rh	$4d^8 5s$ $^4F_{9/2}$	369.24	$4d^8 5p$ $^4D^0_{7/2}$	675.24	$4d^8 6s$ $^4F_{9/2}$	561.7	$4d^8 16*$ p $(^4D^0_{7/2})$
Ru	$4d^7 5s$ $^5F_5$	392.60	$4d^6 5s5p$ $^7D^0_4$	633.05	$4d^7 6s$ $^5F_5$	567.3	$4d^7 16*$ p $(^5F^0_4)$

The wide use of platinum for production of new types of fuels can result in a substantial increase of Pt levels in the environment. Being a good catalyzer of many organic processes, platinum can become a rather dangerous element for all living organisms, and especially for humans. Therefore, establishing the baseline Pt content in the human body is highly important. When determining Pt in human blood, the blood was mixed with an anticoagulant to make possible multiple use of the same blood sample. A 20 $\mu\ell$ aliquot was pipetted into the graphite crucible and dried at 110°C for 10 min. The crucible was then inserted into the atomizer, and after evacuating the atomizer, was heated in a smooth, stepped manner to 2200°C. The Pt ion signal was recorded starting from 1800°C. The ion signal was integrated over a rather wide time gate to include masses 194 to 196. The background ions of hydrocarbon radicals formed by incomplete atomization of the blood matrix was also recorded in this gate. To obtain a reliable value of the selective ion signal, ten measurements were carried out when λ_2 was on resonance with the Pt atomic transition. In the next ten measurements λ_2 was off-resonance. The true selective signal was the difference between the first average value and the second one. Only three measurements were necessary to obtain similar results when the double-pulse field ionization technique was used to suppress background ions. The average value of the Pt concentration was 1.2 ± 0.4 ng/g. When a two-stage atomizer was used, the value increased to 1.7 ng/g, probably due to the improved atomization efficiency of Pt compounds. The concentration values obtained are much higher than expected, and could be explained by a high Pt content in the anticoagulant, or other contamination in the laboratory.

One of the most interesting tasks in contemporary geochemistry and paleontology is the reconstruction of the geological and biological histories of our planet. Both histories have not been smooth. There were stages of dramatic changes, and stages of slow development.

All these have been reflected in the development of life on earth. It is well known that during the 600 million years of the existence of life, about ten great biological crises (extinctions) have taken place in which the variety of living organisms has substantially decreased in a rather short period of time. Some hypotheses based on the results of geochemical and paleontological studies of biostratigraphic boundaries have been suggested to explain the reason for such extinctions. Only two of them appeared to be most consistent: the impact of a large cosmic body, or temporary volcanic super-activity. Both events could cause mass extinctions of biological species.

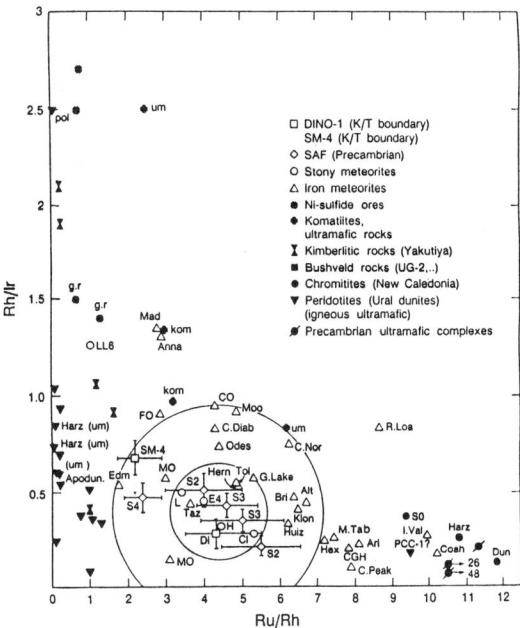

Figure 1 Rh/Ir versus Ru/Rh concentration ratios for different types of crustal rocks (black figures), iron and stony meteorites, synthetic standards containing K/T boundary material (DINO-1, SM-4), and samples analyzed (white figures).

Our geochemical study of sedimentary rocks belonging to the Cretaceous-Tertiary (K/T) boundary with the age of about 65my (the last known large biological crisis) has shown (Bekov et al, 1986b, 1989) that the Rh/Ir concentration ratio in those rocks is closest to the ratio for meteoritic material, and strongly suggest the correctness of the impact hypothesis. The Pt-group elements appeared to be excellent indicators of impact events because extraterrestrial materials are rather rich in Pt metals, whereas their concentrations in surface terrestrial rocks are several orders of magnitude lower. The Rh/Ir ratio can not be so informative for earlier geological periods because some very old rocks are characterized by Rh/Ir ratios similar to meteoritic material. The discrimination of geochemical anomalies in such rocks can be made using the concentration ratios between other Pt group elements. We have used this approach in the investigation of one of the most ancient known deposits belonging to the Precambrian with an age of about 3.5 billion years. Figure 1 presents the concentration ratios Rh/Ir (y axis) and Ru/Rh (x axis) for the majority of known types of crustal rocks, iron and stony meteorites, and samples under study collected in South Africa. A considerable fraction of the data on terrestrial rocks and meteorites has been taken from the literature. The Ir concentrations have been determined in natural samples by neutron activation analysis using a unique spectrometer at Lawrence Berkeley Laboratory. The TARIS technique has been used to determine Ru and Rh concentrations in the samples after their chemical decomposition. As one can see, the meteoritic material, synthetic standards prepared from the K/T boundary sediments, and the Precambrian samples group rather tightly within two circles centered at Ru/Rh = 4.5 and Rh/Ir = 0.38 (see Figure 1). The highest correlation of the Precambrian samples is with the stony meteorites. Thus, the proportions of Pt metal concentration can help establish not only the origin of ancient geological or biological phenomenon, but also the type of extraterrestrial body in the case of an impact.

The establishment of a correlation between the concentration of Pt group elements in rocks can be extremely useful for the study of their rock metamorphism. The results of this study can have a direct practical use in prospecting for deposits of rare elements, and Pt metals in particular. As is known from the literature (i.e., Groves et al., 1987) geological zones with increased platinum content sometimes also have imbedded regions of increased diamond content, thereby drawing attention to Pt metals as possible additional geochemical criteria for prospecting diamond-bearing rocks. The main diamond-bearing rock formations are kimberlite and lamproite pipes. We have investigated the Ru and Rh content in kimberlites and abyssal xenoliths from different diamond-bearing regions of Siberia. When analyzing the kimberlites, large samples (5-10g) were available, and the fire-assay technique in a copper alloy has been used to collect Pt metals. When analyzing small samples (1-10mg) of monominerals and xenoliths, the direct TARIS analysis of natural samples has been used without any chemical treatment. Table 2 contains the averaged data on Ru and Rh concentration in various types of rocks taken from diamond rich and poor kimberlitic pipes. The tentative conclusion to be drawn is that the average concentrations of Ru and Rh are different for areas with low and high diamond content even though the rock types are the same. To establish the distinct correlation between Pt metal concentrations (or their proportions) and diamond content of kimberlitic pipes, it is necessary to carry out many more analyses for better statistics, to increase the number of Pt elements to three or even four, and to use rock samples from diamond-bearing pipes from other parts of the globe.

Table 2. Ru and Rh concentrations (ng/g) in kimberlitic pipes

Rock Type	High Diamond Content Pipe		Low Diamond Content Pipe	
	Ru	**Rh**	**Ru**	**Rh**
Kimberlites	0.5-1.6	<0.5	1.3-2.8	0.5-1.2
Xenoliths	2.0-7.3	1.6-3.3	5.2-12.0	1.8-14.0

ACKNOWLEDGEMENTS

The authors gratefully acknowledge the long-term encouragement of VS Letokhov at the Institute of Spectroscopy, the stimulating cooperation with W Alvarez at the University of California in Berkeley, and AW McMahon at Harwell Laboratory, UK. The TARIS acronym was suggested by N Thonnard of Atom Sciences, Inc. to whom GIB is very thankful for help and useful discussions.

REFERENCES

Bekov GI, Radayev VN and Letokhov VS 1988a Spectrochim. Acta 43B 491.
Bekov GI, Radayev VN, Letokhov VS, Badyukov DD and Nazarov MA, 1988b Nature 332 146.
Bekov, GI and Letokhov VS, 1989 Resonance Ionization Spectroscopy 1988 ed TB Lucatorto and JE Parks (Bristol: Inst. of Physics) pp 331-36.
Bekov GI, Kolpakov IV, Radeav VN and Veselov VA 1991 Resonance Ionization Spectroscopy 1990 ed JE Parks and N Omenetto (Bristol: Inst. of Physics) pp 265-70
Groves DJ, Ho SE and Rock MS 1987 Geology 15 801.

Inst. Phys. Conf. Ser. No 128: Section 6
Paper presented at RIS 92, Santa Fe, NM, USA, 24–29 May 1992

High-precision thorium RIMS for geochemistry

B L Fearey, B M Tissue, J A Olivares, G W Loge, M T Murrell, and C M Miller[†]

Isotope Sciences Group, INC-6, †Nuclear Chemistry and Analysis Group, INC-13
Los Alamos National Laboratory, Los Alamos, NM 87545

ABSTRACT: We present recent results in the development of resonance ionization mass spectrometry (RIMS) using continuous-wave lasers to measure thorium isotope ratios of importance in geochemistry and geochronology. The ionization efficiency, precision, accuracy, and biases in $^{230/232}$Th ratios measured by cw RIMS using broad-band and narrow-band lasers are discussed. Spectra of ^{229}Th, which potentially can be used as an internal standard, are also presented.

1. INTRODUCTION

Isotope ratio measurements of $^{230/232}$Th are used in uranium-series disequilibrium dating of geological samples over the last 350,000 years (Ivanovich and Harmon, 1982). Characterizing geological materials over this time scale is critical to the understanding of recent dynamic geological processes, which are important for assessing geological hazards to nuclear waste repositories, and for studying man's influence on the environment. Uranium-series dating originally utilized alpha spectrometry, which required large sample sizes and provided poor precision. Mass spectrometric methods have greatly reduced the measurement time and the necessary sample sizes, as well as increased the precision of the ratio measurement (Goldstein, *et al.* 1989 and references therein). Our goal has been to further reduce sample size requirements by using the higher ionization efficiency achievable for thorium in cw RIMS compared to thermal ionization mass spectrometry (TIMS). Analysis of smaller sample sizes allows use of the internal isochron method with mineral separates, thereby increasing the precision and accuracy of the dating method (Allegre and Condomines 1976).

In this paper we report recent results on $^{230/232}$Th ratio measurements using cw RIMS. The main concerns addressed are: (1) precision and accuracy in the ratio measurements, (2) laser-induced ratio biases and (3) isotopic selectivity. We compare two approaches, use of a broad-band laser to simultaneously ionize both isotopes, and use of a narrow-band laser which is tuned to each isotope in turn. A broad-band laser should provide more efficient ionization than a narrow-band laser since its bandwidth encompasses the Doppler width and isotope shift. Since the ^{230}Th and ^{232}Th isotopes have an even number of neutrons and zero spin, there is no hyperfine splitting. The power and frequency stability of a broad-band laser is typically not as good as that of a narrow-band laser and, hence, can introduce a higher level of noise in the ratio measurements. The effectiveness of an optogalvanic frequency-locking method to reduce this problem is under evaluation. A narrow-band laser has the advantages of being actively stabilized and providing isotopic selectivity. Isotopically-selective ionization with a narrow-band laser increases the dynamic range of ratio measurements, under our experimental conditions, by about a factor of thirty, (depending on the isotope shift and the degree of saturation broadening). A high-resolution-spectroscopic study of saturation-broadening effects in thorium RIMS spectra is presented elsewhere in this volume (Tissue, *et al.* 1992).

2. EXPERIMENTAL

A description of the cw-RIMS technique and apparatus as applied to thorium can be found in Fearey, *et al.* (1991) and Johnson, *et al.* (1992). A comprehensive listing of many thorium transitions can be found in Engleman and Palmer (1983). We have studied the two most efficient RIMS transitions at 26036.35 and 26113.27 cm^{-1} (transitions are referred to by the ^{232}Th energies in vacuum). Resonant excitation was provided by either a frequency-doubled Ti:sapphire laser (Coherent model 899-21) or a standing-wave dye laser (Coherent model 599). The bandwidth of the Ti:sapphire laser was *ca.* 2 MHz, while the bandwidth of the dye laser was *ca.* 1 cm^{-1}. A laser power controller maintained laser power at a constant level. The ionization process was enhanced with an additional Ar^{+} laser operating in the ultraviolet (333-363 nm). The resonant and enhancement lasers were overlapped using a dichroic beamsplitter and were focused into the mass spectrometer source region with a 15-cm focal-length quartz lens. Samples were prepared by depositing solutions of thorium with a graphite slurry overcoat onto a rhenium filament for atomization in the source chamber. The pressure in the ion source region was typically $< 1 \times 10^{7}$ torr. Ion dispersion and detection was accomplished through a magnetic-sector mass spectrometer with an electron multiplier tube and pulse counter detection system. A PC computer controlled the mass spectrometer and the Ti:sapphire laser frequency, and collected data from the pulse counter.

3. RESULTS AND DISCUSSION

3.1 RIMS Using a Broad-Band Laser

The broad-band dye laser excites both isotopes simultaneously but not necessarily equally. Experimentally, the laser is typically tuned to the peak of a ^{232}Th line, unfortunately, this procedure introduces a large bias in the ratio measurement. Table 1 shows some of the ratios obtained for the 26036.35 and 26113.27 cm^{-1} transitions for dye laser powers of 15 and 20 mW, respectively, with enhancement uv laser power of 4.9 W.

Table 1. $^{230/232}$Th ratio measurements using the broad-band laser.

Method	Resonant laser frequency (cm^{-1},vac.)	Isotope shift (cm^{-1})	$^{230/232}$Th $\pm$ 1 std dev (x10^{-4})
TIMS	---	---	4.096 $\pm$ 0.009
RIMS	26036.35	-0.5194	1.600 $\pm$ 0.042
RIMS	26113.27	-0.3926	2.370 $\pm$ 0.081

The difference in the bias in the ratio for the two transitions (compared to the TIMS result) is due primarily to the difference in the isotope shift. The RIMS signal changed with laser power, although less than linearly, indicating the transitions were not fully saturated at our laser powers. The bias in the measured ratio using the 26036.35 cm^{-1} transition did not depend on changes in dye laser power or sample temperature. However, it did depend on the dye laser operating conditions, dye laser frequency and on the percentage of thermally-generated ions to laser-produced ions (1-20% depending on sample size and the conditioning of the sample filament). The precision of the ratio measurement depended primarily on laser power fluctuations and drift of the laser wavelength. The power fluctuations can be damped with the laser power controller and the frequency drift can be eliminated by externally locking the laser frequency onto an atomic resonance. To frequency-lock the laser a portion of the beam can be split off through a chopper into a hollow-cathode lamp. The reference signal from the chopper and the optogalvanic signal from the hollow-cathode lamp is then fed into a lock-in amplifier. The lock-in output is used to actively control piezoelectric transducers affixed to the birefringent filter tuning element in the laser (David and Gagne 1990). This frequency-locking method was demonstrated using a lutium lamp to measure the stability of lutium RIMS signal and precision in $^{175/176}$Lu ratios. The frequency locking method maintained a constant frequency over time, and improved the precision and reproduciblity of the $^{175/176}$Lu isotope ratios by approximately a factor of three. The internal precision of thorium ratio measurements using the laser power controller and frequency locking is expected to be 1% (2σ) or better. The reproducibility of the bias in the ratio measurement will depend on the reproducibility of the laser conditions (spatial mode, bandwidth, and longitudinal mode structure) and on the ratio of thermally-generated ions to RIMS-generated ions. Accurate ratio measurements will require careful control of experimental conditions to minimize day-to-day changes.

3.2 RIMS Using a Narrow-Band Laser

In the narrow-band thorium RIMS work, the Ti:sapphire laser was scanned by computer prior to each ratio set and centered on the thorium resonance for each isotope. A summary of the $^{230/232}$Th ratio measurements for a thorium standard ($^{230/232}$Th $\cong 4 \times 10^{-4}$), 10 ng loading, obtained under various conditions is presented in Table 2. The overall RIMS signal level was essentially the same for both the narrow-band and broad-band lasers, indicating that the saturation broadening obtained under the narrow-band laser excitation effectively probed all atoms within the Doppler profile.

Table 2. $^{230/232}$Th ratio measurements using the narrow-band laser.

Method	Resonant laser frequency (cm^{-1},vac.)	Enhancement laser power (W)	$^{230/232}$Th $\pm$ 1 std dev ($\times 10^{-4}$) (# of data points)	
TIMS	--	--	4.096 $\pm$ 0.009	(7)
RIMS	26113.27	--	2.985	(1)
RIMS	26113.27	1.9	3.684 $\pm$ 0.130	(11)
RIMS	26036.35	1.9	3.825 $\pm$ 0.067	(3)
RIMS	26036.35	2.9	3.919 $\pm$ 0.027	(9)
RIMS	26036.35	5.2	4.080 $\pm$ 0.035	(5)

The bias in the $^{230/232}$Th ratio (compared to the TIMS measurement) was larger with the resonant laser at 26113.27 cm^{-1} than at 26036.35 cm^{-1}. The large bias at 26113.27 cm^{-1} is due to the proximity of an autoionizing level which clearly is more resonant with ^{232}Th than with ^{230}Th (Johnson, et al. 1992). The bias can be reduced by adding the enhancement laser but was still not as small as when using the 26036.35 cm^{-1} ^{232}Th transition. The source of the bias for the 26036.35 cm^{-1} transition is not well defined, but can be effectively eliminated by increasing the power of the uv enhancement laser. The Ti:sapphire laser power was varied over two orders of magnitude (0.2 and 20 mW) and had no effect on the bias in the ratio. The internal precision of any given set of RIMS measurements matched the internal precision of TIMS ($\leq 0.5\%$, 2σ). The standard deviations in Table 2 indicate the day-to-day reproduciblity of the measurement. Some of the scatter in the RIMS measurements is due to thermal fractionation of the isotopes from the rhenium filament (freshly-loaded filaments were not always used for RIMS as was done for the TIMS measurements). Further development of the sample-heating procedure specifically for RIMS is expected to eliminate this source of error and improve the reproducibility.

3.3 Spectra of ^{229}Th

The isotope ^{229}Th is used as an internal standard in isotope-dilution TIMS ratio measurements. Using ^{229}Th in RIMS ratio measurements may not be straightforward since hyperfine splitting in the ^{229}Th spectra could make the RIMS signal very sensitive to small changes in laser frequency and power. Only single peaks were observed in the ^{229}Th spectra at 26287.7 and 26113.9 cm^{-1}

(experimentally-determined isotope shifts of -0.65 and -0.62 cm^{-1} from the 26287.05 and 26113.27 cm^{-1} ^{232}Th transitions. Figure 1 shows the 26113.9 cm^{-1} spectra by (a) conventional RIMS and (b) by Doppler-free spectroscopy using counterpropagating laser beams (*i.e.*, probing for saturation (Lamb) dips). The Doppler-free spectral line is broadened and flattened, but shows no apparent features indicative of underlying hyperfine lines. The evidently simple ^{229}Th RIMS spectra makes it potentially useful as an internal standard for isotope-dilution RIMS measurements. However, further study is needed to determine if the ^{229}Th spectra are as simple as they appear, or if some peaks are not observed due to the experimental conditions. For the ^{229}Th 26287.7 cm^{-1} spectrum (F'=5/2←F"=7/2) three transitions (out of eight) could disappear for certain experimental conditions (*i.e.*, for certain polarizations and laser powers) due to optical pumping into inaccessible m_F states (Bushaw, 1989). The 26113.9 cm^{-1} transition (F'=7/2←F"=7/2) can not undergo m_F optical pumping, however, calculations of hyperfine level positions show that the observed spectra in Figure 1 could be produced if the ground and excited states have nuclear magnetic hyperfine splitting constants, A, that are nearly equal in magnitude and equal in sign.

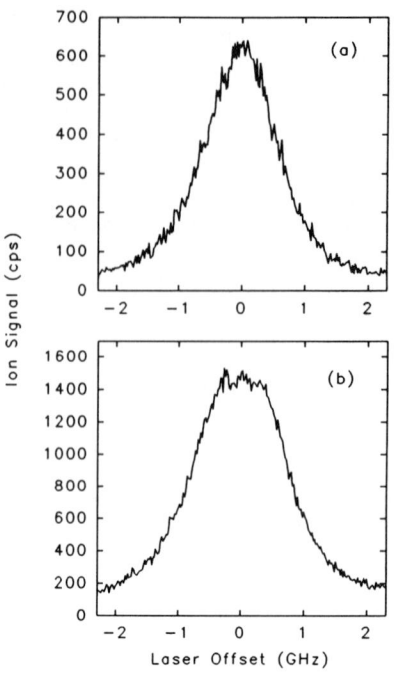

Figure 1 ^{229}Th RIMS spectra at 26113.9 cm^{-1} (a) single beam, (b) Doppler-free.

4. CONCLUSIONS

The precision of ratio measurements using the broad-band dye laser could be greatly improved with a laser power controller and external frequency locking. However, the precision will still be a major concern versus using the narrow-band Ti:sapphire laser because of potential mode structure effects. More importantly, the Ti:sapphire laser had a negligible bias in the ratio so that external effects, such as thermal ionization, do not need to be corrected. Further work is planned; 1) to determine the sensitivity of the ratio bias on the enhancement laser parameters, 2) to ensure the day-to-day reproducibility of the ratio measurements, and 3) to determine the effect of saturation broadening on the ionization selectivity in very-high dynamic range ratio measurements.

5. REFERENCES

Allegre C J and Condomines M 1976 *Earth Planet. Sci. Lett.* **28** 395
Bushaw B A 1989 *Prog. Analyt. Spectrosc.* **12** 247
David E and Gagne J-M 1990 *Applied Optics* **29** 4489
Engleman Jr. R E and Palmer B A 1983 *J. Opt. Soc. Am.* **73** 694
Fearey B L, Johnson S G, Nogar N S, Murrell M T and Miller C M 1991 in *Resonance Ionization Spectroscopy 1990* Parks J E and Omenetto N, eds. (Bristol: Institute of Physics) pp 311
Goldstein S J, Murrell M T and Janecky D R 1989 *Earth Planet. Sci. Lett.* **96** 134
Ivanovich M and Harmon R S, eds. 1982 *Uranium Series Disequilibrium: Applications to Enviromental Problems* (Oxford, Claredon Press)
Johnson S G, Fearey B L, Miller C M and Nogar N S 1992 *Spectrochim acta* accepted
Tissue B M, Miller C M and Fearey B L 1992 *elsewhere in this volume*

Inst. Phys. Conf. Ser. No 128: Section 6
Paper presented at RIS 92, Santa Fe, NM, USA, 24–29 May 1992

213

Determination of uranium at trace levels in nuclear scintillators

P. Benetti, G. Cecchet, M. Cola, M. Rossella and F. Sigon*

University of Pavia and INFN, sezione di Pavia - (Italy)
*Enel-Crtn, Milano (Italy)

ABSTRACT: RIMS-TOF technique, coupled with the laser ablation of the target, has been used for the determination of uranium traces in TMB; the samples were electrochemically deposited on Re substrate.
UO_2^+ turned out to be the measured form with a nominal very good LOD. However the purity of chemicals used and contaminations ascribed to the handling, currently set severe limits in the actual results.

1. INTRODUCTION

In the past years, RIMS has been claimed to be a powerful analytical technique and several very interesting limits of detection (LOD) were presented. These data were also supported by detailed discussions of the different steps of the process, i.e. atomization, photoionization, ion collection and detection etc. However, often the given LOD are extrapolated values, starting from much larger quantities.
In the present work our attempts in measuring low concentrations of uranium in real samples using laser ablation-RIMS technique will be described.

2. EXPERIMENTAL SET-UP

The experimental set-up is shown in Figure 1. For the following measurements only one dye laser (Lambda Physik LPD 3000) has been used (one colour, multisteps photoionization), at 591.5 nm, with an energy pulse of 6 mJ and a laser spot of about 3 mm in the interaction region. The ionizing wavelength has been carefully tuned through the so called "fast optogalvanic effect" (Benetti 1987) generated in an uranium loaded hollow cathode lamp. The laser nominal bandwidth is 6 GHz.
The sample is ablated by the second harmonic of a Q-switched Nd-Yag laser (spot 0.3 mm, energy up to 50 mJ/cm^2). By means of a photodiode, this laser acts as the master trigger for the whole system (excimer and digitizing electronics).The optimum delay between Nd-Yag and dye laser has been found, in our experimental conditions, to be 3.5μs; the time jitter falls in the negligible 20ns range.
The TOF is a home made Wiley-McLaren unit, 2m long, having a measured resolution of 300 at mass 270.
The transient digitizer is the CAMAC based LeCroy TR8828C operating in the integrating mode. Using the 3968 Kinetic Systems Controller which has a DEC rt-VAX 300 on board, it is possible to record up to 80μs each shot, at 10 Hz repetition rate; setting an initial delay of 10μs, this allows for acquisition of masses up to 900 m/z.
During the measurement, the vacuum is better than 10^{-7} torr.

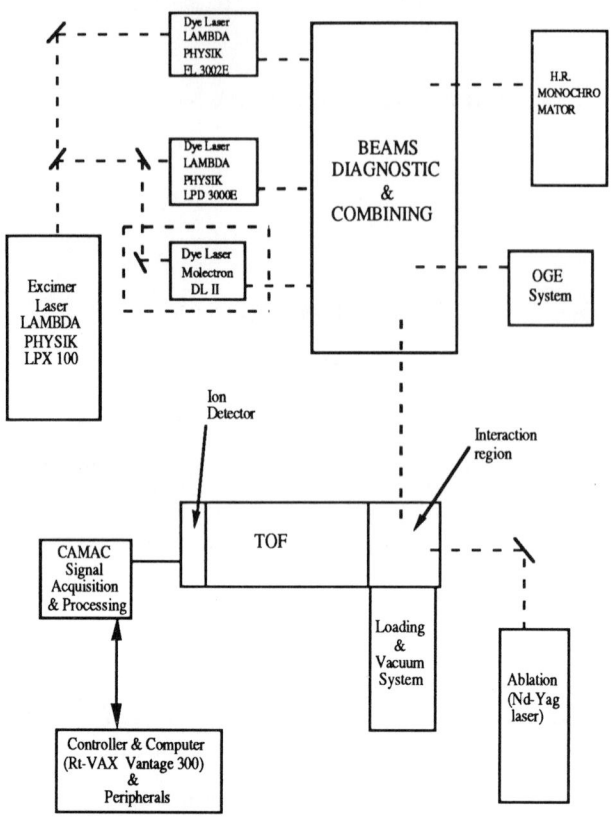

Fig. 1. RIMS-TOF experimental set-up

3. SAMPLES PREPARATION AND RESULTS

As previously said, the final aim of this work is to measure the content of uranium in threemethylborate (TMB), a liquid nuclear scintillator, which is a volatile compound easily damaged by air humidity. Accordingly, as a first approach it has been decided to electrochemically deposit its uranium content on a thin Re substrate. However, preliminarly we have tested some electrodeposited acqueous standards prepared from UO_2Cl_2, processing 0.5 ml of solution. A typical result obtained is reported in Figure 2, where TOF spectrum doesn't show peaks of atomic ions, but fortuitously ionized UO_2^+ shows up.

Although aware of possible contamination in real analysis, we attempted to chemically reduce the deposit, for instance by heating it in a reducing environment or according to other procedures described in the literature (Elewady 1986)].

These attempts have been done in order to obtain a common chemical basis, since the speciation of uranium in TMB is unknown. As shown in Figure 3, the reductions have been only partially successful; UO^+ now exceeds UO_2^+, and U^+ is just coming out.

Because of the unexpected troubles, it has been decided to insist with acqueous standards instead of TMB, and we tried also to get a calibration curve, as reported in Figure 4. A signal at no added uranium (blank) is clearly present.

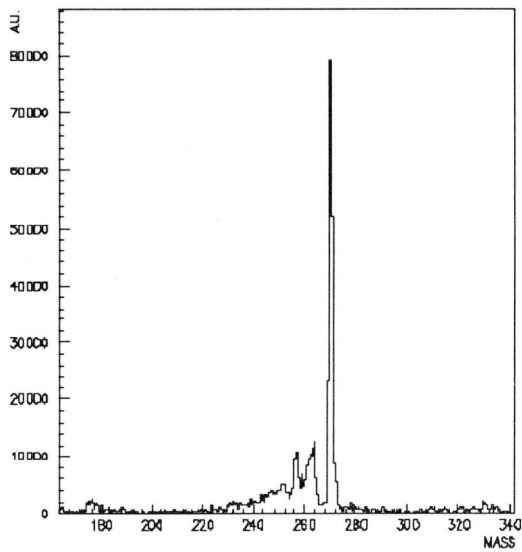

Fig. 2.   TOF spectrum obtained from a part (10^{-2}) of 100 pg of U
electroprocessed. Laser shots : 5000.

As a matter of facts, although from the results of Figure 2 the extrapolated LOD (3σ) lies in the low fg range (nominally ppq concentrations), the processing of standards below pg levels has been found useless because a comparable content of U in the electrolytic solution itself (5pg in 0.5ml, corresponding to 10 ppt); this solution contains NH_4Cl 1.5 M and HCl 1.5 M. Any attempts to reduce this contamination, for example doing a preliminary electrolysis on the reagents, were up to now uncertain and we found this contamination both systematic and erratic. This is consistent with the fact that the best available water itself contains 10^{-2} ppt of uranium, while ultrapure reagents are quoted for $\leq$ 5 ppt; maybe special or customer repurified chemicals could do better. Furthermore the results were quite fluctuating from sample to sample, even operating under laminar flow hood and clean environment. This behaviour affects either standards and blanks; it suggests that the risk of contamination is rather frequent, and, as expected, physico-chemical operations must be avoided as much as possible.A further example of possible contamination is given by the results obtained in an old sample of TMB, which totalizes 0.2 ppb of uranium. Unfortunately for the moment freshly prepared samples were not available to RIMS and have been independantly analyzed only with a HR ICP-MS installed at JRC in Ispra (Italy); these samples gave orders of magnitude lower values.
Afterwards the results obtained with the two instruments have been compared, still as acqueous standards. It turns out that, assuming as reference the HR ICP-MS, the samples analyzed by RIMS provide, on the average, values in agreement within a factor of 2; more precisely, it seems that the yield of electrodeposition is around 50%.

4. CONCLUSIONS

The calculated limits of detection achievable with RIMS TOF technique are excellent although questionable in real analytical measurements. In the present work, for instance, it has been found that the purity grade of chemicals used together with the handling of the samples are a strong limiting factor and vanish the potential advantages of the technique.

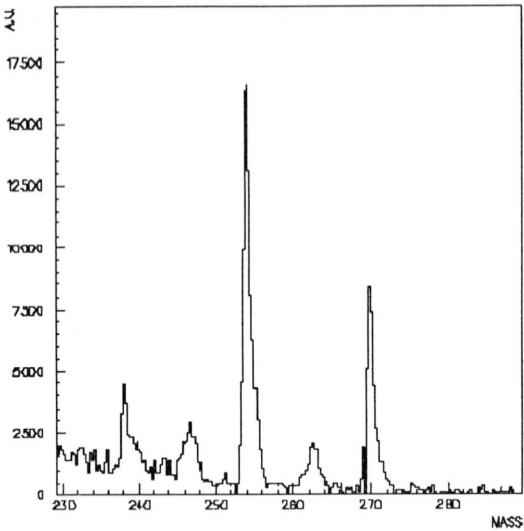

Fig. 3. TOF spectrum resulting from electrodeposition and subsequent chemical reduction.

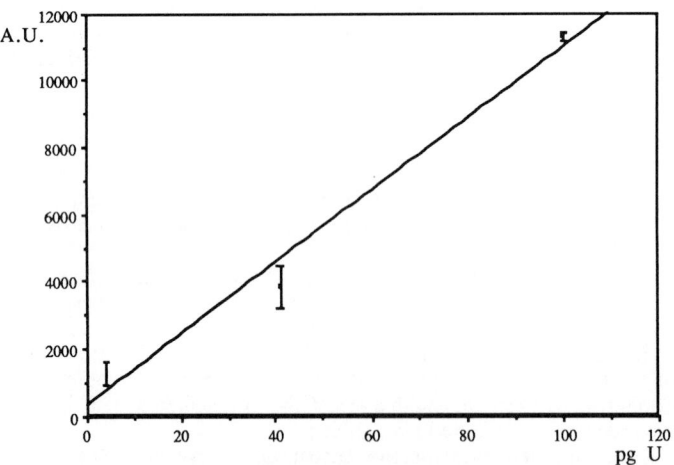

Fig. 4. Calibration curve for UO_2^+, from U standard solutions

Of course these drawbacks are not hanging when artificial isotopes or no chemicals are involved. Our work is currently in progress towards this last direction.

References

Benetti P , Tomaselli A and Zampetti P 1987 SPIE Proceed. 701 496
Elewady Y A 1986 J.Radioanal. Nucl. Chem. Letters 103 207

Inst. Phys. Conf. Ser. No 128: Section 6
Paper presented at RIS 92, Santa Fe, NM, USA, 24–29 May 1992

Microanalysis of gold in minerals using time-of-flight-sputter initiated resonance ionization spectroscopy (TOF-SIRIS)

S.L. Wang, J.H. Tian, W.Y.Ma, K.L.Wen, D.Y.Chen

Laser Single Atom Detection Laboratory, Department of Physics, Tsinghua Univesity, Beijing 100084, China

ABSTRACT: A technique, based on sputter initiated resonant ionization spectroscopy and combined with time of flight mass spectroscopy(TOF—SIRIS), has been recently developed for miroanalysis of mineral samples and other solid materials. Its sensitivity limit now has been reached 40ppb. With this technique, some research on finding out fine discrete mineral grains of gold at low concentrations trapped in pyrite and vertical grain—size analysis of gold have been done. These information are helpful for geologists to evaluate the inborn state of gold in mineral samples.

1. INTRODUCTION

It has been known that gold is usually presented in pyrite as discrete mineral grains. It is important to investigate the shape, size and composition of the discrete mineral grains. Electron probe micro—analysis (EPMA) is very mature and is routinely used to analyze the invisible gold in the discrete mineral grains. However, EPMA can only analyze the major constituents exposed to the sample surface. For minor constituents analysis and depth profiling an ion probe and secondary ion mass spectrometry (SIMS) can be used. But unfortunately, SIMS is difficult to quantify most measurements because the formation of secondary ions is strongly influenced by electric effects arising from the sample matrix. In addition, SIMS usually abound with mass interferences especially for mineral sample analysis.

In order to minimize matrix effects and mass interferences, we have established an apparatus using tunable lasers for stepwise resonance ionization of the neutral atoms sputtered from the sample. This technique will be demonstrated to be an highly sensitive microanalysis tool because of the using of high efficient resonance ionization spectroscopy (RIS) process.Mass interferences can be reduced to be eliminated by the high selectivity of RIS. Detection of the neutral atoms should dramatically reduce the matrix effects and improve the prospects for quantitative analysis. (Kimock et al, 1984)

This technique allows direct analysis of mineral sample without complicated preparation. It can search for fine discrete mineral grains trapped in mineral sample by sputtering to peel off the sample. We have done some research on vertical grain—size analysis of gold.

2. APPARATUS

The TOF—SIRIS apparatus was designed and built by ourselves. A schematic representation of this apparatus is shown in Fig.1.

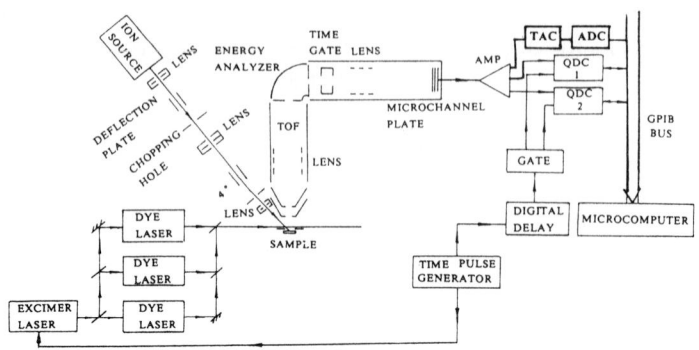

Fig.1   Diagram of TOF–SIRIS apparatus

A primary beam of Ar^+ ions is generated by a duoplasmatron ion source. The kinetic energy of the primary beam can be varied from several hundred ev to 16KeV. After extraction from this source, the ion beam is focused onto a sample in the analysis chamber by a series of Enzel lens and passed through differential pumping system so high vacuum condition can be maintained in the analysis chamber. The analysis chamber allows precision sample manipulation for X–Y–Z and rotation around Z–axis. The ion beam is pulsed at 1–100Hz (defined by the laser) by deflection it through a 150μm diameter aperture with a 440v pulse to a set of parallel plates. Typically, the spot size of the primary ion beam (at 16 KeV beam energy) is less than 20μm in diameter. The primary beam is incident to the sample with an 45 ° angle. By sputtering of the pulse primary ion beam, a distribution of atoms vapor localized in space a few millimeters above the sample is produced. Lasers with appropriate wavelengths selectively ionize the desired atoms (e.g. Au) in the vapor. The produced RIS ions are subsequently extracted and accelerated into a TOF analyzer and with a MCP detector.

The RIS process is carried out by three dye lasers (FL3002E, FL3002EC of Lambda–Physik, and another dye laser made by ourselves) pumped by an eximer laser (model 202EMG, Lambda–Physik). The eximer laser can produce the laser pulse of 400mJ energy and 308nm wavelengths with a repetition rate up to 150Hz. By pumping dye lasers and frequency doubling, wavelength range from red to ultraviolet can be obtained.

Two modes of data acquisition is used, singe ion counting with pulse height discrimination using TAC combined ADC multichannel analyzer and analogous current pulse signal converting to digital by QDC (charge digital convertor).

The vacuum system is of a conventional bakable high–vacuum construction. The vacuum environment in analyzer chamber and TOF is 10^{-7} Torr. Since TOF–SIRIS is not strictly a surface analysis technique, the UHV capability may seem superfluous, however although a vacuum environment in 10^{-7} Torr region in chamber is generally adequate.

3. RESULTS AND DISCUSSIONS

3.1 Background suppression method

In our experiments there are two potential sources of background: Secondary ions (SIMS ions and nonresonant photoions. The photoions could be easily reduced by optimal choice of RIS scheme, ionizing atoms resonantly through autoionization state. The SIMS ions are much more important background of TOF–SIRIS and difficult to be rejected because of its wide energy distribution. However, as SIMS ion is created in different region from the RIS ion, it can be reduced by potential barrier and electrostatic energy analyzer.

The potential barrier is created by an ion extraction electrode system as shown schematically in Fig.2. The saddle point of the potential barrier is below the point at which RIS ions are created. So it can suppress most of the SIMS ions with low energy. The residual SIMS ions which have enough energy to get over the barrier can be reduced by the cylindrical electrostatic energy analyzer. The total rejection factor is about 8 orders of magnitude.

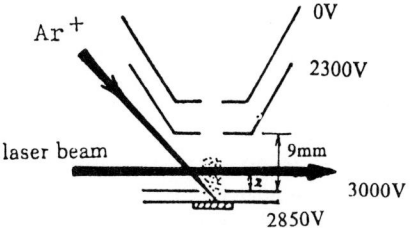

Fig.2 extraction / acceleration electrode system

3.2 Sensitivity

In order to improve the sensitivity of TOF–SIRIS, we have been using three–step resonant ionization through autoionization state for gold(as shown below), instead of two–step resonant excitation with nonresonant ionization through continuum state. The enhancement of ionization efficiency of gold is about 2–3 orders of magnitude. The nonresonant photoions are reduced.

$$5d^{10}6s^2S_{1/2} \xrightarrow{243nm} 5d^{10}6p\ ^2P_{3/2} \xrightarrow{479nm} 5d^{10}6d\ ^2D_{5/2} \xrightarrow{588nm} \text{Autoionization states}$$

The ore sample was used to demonstrate the present detection limit of TOF–SIRIS, which was supplied by Institute of Ore Deposit, Chinese Academy of Geological Science. Gold is present in pyrite as discrete mineral grains. The pyrite sample was pulverized to $500\mu m$ grains. Those gold rich grains were choose, mounted in room temperature polymerizing epoxy resin, polished to less than $1\mu m$ relief, and coated with approximately $200 Å$ of carbon. The assay value of the bulk pyrite ore is 86ppm.

First, ten arbitrary points on the ore sample were investigate to show the gold distribution in the pyrite sample. The result of this experiment is shown in Table 1.

Table 1

points	1	2	3	4	5	6	7	8	9	10
counts*	840	826	817	841	845	829	865	835	833	800

* before corrected by Possion statistics

It is obviously that gold is well–distributed in these discrete mineral grains.

Then the experiment was performed to investigate the preset sensitivity limit of TOF–SIRIS. Data were collected for 1000 laser pulses at 10Hz for a time of 100 second. The current of the primary ion is about $2–3\mu A$. The pulse width of the primary ion is $3\mu s$. When three lasers were tuned to ionize gold resonantly, the signal counts is 14176. Then detuning the first laser (243nm) $2 Å$, the counts (i.e. background noise) averaged about 2. Requiring a measurement to be 3 sigma above the background would establish the detection limit at 36ppb.

3.3   Vertical grain–size analysis of gold

A set of pyrite samples collected from Shandong Province of China were also supplied by Institute of Ore Deposit, Chinese Academy of Geological Science. The pyrite sample was pulverized to different size grains: 224mm, 140μm, 112μm, < 112μm. In fact, most of grains of the last group are about 60μm in size. Those grains of each group were partitioned and mounted in eight holder on one sample plate in room temperature polymerizing epoxy resin, polished to less than 1μm relief, and coated with approximately 200Å of carbon. The assay value for gold in these samples is 130ppm.

The first group (224μm) and the last group (< 112μm) were investigated. Three arbitrary points on samples in each holder were measured to search for gold discrete mineral grains. In 224μm grains only two gold discrete mineral grains have been discovered at two points after sputtering. The vertical grain–size of gold is estimated to about ten to twenty micrometer (as shown in Fig.3) . In grains of less than 112μm, it is easily to find fine discrete mineral grains, nearly at all measured points (21 points in 24 points). A typical result is also shown in Fig.3. The vertical grain–size of gold in these grains is about 2–6μm.

4. CONCLUSION

Several experiments have been performed to demonstrate the sensitivity limit of TOF–SIRIS to microanalysis at 20μm level for mineral samples and vertical grain–size analysis of gold in pyrite. The microanalysis sensitivity limit for minerals has been reached 36ppb. By rejecting the scattering SIMS ions and increasing the energy of second dye laser (479 nm) to saturate RIS process, the sensitivity should be possible to reach ppb level.

In order to investigate finer discrete mineral grains and microparticles of gold an liquid metal ion source (Ga$^+$) and a more precision sample manipulation system for X–Y using piezoelectric device has been underway during this time. A microscopic observing system with CCD camera and TV monitor for watching the sample with micrometer resolution will be installed.

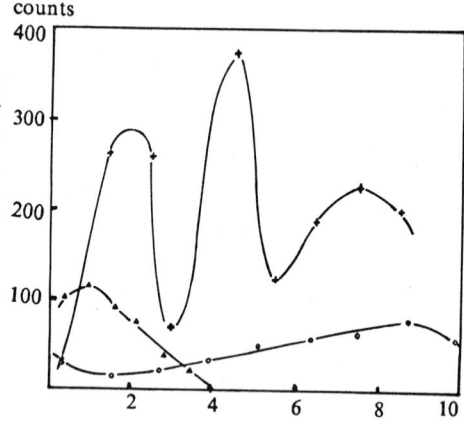

Fig.3   vertical grain size of gold.
(+)($\bigcirc$) discrete mineral grains in 224μm grains
($\triangle$) discrete mineral grains in grains less than
    112μm

References

Kimock F M, Baxter J P, Pappas D L, Kobrin P H, Winograd N, 1984, Anal, Chem., 56, 2782.

Inst. Phys. Conf. Ser. No 128: Section 6
Paper presented at RIS 92, Santa Fe, NM, USA, 24–29 May 1992

Trace analysis of neptunium with resonance ionization mass spectroscopy (RIMS)

J. Riegel, F. Albus*, F. Ames*, R. Deißenberger, G. Herrmann, H.-J. Kluge*, S. Köhler,
P. Sattelberger, F. Scheerer*, N. Trautmann, F.-J. Urban, H. Wendeler
Institut für Kernchemie and *Institut für Physik, Universität Mainz, D-6500 Mainz,
Fed. Rep. Germany

ABSTRACT: Resonance ionization mass spectroscopy (RIMS) is a very element-selective and sensitive analytical technique for the detection of trace elements. This method is based on the stepwise excitation and ionization of atoms with resonant laser light and followed by mass analysis. Our facility for RIMS consists of three dye lasers, pumped by two copper vapor lasers, and a linear time-of-flight spectrometer. For trace analysis of neptunium several two- and three-step excitation schemes have been investigated for maximum detection efficiency. With one of the schemes an overall efficiency of 3×10^{-8} was reached resulting in a detection limit of 4×10^8 atoms (160 fg) of neptunium. Furthermore, the first ionization potential of neptunium was measured to be $I = 6.2656(4)$ eV.

1. INTRODUCTION

Ultrasensitive detection methods with unambiguous element and in some cases isotope assignment have gained more and more importance in trace analysis. Such techniques are required, e.g., for studies of the ecological behavior of radiotoxic elements, which are present in the environment mainly as a result of global fallout from atmospheric nuclear weapons tests and releases from nuclear facilities. So far mostly radiometric methods have been applied for the determination of actinides. In the case of ^{237}Np, the ecologically most important isotope of neptunium, a detection limit of 4×10^{10} atoms can be reached by α-spectroscopy at a counting time of 1000 min. With the availability of powerful lasers yielding very high photon intensities at small spectral bandwidth resonance ionization mass spectroscopy (RIMS) has been established in the last few years as a powerful, alternative technique for fundamental research as well as for analytical applications (Hurst 1988, Letokhov 1987).

2. EXPERIMENTAL SET-UP

Our experimental set-up for resonance ionization mass spectroscopy consists of a high-repetition-rate pulsed laser system, a time-of-flight spectrometer and a data aquisition system (Ruster et al. 1989). Three tunable dye lasers (Lambda Physik, mod.2001) are simultaneously pumped by two copper vapor lasers (Oxford Lasers, mod. Cu-40). With a pump laser power of 30 W for each copper vapor laser, a pulse repetition rate of 6.5 kHz

Table 1: Two- and three-step excitation schemes for resonance ionization of neptunium. The J values are taken from Fred et al. (1977) and Worden et al. (1979).

No.	λ_1 (nm)	E_2 (cm^{-1})	J_2	λ_2 (nm)	E_3 (cm^{-1})	J_3	λ_3 (nm)
1	618.86	16154.3	9/2	576.63	33491.8	-	591.8 - 557.4
2	607.39	16459.3	9/2	585.73	33527.7	-	571.2 ; 581.1
3	587.81	17007.7	13/2	605.17	33527.7	11/2	571.2 ; 581.1
4	587.81	17007.7	13/2	610.93	33371.8	11/2	575.88
5	587.81	17007.7	13/2	612.29	33335.4	-	575.97

No.	λ_1 (nm)	E_2 (cm^{-1})	J_2	λ_2 (nm)
6	314.64	31773.8	13/2	$\leq$ 540
7	312.91	31949.4	9/2	$\leq$ 539
8	311.82	32061.1	11/2	542.99 - 536.11

and a pulse width of 30 ns, conversion efficiencies of up to 25 % have been obtained in a wavelength range from 520 nm to 850 nm. The bandwidth of the light emitted from the dye lasers varies between 3.5 GHz and 9 GHz depending on the dyes used and it can be reduced to 1 GHz by means of an intracavity etalon.

The dye laser beams are deflected into the interaction zone with the atomic beam in the time-of-flight spectrometer with the help of prisms or they are coupled into quartz fibres and focused by lenses into the interaction region, in order to achieve a homogeneous beam profile and an optimal spatial overlap of the beams. The samples are deposited by electrolysis on a rhenium foil and covered with a thin rhenium layer. An atomic beam is generated by heating this filament up to 1700 - 2000° C. The ions produced are accelerated to 2.9 keV by means of two grids and are counted in a channel plate detector after a drifting length in a field free region over 2 m. Mass spectra are recorded by time-to-pulse height conversion and multichannel pulse height analysis. The ion count rates for different masses as a function of the laser wavelengths are recorded with a LSI 11/23 microcomputer, which also controls the dye lasers.

3. MEASUREMENTS AND RESULTS

Starting from the ground state $5f^46d^17s^2$ $^6L_{11/2}$ several two- and three-step excitation schemes of neptunium were investigated for maximum efficiency. They are listed in Table 1. ^{237}Np with a nuclear spin of I=5/2 and an atomic angular momentum of J=11/2 shows large hyperfine structure (HFS) splitting. With a laser bandwidth of about 1 GHz the splittings have been measured in different optical transitions. From the viewpoint of sensitiv excitation paths with large HFS splitting should be avoided. As a result, even with broadband laser light it is not possible to cover the whole range, the overall detection efficiency decreases. Excitation scheme no. 4 (Table 1) shows the smallest HFS splitting. For this case the total splittings of the transitions have been measured to be 15 GHz in λ_1 and 18 GHz in λ_2. Figure 1 shows the result of a two-dimensional scan with a laser bandwidth of 1 GHz in λ_1 and λ_2.

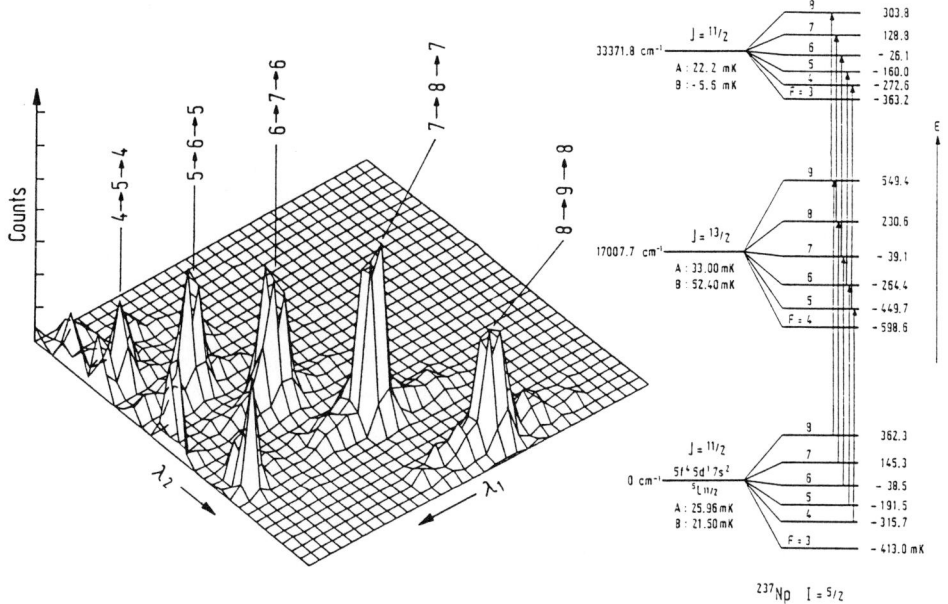

Figure 1: Two dimensional scan of λ_1 and λ_2 showing the HFS splitting of ^{237}Np. The HFS splitting factors A and B of the state at 33371.8 cm^{-1} were determined by this measurement.

We define the overall efficiency of the instrument as the ratio of the number of ions counted in resonance to the number of atoms placed on the filament. Measurements were performed both with synthetic samples of known ^{237}Np content (10^{12} atoms) and with samples after chemical separation from an aqueous medium. By use of excitation scheme no. 4, the highest overall detection efficiency of 3×10^{-8} has been reached, which corresponds to a detection limit of 4×10^8 atoms (160 fg) of ^{237}Np.

According to the classical saddle point model, the energy of the ionization threshold in our experimental set-up depends on the constant electric field strength E in the interaction region. Using excitation scheme no. 8 the wavelength of the ionizing laser was scanned in the region of the ionization threshold, increasing the total excitation energy $h(\nu_1 + \nu_2)$. Several mesurements were done with various electric field strengths. The ionization threshold at a given electric field strength is marked by a strong increase in the ion signal. The ionization energies obtained with different electric field strengths are plotted versus the square root of E in Figure 2. Linear extrapolation to E = 0 V/cm gives an ionization potential of I = 50535(3) cm^{-1} $\simeq$ 6.2656(4) eV which is in excellent agreement with the value obtained by Worden and Conway (1979) by Rydberg series convergence measurements (I = 6.2657(5) eV).

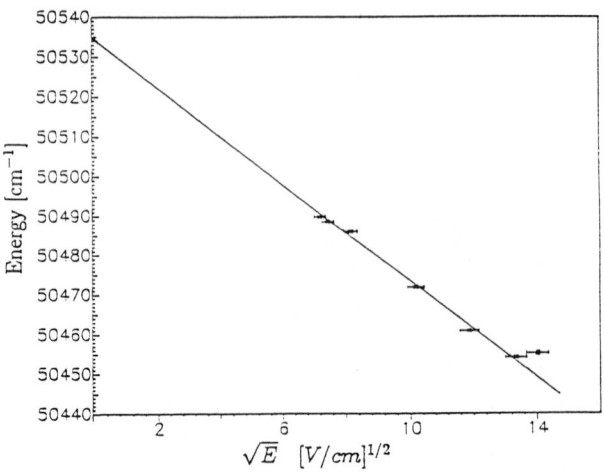

Figure 2: Effective ionization energies for different electric-field strenghts (E) in the interaction region plotted versus the square root of E (points). Linear extrapolation to E = 0 yields the first ionization potential of neptunium to be I = 50535(3) cm^{-1} $\simeq$ 6.2656(4) eV

4. CONCLUSIONS

Our experiments demonstrate that RIMS is a very sensitive method for trace analysis. For neptunium a detection limit of 4×10^8 atoms (160 fg) was reached, which is two orders of magnitude better than the value obtained by α-spectroscopy. This allows the detection of neptunium in the environment in smaller samples. The high sensitivity also offers the possibility of doing spectroscopy with a very small number of atoms, as demonstrated by hyperfine structure splitting measurements of ^{237}Np with 10^{12} atoms. The onset of ionization as a function of the electric-field strength within the interaction region makes it possible to determine the first ionization potential of neptunium to be I = 50535(3) cm^{-1} in good agreement with literature value. For the transplutonium elements the ionization limits are not known very accurately. We therefore plan to apply this method to these elements for which only small quantities are available.

This work was funded by the Bundesministerium für Umwelt, Naturschutz und Reaktorsicherheit under contract number StSch 4020.

References:
Fred M et al 1977 J. Opt. Soc. Amer. **67** 7.
Hurst G S, Payne M G 1988 Principles and Applications of Resonance Ionization Spectroscopy, Adam Hilger Ltd, Bristol.
Letokhov V S 1987, Laser Photoionization Spectroscopy, Academic Press, London.
Ruster W et al 1989 Nucl. Instr. Meth. **A281** 547.
Worden E F, Conway J G 1979 J. Opt. Soc. Amer. **69** 733.

Inst. Phys. Conf. Ser. No 128: Section 6
Paper presented at RIS 92, Santa Fè, NM, USA, 24–29 May 1992

225

Collinear resonance ionization spectroscopy for the determination of strontium-90 and strontium-89 in environmental samples

L. Monz, R. Hohmann, H.-J. Kluge, S. Kunze, J. Lantzsch, E.W. Otten,
G. Passler, P. Senne, J. Stenner, K. Stratmann, K. Wendt, K. Zimmer,
G. Herrmann, N. Trautmann, K. Walter

Institut für Physik and Institut für Kernchemie,
Johannes Gutenberg-Universität, D-6500 Mainz, Fed. Rep. Germany

ABSTRACT: An experimental method for the sensitive and selective detection of the radionuclides ^{90}Sr and ^{89}Sr in environmental samples has been worked out. The technique allows the quantitative detection of about 10^8 atoms of these isotopes in the presence of 10^{18} atoms of stable strontium. The sample is ionized and the ions are accelerated to an energy of 50 keV. In a conventional mass spectrometer the isotope under investigation is enriched by a factor of $>10^5$. The ions are neutralized and the resulting atoms are selectively excited to a Rydberg state by narrow-band cw laser light in collinear geometry and detected after field ionization. A total isotopic selectivity of S $>10^{11}$, an overall efficiency of $\varepsilon \approx 10^{-6}$ and a detection limit for ^{90}Sr of $5 \cdot 10^8$ atoms have presently been achieved.

1. INTRODUCTION

The radioisotopes ^{90}Sr ($T_{1/2}$ = 28.5 a) and ^{89}Sr ($T_{1/2}$ = 50.5 d) are produced by fission of ^{235}U and other actinide nuclides. The nuclide ^{90}Sr is one of the most dangerous fall-out products as it is accumulated in human bones and causes a high local radiation dose. Since the quantitative radiochemical detection of the pure β-emitters ^{89}Sr and ^{90}Sr is difficult and time consuming, a new approach to fast detection by laser spectroscopy is presented in the following.

As laser spectroscopy is sensitive to both the stable and the radioactive isotopes, the ratio of concentration [^{nat}Sr]/[$^{89/90}$Sr] must be considered. In a typical air sample one expects about 10^{18}-10^{19} atoms ($\approx$ 100 μg) of natural strontium (mainly ^{88}Sr) per 1000 m^3. The ^{90}Sr concentration in air after the Chernobyl accident was about 10^8 atoms/1000m^3 in the northern part of Germany. Hence a selectivity of S $>10^{11}$ is required for adequate suppression of ^{88}Sr.

The optical selectivity is given by the frequency spacing between the resonances of two isotopes of concern and the natural line width of the resonances. In the case of strontium the required selectivity cannot be reached just by use of the isotopic shift since it is too small. Therefore collinear excitation of a fast atomic beam is used to produce a suffiently large velocity-dependent Doppler shift. Hereby the line spacing is artificially enlarged by about to orders of magnitude (Kaufman 1976). Furthermore this technique results in a sizeable narrowing of the Doppler width and thus to more efficient use of cw laser light. To achieve high sensitivity this method can be combined with resonance ionization spectroscopy (Kudriavtsev 1982).

2. EXPERIMENTAL

The experimental set-up is shown in Figure 1. After chemical extraction, the strontium sample is introduced into an ion source. Subsequent to atomization the atoms are ionized on a hot tantalum surface (T $\approx$ 2400 K) and the ions are accelerated to about 50 keV. A first suppression of the stable isotopes is achieved in a 60° sector magnet. After deflection of the ions by 10° to collinearly overlap the laser beam, neutralization is carried out in a charge exchange cell containing cesium vapor at a pressure of 10^{-3} $mbar$. Here the long-lived $5s5p\ ^3P_{0,1,2}$- and $5s4d\ ^3D_{1,2,3}$- states of strontium are predominantly populated by quasi-resonant charge exchange.

The isotope under investigation is excited to a high-lying Rydberg state by cw laser light in a field-free region. Then the Rydberg atoms are field-ionized and detected by a channeltron (Fig. 1). Background events are strongly suppressed by the following techniques: three ion deflectors, located along the beam path behind the charge exchange cell are used to clean the atomic beam of remaining ions and of Rydberg atoms produced by collisions in the charge exchange cell. Ions produced along the beam path by collisions with residual gas molecules are eliminated by energy selective detection: In the longitudinal ionization field the laser-excited Rydberg atoms are field-ionized at a well-defined potential. Therefore they experience an additional acceleration in contrast to the background ions. Hence the two species can be efficiently separated by electric-field deflection (Schulz 1991).

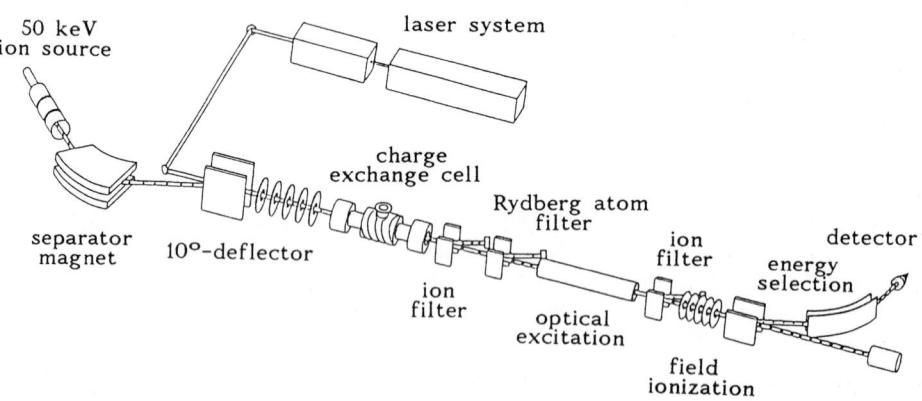

Fig. 1: Experimental set-up for the trace detection of ^{90}Sr and ^{89}Sr

3. RESULTS

The total isotopic selectivity is the product of the selectivity of the magnetic mass separation and of the collinear laser excitation. Both contributions can be investigated separately. Figure 2 shows the ion current as a function of the magnetic field strength. The ^{88}Sr signal is suppressed by a factor of about 10^5 at mass 89 and about 10^6 at mass 90. Contaminations of the sample or the ion source by ^{89}Y and ^{90}Zr are responsible for the small peaks at these mass values.

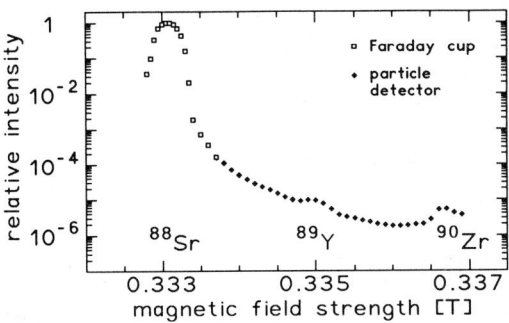

Fig. 2: Selectivity of the mass separator

The isotopic selectivity of the laser excitation has been investigated in the transition 5s5p $^3P_2 \rightarrow$ 5s21s $^3S_1$ (Fig. 3). On the lower-frequency side of the resonance the intensity decreases to about 10^{-4}, on the opposite side to $\langle 10^{-5}$ of the peak intensity of ^{88}Sr which is the limit due to background events. The asymmetric line shape is understood by taking into account the Stark effect in the electric fields of the deflectors and the field-ionization region. With anti-parallel atomic and laser beams

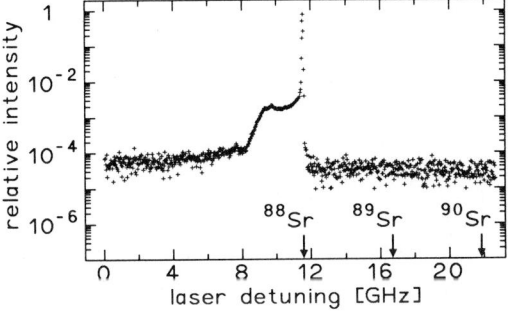

Fig. 3: Optical selectivity in the 5s5p $^3P_2 \rightarrow$ 5s21s $^3S_1$ ($\lambda \approx 326\,nm$) transition of Sr

the ^{89}Sr and ^{90}Sr resonances are located on the higher-frequency side of the ^{88}Sr resonance. Thus the Stark effect does not influence the ^{88}Sr/^{90}Sr and ^{88}Sr/^{89}Sr optical selectivity, which was found to be $\rangle 10^{11}$ and $\rangle 10^{10}$ (mass separation included), respectively.

The overall efficiency of the detection system is determined by the following factors:

* Production of the 50 keV ion beam	$\approx 2\,\%$
* Transmission of the apparatus	25–30 %
* Charge exchange	50 %
* Population of the 5s5p $^3P_{0,1,2}$ and 5s4d $^3D_{1,2,3}$ states	$\approx 5\,\%$
* Optical excitation into a Rydberg level	0.1–10 %
* Field ionization and ion detection	90 %
Total	$1 \cdot 10^{-7} - 1 \cdot 10^{-5}$

The efficiency of the ion source is limited by the ionization probability of tantalum used so far for surface ionization. While the transmission of the mass separator is about 90 %, the corresponding value for the subsequent apparatus does not exceed 30 %. These losses have to be accepted because the beam can be manipulated by ion optics only prior to entering the charge exchange cell.

The efficiencies of different single-step excitation schemes have been investigated. For a quantitative analysis, the population of the initial state, the transition probability to the Rydberg level, the lifetimes of the states involved, and the available laser power must be considered (Wendt 1992). The transition 5s4d $^3D_2$ → 5s 23f $^3F_3$ has proved to be best suited as the wavelength (364.8 *nm*) can be directly obtained with high power from a single-mode argon ion laser. With this excitation scheme an overall efficiency of the set-up of about $5 \cdot 10^{-6}$ has been realized.

Background count rates were determined to be $2 \cdot 10^{-8}$ events per incoming atom in the atomic beam. While collisionally produced ions are sufficiently suppressed by the filter deflectors and the technique of energy-marking in the field ionization region, the main component of the background is caused by Rydberg atoms produced in the charge exchange process. The total background has been found to be typically 300 counts per sample.

The presently achieved detection limit is $5 \cdot 10^8$ atoms of ^{90}Sr. As an example Fig. 4 shows the result of a measurement of $1 \cdot 10^9$ atoms of ^{90}Sr in the presence of $5.6 \cdot 10^{13}$ atoms of ^{84}Sr and $1 \cdot 10^{16}$ atoms of ^{88}Sr. This analysis took about one hour (chemical preparation of the sample not included).

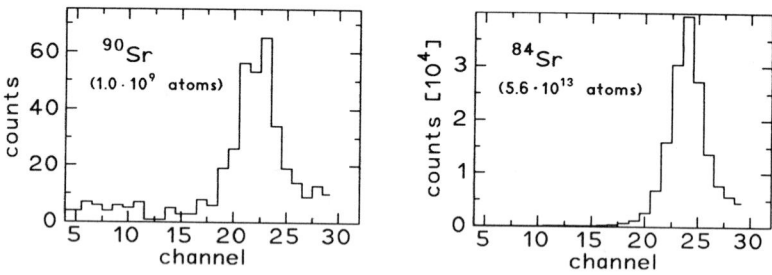

Fig. 4: Optical spectra of 10^9 atoms of ^{90}Sr and $5.6 \cdot 10^{13}$ atoms of ^{84}Sr obtained by collinear resonance ionization spectroscopy

4. OUTLOOK

For the near future a further improvement of the detection limit by about one order of magnitude is envisaged and first measurements of the odd-A isotope ^{89}Sr are planned. The use of ^{85}Sr as tracer isotope is intented for quantitative analysis of environmental samples.

This work has been funded by the Deutsches Bundesministerium für Umwelt, Naturschutz und Reaktorsicherheit under contract number St Sch 4020

REFERENCES:
Kaufman S L 1976 *Opt. Comm.* **17** 309
Kudriavtsev Y A and Letokhov V S 1982 *Appl. Phys.* **B29** 219
Schulz Ch et al. 1991 *J. Phys B* **24** 4831
Wendt K et al. 1992 *contribution to this conference*

Inst. Phys. Conf. Ser. No 128: Section 6
Paper presented at RIS 92, Santa Fe, NM, USA, 24–29 May 1992

229

Measurement of isotopic ratios by resonance ionization mass spectrometry in the presence of optical isotope shifts

R. K. Wunderlich*, G. J. Wasserburg, I.D. Hutcheon and G.A. Blake

Division of Geological and Planetary Sciences, California Institute of Technology, California, CA 91125, *Physikalisches Institut Universität Augsburg, W-8900 Augsburg, FRG

ABSTRACT: The effect of optical isotope shifts on Os and Ti isotopic ratio measurements by resonance ionization mass spectrometry has been investigated. Conditions with regard to laser bandwidth and intensity and the reproducibility of laser wavelength setting have been evaluated wich allowed Os and Ti isotopic ratios to be obtained with an accuracy better than 0.4%.

1. INTRODUCTION

The determination of isotopic ratios by Resonance Ionization Mass Spectrometry (RIMS) requires a thorough characterisation of laser induced isotope effects. Besides the widely discussed odd-even effects (Fairbanks et al.1989,Lambropoulos et al.1989, Payne et al. 1991, Wunderlich et al 1992 a,b) there have also been reports of laser induced shifts in even-even isotopic ratios (Joung et al.,1990 Miller et al.1990). Both effects complicate the application of RIMS to accurate measurements of isotopic ratios in geo- and cosmochemistry (Spiegel et al. 1992).
In this study we focus on the effects of the optical isotpe shift (IS) on the measurement of even-even isotope ratios of Os and Ti in order to evaluate the conditions which allow the measurement of isotopic ratios by RIMS with high accuracy

2. WAVELENGTH TUNING EFFECTS AND MASS FRACTIONATION

In the presence of optical isotope shifts the ionization efficiency of different isotopes depends on parameters such as laser bandwidth and intensity the atomic absorption width and the size of the IS. This results in a dependence of measured isotpic ratios on the precise laser wavelength tuning and intensity. By precise we mean that measured isotope ratios may change on a scale smaller then the laser bandwith with laser tuning. In addition, mass spectrometric analyses are always is characterized by pure mass fractionation which cause a shift in measured isotopic ratios as compared to the isotopic abundances present in the sample. This pure mass fractionation can be corrected by the application of a fractionation 'law':

$$R^{ij}_M \times f(\alpha) = R^{ij}_S$$

where R^{ij}_M is the measured ratio of ion currents at masses i and j, R^{ij}_S is the isotope ratio in a reference standard and $f(\alpha)$ is an empirical function of the form $(1 + n\alpha)$, $(1 + \alpha)^n$ or $\exp(n\alpha)$ with n= i-j. If at least two isotopes in the sample, i and j, are of known standard abundance the mass fractionation per mass unit, α, can be dertermined from measurements of R^{ij}_M. Pure mass fractionation is characterized by a constant, mass independent α. In order to obtain accurate isotopic ratios from RIMS measurements it is thus important to separate the shifts in isotopic ratios originating from pure mass fractionation from the effects of laser wavelength tuning.

The differences in the ionization probability of the even isotopes arise from the dependence of the effective absorption cross section of isotope i, $\sigma_A{}^i$,on the detuning of the laser center wavelength from the absorption maxima of isotope i. The discrete transition of isotope i is saturated if:

$$F_L \times \sigma_A{}^i \times t_P \gg 1$$

where F_L and t_P are the total flux of laser photons and pulse duration respectively. If this condition can be met simultaneously for isotopes i and j in a certain wavelength tuning interval, then $R^{ij}{}_M$ will be constant in that interval. The value of $R^{ij}{}_M (\lambda) = $ const. is then considered to represent the purely mass fractionated value of that isotopic ratio. If the pair i,j has the largest IS, all other isotopes can be measured simultaneously. Alternatively, if the different isotopes are spectrally well resolved, the purely mass fractionated ratios can be obtained from measurements of the ion current at the individual isotopic ionization maxima. This procedure however considerably complicates analytical applications.

3. EXPERIMENTS AND RESULTS

The experiments were performed with a thermal atom source in a modified quadrupole mass spectrometer. Os and Ti were ionized with 1+1 photoionization schemes with resonance transitions near 40000. cm^{-1}. Measurements were performed with a pulsed laser system. Details of the experimental set up have been described elswhere (Wunderlich et al. 1992 a and b).

Fig.1 shows the laser wavelength dependence of the ^{190}Os/^{188}Os isotope ratio, expressed as the per mil deviation from its standard reference value (δ -values) (Creaser et al. 1991). In this experiment the ratio of the laser bandwidth $\Delta\omega_L$ to the optical isotope shift of the transition ΔT was $\Delta\omega_L/\Delta T = 15$ ($\Delta\omega_L = 0.75$cm^{-1}, ΔT 0.05cm^{-1}). The measured isotope ratio is constant within a wavelength tuning interval of ≈ 0.4 cm^{-1}. For $\Delta\omega_L/\Delta T \approx 8$ the width of the plateau was reduced to 0.20 cm^{-1}. Similar experiments with Lu isotopes (Miller etal. 1990) performed at much lower laser intensity and with $\Delta\omega_L/\Delta T \approx 10$ did not reveal a plateau in the measured isotope ratios. The constancy ofhe measured isotope ratio is a result of the

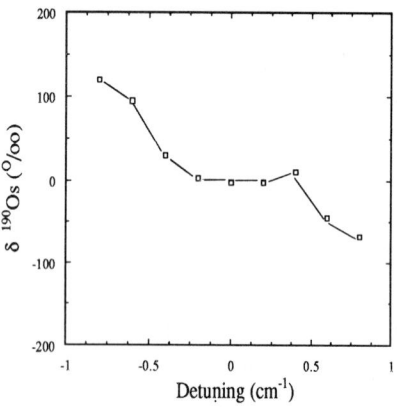

Fig. 1: Wavelength dependence of the ^{190}Os/^{188}Os ratio. J=4, 40361 cm^{-1}

strong power broadening of the resonance, following the arguments given above. The power broadend width of the resonance was estimated as 0.8cm^{-1}. The mass fractionation factor per mass unit, α, is then calculated from the value of the ^{190}Os/^{188}Os ratio measured in the region $R^{ij}(\lambda)=$const. Table 1 shows the α-values obtained for other Os isotopes with the laser tuned in the center of the region ^{190}Os/^{188}Os(λ) = const.. Within the precision of the experiment the α-values are indistinguishable , showing that these isotopic ratios differ from those present in the sample only by pure mass factionation and can thus be simultaneously measured with high accuracy. Using the condition ^{190}Os/^{188}Os(λ) = const. for laser wavelength tuning tuning we obtained in a series of different experiments a reproducibility of the ^{192}Os/^{188}Os ratio of 0.3 %. After mass fractionation correction ithe absolute value of the 192/188 ratio was indistinguishable from the standard ratio.

The width of the region $R^{ij}(\lambda)$=const. directly shows the the required accuracy of laser wavelength setting for a given set of laser parameters required to obtain reproducible isotope ratio measurements. For the case shown in Fig.1 this accuracy is easily obtained with e.g. a pulsed wave meter or a well calibrated dye laser.

The importance of power broadening is also demonstrated in the power dependence of even-even isotope ratios shown in Fig.2. The ^{190}Os/^{188}Os ratio was measured with $\Delta\omega_L/\Delta T$=7 at the maximum of the ^{190}Os ion signal and for a detuning of 0.08 cm^{-1}. It is seen that in particular for lower pulse energies the effects of a small laser detuning can be very large. Both curves converge at higher laser intensities. The maximum laser intensity in this experiment was 5×10^7 Watt/cm^2.

Isotopic ratio measurements of Ti proved more difficult due to the larger IS approaching the closed neutron shell with magic netron number 28 in ^{50}Ti. Compared to Os, the IS has a different sign. Optical isotope shifts were measured for different resonance transitions of the 3d^{3}4p and 3d^{2}4s4p excited state electron configuration (Wunderlich et al. 1992 b). Fig.3 shows the wavelength dependence of the ^{50}Ti/^{46}Ti ratio using the 39715.5 cm^{-1} (J=3, 3d^{3}4p configuration) resonance transition together with the ion signals at masses 50 and 46. In this experiment we had $\Delta\omega_L/\Delta T$=3 and $\Delta T(50$-$46)$=0.15 cm^{-1}. It is obvious that in this case isotopic ratios can *not* be measured with a single laser wavelength setting. For resonance states of the 3d^{2}4s4p electron configuration which have smaller IS, $\Delta T(50$-$46)$=0.06 cm^{-1}, it was possible to obtain a wavelength tuninginterval with ^{50}Ti/^{46}Ti(λ)=const. Very good agreement between isotopic ratios measured in this region $R^{ij}(\lambda)$=const. and ratios obtained from measurements of the ion current at the ionization maxima of the individual isotope for 3d^{3}4p resonance states, have been obtained. This agreement justifies the

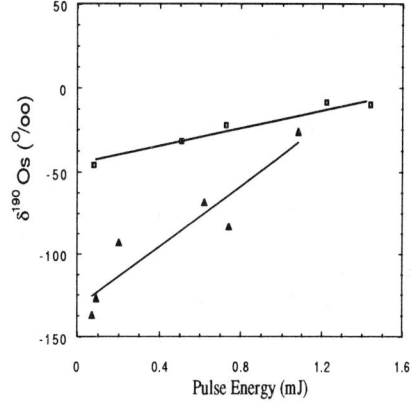

Fig.2: Power dependence of the ^{190}Os/^{188}Os ratio. Squares: zero detuning, triangles: detuning 0.08 cm^{-1}.

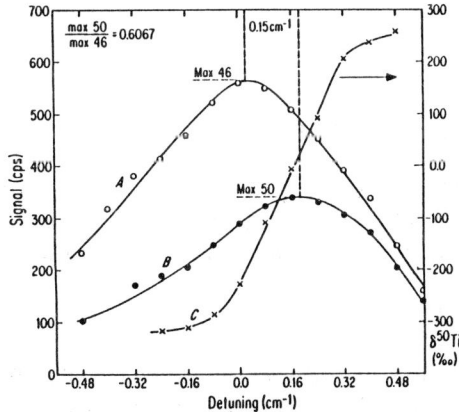

Fig.3: Wavelength dependence of the ^{50}Ti/^{46}Ti ratio (left) and ion signals at masses 46 and 50 (right)

identification of the ratio measured in the region ^{50}Ti/^{46}Ti(λ)=const. with the purely mass fractionated value of that ratio. The α-values obtained in a series of measurements are summarized in Tab.1. Within experimental precision the values are identical and it is again seen that the measured isotope ratios are characterized by pure mass fractionation. After application of mass fractionation correction, the measured ^{50}Ti/^{46}Ti and ^{48}Ti/^{46}Ti are indistinguishable from the standard reference values (Niederer et al.).

Table 1: Mass fractionation per mass unit for Os and Ti ratios measured in the region $R^{ij}(\lambda)$=const.

Os Ratios	$\alpha \times 10^3$	Ti Ratios	$\alpha \times 10^3$
192/188	7.0 (2)	50/46	25 (2)
190/188	6.5 (3)	48/46	24 (7)
192/190	8.0 (3)		

3. CONCLUSIONS

We have shown that under selected conditions even-even isotopic ratios differing from the 'true' ratios only by mass dependent fractionation can be measured by RIMS. Application of the mass fractionation correction then yields very good agreement with the standard ratios. A large laser bandwidth and strong power broading allow measurement at a single laser wavelength setting. This approach is important for analytical applications as compared to measurements of the ion intensity at the individual isotopic peaks which requires careful measurement of the ionization lineshapes of all isotopes. The accuracy of 0.3% of our best Os measurements suggests that it will be intersting to infer the absolute accuracy which can be obtained in this type of RIMS experiment. The importance of power broadening suggests that the intensity distribution in the laser beam and the focussing conditions may become limiting factors for the accuracy at a higher level of precision. The results given here, together with our measurements of the odd even effect in Os and Ti, show   that it is possible to obtain isotopic ratios by RIMS with an absolute accuracy of 0.4% or better.

ACKNOWLEDGEMENTS

This work was supported by DOE grant DE FG03-88ER-1351 and NSF grant EAR-8816936 to G. J. Wasserburg. The laser system was obtained through support from the Packard and Sloan Foundations and NASA grant NAGW -1955 to G. A. Blake.

REFERENCES

Creaser R A, Papanastassiou D A and Wasserburg G J 1991 *Cosmochim. Acta* **55** 397
Fairbank W M Jr., Spaar M T, Parks J E and Hutchinson J M R 1989 *Phys. Rev.* **A40** 2195
Lambropoulos P and Lyras A 1989 *Phys. Rev.* **A40** 2199
Miller C M, Fearey B L, Palmer B A and Nogar N S 1988 *Resonance Ionization Sectroscopy 1988,* eds.
    J E Parks and T B Lucatorto Institute of Physics Conf. Series **94**, (Bristol UK) 297-301
Niederer F R, Papanastassiou D A and Wasserburg G J 1981 *Geochim Acta* **45** 1017
Payne M G, Allman S L and Parks J E 1991 *Spectrochim. Acta* **46B** 1439
Spiegel D R, Pellin M J, Calaway W F, Burnett J W, Coon S R, Young C E, Gruen D M
    and Clayton R N 1992 *Anal. Chem.* **64** 469
Wunderlich R K, Hutcheon I D, Wasserburg G J and Blake G A 1992 a *Int J. Mass
    Spectrom. and Ion Processes,* in press
Wunderlich R K, Hutcheon I D, Wasserburg G J and Blake G A 1992 b submitted
    Anal. Chem.
Young Y P, Shaw R W, Goeringer D E, and Smith D H 1988 *Resonance Ionization Sectroscopy
    1988,* eds. J E Parks and T B Lucatorto Institute of Physics Conf. Series **94**, (Bristol UK) 367-371

Inst. Phys. Conf. Ser. No 128: Section 6
Paper presented at RIS 92, Santa Fe, NM, USA, 24–29 May 1992

233

Resonance ionization mass spectroscopy of plutonium with a reflectron time-of-flight mass spectrometer

F-J Urban, R Deißenberger, G Herrmann, S Köhler, J Riegel, N Trautmann, H Wendeler
Institut für Kernchemie, Universität Mainz, D - 6500 Mainz, Fed. Rep. Germany
F Albus, F Ames, H-J Kluge, S Kraß, F Scheerer
Institut für Physik, Universität Mainz, D - 6500 Mainz, Fed. Rep. Germany

ABSTRACT: Trace amounts of plutonium are detected by means of resonance ionization mass spectroscopy (RIMS). A three-step excitation scheme leading to an autoionizing state is used for the detection of trace amounts of ^{239}Pu down to 10^7 atoms. The isotope shifts (IS) of different plutonium isotopes were measured for several excitation schemes. Taking into account the isotope shifts a good reproducibility of the isotopic abundances is obtained with RIMS. In order to increase the detection efficiency a reflectron time-of-flight (TOF) mass spectrometer was built and tested with gadolinium.

1. INTRODUCTION

Resonance ionization mass spectroscopy is used for the detection of trace amounts of plutonium in environmental samples. From the fall-out of nuclear weapon tests trace amounts of plutonium are present in the environment. Thus a sensitive and element-selective analysis is required. Because the isotopic ratios of plutonium produced in nuclear weapon tests differ from those in reactors, the measurement of isotopic abundances is essential for determining the origin of plutonium samples.

2. EXPERIMENTS AND RESULTS

2.1. Experimental set-up

The experimental set-up consists of three tunable dye lasers which are pumped by two copper vapour lasers operating at a repetition rate of 6.5 kHz. The light of the dye lasers is coupled via mirrors or quartz fibres into the interaction zone, which is located at the starting point of a TOF mass spectrometer. There the laser beams interact with an atomic beam of plutonium produced by evaporation from a heated rhenium filament, where the plutonium has been deposited electrolytically. The atoms are photo-ionized in a three-step, three-colour resonant scheme. The third step leads to an autoionizing state above the ionization limit. The ions are accelerated by electric fields and mass-analyzed in the TOF mass spectrometer. A detailed description of the experimental set-up is given in Ruster et al (1989).

2.2. Isotope shift and hyperfine splitting of plutonium

The excitation scheme with $\lambda_1 = 586.49$ nm, $\lambda_2 = 665.57$ nm and $\lambda_3 = 577.28$ nm was found to be best suited for sensitive detection. The isotope shifts were measured in all three steps for the plutonium isotopes with mass numbers 240, 241, 242 and 244. These data are important for the determination of isotopic abundances. An intra-cavity etalon was placed in the dye lasers for reducing their bandwidths from usually 5 GHz to 1 GHz. Figure 1 shows the ion signal for the plutonium isotopes 240-244 as a function of the detuning of the first wavelength λ_1 at 586.49 nm with $\lambda_2 = 665.57$ nm and $\lambda_3 = 577.28$ nm kept fixed. Due to its nuclear spin of $I = 7/2$, ^{241}Pu shows hyperfine splitting. Isotope shifts of similar size were found for λ_3, whereas the isotope shifts in λ_2 are in the same range as the laser bandwidth. The measured isotope shifts are in good agreement with data published by Blaise (1986).

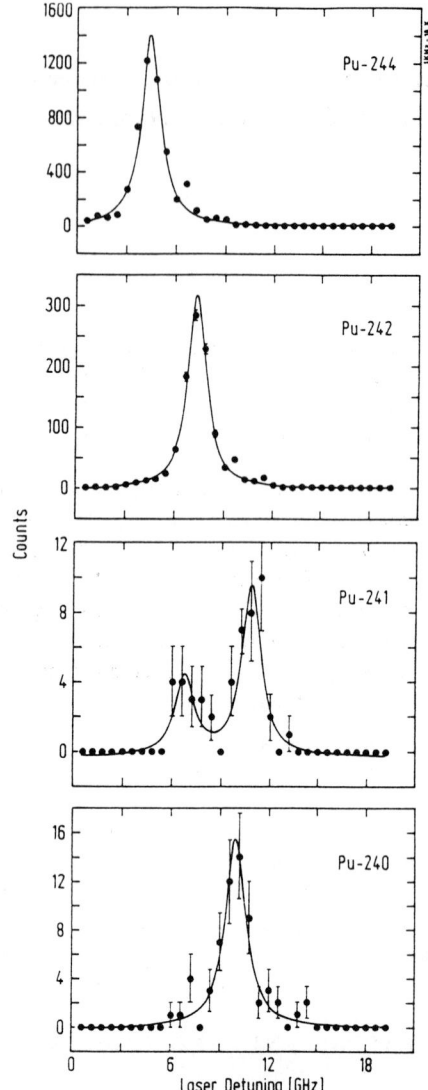

Fig. 1. Isotope shift and hyperfine splitting of plutonium isotopes in the first excitation step at $\lambda_1 = 586.49$ nm.

2.3. Isotopic abundances

A plutonium sample containing a known amount of the Pu-isotopes 240, 241, 242 and 244 was measured several times in order to test the reproducibility of the isotopic analysis with our set-up. The data are shown in Table 1. λ_1 and λ_3 were tuned over the resonances of the isotopes, whereas λ_2 was fixed at 665.57 nm during the measurement. The results for the isotopic ratios agree quite well with each other and are in good agreement with a mass spectrometric determination.

2.4. Reflectron TOF spectrometer

So far a detection efficiency of about 10^{-6} for plutonium has been reached in a set-up with a linear TOF spectrometer. The efficiency is mainly limited by the poor spatial overlap of the atomic beam with the laser beams. It depends on the distance between the filament and the interaction region, which was 4 cm in the set-up used so far.

In a new TOF-spectrometer this distance is reduced to 4 mm. Here the filament is placed behind the end plate of the two-stage acceleration region. The atoms drift through a hole in the plate, which is 1 mm thick, and interact with the laser beams immediately behind it. The spatial overlap is thereby increased by two orders of magnitude.

In this set-up the direction of the thermal atomic beam evaporated from the filament is parallel to the flight direction of the accelerated ions.

Table 1. Isotopic abundances determined in a plutonium sample. The wave lengths of the first and third excitation step were scanned in order to assure equal excitation probability for each isotope. λ_2 was kept fixed. The known abundances were obtained by mass spectrometry.

Isotope	^{240}Pu	^{241}Pu	^{242}Pu	^{244}Pu
	2.3	0.2	8.2	89.3
	2.3	0.5	8.2	89.1
	2.1	0.4	8.6	88.9
	2.6	0.6	11.2	85.6
	2.8	0.5	11.6	85.1
	2.8	0.5	10.9	85.7
	2.4	0.3	9.6	87.7
	2.4	0.3	9.2	88.1
	2.4	0.5	7.3	89.8
	1.9	0.5	8.1	89.5
	2.0	0.4	9.8	87.8
	2.8	0.4	9.0	87.9
	2.1	0.2	8.8	89.0
Average	2.4	0.4	9.3	87.9
σ	0.3	0.1	1.3	1.6
Mass Spec.	2.7	0.4	9.0	87.8

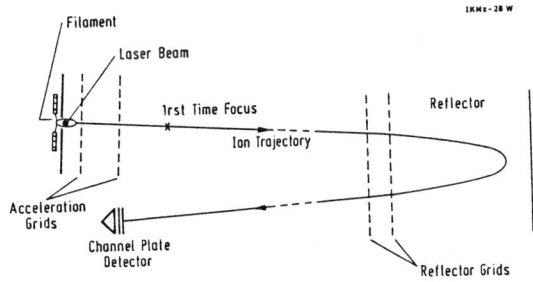

Fig. 2. Scheme of the reflectron time-of-flight mass spectrometer

The mass resolution of the TOF spectrometer is reduced by the initial thermal velocity distribution of the atoms and the spatial spread of the starting point due to the laser beam diameter of 5 mm. Furthermore UV photons are emitted from the filament at temperatures up to 2300 °C. They can cause background events in the channel plate detector, if it faces the filament at the end of the drift tube. Therefore an ion reflector must be added. In Figure 2 the TOF spectrometer is shown schematically. The ions are accelerated to 2.8 keV by a two-stage electric field and are bunched in a first time focus 5 cm behind the second grid. There the duration of the ion bunch is determined mainly by the initial ion velocity spread, which amounts to $\delta t=17$ ns for Pu-ions with an energy of 2.8 keV. With a two-stage ion reflector a similar spread of the time focus is obtained at the detector position. A time resolution better than 4000 is expected for a flight time of 68 μs. This corresponds to a mass resolution of M/ΔM(FWHM)=2000 with a field free drift length of 2 m and a reflector angle of 4.5°.

Our set-up was tested with an atomic beam of gadolinium. Laser beams with λ_1=561.79 nm, λ_2=635.17 nm and λ_3=613.28 nm were used for resonance ionization. Figure 3 shows a TOF spectrum obtained in this way. The mass resolution is M/ΔM(FWHM)=1100.

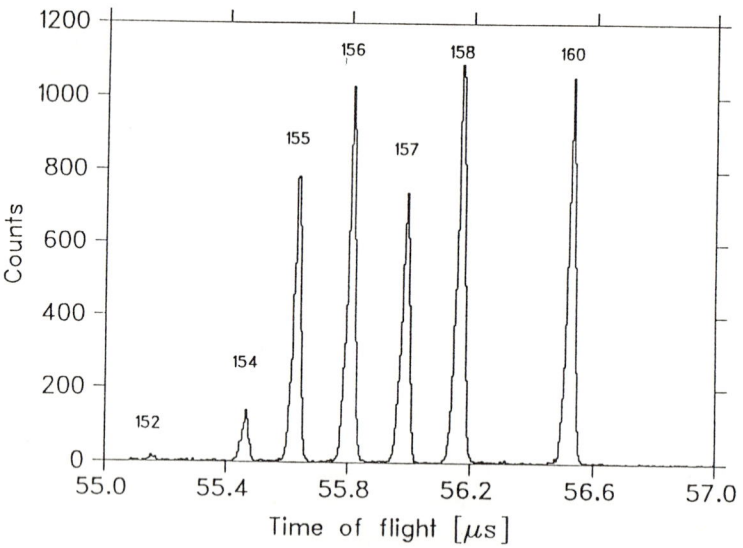

Fig. 3. Time-of-flight spectrum of gadolinium obtained by RIMS with a reflectron TOF mass spectrometer

REFERENCES:

Blaise J, Fred M and Gutmacher R 1986 *J. Opt. Soc. Am. B* **3** 403

Ruster W, Ames F, Kluge H-J, Otten E-W, Rehklau D, Scheerer F, Herrmann G, Mühleck C, Riegel J, Rimke H, Sattelberger P and Trautmann N 1989 *Nucl. Instr. and Meth. A***281** 547

Inst. Phys. Conf. Ser. No 128: Section 6
Paper presented at RIS 92, Santa Fe, NM, USA, 24–29 May 1992

Shoestring resonance ionization mass spectrometry (SSRIMS)

J Zoller, R Lewis, G Rothschopf and R Estler

Department of Chemistry, Fort Lewis College, Durango, Colorado 81301

A Time-of-Flight (TOF) mass spectrometer for use in Resonance Ionization Mass Spectrometry (RIMS) studies has been constructed in the undergraduate environment using a design that minimizes instrumentation and costs. Accordingly, 2+1 (photons to resonance + photons to ionize) ionization schemes driven by a nitrogen pumped dye laser have been explored. In Ni, for example, several low lying states provide ionization pathways as well as the opportunity to measure the electronic temperature of the probed atoms. This detection scheme is being investigated for future use as a probe in laser ablation mechanistic studies.

1. INTRODUCTION

As the important role of trace metals in biological functions becomes well documented, both as nutrients and toxins, the need for selective and sensitive detection of these trace elements becomes increasingly important.

To address some aspects of this detection, we have constructed within the undergraduate institution environment an instrument capable of pursuing Resonance Ionization Mass Spectrometry (RIMS) studies while under the constraints posed by very limited resources (i.e., ShoeString RIMS, SSRIMS). Although simple by many laboratory standards, the instrumentation is well suited to develop sensitive and selective pulsed laser ionization schemes for selected systems as well as to study and characterize the laser/surface interactions that occur during the laser ablation process.

A significant fraction of the published work in RIMS analysis describes the use of conventional thermal sources to vaporize and atomize the sample. Multistep laser photoionization is then followed by mass spectral separation and detection. Although numerous ionization processes have been proposed and demonstrated, most analytical work to date has utilized simple 1+1 (photons to resonance + photons to ionize) ionization schemes, using the fundamental or frequency-doubled output from a single laser. This bias is due in part to two factors: the simplicity and ease of modeling for this ionization process, and a commonly held belief that n-photon resonances (n>2) are difficult to drive efficiently. The potential advantages derived from the use of n-photon resonances include minimal laser hardware (since the fundamental from a single dye laser is normally sufficient to effect ionization) and the potential for Doppler-free excitation. Calculations have indicated that with modest spectral densities, a representative two-photon transition can be saturated.

Several general observations can be made regarding the utility of the 2+1 ionization process for analytical resonance ionization studies. First, the use of three photons in the ionization process allows relatively broad sprectral coverage with a single dye. For example, the nitrogen laser pumped dye C540A has a useful tuning range of 525 nm < λ < 580 nm, thereby providing possible ionization of metal atoms with ionization potentials up to 7.3 eV. In addition, the range of intermediate states that can be accessed doubles, relative to 1+1 type processes: using the example cited above, the range of accessible states via one-photon resonances is ~2.36-2.14=.22 eV, while for a two-photon resonance, the range is 4.72-4.28 = .44 eV. Second, the use of UV dyes (e.g., laser dye BBQ) allows atoms with ionization potentials up to 10.1 eV to be ionized without frequency doubling. Last, the resonant excitation process is not normally over-saturated, as can be (and often is) the case with one-photon resonant excitation. This allows better spectral definition and higher isobaric discrimination. Although selection of 2+1 ionization schemes is not always possible due to the resonant states available for a particular atom versus dye selection, their use can offer advantages over 1+1 routes.

2. EXPERIMENTAL

The instrumentation design has emphasized 'standard' components wherever possible, minimized the number of components and maximized the number of possible instrument configurations for a number of different types of investigation. (For example, this instrument is also used for the molecular weight determinations of biopolymers in Matrix-Assisted Laser Desorption Mass Spectrometry). Figure 1 provides an overview of the instrumentation.

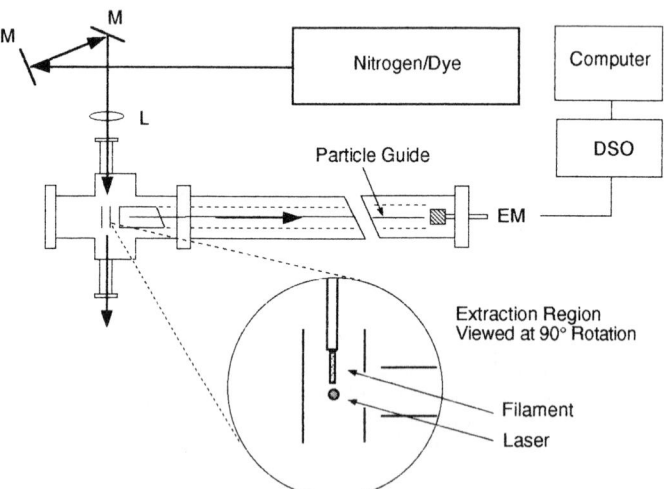

Figure 1. Experimental apparatus constructed for these studies: DSO, Digital storage oscilloscope; EM, electron multiplier; L, lens; M, dielectric mirror.

The vacuum chamber for the linear time-of-flight mass spectrometer has been fabricated from standard (2.75 inch) metal gasket vacuum hardware and provides a 1-m flight path length. The pumping system consists of a single, baffled 4-inch diffusion pump backed by a mechanical pump. The extraction region ion optics are of traditional design and allow the placement of a resistively heated filament within this region from either of two mutually perpendicular directions. In order to maintain the absolute highest transmission efficiency of ion transport from extraction field to detector, we have incorporated the use of an electrostatic particle guide (Abbe *et al*, 1972). The ease of use of this device (consisting of a biased central wire within the flight tube) in terms of additional powers supplies, wiring, etc., versus mutually perpendicular steering plates greatly simplifies the construction and operation of the instrument. Charged particle detection is accomplished via a Ceratron electron multiplier followed by a 300 MHz preamplifier.

Ionization within the extraction region is effected by the output pulses of a nitrogen pumped dye laser combination propagating perpendicular to the time-of-flight ion axis. The 100-200 μJ output pulses of this system are directed into the mass spectrometer using high reflectivity dielectric mirrors and are brought to a focus using a 100-mm focal length lens. Calculated spot diameters above the filament source for the wavelengths of interest are on the order of 10μm providing irradiances of approximately 10^9-10^{10} W/cm^2.

Signal capture/analysis/display is accomplished using a fast digital storage oscilloscope (DSO, Tektronix 2440) and computer (Macintosh SE/30). Communication between these devices is maintained via a GBIP bus controller. BASIC software provides total software control of the DSO. Repetitive file transfers from the DSO maintain total accumulation (and averaging) of the signal. Transfer rates of up to 30 Hz (equal to the fastest laser system repetition rate used) have been tested while operating the DSO in a fast-dump mode. Averaging is performed with the computer rather than the DSO itself in order to increase the limited digitization range provided by the DSO (8 bit), as well as to provide discrimination and elimination of aberrant data capture and transfer. Software allows for near real-time display of data and file storage for later analysis.

3. RESULTS AND DISCUSSION

While the Cu system was used to test the initial operation of the instrumentation via the 2-photon resonance at 463 nm (Apel *et al*, 1987), our initial work on the Ni system is presented below.

Nickel has been investigated both for its significance as a toxin as well as for its interesting spectroscopy. Trace Ni detection in traditional mass spectrometry can be complex due to a number of factors including the interference of the P_2^+ molecular ion in samples of biological (tissue) origin.

Most of the prior resonance ionization schemes for nickel have involved 1+1 processes (Moore *et al* 1984; Saloman, 1991). A partial Grotian diagram for nickel as well as the ionization scheme used in this work in shown in Figure 2.

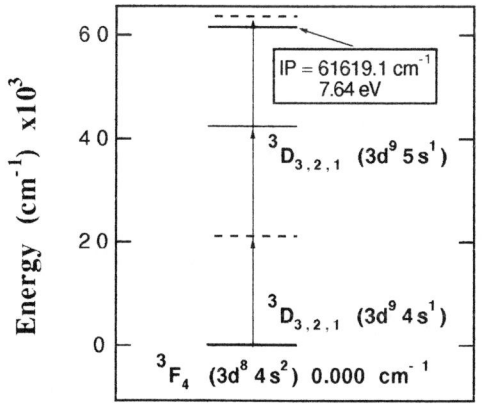

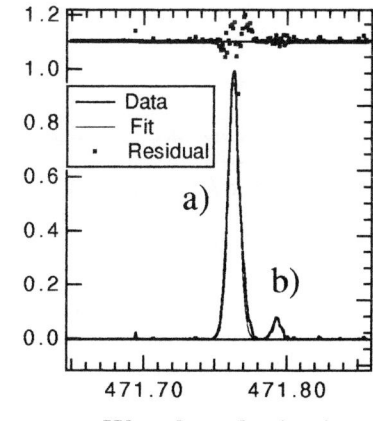

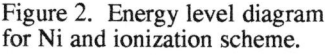

Figure 2. Energy level diagram for Ni and ionization scheme.

Figure 3. Ni (2+1) ionization spectrum near 471 nm originating from the a) $^3D_3$ state b) $^3D_1$ state.

Although nickel may be ionized from its ground state $(3d^8\ 4s^2,\ ^3F_4)$ in a 2+1 ionization, three low lying electronic states provide alternate ionization schemes. The $3d^9\ 4s^1,\ ^3D_3$, $^3D_2\ ,^3D_1$ at 204, 880, and 1713 cm^{-1} above the ground state are accessible at the temperatures involved during evaporation from a filament and those possible during laser ablation (currently being incorporated into our instrument). Two-photon transitions to the $3d^9\ 5s^1,\ ^3D$ manifold of states provide the resonant ionization pathways. The serendipitous close proximity in wavelength of two of these resonances, as illustrated in the Figure 3, provides a 'thermometer' for the detected Ni atoms. Since Fassett *et al* (1983) have demonstrated the thermodynamic equilibrium for Ni atoms evaporated from thermal sources, resonance intensity ratios measured from the Gaussian fits of the two resonances (also shown in Figure 3) at know temperatures allow determination of a temperature calibration factor. This factor would include 2-photon line strengths as well as experimental variables such as laser power differences at the two wavelengths. The wavelength resonance scan of these transitions shown in Figure 3 was recorded at a filament source temperature of approximately 1000K. Since the overall efficiency of detection is partially determined by the electronic partitioning present for a particular analyte, investigating this partitioning for pulsed-sample introduction, i.e., laser ablation, becomes increasingly important. We are currently studying these resonance signal ratios as a function of atom source temperatures as well as laser power and bandwidth for future use in laser ablation studies.

4. CONCLUSIONS AND ONGOING WORK

Relatively simple instrumentation can be constructed with limited resources which is quite suitable for many RIMS investigations. Using this instrumentation, we have initiated filament studies necessary to identify viable ionization schemes of analytes of interest in a number of systems (Cu and Ni illustrated here). Future work includes the integration of laser desorption sampling within this instrument to improve the duty cycle use of analyte as well as to study the fundamentals of laser-material interactions by diagnosing the evolution and speciation of evaporated material and the energy distributions (kinetic and internal) of the evaporated atoms, molecules and fragments.

5. ACKNOWLEDGEMENTS

The authors would like to acknowledge the support of the National Science Foundation (Grant CHE-9100426) and that of the RIS-92 program committee for its encouragement of undergraduate participation in this conference.

6. REFERENCES

Abbe J C, Amiel S and MacFarlane R D 1972 *Nucl. Instrum. and Meth.* **102** 73

Apel E C, Anderson J E, Estler R C, Nogar N S and Miller C M 1987 *Appl. Opt.* **26** 1045

Fassett J D, Moore L J, Travis J C and Lytle F E 1983 *Int. J. Mass Spectrom. Ion Proc.* **54** 201

Moore L J, Fassett J D and Travis J C 1984 *Anal. Chem.* **56** 2770

Saloman E B 1991 *Spectrochimica Acta* **46B** 319

Inst. Phys. Conf. Ser. No 128: Section 7
Paper presented at RIS 92, Santa Fe, NM, USA, 24–29 May 1992

Towards laser photoelectron resonance ionization spectromicroscopy (RISM)

V.S. Letokhov and S.K. Sekatskii

Institute of Spectroscopy, Russian Academy of Sciences, 142092 Troitsk, Moscow Region, Russia

ABSTRACT: The first experimental results are presented on the resonance photoionization of impurity ions on the surface of crystals. The effect discovered opens the way for realization of laser photoelectron spectromicroscopy capable of a subwavelength spatial resolution.

1. INTRODUCTION

The various laser spectroscopic methods, resonance ionization spectroscopy [1] in particular, make it possible to improve drastically the main characteristics of optical spectroscopy, namely:

1°. Spectral resolution, to be capable of covering the sub-Doppler range.

2°. Sensitivity, to be capable of detecting single atoms and a few molecules.

3°. Time resolution, to be capable of covering the femtosecond range.

4°. Detection selectivity, to be capable of detecting trace atoms, isotopes, and molecules.

The present paper is concerned with the possibility of improving radically, with the aid of laser radiation, one more spectroscopic characteristic –

5°. Spatial resolution, to be capable of covering the sub-wavelength range, down to atomic-molecular resolution, along with spectral resolution. It is essentially a matter of the creation of the method of laser resonance ionization spectromicroscopy (RISM).

The idea of RISM was put forward quite long ago [2], simultaneously with those of resonance ionization spectroscopy (RIS) and resonance ionization mass spectrometry (RIMS) (see [1, 3]). The possibility of implementing RISM is based on the following two key factors: (1) the photoionization fragmentation site (photoion of photoelectron detachment site) is localized with an accuracy much better than the photoionization laser wavelength and (2) the various microscopic techniques (projective [4] or electron/ion lens [5]) make it possible to observe the ejection site of a charged particle with a high magnification.

To realize the RISM idea, one should be able to ionize in a resonance manner suitable chromophores on a surface, i.e., selectively to detach by means of a laser pulse either an electron or an ion from the surface without causing damage to the host matrix containing the absorbing chromophores. We conducted a series of experiments to study the laser-pulse detachment of ionized chromophores from large molecules adsorbed on a surface. The formation of photoions upon irradiation of a surface with powerful UV laser pulses was revealed in [6-9] where it was demonstrated that the production of photoions resulted from the MPI of neutral molecules undergoing laser-induced thermal desorption. To avoid this effect, the laser pulse duration, i.e., the time of deposition of the required energy in the chromophore, τ_{depos}, must satisfy the following obvious condition

$$\tau_{depos} << \tau_{detach} \simeq <a>/<v> << \tau_{transf} \qquad (1)$$

where τ_{detach} is the time it takes for a photoion with an average size of $<a> \simeq 1$ Å to move with an average velocity of $<v> \simeq 10^4$ cm/s for a distance of the order of $<a>$ from the projector tip surface and τ_{transf} is the time of excitation transfer from the chromophore to the other portions of the molecule. Since $\tau_{detach} \simeq 1$ ps and $\tau_{transf} \simeq 0.1$-1 ps, it is obvious that to effect the MPI of molecular chromophores directly on the surface necessitates the use of femtosecond laser pulses.

Successful experiments on the femtosecond MPI mass spectrometry of biomolecules on a surface were performed in [10, 11]. However, following the laser detachment of an ionized chromophore, the molecule may nevertheless get desorbed from, or at least move over the laser-heated surface. The possible ways to localize the molecule on the surface were discussed in [12].

To overcome this difficulty, it seems useful to realize first the photoelectron RISM version [2], wherein it is sufficient to eject photoelectrons from the chromophores being excited by the laser radiation. Such chromophores may be impurity centers (impurity ions) in glass and crystals. Impurity ions are in part located near the surface (1-10 Å) and can, in principle, repeatedly emit photoelectrons under excitation, while remaining rigidly localized in the host matrix.

2. PHOTOELETRIC EFFECT FOR IMPURITY IONS IN CRYSTALS

Consider a crystal matrix doped with ions of some metal having an incomplete d- or f-electron shell. In such a system, there arises a ladder of energy levels belonging to different transition metal ion configurations, and a stepwise resonance laser excitation of the implanted ions is possible, as demonstrated by numerous experiments [13 and references cited therein]. If an energy exceeding the photoelectric effect threshold of the crystal matrix is imparted to the ions, electrons with an energy above the vacuum level may then be formed in the matrix as a result of the highly excited ions imparting their energy to the matrix. Such electrons may escape from the matrix into the vacuum (photoelectric effect) with a certain probability. For some media, for example, $MeF_2 : TR^{2+}$ where Me = Ba, Sr, or Ca and TR^{2+} is a bivalent rare earth ion, photoelectric effect is also possible as a result of the direct photoionization of the impurity ion, $TR^{2+} \rightarrow TR^{3+} + e^-$, upon absorption of one or two visible or UV photons (see [14, 15] for the data on the energy level structure of MeF_2: TR^{2+}). Photoconduction (internal photoelectric effect) due to the absorption of light by the ions TR^{2+}, TR^{3+}, and Ti^{3+} implanted in a matrix was already observed earlier [14-16]. Photoelectric effect due to resonance stepwise excitaion can be expected to occur in such media as well.

2.1. Doubly Ionized Impurity Atoms

The ionization potentials of free doubly ionized atoms of rare earth elements, TR^{2+}, are high and range between 20 and 25 eV (Fig. 1, left). Consequently, to effect selective stepwise photoionization of such ions requires 7-12 laser quanta 2-3 eV in energy, which seems impracticable. But the situation changes cardinally for bivalent rare earth ions implanted in ionic crystals, first of all such as CaF_2, SrF_2 and BaF_2 (Fig. 1, right). This is due to the fact that the rare earth ions which replace in these crystals the ions Ca^{++}, Sr^{++}, and Ba^{++} possess a very high potential energy E_M (the Madelung energy) due first of all to the electric field of the neighboring ions: 19.94 eV, 18.78 eV, and 17, 59 eV, respectively [17]. For this reason, the ionization potential of the ions TR^{2+} is in that case greatly reduced:

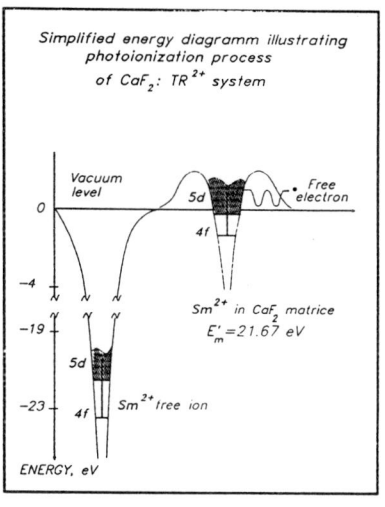

Simplified energy diagramm illustrating photoionization process of CaF_2: TR^{2+} system

$$I_p^{cryst} = I_p^{free} - E_M - \Delta E_M - \Delta E_{pol}, \quad (2)$$

where the term ΔE_M allows for the change of the Madelung enegry due to the crystal lattice deformation resulting from the replacement of a lattice ion (e.g., Ca^{++}) by a TR^{2+} ion, and the term ΔE_{pol}, for the matrix polarization due to the removal of an electron from a bivalent ion after ionization. The detailed analysis [15] shows that for the CaF_2: TR^{2+} systems, the ionization energy of a bivalent ion must fall within the range 1-5 eV, and this was confirmed in the experiments [14, 15] on the photoconduction of such crystals.

Fig. 1. Simplified energy diagram illustrating the photoionization of doubly ionized atoms: left – a free Sm^{2+} ion; right – a Sm^{2+} ion in a CaF_2 crystal (E_m' is the corrected Madelung energy).

The threshold of the external photoelectric effect associated with the escape of an electron from the crystal into the vacuum differs from that of the internal photoelectric effect (photoconduction) by the amount of the electron affinity χ of the CaF_2 crystal, which is very small: $\chi \simeq 0$ eV [18]. For this reason, external photoelectric effect can be expected to occur in the CaF_2:Sm^{2+} crystal upon absorption of a single laser quantum with an energy of $\hbar\omega > 2$ eV.

2.2. Triply Ionized Impurity Atoms

The situation with the trivalent rare earth ions TR^{3+} is different. The ionization potential of the free ions TR^{3+} ranges between 36 and 46 eV, i.e., even if the Madelung energy is taken into account, no less than 6 quanta with an energy of 2-3 eV is required to effect "direct" photoionization the type of $TR^{3+} \longrightarrow TR^{4+} + e$ (Fig 2, left). Nevertheless, selective stepwise laser photoelectric effect can be realized in this case as well. If one imparts to a TR^{3+} ion an energy of E higher than the

photoelectric effect threshold of the
crystal matrix,   (the ion in that case
can be in a relatively long-lived
metastable state!), the excitation energy
can be transferred from the TR^{3+} ion to
the matrix (Fig. 2, right). As a result
the ion drops back to the ground state (or
some low lying one) and a free electron
with an energy above the vacuum value
appears in the matrix. Later on this
electron can escape from the crystal into
the vacuum (photoelectric effect). Such an
energy transfer can be considered an
Auger-type relaxation of the electronic
excitation of rare earth ions in crystal
matrices, and such relaxation processes
have recently become the object of
extensive studies [19]. Of course, for
stepwise photoelectric effect to occur in
that case, it is desirable to select
crystals with as low as possible
photoelectric effect threshold.

3. EXPERIMENTS WITH $CaF_2 : Sm^{2+}$ AND $ZrO_2 : Nd^{3+}$

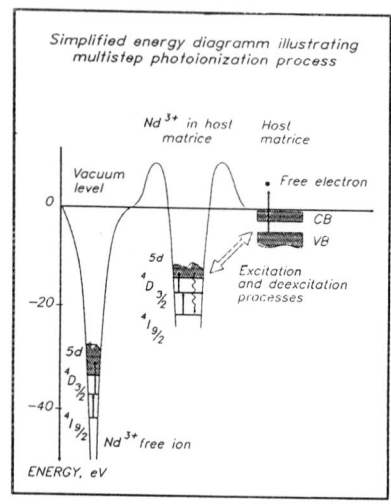

We conducted experiments with a series of
crystals doped with both doubly and triply
ionized atoms, in which we managed to
observe both of the external photoelectric
effect mechanisms considered above.

3.1. $CaF_2:Sm^{2+}$

We studied external photoelectric effect
in the course of laser irradiation of the
$CaF_2:Sm^{2+}$ crystals (the concentration of
the Sm^{2+} ions 0.04 mol. %), nominally pure
CaF_2 crystals, $CaF_2: Sm^{2+}$ crystals

Fig. 2. Simplified energy
diagram illustrating the
photoionization of triply
ionized atoms: left – a free
Nd^{3+} ion; center – a Nd^{3+}
ion in a crystal matrix;
right – energy transfer from
a Nd^{3+} ion to the crystal
matrix, followed by its
ionization.

(0.3 mol. %), and $CaF_2:Tm^{2+}$ crystals (the concentration of the Tm^{2+} ions
around 0.02 mol. % as estimated by absorption).

The experiments were conducted with both a CW argon laser (a group of lines
in the range 458–528 nm), whose intensity on the sample surface amounted to
2.6 W, and a pulsed copper-vapor laser (510.6 and 578.2 nm, pulse
repetition frequency 10^4 Hz) with up to 2 W average power at each line.
The laser radiation was split by means of a diffraction grating into
separate spectral lines, and studies were made both on these individual
lines and on all the lines with the exception of the UV ones.

Experiments with both the argon and the copper-vapor laser revealed a
substantial difference in photoelectron yield between the $CaF_2:Sm^{2+}$ sample
and all the rest of the samples mentioned above (see Fig 3). As can be
seen, the photoelectric effect in this case is of a linear character, and
the photoelectron yield for the doped $CaF_2:Sm^{2+}$ sample is about 1.5×10^2
times that for the nominally pure CaF_2. The photoelectron yield (also
linear) for $CaF_2:Sm^{3+}$ and $CaF_2:Tm^{2+}$ samples was on the whole close to that

for nonactivated CaF_2 samples and much lower than the photoelectron yield for the $CaF_2:Sm^{2+}$ sample.

These results agree well with the photoconductivity measurements for the $CaF_2:TR^{2+}$ crystals: according to [14, 15], the photoconduction threshold for CaF_2: Sm^{2+} is 1.7 eV, and for $CaF_2: Tm^{2+}$, it is 2.75 eV; the latter value corresponds to a shorter laser wavelength than those used in our investigations.

3.2. $ZrO_2:Nd^{3+}$

We investigated various activated crystals (neodymium-doped silicate glass, $YAG:Nd^{3+}$, $CaMoO_4: Nd^{3+}$, $CaF_2: Nd^{3+}$, $ZrO_2: Nd^{3+}$) and laser irradiation conditions (irradiation of samples with nanosecond pulses from a nitrogen laser at λ = 337 nm, a nitogen-laser-pumped dye laser at λ = 520-600 nm, a copper-vapor laser at λ = 510.6 and 578.2 nm, and neodymium-doped glass laser at λ = 532 and 355 nm (harmomnics), and also simultaneous irradiation with several of these lasers). Stepwise laser resonance photoelectric effect was detected in the $ZrO_2:Nd^{3+}$ sample (the Nd^{3+} ion concentration was 0.3 at. % and the ZrO_2 crystal was stabilized with 16% gadolinium oxide) irradiated with the

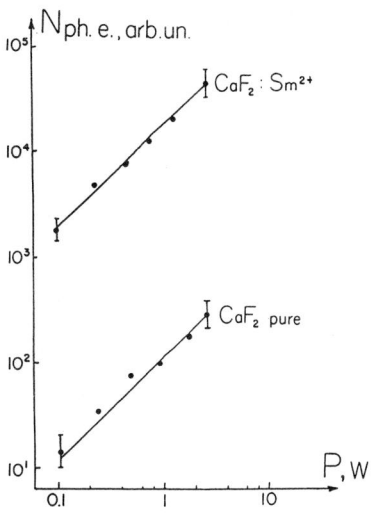

Fig 3. Photoelectron yield (arbitrary units) for CaF_2: Sm^{2+} and pure CaF_2 as a function of the argon laser intensity.

third-harmonic pulse of the neodymium-doped glass laser (λ = 355 nm, pulse half-width 15 ns, pulse energy up to 4 mJ). The experimental data are presented in Fig. 4. The relationship between the photoelectron yield $N_{ph.e}$ and laser intensity P in the range $P \leqslant 10^6$ W/cm^2 is linear. Such a relationship was observed earlier for some samples at about the same laser intensities and wavelengths (see, for example, [21]). It is most probably explained by the photoionization of impurities, surface defects, and so on. In the range $P \geqslant 10^6$ W/cm^2, this relationship is quadratic, and to our view, can be explained by the two-step resonance photoelectric effect involving the implanted ions Nd^{3+} (the matrix itself is transparent at λ = 355 nm) in the following sequence of transitions: $^4I_{9/2} \xrightarrow{---} ^4D_{3/2}$ (cross section $\sigma \simeq 10^{-19}$ cm^2) and then $^2P_{3/2} \dashrightarrow 5d4f^2$ ($\sigma \lesssim 10^{-17}$ to 10^{-16} cm^2), with a fast, subnanosecond relaxation occurring between the $^4D_{3/2}$ and $^2P_{3/2}$ levels (see Fig. 5). A similar stepwise photoexcitation of the Nd^{3+} ions in YAG and YGaG matrices and neodymium-doped glass was earlier observed in [22].

The relationships between the photoelectron yield $N_{ph.e}$ and the laser intensity P for the other samples studied, nonactivated ZrO_2 samples included, differ materially from that for the $ZrO_2:Nd^{3+}$ sample considered above. For these samples, the threshold marking the occurrence of the quadratic dependence of $N_{ph.e}$ on P under irradiation at λ = 355 nm is higher by one or two orders of magnitude than the corresponding threshold for the $ZrO_2:Nd^{3+}$ sample. Accordingly, the photoemission current at a laser intensity of $P \simeq 10^7$ W/cm^2 is in this case one or two orders of magnitude smaller. By contrast, when irradiating $ZrO_2:Nd^{3+}$ and pure ZrO_2

samples at wavelengths of 337 and 532 nm, the $N_{ph,e}$ -P relationships are close to one another.

4. POTENTIALITIES OF PHOTOELECTRON RISM

The first experimental results bear witness to the occurrence of stepwise resonance photoelectric effect in the above media, hence to the possibility of imaging them by photoelectron microscopy means with a spatial resolution about a few nanometers (see the estimates in [1], Ch. 12]. An interesting potentiality of photoelectron microspectroscopy is the imaging of small-size irregularities in dielectric samples. The experimental results obtained on the breakdown of transparent dielectrics in laser fields with an intensity of $I \geqslant 10^9$ W/cm^2 are usually explained by the absorption of light by small-size (typically 10-100 nm across) irregularities in the medium [23]. It has also been known that the photoemission current caused by the irradiation of a dielectric with light of an intensity much lower than the photodamage threshold value ($I \geqslant 10^6$ W/cm^2) contains not only the undelayed component due to single- or multiple-photon photoelectric effect, but also a delayed component associated with thermionic emission from the absorbing irregularities or with other processes directly related to the absorption of light by these irregularities [23].

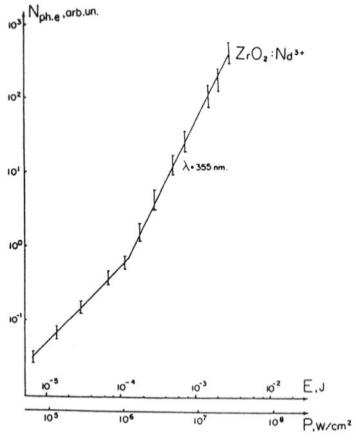

Fig. 4. Photoelectron yield as a function of laser pulse energy when irradiating the ZrO$_2$: Nd^{3+} sample with the third-harmonic pulses from a neodymium-doped glass laser.

Observing in a photoelectron microscope the cathode image produced by such "delayed" electrons can help directly to image the absorbing irregularities or defects in transparent materials. Our estimates show that the use in this case of a copper-vapor laser, or dye lasers pumped by such a laser, holds much promise, for its pulsed power output and pulse repetition frequency are high enough to provide for an acceptable intensity and pulse-period-to-pulse-duration ratio. Possibilities provided by the use of resonance photoelectric effect in the case of photoelectron microscopy exist also for an electron projector wherein electrons (ions) are emitted from a tip of a very small radius of curvature, r, placed at the projector center, and projected onto a screen of radius R at a high magnification of the order of R/r under the action of a radial electic field [4]. If the projector tip here is manufactured from a suitable material and made to exhibit resonance photoelectric effect under irradiation in a strong electric field, one can image absorbing impurity ions, color centers, and small-size irregularities located near the projector tip surface. But in that case it will be necessary to develop a method to manufacture small-radius projector tips from the above materials and make them electrically conductive to ensure effective operation of the electron projector.

REFERENCES

1. V.S. Letokhov, *Laser Photoionization Spectroscopy* (Academic Press, Orlando, 1987).
2. V.S. Letokhov, *Kvantovaya Elektron.* 2, 930 (1975) (in Russian).
3. V.S. Letokhov, in: *Tunable Lasers and their Applications*, ed. by A. Mooradian, T.Jaeger and P. Stokseth (Springer-Verlag, Berlin, 1976).
4. E.W. Müller and T.T. Tsong, *in: Progress in Surface Sciences* (Pergamon Press, Oxford, 1973), Vol. 4.
5. O.H. Griffith and G. Rempfer, *Adv. Opt. Electr. Microsc.* 10, 269 (1987).
6. V.S. Antonov, V.S. Letokhov, and A.N. Shibanov, *JETP Lett.* 21, 471 (1980); *Appl. Phys.* 25, 71 (1981).
7. V.S. Antonov, V.S. Letokhov, Yu.A. Matveetz, and A.N. Shibanov, *Laser Chemistry* 1, 37 (1982).
8. S.E. Egorov, V.S. Letokhov, and A.N. Shibanov, in: *Surface Studies with Lasers*, ed. by F. Ausseneg, A. Leither, M.E. Lippitsch (Springer-Verlag, Berlin, 1983), p. 156.
9. S.E. Egorov, V.S. Letokhov, and A.N. Shibanov, *Sov. J. Quant. Electr.* 121, 1393 (1984).
10. S.V. Chekalin, V.V. Golovlev, A.A. Kozlov, Yu.A. Matveetz, A.P. Yartzev, and V.S Letokhov, *J. Phys. Chem.* 92, 6855 (1988).
11. S.V. Chekalin, V.V. Golovlev, Yu.A. Matveets, A.P. Yartsev, V.S. Letokhov, C. Seidel, J. Wolfrum, and K.V. Greulich, *IEEE Journ. of Quantum Electronics* 26, 2158 (1990).
12. V.S. Letokhov, in: *Laser Spectroscopy IX*, ed. by M.S. Feld, J.E. Thomas and A. Mooradian (Academic Press, Boston, 1989), pp. 495-499.
13. A.A. Kaminskii, B.M. Antipenko: *Mnogourovnevye functsional'nye skhemy kristallicheskikh lazerov* (Many-level functional schemes of crystalline lasers) (Nauka, Moscow, 1989) (in Russian).
14. C. Pedrini, D.S. McClure, C.H. Anderson: *J. Chem. Phys.* 70, 4959 (1979).
15. C. Pedrini, F. Rogemond, D.S. McClure: *J. Appl. Phys.* 59, 1196 (1986).
16. S.A. Basun, A.A. Kaplyanskii, V.K. Sevast'yanov *et al.*: *Fiz. Tv. Tela* 32, 1898 (1990) (in Russian) [Sov. Phys. Solid State].
17. R.T. Poole, J. Szajman, R.C.G. Leckey, J.G. Jenkin, and J. Lisegang, *Phys. Rev.* 12B, 5872 (1975).
18. J.M. Langer, *Post. Fiz.* 31, 435 (1980).
19. R. Boyn, *Phys. Stat. Solidi (b)* 148, 11 (1988).
20. V.S. Letokhov, S.K. Sekatskii, S.B. Mirov, *Pis'ma Zh. Eksp. Teor. Fiz.* 54, 437 (1991) (in Russian) [JETP Letters].

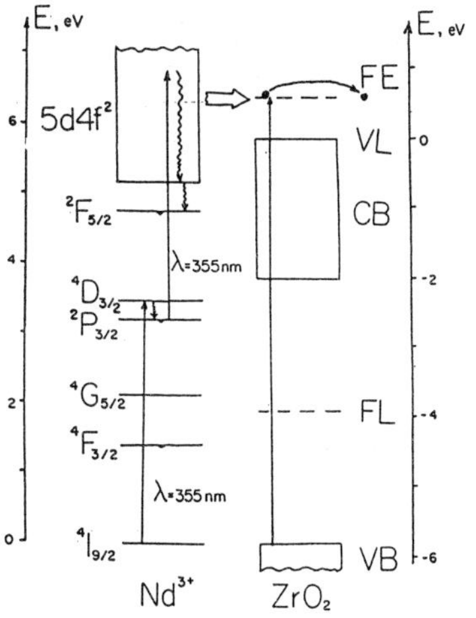

Fig. 5. Illustrating laser stepwise photoelectric effect. Shown on the left are some energy levels of the Nd^{3+} ion in the matrix; the ground state energy is taken as the zero energy level. On the right is a simplified band structure diagram of the matrix: *VB* - valence band; *CB* -conduction band; *FL* - Fermi level; *FE* - free electron; *VL* - vacuum level. The energy transfer process is indicated by the broad arrow.

21. E.M. Logothetis, P.L. Hartman: *Phys. Rev.* **187**, 460 (1969).
22. G.J. Quarles, G.E. Venikouas, R.C. Powell: *Phys. Rev.* **B31**, 6935 (1985).
23. E.F. Lazneva, *Lazernaya desorbtsiya* (Laser desorption) (Leningrad, State Univ. Publ. House, Leningrad, 1990) (in Russian).

Inst. Phys. Conf. Ser. No 128: Section 7
Paper presented at RIS 92, Santa Fe, NM, USA, 24–29 May 1992

Measurement of ultrafast surface processes using laser photoionization

D. von der Linde and H.- P. Ludescher

Institut für Laser- und Plasmaphysik, Universität Essen, D-4300 Essen 1, Germany

ABSTRACT: A pump-probe scheme for ultrafast spectroscopy using laser photoionization is discussed. Picosecond time-resolved measurements of thermal desorption from laser-heated GaAs surfaces are presented.

1. INTRODUCTION

Due to impressive progress in the development of tunable lasers it is nowadays possible to photoionize atoms or molecules in an extremely well controlled way. By tuning lasers into resonance with suitable energy levels species-selective and state-selective ionization can be achieved. Also, powerful techniques of mass spectroscopy and electron spectroscopy for the detection and analysis of the photoproducts are available. Because of these attractive features a considerable number of novel experimental methods have emerged during the recent past in which laser photoionization plays a key role (Hurst and Payne 1988, Letokhov 1987).

In this contribution we will discuss the use of laser photoionization to probe the time evolution of very fast transient processes. We believe that a promising application of this technique is the study of the dynamics of adsorbate-substrate interaction in surface science. Quite a number of fundamental processes of this category are expected to take place on a picosecond or femto-second time scale. The study of these processes requires experimental techniques providing (i) high time resolution, and (ii) high sensitivity and selectivity, the latter because the response of a relatively small number of surface particles must be measured.

2. ULTRAFAST PHOTOIONIZATION SPECTROSCOPY

With the advent of picosecond and femtosecond laser pulses the time resolution of optical measurements has dramatically improved. For ultrafast spectroscopy indirect measuring techniques based on pump-probe schemes have been devised. A first laser pulse (pump pulse) is used to excite the process of interest, and a second laser pulse (probe pulse) to interrogate its subsequent temporal evolution. The time resolution of this method is determined by the laser pulse duration and not by the resolution of the detectors. This basic pump-probe scheme has been applied with great success in a variety of fields (Kaiser 1988). Pump-probe schemes with laser photoionization for the interrogation step appear to be well adapted for studying certain

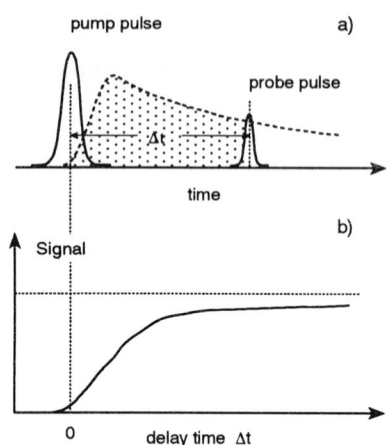

FIG. 1: (a) Timing of pump and probe pulses, and desorption current (dotted curve); (b) photoion signal from the detector versus delay time.

types of ultrafast surface processes. To illustrate the principle let us consider a laser-induced surface desorption process. It is assumed that the first laser pulse strikes the surface and starts the release of some adsorbed material. The time evolution of the process is measured using a second laser pulse to probe the instantaneous concentration of the desorbed products above the surface. This can be done by choosing the probe pulse frequency (or frequencies) in such a way that the desorbed free particles interacting with the probe pulse are photoionized. The photoions are detected by means of a suitable particle detector, for example, a mass spectrometer.

Figure 1a is a schematic illustration of the timing of the laser pulses and the assumed temporal evolution of a desorption product. Figure 1b shows the resulting photoion signal of the detector as a function of Δt. The signal is proportional to the time integral of the particle current emerging from the surface.

A necessary requirement of this scheme is that the probe pulse must spatially overlap the entire cloud of desorbed particles, implying that the probe beam must also strike the surface. There are two important consequences of this requirement (von der Linde 1989): (i) Optical interference of the incident and the reflected beam can lead to a spatially non-uniform intensity distribution near the surface. (ii) The energy of the probe pulse must be kept low enough that possible surface excitation by the probe pulse (e.g. surface heating) is negligible.

Examination of the Fresnel formulas indicates that in most cases the incident and the reflected wave are practically out of phase at the surface. The resulting destructive interference leads to a spatial pattern with an intensity minimum at z=0, and a first maximum at $z=\lambda/4\cos\theta$, where z is the distance from the surface, and θ is the angle of incidence. Clearly such a situation is undesirable because in this case the measured variation of the photoionization signal would not be determined by the kinetics of the desorption process alone, but also by the velocity distribution of the particles and the spatial distribution of the probe field. A spatially uniform distribution can be obtained when p-polarized probe pulses and an angle of incidence of 45 degrees are used. In this case the incident and the reflected fields do not interfere because the electric field vectors are orthogonal. A disadvantage of this configuration is that a relatively large fraction of the probe light is transmitted and eventually absorbed by the substrate.

An interesting alternative probe configuration is possible with highly reflective metallic substrates. In this case the phase change upon reflection of p-polarized light is relatively small up to rather large angles of incidence, and *constructive* interference is obtained, which leads to an intensity maximum at the surface and a fourfold enhancement compared with the incident intensity (Gadzuk 1987).

3. TIME RESOLVED THERMAL DESORPTION

To demonstrate the feasibility of the method we have studied thermal desorption of Ga and a few other species from laser-heated GaAs surfaces. Thermal desorption is a relatively slow process. The characteristic time scale is roughly 100 ps. For the rapid heating of the GaAs surfaces we used laser pulses of 25 ps duration at λ=532 nm. These pulses were obtained from the frequency-doubled amplified output of an actively and passively mode-locked Nd-YAG laser. The energy fluence of the heating pulses was typically 30 mJ/cm^2 or less. Under these circumstances the surface reached a maximum temperature of about 1000 K, which is much less than the melting temperature of GaAs (1513 K). The threshold fluence for melting has been measured (von der Linde 1989) to be 45 mJ/cm^2 (s-polarization, 45 degrees angle of incidence).

Two different photoionization schemes were compared:
(i) *Fixed frequency probe pulses:* The probe wavelength was 266 nm (hv=4.66 eV, fourth harmonic of the Nd-YAG laser). In this case two-photon ionization of Ga is possible (ionization potential I=6.0 eV). We note that there is an accidental resonance with a $^2P_{1/2} \rightarrow ^2S_{1/2}$ transition, but the oscillator strength is rather small. For the ionization of As, on the other hand, a three-photon process is required (I=9.79 eV).

(ii) *Tunable frequency, two-color probe pulses* : The two wavelenths could be tuned around λ_1 = 588 nm and λ_2 = 294 nm (hv$_1$+hv$_2$=6.33 eV). The tunable probe pulses were obtained from a synchronously mode-locked dye laser which was pumped by a fraction of the (frequency-doubled) output pulse train of Nd-YAG laser. The dye laser pulses (588 nm) were amplified and frequency-doubled in a KDP crystal (294 nm). Synchronous pumping ensured that the two-color probe pulses and the heating pulses (532 nm) were well synchronized. The UV part of the probe pulses could be tuned in and out of resonance with the $^2P_{3/2} \rightarrow ^2D_{5/2}$ transition of Ga at 294.3 nm for resonant two-photon ionization of the Ga atoms.

The time-resolved thermal desorption experiments were performed in an ultrahigh vacuum chamber at a base pressure of about 3x10^{-10} mbar. The time delay between pump and probe pulses was controlled by means of an optical delay line. The pump beam and the probe beam were made collinear and focussed onto the surface of the sample at an angle of incidence of 45 degrees. The samples were raster-scanned to avoid surface erosion effects that can result from multiple exposure. We used undoped, optically polished (100) GaAs wafers, some of them treated in an oxygen atmosphere. The GaAs crystals could be sputtered with 3 keV Ar$^+$ ions and thermally annealed at temperatures up to 1000 K. However, surface diagnostic equipment for a detailed characterization of the crystal surfaces was not available.

4. RESULTS

First we will discuss some results obtained with fourth harmonic probe pulses (266 nm). A typical mass spectrum of a weakly tempered (550 K for 10 minutes) GaAs sample is shown in Fig. 2. Gallium with the two isotopes ^{69}Ga and ^{71}Ga is readily identified. Atomic arsenic is not observed, probably because of its relatively high ionization potential. However, arsenic cluster ions can be seen, and some oxide radicals, e.g. AsO$^+$ and Ga$_2$O$^+$. There are many other species that cannot be clearly identified. The observed number of lines and their strength is greatly

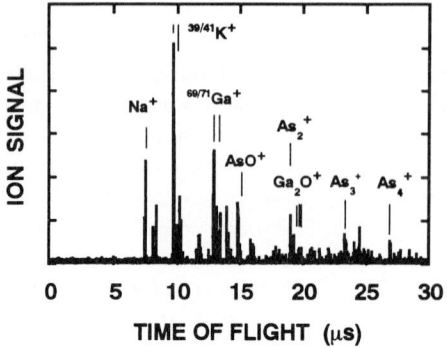

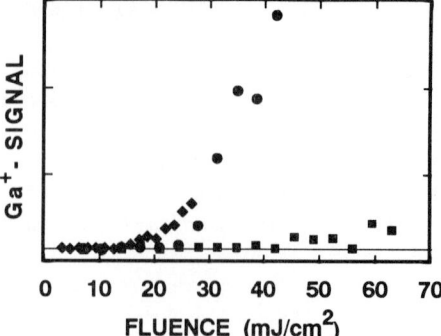

FIG. 2: Mass spectrum from GaAs for a 266 nm probe pulse (Δt=75 ps).

FIG. 3: Ga$^+$ versus fluence. Squares: pump only; diamonds: probe only; full circles: pump and probe (Δt=75 ps).

dependent on the type of surface treatment and the particular photoionization scheme.

For the pump-probe scheme to work it must be shown that the ions are indeed produced by photoionization from desorbed neutrals. It turned out that quite a number of ion species are observed with the pump pulse or the probe pulse alone, indicating that other ionization processes compete with the desired photoionization by the probe pulse. For example, Na$^+$ and K$^+$ ions were observed for almost any combination of pump or probe pulses.

The situation for Ga is illustrated in Fig. 3. Shown is the Ga$^+$-signal as a function of the energy fluence of the laser pulses. The Ga$^+$-signals measured with pump pulses only (532 nm) and probe pulses only (266 nm) are given by the squares and the diamonds, respectively. It was observed that the signals increased rapidly near the threshold fluence for melting ($\approx$20 mJ/cm^2 for 266 nm). The full circles show the Ga$^+$-signal when both pump *and* probe pulses are present. The delay time of the probe pulses was 75 ps. The pump fluence was varied, while the probe pulses were kept constant at approximately 10 mJ/cm^2. These data show that the Ga$^+$-signal could be increased by as much as a factor of ten when the UV probe pulses were applied.

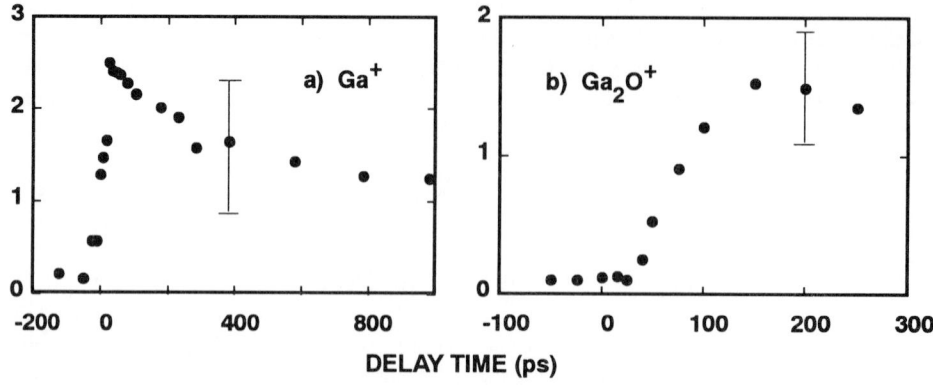

FIG. 4: Ga$^+$ and Ga$_2$O$^+$-signal as a function of probe pulse delay.

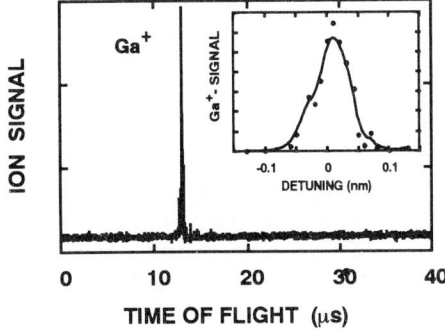 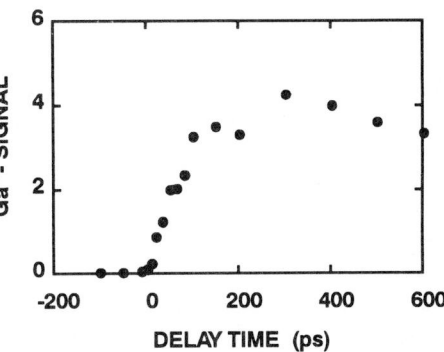

FIG. 5: Mass spectrum from GaAs for resonant two-photon ionization. Insert: Ga⁺-signal versus detuning of the probe pulse.

FIG. 6: Ga⁺-signal as a function of probe pulse delay time for resonant two-photon ionization.

Next we present some results of time-resolved measurements. Two instructive examples are shown in Fig. 4 for Ga^+ and Ga_2O^+. A sharp increase of the signal near $\Delta t=0$ was measured for Ga^+. A maximum was observed around $\Delta t=30$ ps, followed by a decay to a constant level over a few hundred picoseconds. On the other hand, the Ga_2O^+-signal showed quite a different behaviour. In this case a step-like, monotonic increase of the signal with a rise time of about 80 ps was observed.

On the basis of the simple model discussed in Section 2, a signal maximum near $\Delta t=0$ is not expected (see Fig. 1). We have observed that detailed shape of the Ga^+-signal was strongly dependent on sample preparation. By different treatments of the surface the ratio of the signal maximum to the constant level for large Δt could be varied. For example, for sputtered and thermally annealed clean surfaces the initial maximum was absent, and a step-like increase similar to the Ga_2O^+-signal was measured.

A great advantage of the two-color resonant photoionization scheme is that the energy fluence of the probe pulses could be substantially reduced and yet photoion yields much higher than those obtained with 266 nm probe pulses were measured. In the resonant two-photon ionization experiments the fluence of the visible part of the probe pulse was 5 mJ/cm², a factor of ten below the melting threshold of GaAs, whereas the UV part had a fluence of only about 0.2 mJ/cm². Figure 5 shows a typical mass spectrum for resonant two-photon ionization of Ga. As expected, only Ga^+ is observed. Sometimes Na^+ and K^+ ions produced by the pump pulse are also seen, but Ga^+ is clearly dominant. The insert depicts the Ga^+-signal as a function of wavelength detuning from the exact resonance. The width of the tuning curve of about 0.1 nm curve is determined by the spectral width and, to some extent, by pulse-to-pulse fluctuations of the spectrum of the probe pulses.

The Ga^+-signal as a function of delay time for resonant photoionization is shown in Fig. 6. A monotonic increase of the signal with a rise time of approximately 100 ps is observed. We note, however, that in this experiment the intensity distribution of the UV part of the probe pulse was not uniform because of optical interference effects.

5. DISCUSSION AND CONCLUSIONS

Our experimental results provided clear evidence of ion generation by a two-step process: desorption of surface material caused by the pump pulses, and photoionization due to interaction with the probe pulses. Thus it has been shown that the basic pump-probe scheme works.

However, in the experiments with GaAs the ion yield produced by the pump pulses alone and the probe pulses alone was not insignificant. The dynamic range of the time-resolved measurements was approximately a factor of ten, limited by these background ion signals. The ionization mechanisms responsible for the background have not yet been clarified, but the observed increase near the melting threshold indicates that thermal effects may play a role.

The background problem can be overcome and the dynamic range greatly improved with a suitable choice of the photoionization process for probing. We have shown that with resonant two-photon ionization the necessary energy fluence of the probe pulse can be greatly reduced and undesired heating avoided. From this point of view single photon VUV ionization may also provide some advantages.

It appears that the relatively high energy fluence that was necessary for probing with 266 nm offers a possible explanation of the unexpected results of the time-resolved measurements of Ga. Detailed model calculations taking into account surface heating, first order thermal desorption, and photoionization of the desorbed neutrals indicate that the photoion signal should increase monotonously with a rise time of about 100 ps. The results for Ga_2O^+ (FIG. 4) and for Ga^+ from GaAs samples that had been tempered, sputtered, and annealed (not shown here) are in good agreement with such a model. The unexpected maximum of the Ga^+-signal near $\Delta t = 30$ ps on untreated samples and weakly tempered samples could be due to thermal ionization of relatively weakly bound Ga-species, when surface heating by the probe pulses is not negligible. However, other possible explanations of the effect could be given, for example, photoions produced out of thermally excited states of surface-bound Ga species.

Interesting extensions and new applications of ultrafast photoionization spectroscopy are conceivable. State-selective probing of desorbed molecular species and measurement of the photoelectrons could provide much more detailed information on the dynamics of surface processes. Experiments with femtosecond pulses are expected to be very promising. While picosecond pulses resolve the evolution of the final products of desorption, femtosecond time-resolution is likely to reveal the transition states in a desorption process and other interesting short-lived excited states.

REFERENCES

Gadzuk J W 1987 in *Vibrational Spectroscopy of Molecules on Surfaces* eds Yates J T and Madey T E (New York: 1987)
Hurst G S and Payne M G 1988 *Principles and Applications of Resonance Ionization Spectroscopy* (Bristol: Adam Hilger)
Kaiser W ed 1988 *Ultrashort Laser Pulses and Applications* (Heidelberg: Springer Verlag)
Letokhov V S 1987 *Laser Photoionization Spectroscopy* (Orlando: Academic Press)
von der Linde D and Danielzik B 1989 *IEEE J. of Quant. Electr.* **25** 2540

Inst. Phys. Conf. Ser. No 128: Section 7
Paper presented at RIS 92, Santa Fe, NM, USA, 24–29 May 1992

Resonance ionization mass spectrometry of device materials

S. W. Downey, A. B. Emerson, R. F. Kopf and E. F. Schubert

AT&T Bell Laboratories, 600 Mountain Ave., Murray Hill, NJ 07974

ABSTRACT: Resonance ionization mass spectrometry (RIMS) is used to depth profile devices and device substrates of exceedingly complicated structures which are difficult to characterize by conventional analytical techniques. Of particular interest in many advanced devices is the location of non-metal dopants such as carbon. The high energy levels and ionization potential of non-metals makes RIMS of C difficult. Carbon-doped devices are compared with Be doped devices. RIMS profiles are assoiciated with improved device performance.

1. INTRODUCTION

We have been using resonance ionization mass spectrometry (RIMS) as a sputtered neutrals probe for compound semiconductor depth profiling (Downey, et. al., 1990). RIMS is a form of so called "post-ionization" sputtered neutral mass spectrometry (SNMS). Because atoms are often the predominantly sputtered product, detection sensitivity is enhanced and matrix effects, which plague secondary ion mass spectrometry (SIMS), are reduced. Ideally, only one element is ionized at a time, reducing potential isobaric background interferences. In this work, depth profiling RIMS is shown to be an effective technique for quantitative analysis of dopants at or near interfaces in layered III-V heterostructures. In particular, diffusion of carbon in doped $Al_xGa_{1-x}As$ is compared with Be doped material.

2. EXPERIMENTAL

The RIMS/SIMS instrument has been described and calibrated previously (Downey and Emerson, 1991). SIMS ions are produced with O_2^+ primary ions and detected with pulse counting, electronic gating and physical apertures to improve depth resolution. Xe^+ sputtering is used for RIMS experiments to enhance the atom yield. Both techniques are thus easily compared. The incident beam energy is 2-6 keV and currents up to 1 μA are contained in a spot of about 100 μm (FWHM) diameter. The flat-bottomed portion of the craters is 250 to 300 μm. Sample high-voltage bias and time-gated detection are used to select the RIMS ions while rejecting SIMS ions. In the RIMS mode, the output of a Daly (analog) detector is monitored with a gated integrator and computer as a function of time. One crater is required for each element depth profiled. Profilometry is performed on the craters to obtain depth information.

An excimer-pumped, frequency-doubled pulsed dye laser is used to create RIMS ions. In this work, the wavelengths used are, Al (236.7 nm), As (235.0 nm), Be (234.8 nm), C (245.9 nm) and Si (243.5 nm). The laser light has intensity of more than 10^7 W cm^{-2}, and is

(245.9 nm) and Si (243.5 nm). The laser light has intensity of more than 10^7 W cm^{-2}, and is focused into the sputtering region about 100 μm in front of the sample to optimize detection. The laser is synchronized to the rastered primary ion beam. A digital delay generator fires the laser (40 Hz) when the primary beam is in the center of the crater for optimal depth resolution.

A variety of Al$_x$Ga$_{1-x}$As samples grown by molecular beam epitaxy (MBE) are examined with RIMS. The MBE is performed with an elemental source Intevac Gen II system at 580°C on (100) oriented, 2 inch undoped or n$^+$ doped (Si ≈1 x10^{18} cm^{-3}) GaAs substrates, with substrate rotation. Carbon doping was obtained from a heated graphite filament source as described previously (Malik, et. al, 1988). Growth rates and layer thickness are monitored by high energy electron diffraction (RHEED) intensity oscillations (Kopf, et. al., 1991).

Two types of structures were grown. Si, Be, and C dopant segregation was studied, using simple test structures, each consisting of four layers: 1000Å GaAs, 1000Å AlAs, 1000Å GaAs, and 1000Å AlAs, grown on undoped substrates. Both the GaAs and AlAs growth rates were 0.5 μm/hr, and the entire structure was uniformly doped by maintaining a constant dopant flux during growth. These test structures were doped with Si; 3.0 x 10^{18} cm^{-3}, Be; 2.0 x 10^{19} cm^{-3}; and C; 2.0 x 10^{19} cm^{-3}, respectively. Distributed Bragg reflector (DBR) structures, designed for a peak reflectivity at 850 nm, with 19 periods of bilayers, were grown on n$^+$ substrates. The DBRs are multilayer structures of AlAs and Al$_{0.14}$Ga$_{0.86}$As which are nominally uniformly doped to 5.0 x 10^{18} cm^{-3} with either Be or C. Compositional changes of the layers can occur in stepwise fashion or by linear or parabolic changes in Al and Ga concentration (Kopf, et. al., 1992a).

3. RESULTS AND DISCUSSION

Figure 1 shows the RIMS depth profiles of the Be-doped DBR. The Al trace indicates the layers' locations as well as the parabolic grading of the Al/Ga ratio. Note that this is raw data without any corrections for the changing matrix. Although the As concentration is constant in these materials, the As signal is modulated thus also marking the layers. This is due to slight sputter rate changes and some molecular formation. The sputter yield from GaAs is greater than that of AlAs under the present conditions, giving rise to larger signals, and AlAs is noted to sputter in significant quantity (Downey, et. al., 1992). The most remarkable feature in Fig. 1 is the Be data. Although the doping was intended to be constant throughout this structure, the Be concentration is seen to be lower (5 x 10^{17} cm^{-3}) in AlAs than in the Al$_{0.14}$Ga$_{0.86}$As, and spikes of Be "pile-up" are noted at the interfaces. Movement of Be occurs in the MBE growth direction (forward), out of the AlAs until a suitable Al/Ga composition is reached for its incorporation into the growing crystal. SIMS profiles of the structure produce intractable results due to matrix effects for Be associated with the parabolic change in Al/Ga composition.

Beryllium diffusion out of AlAs in the DBR is a significant drawback to this structure because in certain applications, the DBR must be both conductive and reflective. Here the AlAs resistivity is too high because the maximum obtainable dopant concentration is too low. From these data, it was determined to substitute carbon for Be as the DBR dopant and determine if C diffusion is problematic.

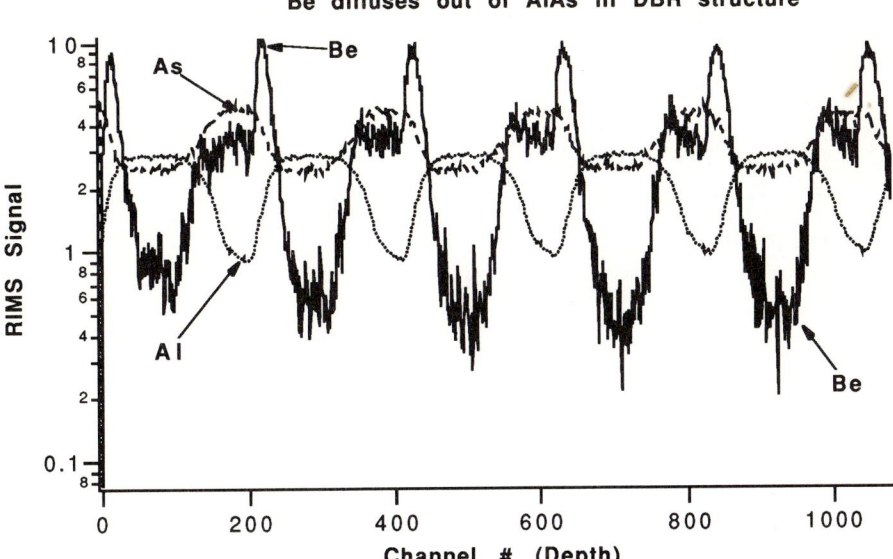

Figure 1. RIMS profiles of five periods of AlAs/Al$_{0.14}$Ga$_{0.86}$As DBR. Sputtering was performed with 6 keV Xe$^+$. The total sputtered depth is 0.83 μm.

RIMS of C is difficult because the ionization potential is 11.2 eV, which is high enough to require absorption of three near ultra-violet photons. Resonant post-ionization of sputtered C atoms has been demonstrated to require > 10^8 W/cm^2 to saturate the ionization process at 280 nm (Gelin, et. al., 1990). In the present work at 245.9 nm, the 2p^2 ^{3}P$_J$ → 4p ^{3}P$_{J'}$ transition is used, involving the simultaneous two-photon absorption (2+1) process. We found that the intensity required to saturate the overall process was also in excess of 10^8 W/cm^2. Figure 2 shows the depth profiles of the C doped multilayer test structure. The C profile follows the As profile, indicating that sputter rate changes are mostly responsible for different signal levels from the two materials. A large amount of surface carbon contamination is obvious, but no spike-like diffusion to the heterointerfaces are evident as in the Be-doped case. The Si doped sample was nearly identical to the C doped sample, except no contamination spikes were present at interfaces.

The sensitivity for C in the RIMS instrument is reduced relative to Be because the ground state of C is split into nearly degenerate ^{3}P$_{J=0,1,2}$ sub levels which are not simultaneously addressed by a single laser wavelength whereas the Be ground state is the unsplit ^{1}S$_0$. If the sub levels are populated in accordance to their degeneracies, only about one half of the sputtered C resides in the J=2 level. Moreover, the C signals were not saturated with the available laser intensity. Even so, the detection limit for C in these materials is about 1 x 10^{18} cm^{-3} (with about 0.5 μA of Xe$^+$) which is adequate for C-doped DBR diagnostics.

The series resistance of DBRs, doped with C, exhibits a factor of two lower than their Be doped counterparts (Kopf, et. al., 1992b). As in Fig.2, RIMS profiles of C-doped test structures of GaAs/AlAs show no detectable movement of carbon. These results are consistent with known diffusivities of C and Be in AlAs and GaAs (Abernathy, et. al., 1989). Carbon is about three orders of magnitude less mobile than Be in these materials.

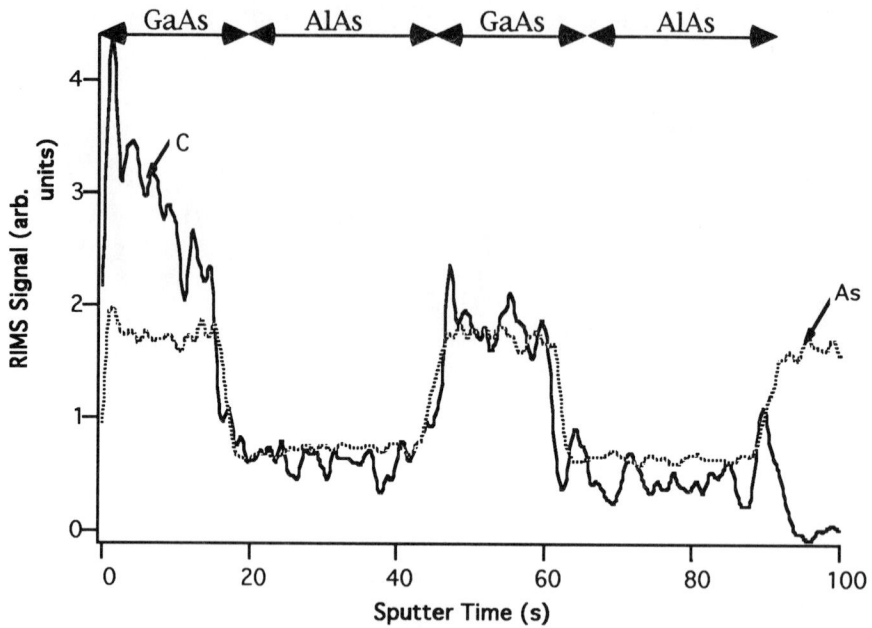

Figure 2. RIMS profiles carbon-doped GaAs/AlAs test structure. Sputtering was performed with 6 keV Xe^+. C spikes are seen at the surface and substrate interface.

4. CONCLUSIONS

Resonance ionization mass spectrometry is used to measure Si, Be and C dopant profiles in GaAs/AlAs and AlAs/AlGaAs DBR structures grown by MBE. When $Al_xGa_{1-x}As$ doped with Si and C, the targeted dopant profiles were readily obtained. However, when the structures were doped with Be, we observed significant redistribution of the dopant, resulting in an accumulation of Be at heterointerfaces. These results show that the high series resistance in DBR structures is due in part to segregation of Be during growth, and a much lower solubility limit for Be in AlAs than GaAs. These problems can be avoided by doping the structure with C.

5. REFERENCES

Abernathy, C R, Pearton, S J, Caruso, R, Ren, F and Kovalchick, J 1989 *Appl. Phys. Lett.* **55** 1750

Downey, S W, Emerson, A B, Kopf, R F and Kuo, J M 1990 *Surf. and Int. Anal.* **15** 781

Downey, S W and Emerson, A B, 1991 *Anal. Chem.* **63** 916

Downey, S W, Emerson, A B and Kopf, R F 1992 *Nucl. Instr. and Meth.* **B62** 456

Gelin, P, Barthe, M F, Debrun, J L, Gobert, O, Gibert, T, Inglebert, R L, and Dubreuil, B 1990 *Nucl. Instr. and Meth.* **B45** 580

Kopf, R F, Kuo, J M, and Ohring, M 1991 *J. Vac. Sci .and Technol.* **B9** 1920

Kopf, R F, Herman, M H, Schnoes, M L, Perley, A P, Livescu, G and Ohring, M 1992a *J. Appl. Phys.* **71** in press

Kopf, R F, Schubert, E F, Downey, S W and Emerson, A B 1992b *Appl. Phys. Lett.* submitted

Malik, R J, Nottenberg, R N, Schubert, E F, Walker, J F and Ryan, R W 1988 *Appl. Phys. Lett.* **53** 2661

Inst. Phys. Conf. Ser. No 128: Section 7
Paper presented at RIS 92, Santa Fe, NM, USA, 24–29 May 1992

259

Prospects for submicron molecular imaging with ion beams and lasers

N Winograd, Y Zhou, M Wood and S Lakiszak

Pennsylvania State University, Department of Chemistry, 152 Davey Laboratory, University Park, PA 16802 USA

S Mullock

Kore Technology Ltd, Cambridge Science Park, Cambridge CB4 4FX, UK

ABSTRACT: We have combined energetic particle bombardment with MPRI as a tool for analysis of atoms and molecules adsorbed on solid surfaces. The selectivity of the postionization process and the high ionization efficiency yield excellent detection limits for trace levels of surface atoms and molecules. In this work, we examine the use of a liquid metal ion source for surface studies. This source offers the unique property that it can be focused to a spot size of smaller than 400 Å. The flux of incident ions, however, is extremely low, necessitating the use of MPRI to enhance detection limits.

1.   INTRODUCTION

Energetic particle bombardment of surfaces followed by MPRI detection of desorbed species has become a powerful new surface characterization methodology since its inception 10 years ago (Winograd et al 1982) (Parks et al 1983). The experiments have found several arenas where they have contributed significant new science. Initial studies focused on the high selectivity and ionization efficiency of MPRI detection for desorbed atoms. These properties extended the analytical utility of existing surface analysis techniques such as secondary ion mass spectrometry (SIMS) where large matrix ionization effects are problematic. Detection limits have ultimately reached to less than 200 impurity atoms over 1 cm^2 of the target surface (Pappas et al 1989). A second area has involved the use of MPRI to measure the energy and angle distributions of desorbed atoms by taking advantage of the spatial distribution of the sputtered flux. These studies have provided the first measurements of the trajectories of desorbed species which can then be directly compared with molecular dynamics computer simulations of the ion/solid interaction (Garrison et al 1988). Recently, this approach has even been successful for describing species desorbed in metastable excited states (Bernardo et al 1992). Finally, there is now considerable evidence to suggest that MPRI can be effectively used to detect molecular species desorbed from surfaces. So far, a variety of aromatic molecules, amino acids and simple polymers have been detected using a simple 1+1 ionization scheme with detection limits in the attomole range (Hrubowchak et al 1991). At this point, then, MPRI detection has offered a new window for elucidating the fundamentals of the ion/solid interaction and for expanding the applications of ion beam methods.

Let us consider here the high sensitivity of MPRI detection. To detect a small amount of an impurity element in a background matrix, for example, 1 part in 10^9, it is of course necessary to guarantee that enough of the impurity atoms reach the laser beam to ensure a statistically accurate analysis. In the past, this has been accomplished by using a high current primary ion source which has a relatively large area of several mm in diameter. For this case, if the primary ion current, I_p, is 100 μA/cm^2, is pulsed on for τ_p of 3 μsec at a 30 Hz repetition rate, and the yield, Y, of sputtered atoms per incident ion is ~3, there will be 2 x 10^{10} atoms /s placed in the gas phase above the target. Assuming a 10% detection efficiency, then, an impurity present at 1 part in 10^9 would yield 600 counts after 5 minutes of analysis. These types of numbers have been demonstrated in real applications by a number of groups.

Where are the applications for such high sensitivity measurements? As noted, detection limits for atoms can reach 200/cm^2 (Pappas 1989). For molecules, preliminary studies have already shown that 10^6 molecules/cm^2 can be achieved (Hrubowchak 1991). Clearly, there are potential uses for such techniques in materials science and semiconductor processing where quality control is a critical factor. These avenues are being vigorously pursued.

Another important direction for this work involves imaging a small number of atoms or molecules from a well–defined spot. Using a liquid metal ion gun (LMIG), for example, it is possible to probe an area of less than 400 Å x 400 Å. The analytical difficulty arises when considering the number of atoms or molecules that are present in the surface layer of an area of this magnitude.

As seen in Table I,depending on size, there are roughly 40,000 molecules or 250,000 atoms in a spot of 1000 Å x 1000 Å. Hence, the need for ultrasensitive techniques is obvious for imaging experiments, particularly in the submicron regime.

There is another important issue associated with this experiment. For the earlier trace analysis studies, I_p ~ 100 μA. For the LMIG operating under conditions of maximum spatial resolution, I_p = 60 pA into a 400 Å spot. (This value can be increased to 500 pA with a concomitant loss of spatial resolution). For τ_p = 500 ns operated at a 30 Hz repetition rate, there will be only 15,000 atoms/s overlapping with the photon field, even though I_p = 4 Amp/cm^2! The conclusions from these numbers are that the LMIG is good enough to yield measurable signals from

Estimation of Number of Molecules/Pixel

Imaged Area	Pixel Size	Pixel Area	Molecules/pixel molecular size 5 Å x 5 Å	Atoms/pixel
100 μm	10 μm x 10 μm	10^{-6} cm^2	4 x 10^8	2.5 x 10^9
10 μm	1 μm x 1 μm	10^{-8} cm^2	4 x 10^6	2.5 x 10^7
5 μm	5000 Å x 5000 Å	2.5 x 10^{-9} cm^2	1 x 10^6	6.25 x 10^6
1 μm	1000 Å x 1000 Å	1 x 10^{-10} cm^2	40,000	2.5 x 10^5
0.2 μm	200 Å x 200 Å	4 x 10^{-12} cm^2	1600	10,000

small areas. It is not good enough, however, to perform analysis of trace levels of impurities that may be present within these spots. In the remainder of this paper, we present our preliminary experiments aimed toward demonstrating the applicability of ion–beam induced desorption using an LMIG to obtain submicron resolved, chemical specific images.

2. EXPERIMENTAL SETUP

A generic description of the apparatus has been described previously, although there are a number of changes and additions needed for these experiments (Pappas et al 1989). The basic instrument is a Kratos TOF–SIMS reflectron–based mass spectrometer equipped with a FEICO LMIG. The analyzer has a nominal mass resolution of 1 part in 10,000 amu with transmission of about 50%. The UHV system is ion–pumped except for the sample inlet system which is pumped by a turbomolecular pump. This pump must be turned off during imaging experiments to avoid vibrations. The laser beam is constricted to a diameter of about 1.5 mm and positioned less than 0.5 mm above the target surface. Indium is used as a model system for this work. The MPRI scheme utilizes 303.9 nm to pump the $^2D_{3/2}$ state and 607.9 nm to reach the ionization level. The laser system is identical to that described previously (Pappas et al 1989).

Pulsing of the LMIG to optimize the spatial overlap between the sputtered species and the laser beam presents a number of subtle challenges and opportunities. The pulsing is typically carried out by applying a deflection voltage to a set of blanking plates, preventing the ion beam from exiting through an aperture. When the proper voltage pulse is applied to the blanking plate, the beam exits through the aperture and strikes the target. For the short pulse–length required for SIMS experiments (~5 ns), the movement of the beam across the aperture causes considerable broadening of the beam spot on the target. In our case, however, the optimal τ_p is 100 times longer or 500 ns. These broadening artifacts are then insignificant. Moreover, the longer τ_p partially makes up for the relatively low repetition rate of the laser experiment (30 Hz) vs the SIMS experiment (5 kHz). An example of the spatial resolution of our system obtained by collecting all the secondary electrons emitted from the sample is shown in Figure 1.The edge of the 10 μ core is defined to within 2 pixels or 400 Å.

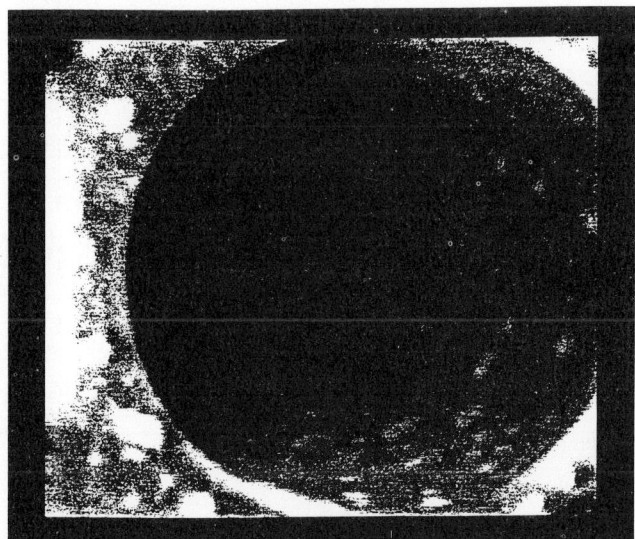

Figure 1. Secondary electron image of a 10 μ fiber in a 12 μ x 10 μ field. Each pixel measures 20 nm^2.

3.   RESULTS AND DISCUSSION

The partial TOF spectrum obtained by bombarding In foil with 25 keV Ga$^+$ ions from the LMIG is shown in Figure 2. The spectrum was obtained using a single–stop time–to–digital– converter (TDC) with 1.25 ns time registration. The incident ion beam current was arbitrarily attenuated so that less than one ^{113}In$^+$ ion reached the detector for each laser shot. The expected isotope ratio of ^{113}In/^{115}In = 0.04 is not observed since there is more than one ^{115}In$^+$ ion per laser shot and the detector is not responding to these multiple hits. The time width of the ^{113}In peak is 11 ns and the flight time is 60,321 ns. These results show that there is an adequate signal generated from the LMIG and that the TOF–reflector yields a mass resolution m/Δm of ~2800. Note that most of this width arises from the 7 ns laser pulse width and that the analyzer is almost completely compensating for the energy spread of the ionized In atoms, estimated to be ~300 eV.

The expected useful yield of In atoms detected in this experiment is easily estimated. The design parameters of the analyzer are aimed to yield about 50% collection efficiency. The laser extraction optics can draw at least 50% of the ionized atoms into the analyzer. And finally, we estimate that about 50% of the sputtered flux actually overlaps with the ionizing laser. Assuming 100% ionization efficiency, then, we expect to be able to detect ~12% of the sputtered In atoms. Using a 300 ns, 60 pA LMIG source, we expect to produce 338 In atoms per laser shot. It is not yet feasible to directly measure this many ions with a TDC, although the Kratos system provides a multistop, 10 ns TDC that can register more than one hit for ^{113}In. Using this detector we find 1.3 ^{113}In$^+$ ions per laser shot or ~33 total In$^+$ ions for a measured useful yield of 33/338 $\cong$ 9%.

We have tested the possibility of using the LMIG source to obtain spatially resolved maps of the In signal. As seen in Figure 3, 300 mesh Cu grid was placed over an In foil so as to shadow

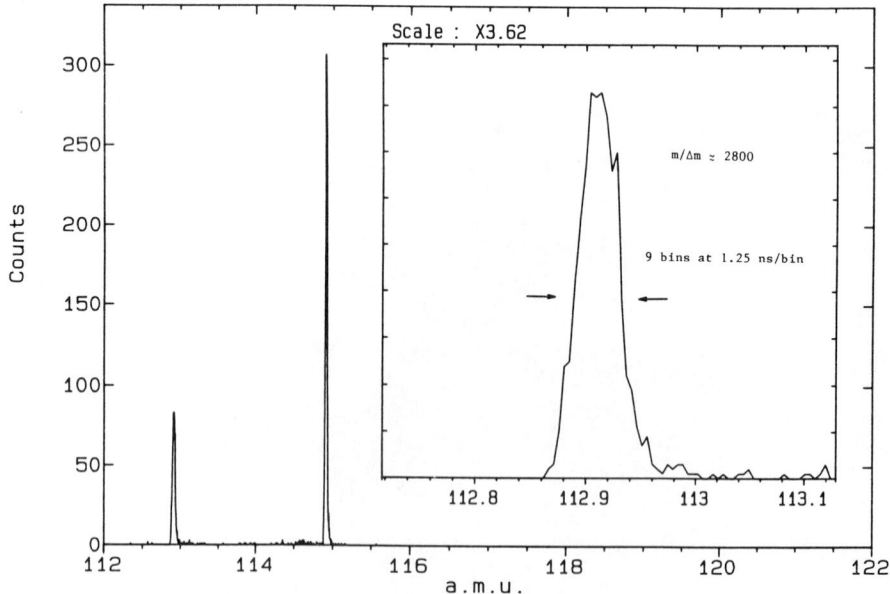

Figure 2. TOF spectrum of In atoms desorbed from In foil and resonantly ionized by a 1.5 mm laser beam.

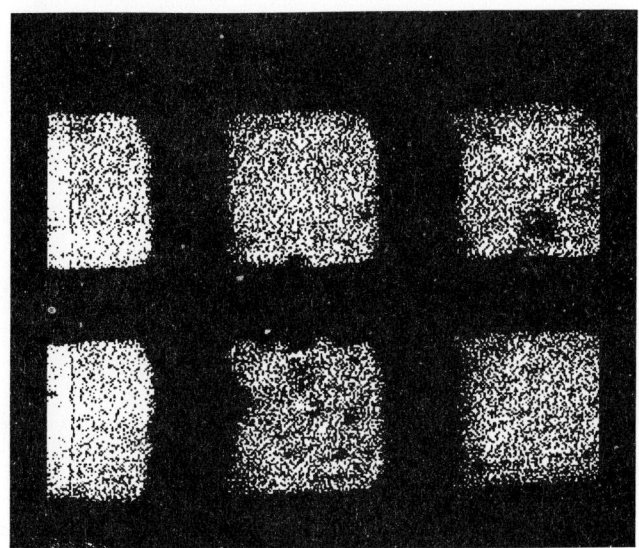

Figure 3. Indium atom image of a copper grid overlaying Indium substrate. Field of view, 150 μ x 130 μ. Each pixel measures 500 nm^2, with the brightest pixels containing 18 counts. ~1 x 10^5 pulses, total dose 2 x 10^7 Ga$^+$ ions or 10^{11} Ga$^+$ ions/cm^2.

specific regions of the In surface. For this image, each pixel measures 5000 Å x 5000 Å and the total field of view is 150 μ x 130 μ. Each pixel was produced by summing the results of two laser shots. The image, which contains about 50,000 pixels was recorded in about 45 min. We estimate that the data acquisition efficiency can be improved by a factor of 4 when using a transient digitizer rather than a multihit TDC. Moreover, if the LMIG is employed using the larger aperture, the count rates improve by an additional factor of 8 with only a small loss in spatial resolution.

Will it be possible to record comparable images for molecules on surface? It is, of course, well–known that ion beams can effectively desorb a wide array of fragile molecules intact, forming the basis for much of SIMS research. Our group has also reported that a number of polycyclic aromatic hydrocarbons (PAH) can be desorbed and resonantly ionized using UV lasers with detection limits near 10^6 molecules (Hrubowchak 1991) per cm^2. In many cases, the signals obtained using laser postionization are several orders of magnitude larger than those obtained using SIMS, indicating that the resonance ionization probability can be as high as 1%. A recent example of a laser postionized PAH is shown in Figure 4. Given the magnitude of the signals observed for In, these data suggest that it will certainly be feasible to detect the molecular ion of these types of molecules using the LMIG and that submicron molecular imaging is indeed a feasible objective.

ACKNOWLEDGMENT

The authors would like to thank the U.S. Department of Energy, the National Science Foundation, the Office of Naval Research and the National Institute of Health for partial support of this work. Penn State University also supplied a significant portion of the equipment, for which we are most grateful.

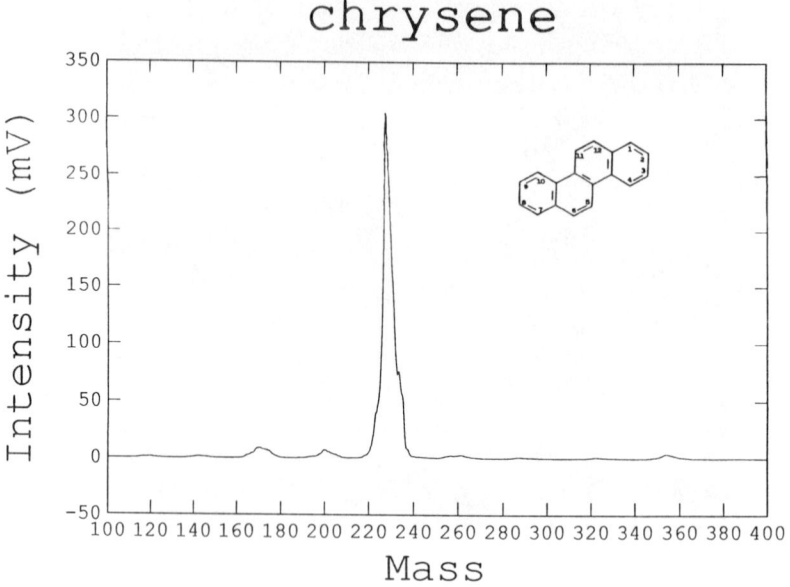

Figure 4. TOF spectrum of a thin film of chrysene ionized by 280 nm laser light after bombardment by 5 keV Ar+ ions.

REFERENCES

Bernardo D, El–Maazawi M, Maboudian R, Postawa Z, Winograd N and Garrison B J 1992 *J. Chem. Phys.* in press
Garrison B J, Winograd N, Deaven D M, Reimann C T, Lo D Y, Tombrello T A, Harrison Jr. D E and Shapiro M H 1988 *Phys. Rev.* **B37** 7197
Hrubowchak D M, Ervin M H, Wood M C and Winograd N 1991 *Anal. Chem.* **63** 1947
Pappas D L, Hrubowchak D M, Ervin M H and Winograd N 1989 *Science* **243** 64
Parks J E, Schmitt H W, Hurst G S and Fairbank Jr. W M 1983 *Thin Solid Films* **108** 69
Winograd, N, Baxter J P and Kimock F M 1982 *Chem. Phys. Lett.* **88** 581

Inst. Phys. Conf. Ser. No 128: Section 7
Paper presented at RIS 92, Santa Fe, NM, USA, 24–29 May 1992

RIMS study of low energy-laser sputtering of metal and semiconductor surfaces

B. Dubreuil, T. Gibert, M. F. Barthe*, J. L. Debrun*

GREMI, CNRS - Université d'Orléans, BP 6759, 45067 Orléans cédex 2, FRANCE

*CERI - CNRS, 3A, rue de la Férollerie, 45071 Orléans cédex 2, FRANCE

ABSTRACT : N_2 laser irradiation (λ = 337 nm) of metal (iron and alloys) and compound semiconductor (InP) surfaces at fluences below the boiling threshold induces surface atom sputtering. The unique sensitivity and selectivity of R.I.M.S. allow to identify the laser-sputtered neutral particles (atoms, molecules), to study their velocity distribution and excitation state and to measure the laser-sputtering yield at very low emission level. These measurements provide information on the irradiated surface and finally on the mechanisms of laser-induced surface sputtering. In the two examples studied, laser surface interaction can be described as a thermal process with the direct evidence of preferential sputtering in the case of InP.

1. INTRODUCTION

Surface sputtering induced by an incident laser beam (vaporization, ablation, desorption) is becoming increasingly important as technologies are developed which utilize this process. For example, laser sputtering is used in connection with mass-spectrometry or optical methods as a sampling technique for materials analysis (Estler et al 1987, Radziemski and Cremens 1989, Arlinghaus et al 1989). Laser sputtering is also used to produce high-quality thin films and to perform surface treatments and etching (Von Allmen 1987, Miller and Haglund 1991).

In the case of film deposition, real-time monitoring and precise control of the process requires the knowledge of the space and energy distributions of the sputtered products as well as their chemical composition as they impinge on the substrate. When laser sputtering is used as an atomization source for materials analysis, one needs to know if the chemical composition of the emitted particle cloud is representative of the surface stoechiometry.

The basic mechanisms involved in low-energy laser sputtering are currently the subject of numerous experimental and theoretical studies. One way to gain insight into the sputtering process itself is to probe the sputtered particles whose composition is expected to reflect the state of the irradiated surface. However, the fundamental processes are often masked by reactions occuring above the surface. For example, due to hydrodynamical effects, the energy/angle distributions and the chemistry of the probed particles can be strongly modified when their near-surface density is high enough to induce efficient elastic and inelastic gas phase collisions (Kelly and Dreyfus 1988).

Many forms of excitation are possible following the initial laser-surface interaction : selective removal of species via electronic transitions in the solid or on surface defects, photodissociation, thermal desorption. But, to have any hope of understanding the mechanisms in laser sputtering, it becomes necessary to conduct experiments where the laser fluences are near threshold levels.

2. EXPERIMENTAL STUDY

In this context, we have studied the interaction of a N_2 laser beam at 337 nm with metal (Fe and Fe/Ni/Cr alloy) and semiconductor (InP) targets in ultra-high vacuum conditions. The experimental set-up is similar to the one developed to perform material analysis by S.I.R.I.S. (Gelin 1989, Gobert et al 1990) in which the ion gun is replaced by the N_2 laser. The laser was operated at low to moderate fluences (E = 25 - 400 mJ/cm²) near the sputtering threshold. The emitted particles (ions, atoms) were analyzed by mass spectrometry, directly in the case of the ionized species and following resonant photoionization by a pulsed tunable dye laser for the neutrals. The unique sensitivity of RIS detection of neutrals (compared to an ionizer) allowed us to detect the sputtering event at a very low laser fluence and to discriminate amongst the ion and neutral emmission thresholds. Delaying the RIS probe laser shot with respect to the ablating laser pulse, we were able to measure the time-of-flight (TOF) distribution of the neutrals for different conditions. These distributions are compared to theoretical models : half-space Maxwell or Maxwell distributions with a finite center of mass velocity (Kelly and Dreyfus 1988). Great care must be taken in interpreting TOF measurements. For example, neglecting the finite dimensions of the laser probe can give rise to erroneous results even if the beam is spatially very uniform.

2.1 Laser sputtering of iron and alloy samples

Fe° atoms in their ground state were probed by (1+1) resonant multiphoton ionization at λ = 293.7 nm (Gobert 1991).

T.O.F. measurements were performed as a function of laser fluence and distance from the surface. The experimental results are well described over the whole laser fluence and distance range by a pure Maxwell velocity distribution limited to a half-space (Figure 1). Fitting the experimental T.O.F. distributions allows the determination of the kinetic

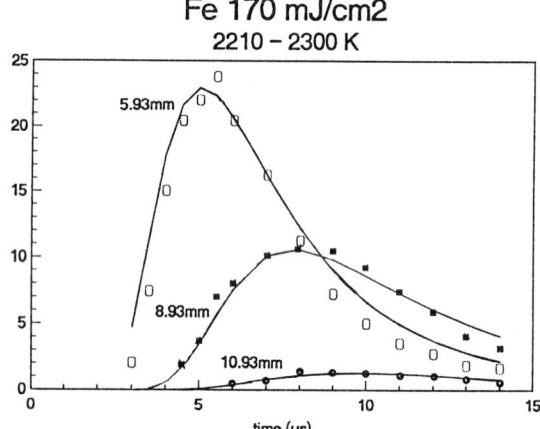

Fe 170 mJ/cm2
2210 – 2300 K

time (µs)

Figure 1. TOF distributions of sputtered Fe atoms (arbitrary unit) for 170 mJ/cm² laser fluence and 3 distances from the surface. Fit by Maxwell distributions with T_k = 2210 – 2300K

In the 25 - 100 mJ/cm² fluence range, Fe° and Fe+ exhibit strongly nonlinear increases in sputtering yields. However the threshold for the neutrals (30 mJ/cm²) is lower than that for the ions (90 mJ/cm²), suggesting separate primary production processes (Gibert 1991). Above 160 mJ/cm², the Fe° ground state population stays nearly constant whereas as a result of the large temperature increase (T > 1500 K), Fe° excited states are populated and Fe^2 molecules are formed in large quantities.

temperature T_k of the sputtered Fe° cloud. The variation of T_k with the laser fluence is shown in Figure 2.

Acquisition of a T.O.F. distribution results from multipulse irradiation of a same site on the surface. This is possible because there is no alteration of the signal during the measurement performed after a short period (a few minutes) corresponding to surface cleaning.

R.I.M.S. probing of the Fe° excited state population distribution (Gobert et al 1991) allows the determination of the atomic excitation temperature T_e. As shown in Figure 2, the T_e and T_k values agree rather well. This result corroborates, as expected, a thermal mechanism for laser sputtering of iron or alloy samples. The laser beam acts as a pulsed heat source and the surface temperature rise can be modelled by heat flow equations (Wedler and Ruhman 1992, Von Allmen 1987, Prokorov et al 1990).

It is interesting to note that the energy characteristics of the emitted neutrals reflect rather well the thermal state of the surface : the measured atom temperatures increase from the ambient at threshold to the melting and boiling temperatures at higher laser energies.

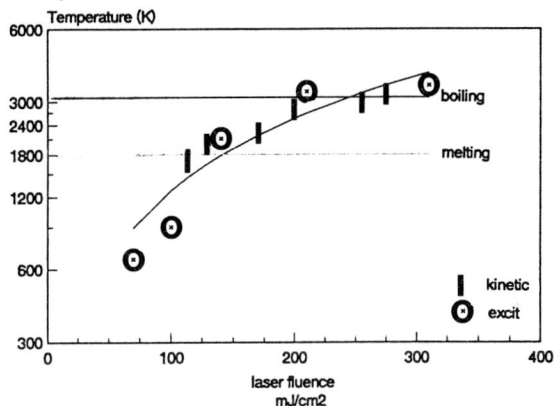

Figure 2. Kinetic and excitation temperatures of laser-sputtered Fe atoms as a function of the fluence. Fit by the results of a 1D heat-flow model (divided by 1.5)

The number of Fe° atoms emitted per laser shot has been estimated by comparison of laser ablation with 12 keV Ar ion sputtering. At 300 mJ/cm² laser fluence (neglecting the Fe_2 contribution), about $2 \cdot 10^7$ Fe atoms and ions are emitted which represent 10^{-4} of a monolayer.

At threshold (25 mJ/cm²), this corresponds to about 1000 Fe atoms, which probably are emitted from defect sites (laser cleaning). Under these conditions, near surface gas phase collisions are indeed negligible and a Knudsen layer is not established near the surface. Consequently, the measured T.O.F. particle distribution reproduces the nascent energy distribution.

2.2 Laser sputtering of InP surface

Pulsed laser irradiation of elemental and compound semiconductors is the object of numerous studies due to its potential application in the micro and optoelectronic industries. In the case of above bandgap laser irradiation, which is the case for irradiation of the InP (E_g = 1.34 eV) with 337 nm photons, absorption is expected to occur via interband transitions generating a large non-equilibrium electron-hole pair population. For nanosecond laser pulses, these highly excited carriers transfer their initial energy to the lattice and recombine according to different pathways which finally lead to bond breaking and particle emission. But, the distinction between thermal and non thermal (electronic) contributions is not clear (Brewer et al 1990, Namiki et al 1990).

In our experiment, In and P atoms and molecules sputtered from a (100) InP surface were detected by R.I.M.S. In° was photoionized at λ = 304.02 nm by a 1+1 photon process (Kimock et al 1983) whereas P° was probed by a 2+1 photoionization process at λ = 301.39 nm (Gobert et al 1990). Whereas the In° ionization scheme is easily saturated, only a 15% ionization efficiency can be expected for P°. Molecular species, particularly P_2 were detected for 200 μJ incident laser energy (350 mJ/cm²).

The laser sputtering yields of In°, P° and In+ exhibit a near exponential increase with the laser fluence from the threshold (60 mJ/cm²) to 300 mJ/cm² with a strong transition around 240 mJ/cm² defining a low energy (LE) and a high energy (HE) domain. The variation in the amount of In° with the number of laser shots on the same site are completely different in these two domains. In the LE domain, In° yield decreases as the sum of two exponentials. The fast and slow components can be attributed (Hattori et al 1991) to the sputtering of surface defects (adatoms and steps) probably resulting from localized absorption on the defect sites followed by bond breaking. Under similar conditions, Yamamoto et al (1989) and Hattori et al (1991) observed no modification of the surface structure. On the contrary, in the HE domain In° intensity increased with the number of laser shots n as In° (∞) (1 - exp (-a n)) up to a nearly constant value. In this case surface modification was observed (Gibert 1991). The P° yield measured under similar conditions shows a complementary evolution as seen in Figure 3. It decreases as : P° (0) exp (-a n) + P° (∞) to a quasi constant value.

This behaviour is quite similar to what is observed in ion sputtering of multicomponent systems and is known as preferential sputtering (Blaise and Coudray 1985). Preferential ion sputtering of the P species from InP leads to a change in the surface composition (Malherbe and Barnard 1991).

In the case of laser sputtering P° and P_2 are preferentially emitted leading after many laser shots to a change in the surface composition. The irradiated surface becomes In-rich up to the level where the change compensates for the difference between In and P sputtering yields.

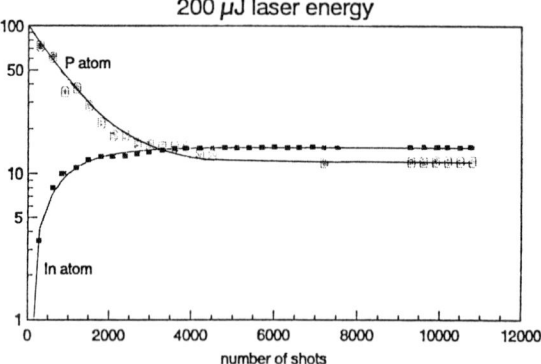

Laser sputtered In and P atoms as a function of the number of laser shots
200 µJ laser energy

Figure 3. In° and P° laser sputtering yields (relative unit) as a function of the number of laser shots (200 µJ laser energy)

This behaviour was revealed in earlier studies (Moison and Bensoussan 1982) of InP laser surface reconstruction and annealing via AES and LEED spectroscopies. Some authors attribute this preferential sputtering to a dimerization reaction occuring at the surface as a result of strong electronic excitation (Namiki et al 1991). This could explain the rather high P_2 R.I.M.S. signal which we observed in the HE domain.

Finally, T.O.F. distributions have been measured for In° and P° in the HE domain. The results together with the fits by Maxwell distributions are shown in Figure 4.

In P atom time–of–flight distribution

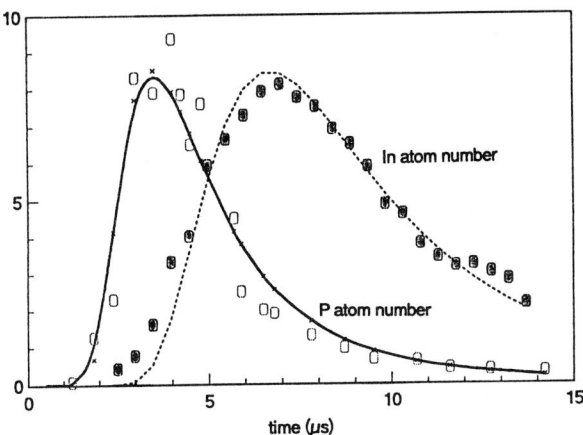

Figure 4. TOF distributions of In and P atoms sputtered from a InP surface by 200µJ energy laser pulses (density measurements). Fits by Maxwell distributions with T_k = 2000K

Two important results can be deduced.
  - The kinetic temperature is the same for In° and P°. This implies that in the laser surface interaction process the same amount of kinetic energy is transfered to the sputtered In° and P° atoms. The velocities of the P° and In° populations are quite different : P° atoms propagate faster than In° atoms. The relative concentration of the two species in the propagating cloud is then strongly dependent on the time of measurement. Simultaneous laser probing of In° and P° in such a mixture using, for example, nonresonant laser ionization of the two species can give rise to erroneous results. Probing at the early time should result in an overestimated P° relative concentration.
  - The kinetic temperature deduced from In° T.O.F. measurements varies from 1400 K at 100 µJ energy to 2000 K at 200 µJ. These values can be compared to the 1340 K boiling temperature of this material.
For low energy laser irradiation, laser sputtering of InP appears to result mainly from localized absorption and excitation of defect sites. Conversely, at moderate energy, interband absorption in the whole absorption volume results, after relaxation of the initial non equilibrium carrier energy, in a surface sputtering (and modification) process which can be well described by a thermal model.

References

Arlinghaus H F, Thonnard N. and Schmitt H W 1989 *Microbeam Analysis* ed P E Russell (San Francisco Press, Inc) pp 180-5

Blaise G and Coudray C 1984 *Le Vide Les Couches Minces* Supplement au n°224 pp 70-106

Brewer P D, Zinck J J and Olson G L 1991 *Laser Ablation - Mechanisms and Applications* eds J C Miller and R F Haglund, Jr. (Berlin Heidelberg : Springer - Verlag) pp 96-105

Estler R C, Apel E C and Nogar N S 1987 *J Opt. Soc. Am. B* 4 281

Gelin P, Debrun J L, Gobert O, Inglebert R L and Dubreuil B 1989 *Nucl. Instrum. Methods B* 40 290

Gibert T 1991 *Ph. D. Thesis* (Université d'Orléans)

Gobert O, Dubreuil B, Inglebert R L, Gelin P and Debrun J L 1990 *Phys. Rev A* 41 6225

Gobert O, Gibert T, Dubreuil B, Gelin P and Debrun J L 1991 *J. Appl. Phys.* 70 7602

Hattori K, Okano A, Nakai Y, Itoh N and Haglund Jr R F 1991 *J. Phys. : Condens. Matter* 3 7001

Kelly R and Dreyfus R W 1988 *Nucl. Instrum. Methods B* 32 341

Kimock F M, Baxter J P and Winograd N 1983 *Surface Sci.* 124 L41

Malherbe J B and Barnard W O 1991 *Surface Sci.* 244 309

Miller J C and Haglund Jr R F (Eds.) 1991 *Laser Ablation - Mechanisms and Applications* (Berlin Heidelberg : Springer-Verlag)

Moison J M and Bensoussan M 1982 *J. Vac. Sci. Technol.* 21 315

Namiki A, Katoh K, Yamashita Y, Matsumoto Y, Amano H and Akasaki I 1991 *J. Appl. Phys.* 70 3268

Prokhorov A M, Konov I, Ursu I and Mihailescu I N 1990 *Laser Heating of Metals* (Bristol : Adam Hilger)

Radziemski L J and Cremers D A (Eds.) 1989 *Laser-Induced Plasmas and Applications* (New York : Marcel Dekker, Inc)

Von Allmen 1987 *Laser-Beam Interactions with Materials* (Berlin : Springer-Verlag)

Wedler G and Ruhman H 1982 *Surface Sci.* 121 464

Yamamoto T, Hattori K, Nakai Y, Itoh N and Szymonski M 1989 *Radiat. Eff. Def. Solids* 109 213

Inst. Phys. Conf. Ser. No 128: Section 7
Paper presented at RIS 92, Santa Fe, NM, USA, 24–29 May 1992

271

Resonance ionization of sputtered atoms—progress toward a quantitative technique

W. F. Calaway,[1] S. R. Coon,[1,2] M. J. Pellin,[1] C. E. Young,[1] J. E. Whitten,[1] R. C. Wiens,[3] D. M. Gruen,[1] G. Stingeder,[4] V. Penka,[5] M. Grasserbauer,[4] and D. S. Burnett[3]

ABSTRACT: The combination of RIMS and ion sputtering has been heralded as the ideal means of quantitatively probing the surface of a solid. While several laboratories have demonstrated the extreme sensitivity of combining RIMS with sputtering, less effort has been devoted to the question of accuracy. Using the SARISA instrument developed at Argonne National Laboratory, a number of well-characterized metallic samples have been analyzed. Results from these determinations have been compared with data obtained by several other analytical methods. One significant finding is that impurity measurements down to ppb levels in metal matrices can be made quantitative by employing polycrystalline metal foils as calibration standards. This discovery substantially reduces the effort required for quantitative analysis since a single standard can be used for determining concentrations spanning nine orders of magnitude.

1. INTRODUCTION

Since the demonstration that atomic selectivities of 10^{19} could be achieved by resonance ionization spectroscopy (Hurst, 1977), the field has rapidly grown as researchers have attempted to exploit this unique capability. One arena where these endeavors have been particularly fruitful is surface analysis where resonance ionization mass spectrometry (RIMS), in combination with ion sputtering, has been employed to analyze for trace impurities on the surfaces and in the bulk of a solid by probing the material ejected into the gas phase during ion bombardment. Here, the selectivity of resonance ionization is used to separate the element of interest from the bulk material that is sputtered and a mass spectrometer is used for noise discrimination and detection. At least three different laboratories have reported detection capabilities below atom fractions of 10^{-9} (1 appb) employing this technique (Young, 1986; Pappas, 1989; Thonnard, 1989; Pellin, 1990).

In addition to selectivity, RIMS of sputtered atoms has the potential for making quantitative measurements with ease. This is derived from the fact that sputtering of clean metals by noble gas ions produces a flux dominated by ground state neutral atoms and the fact that there is a large data base on the rate of removal for most metals due to sputtering. By knowing sputtering yield and instrument transmissions, it is possible to draw quantitative conclusions

*Work supported by the U.S. Department of Energy, BES-Materials Sciences, under Contract W-31-109-ENG-38.

[1]Materials Science/Chemical Technology/Chemistry Divisions, Argonne National Laboratory.
[2]Department of Chemistry, University of Texas at Austin.
[3]Division of Geology and Planetary Sciences, California Institute of Technology.
[4]Institute of Analytical Chemistry, Technical University of Vienna
[5]Siemens AG, Munich

regarding impurity concentrations if the degree of ionization is known. Thus, RIS is unique in that complete ionization of a species is possible without extraordinary efforts.

One of the difficulties associated with quantitative analysis is the development of standards. This is particularly true for techniques that are attempting to extend detection limits to heretofore unobtainable levels. To better define the usefulness of surface analysis by resonance ionization of sputtered atoms (SARISA) for quantitative analysis, a series of samples has been analyzed for various impurities at levels that can be confirmed by other techniques. The experiments that have been completed to date are presented herein.

2. EXPERIMENTAL

RIMS experiments have been performed to determine the concentration of Fe, Cu, and Mo in Si. Pure Si wafers were dipped into solutions containing Cr, Cu, Ni, and Fe in order to produce contaminated samples. These samples were then analyzed by total reflection x-ray fluorescence (TXRF). Each wafer was then split in half, one half being analyzed by secondary ion mass spectrometry (SIMS) and the other half by SARISA. In addition, two targets were prepared by ion implanting 200 keV Mo into Si wafers at doses of $3.0 \times 10^{13}/cm^2$ and $1.0 \times 10^{11}/cm^2$

The TXRF that was used was an Atomika model XSA 8000 located at Siemens, Munich, Germany. The SIMS instrument was the CAMECA IMS 3f located at the Technical University, Vienna, Austria. SARISA is a one-of-a-kind laser postionization secondary neutral mass spectrometer built and operated by Argonne National Laboratory, USA. Both the TXRF and SIMS are commercially available instruments and require no further description. Details of SARISA can be found in past publications (Pellin, 1988; Pellin 1987).

For Cu and Mo in Si, simple two-color/two-photon ionization schemes produce good discrimination between signal and background. In the SARISA analyses, the Cu resonance transition at 327.40 nm ($^2S_{1/2} \rightarrow {}^2P_{3/2}$) was utilized. A dye laser containing DMT and pumped with a KrF excimer laser was employed for the resonance step and a XeCl excimer laser (308 nm) was used to access the ionization continuum. For Mo RIMS, the z^7P_0 state is accessed from the ground state (a^7S_3) by a dye laser tuned to 386.41 nm and then ionized with light from a XeCl excimer laser. For Fe in Si, a three-color/three-photon scheme was employed in order to minimize isobaric interference from nonresonantly ionized Si dimers (Gruen, 1991). This ionization scheme requires three dye lasers tuned to 344.06, 640.00, and 510.0 nm in order to access the $z^5P_3^0$, e^5D_4 and an autoionizing state of Fe, respectively, in a stepwise fashion.

In order to make quantitative determinations, the SARISA instrument was calibrated before and after analysis of the unknown sample. This was accomplished by determining the transmission of the instrument by measuring the signal from a pure metal foil of the element being investigated. For a constant ion current and instrument transmission, this signal, S, from the pure, p, elemental target is then used to calculate the concentration of that element, [M], in an unknown sample, u, through the equation,

$$[M] = \frac{S_u Y_p G_p}{S_p Y_u G_u} \tag{1}$$

where Y is the sputtering yield and G is the gain of the detector. Note that Equation 1 assumes that the sputter yield for the element to be measured and for the matrix containing it are identical and equal to the sputter yield for the uncontaminated substrate.

3. RESULTS AND DISCUSSION

Both SIMS and SARISA have the capability to determine elemental concentrations as a function of depth. This is accomplished by monitoring signal for the element of interest as an ion beam mills into the target. Typical results for the SARISA instrument are shown in Figures 1 and 2 for Cu and Mo, respectively. As can be seen in Figure 1, the Cu is clearly a surface contaminate that falls off with depth. This is in contrast to the Mo sample shown in Figure 2. Here, the Mo depth profile shows the typical shape expected for an ion-implanted target. The concentration maximum at 200 nm in Figure 2 is in agreement with the calculated depth for 200 keV Mo implanted into Si.

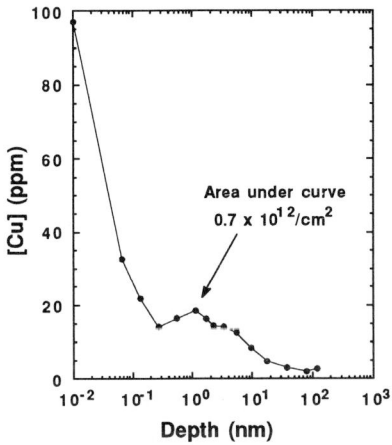

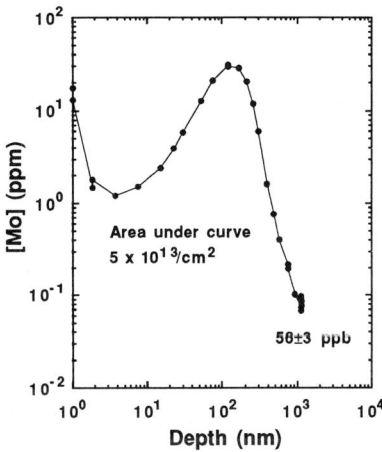

Fig. 1 SARISA analysis showing the Cu in Si concentration as a function of depth. Integration of the data from 0 to 15 nm yields a total Cu content of $0.7 \times 10^{12}/cm^2$.

Fig. 2 SARISA analysis showing the Mo in Si concentration as a function of depth. Integration of the entire curve yields a total Mo content of $5.0 \times 10^{13}/cm^2$.

TXRF instruments are unable to produce concentration versus depth data but rather measure the total concentration of contaminants in the near-surface region. For Si targets, this corresponds to about 15 nm. Thus, TXRF is capable of measuring surface contamination such as shown is Figure 1 but is unable to determine the concentration of implanted species as shown in Figure 2. In order to compare the SARISA measurements with TXRF data, concentration versus depth data must be integrated over the depth of TXRF sensitivity. Similarly, depth profile data of ion implanted standards must be integrated over the entire concentration peak in order to compare the SARISA measured results with the known ion dose. Integrations of SARISA depth profiles have been performed and are listed in Table I along with results from TXRF measurements and from implanted targets of known ion doses. Also listed in Table I are results of the SIMS analyses. Since SIMS data is difficult to quantify, results from these analyses are expressed only as to whether or not real signal above interferences could be observed.

As can be seen in Table I, agreement is reasonably good when various types of instruments are used to analyze the same sample. To date, only Cu analyses have been performed on identical Si samples by SARISA, SIMS, and TXRF. Unfortunately, the SIMS analysis suffered from an isobaric interference that, when suppressed, reduced the sensitivity of the measurement to the point that no Cu could be detected. Only SIMS analyses of Fe and Cr had confirmed signal above levels of isobaric interferences, but the SARISA analyses for these samples are yet to be performed. The agreement between the TXRF and SARISA measurements indicate that either may be used to measure surface contaminations. However, TXRF is unable to furnish the concentration versus depth information that SARISA can produce. In addition, SARISA has been shown to have a detection limit below 1 ppb while maintaining near monolayer resolution (Pellin, 1990).

Table I.   Comparison of impurity concentrations in Si as determined by various analytical methods. All concentrations are given in atoms/cm^2. For SIMS analysis, yes and no indicate whether an element was detected.

Element	SARISA	TXRF	SIMS	Ion Implant
Mo	5.0×10^{13}	-	-	3.0×10^{13}
Mo	2.5×10^{11}	-	-	1.0×10^{11}
Fe	4.0×10^{12}	8.0×10^{12}	-	-
Fe	-	0.6×10^{12}	yes	-
Cu	0.8×10^{12}	1.3×10^{12}	no	-
Cu	-	0.3×10^{12}	no	-
Cr	-	0.3×10^{12}	yes	-
Ni	-	1.1×10^{12}	no	-

A point regarding the accuracy of the SARISA measurements that should be emphasized is the method of calibration as described in the experimental section. Simple metallic foils were used as standards for these measurements, yet the accuracy appears to be good to within a factor of 3. As shown here for Si and observed in our laboratory for other metals, Equation 1 is valid down to at least ppb levels for metallic targets. The only other information required to make quantitative determinations in this manner is knowledge of the sputtering yields for the metal and the unknown substrate, and knowledge of the detector gain curve. Thus, it is possible to perform trace analyses with SARISA without the laborious task of developing and fabricating calibration standards specific to the substrate of interest.

4. REFERENCES

Gruen D M, Calaway W F, Pellin M J, Young C E, Spiegel D R, Clayton R N, Davis A M and Blum J D 1991 *Nucl. Instrum. Methods Phys. Res. B* **58** 505
Hurst G S, Nayfeh M H and Judish J P 1977 *Appl. Phys. Lett.* **30** 229
Pappas D L, Hrubowchak D M, Ervin M H and Winograd N 1989 *Science* **243** 64
Pellin M J, Young C E, Calaway W F, Burnett J W, Jørgensen B, Schweitzer E L and Gruen D M 1987 *Nucl. Instrum. Methods Phys. Res. B* **18** 446
Pellin M J, Young C E and Gruen D M 1988 *Scanning Microsc.* **2** 1353
Pellin M J, Young C E, Calaway W F, Whitten J E, Gruen D M, Blum J D, Hutcheon I D and Wasserburg G J 1990 *Phil. Trans. R. Soc. Lond. A* **333** 133
Thonnard N, Parks J E, Willis R D, Moore L J and Arlinghaus H F 1989 *Surf. and Interface Anal.* **14** 751
Young C E, Pellin M J, Calaway W F, Jørgensen B, Schweitzer E L and Gruen D M 1986 *Inst. Phys. Conf. Ser.* **84** 163

Inst. Phys. Conf. Ser. No 128: Section 7
Paper presented at RIS 92, Santa Fe, NM, USA, 24–29 May 1992

275

Three-dimensional analysis of semiconductors and biological surfaces using SIRIS and LARIS

H. F. Arlinghaus[+], M.T. Spaar[+], N. Thonnard[+], P. Holloway[*], G.W. Kabalka[#], and R.C. Switzer[#]

[+]Atom Sciences, Inc., 114 Ridgeway Center, Oak Ridge, TN 37830
[*]Department of Materials Science and Engineering, University of Florida, Gainesville, FL 32611
[#]University of Tennessee, Knoxville, TN.

ABSTRACT: Resonance Ionization Spectroscopy (RIS) is becoming recognized as an analytical tool to quantitate trace elements in semiconductors and biological materials. Using either Sputter Initiated RIS (SIRIS) or Laser Atomization RIS (LARIS), it is possible to localize with high spatial resolution ultra-trace concentrations of a selected element down to the sub-parts-per-billion level. We present quantitative SIRIS depth profile analysis results yielding a depth resolution of ~2 nm and a dynamic range of $>10^6$ as well as SIRIS/LARIS imaging results of trace elements on semiconductors and biological surfaces with high lateral resolution down to ~5 μm.

1. INTRODUCTION

Sputter initiated Resonance Ionization Spectroscopy (SIRIS) is an analytical technique having extremely high sensitivity, selectivity, dynamic range, and quantitation accuracy. SIRIS also promises good spatial resolution, and freedom from matrix effects on surfaces and at interfaces. Even higher sensitivity in the same analysis time can be achieved if the ion beam is replaced with a laser beam, i.e., Laser Atomization RIS (LARIS) (Arlinghaus, et al. 1991a, b). By use of a separate laser pulse instead of an ion pulse for the atomization process, three, or more, orders of magnitude more material can be released from the sample. Thus, with LARIS, tens of monolayers can be removed with a single atomizing laser pulse while with SIRIS, many bombarding ion pulses are required to remove a fraction of a monolayer. In this paper we will demonstrate the capability of the SIRIS technique to measure concentration profiles in semiconductors with high dynamic range and ultra-high depth resolution. Furthermore, SIRIS/LARIS imaging results of trace elements on semiconductors and biological surfaces will be shown and discussed.

2. EXPERIMENT

The SIRIS/LARIS instrument consist of a pulsed microbeam ion gun, a pulsed flood electron gun, a RIS laser system, an Excimer laser for laser atomization, a high-resolution computer-controlled (x, y, z, ϕ) sample holder, a secondary electron imaging system, a video imaging system for sample observation, a sample interlock system, and a mass spectrometer detection system. Further details of the SIRIS/LARIS instrument can be found in Arlinghaus, et al. (1991a,b). In the SIRIS experiments, the samples were bombarded with a high energy pulsed Ar$^+$ ion beam, while in the LARIS experiments, 248 nm photons were used to irradiate samples. The diameter of the bombarding ion beam can be focussed between a few and several hundreds of microns. A UV achromatic focusing objective was used to focus the laser beam down to 10 μm. A focal spot diameter of 3 μm can be obtained with the current setup by increasing the input atomization laser beam diameter from 3 mm to 10 mm. Secondary ions are removed electrostatically. The remaining neutral particles are selectively ionized by the RIS laser beams and analyzed with a mass spectrometer. With SIRIS/LARIS, charge compensation of insulating samples is achieved using pulsed low energy electrons. Imaging is achieved by either scanning the ion beam over the sample (for very small image areas) or by changing the x and y sample position while the RIS laser beam and the bombarding ion/atomization laser beam position remains fixed. Depth profiles are obtained by (a) scanning the sample with a continuous ion beam to etch the sample to a specific depth and (b) taking data with a pulsed ion beam after a specific number of raster frames in the center of the resulting crater.

3. RESULTS AND DISCUSSION

The depth resolution of a depth profile is best described in terms of the measured width of an interface between two layers. The ideal "true" profile would be rectangular in shape. Since atomic mixing by the sputtering ion beam will cause broadening, the measured profile can be represented by an integrated error function. The interface width is defined as the interval where the intensity drops from 84.13% to 15.87% of maximum signal, equivalent to two standard deviations (2σ) of the error curve. Figure 1 shows the depth resolution obtained for SIRIS measurements of layered GaAs/AlGaAs/GaAs structures with a primary ion energy of 0.5 keV. The change of Al concentration from 84% to 16% for the leading and trailing edge is shown. A depth resolution of ~2 nm is obtained for 0.5 keV ion energy, while for 2 and 7 keV, a depth resolution of approximately 6 and 10 nm was measured.

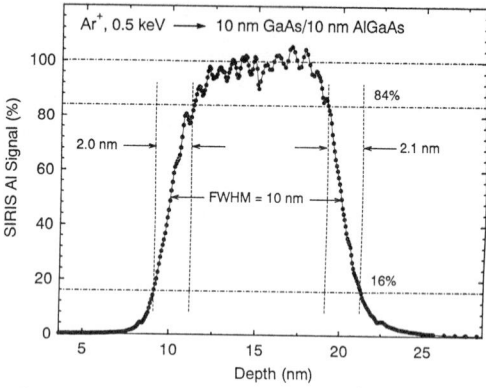

Figure 1. SIRIS depth resolution for Ar$^+$ bombarding energy of 0.5 keV. Note that data points are taken at less than 0.1 nm intervals.

To determine the achievable dynamic range of the current SIRIS instrument, and to investigate its limits, we measured the concentration profile of B implants with various peak concentrations. Figure 2 shows a typical B depth profile obtained from a silicon sample implanted with 70 keV ^{11}B at a dose of 1×10^{16} atoms/cm^2 along with the B implant profile simulation calculated using TRIM89. The simulated curve follows the measured curve for the first 400 nm, and then exhibits a faster decay than the measured curve. The longer decay of the SIRIS profiles is mostly due to atomic mixing effects. The dynamic range is ~2×10^6, which could be further increased if a cleaner or undoped Si wafer would be used for B implantation (detection limit $\ll 2 \times 10^{13}$ atom/cm^3, i.e. <400 ppt).

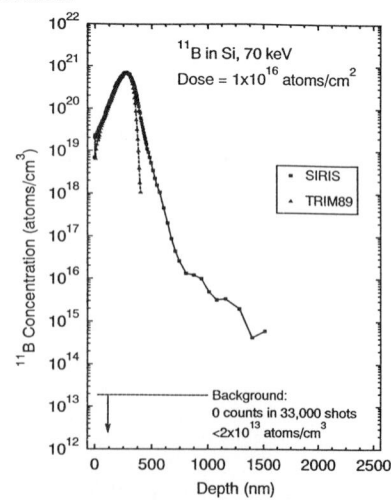

Figure 2. SIRIS dynamic range demonstration using 70 keV, 1×10^{16} B atoms/cm^2 in Si implant sample and a 7 keV Ar$^+$ sputtering beam.

The SIRIS technique can also be used to determine the spatial distribution of trace elements on surfaces. 3D images are generated by taking 2D images at various depths. Figure 3 shows the lateral B distribution of a four-dot mesa-type Si sample implanted with 25 keV [11]B at a dose of 1×10^{15} atoms/cm². The diameter of the dots was ~1 mm. From this 2D image the center and the diameter of the four-dot pattern can be determined. It can be concluded that the SIRIS technique is capable of imaging trace elements in semiconductors and of providing quantitative measurements of dopant and impurity concentrations in semiconductor materials at the 10^{13} to the 10^{21} atoms/cm³ level with high depth resolution and accuracy. If a liquid metal ion gun would be used, images down to 50 nm spatial resolution could be obtained.

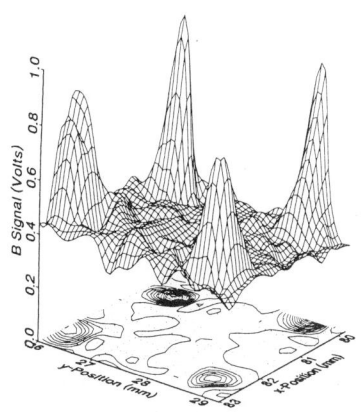

Figure 3. Lateral B distribution of a four-dot mesa-type Si sample implanted with 25 keV [11]B at a dose of 1×10^{15} cm⁻².

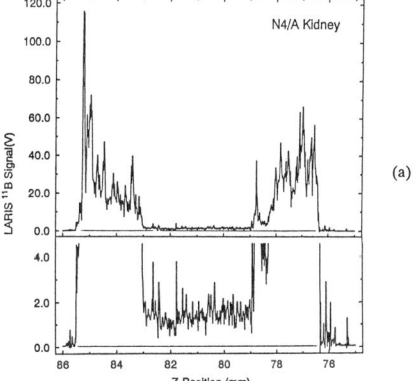

(a)

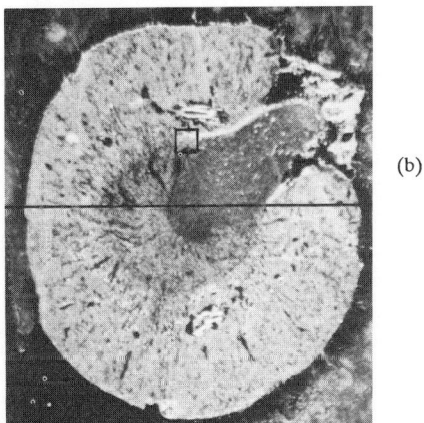

(b)

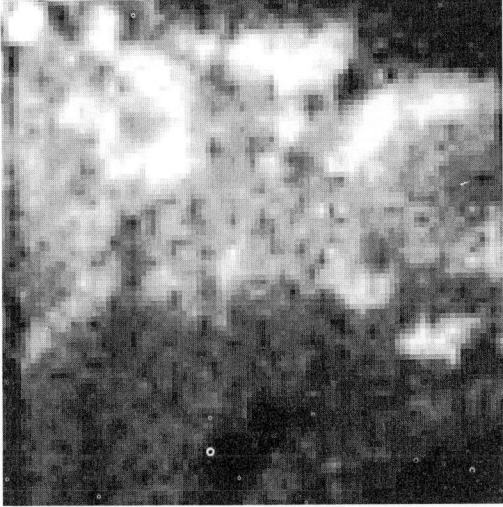

(c)

Figure 4. Boron concentration as a function of position (a), photomicrograph (b), and LARIS image (c) of BSSB infused rat kidney. Data were taken at 10 μm spacing with 30 laser shots per position. Bright areas correspond to high boron concentration. The black horizontal line displays the location of the position-concentration measurement, which has been plotted to the same scale as the optical image.

500 μm

Boron Neutron Capture Therapy (BNCT) is a very promising treatment for Glioblastoma Multiforme and other tumors in the brain (1989). In this therapy, boron, which has been prepositioned by biochemical methods in the tumor, captures a thermal neutron from a beam of thermal or epithermal neutrons and undergoes the $^{10}B(n,\alpha)^7Li$ reaction, producing energetic alphas and Li ions. Since the range of the fission fragments is relatively small (few μm), the clinical efficacy of the therapy is only guaranteed if the boron is positioned in a sensitive area of the tumor cell, such as the cell nucleus. A major technical issue impacting BNCT clinical efficacy is the question of boron concentration and its variability as a function of inter/intracellular location. To date, it has not been possible to accurately determine the location and quantity of boron within an individual cell and to determine its variability. We have used for the first time an imaging LARIS instrument to determine with micrometer-scale lateral resolution boron concentration in thin sections of kidney, liver, brain, and tumor tissues obtained from BSSB ($Na_4B_{24}H_{22}S_2$) infused rats. Position-boron concentration scans and boron-concentration images were generated at a number of different locations on these tissues, resulting in plausible changes in boron concentration as a function of morphology and organ type. Figure 4 shows the boron concentration as a function of position, a photomicrograph, and a LARIS image of BSSB infused rat kidney. The position scan shows that there is a significantly higher boron concentration in the outer, lighter colored region of the kidney as compared to the inner regions. The outer part of the kidney is the region in which filtration is taking place, with a very large mass of blood vessels and a barrier through which numerous electrolytes and metabolites diffuse and exchange. The inner part of the kidney is just a mass of collection tubes that transports the substances extracted from the blood to the bladder. Therefore, it is not unreasonable to have a high boron (presumably BSSB) concentration at this barrier as the kidneys work overtime to remove this foreign substance from the blood. If one assumes that the LARIS response to the boron in BSSB in tissues is the same as to boron in BSH in 5% gelatin, then the ~30 V average signal seen in the outer regions corresponds to ~40 μg B/g and the ~1.5 V average signal in the inner region to ~2 μg B/g. The boron concentration image was taken at the junction between the outer and inner regions of the rat kidney. Notice the detailed structure seen in the image, with "hot spots" in boron concentration. Also notice that there is also structure in the inner (low B concentration) region, with infrequent, but occasional "hot spots."

These data demonstrate that LARIS is capable of quantitative and sensitive imaging of trace element and molecule concentrations in biological samples. Even though the measurements of tissue sections were only intended to demonstrate the feasibility of the technique, the quality of the data is sufficient to make useful biological observations. The LARIS signal was 3000 times higher than the SIRIS signal. In the brain tissue section scans (not shown), good signal-to-noise LARIS boron signals were seen after 30 laser shots with a 10 μm beam at the few ppb level. Therefore, sub-ppm (~100 ppb) sensitivity should be achievable with a 1-2 μm beam and 5 laser shots per point. 3D boron concentration images could be obtained by either slicing tumors or organs into several thin sections and imaging each individual section, or by scanning the entire imaged area several times and comparing the individual frames.

This work was supported in part under various contracts and grants with DOD, DOE and NIH.

REFERENCES

Arlinghaus H F et al. 1991a *J. Vac. Sci. Technol.* **A9** 1312
Arlinghaus H F et al. 1991b SPIE **1435** 26
Proc. of the 3rd International Symp. on Neutron Capture Therapy 1989 *Strahlenther. Onkol.* **165** 213

Inst. Phys. Conf. Ser. No 128: Section 7
Paper presented at RIS 92, Santa Fe, NM, USA, 24–29 May 1992

Laser ablation as a sample atomisation technique for resonance ionisation mass spectrometry

I S Borthwick, K W D Ledingham, C T J Scott and R P Singhal

Department of Physics and Astronomy, University of Glasgow, Glasgow G12 8QQ, Scotland.

ABSTRACT: Laser ablation utilising a Q-switched Nd:YAG laser has been investigated as a technique for atomising solid samples for subsequent analysis by resonance ionisation mass spectrometry. The effects of increasing the ablation laser power and of changing the ablation laser wavelength for the analysis of aluminium are discussed. The analysis of a series of aluminium samples containing manganese and magnesium is reported to demonstrate the trace analysis capability of the technique.

1. INTRODUCTION

Before resonant ionisation mass spectrometry (RIMS) can be applied to the detection of trace quantities of elements in solid samples the sample must be atomised. If RIMS is to be used to analyse small areas on a surface the atomisation technique must be able to be applied selectively to these areas. Laser ablation is such a technique and has been demonstrated to offer sensitivity approaching the part per billion level for the analysis of metals (Beekman and Thonnard 1989, McCombes *et al* 1991). However the technique has not been well characterised and so a detailed study of the laser ablation atomisation of aluminium has been undertaken. This has involved varying both the power and wavelength of the ablating laser and monitoring the temporal development of the neutral atoms liberated in order to assess the optimum operating conditions for the technique.

2. EXPERIMENTAL ARRANGEMENT

These investigations were carried out on the Glasgow RIMS instrument (Towrie *et al* 1990). Samples are mounted on a precision manipulator at the centre of a spherical sample chamber. Sample changeover takes about 10 minutes using a rapid transfer probe. The system is maintained at a pressure of 10^{-9} torr by turbomolecular and diffusion pumps.

The samples were vapourised using a JK HY750 Nd:YAG laser with a pulse duration of about 10ns and a repetition rate of 10Hz. The harmonics at 532, 355 and 266nm were also available and an intercavity aperture provided essentially TEM_{00} mode operation. The beam was focussed using a 30cm lens to give spot sizes of about 100μm, which were measured using exposed Polaroid plates, and was incident on the sample at an angle of $45°$. The average laser power was measured at the window into the chamber using an Ophir 2A-P thermopile.and the stability was monitored by diverting a fraction of the beam onto a Joulemeter.

The post ablation ionising laser system consists of a Spectron SL2Q/SL3A Nd:YAG laser pumping a Spectrolase 4000 dye laser. Frequency doubling crystals can extend the wavelength to 250-350nm. The ionising laser beam was introduced parallel to the surface of the sample stub and a known distance from the surface, usually 3mm. The time delay between the ablation laser and the ionising laser could be adjusted from 0-16µs using a custom made pulse generator. The reflectron time of flight mass spectrometer (Mamyrin *et al* 1973) has an overall length of 3m with a FWHM mass resolution of about 700 for ions of 50amu. The data acquisition system consists of a Lecroy 2261 transient recorder linked to a COMPAQ 386/25 personal computer and records and stores mass spectra and laser pulse energies on a pulse-to-pulse basis.

3. RESULTS AND DISCUSSION

When a high powered laser beam is focused onto a surface, both ions and neutral atoms can be produced. Fig.1 shows the result of increasing the ablation laser power incident on a pure aluminium foil 1mm thick. Line (a) shows the signal resulting from ionising the ablated neutral atoms non-resonantly with ~1mJ of tightly focused 266nm radiation, which was generated by frequency quadrupling the output of the Spectron Nd:YAG laser. This method of ionisation was used extensively in these investigations as it allows several elements to be studied simultaneously, although the signals were not as large as could be obtained using resonant ionisation. Line (b) shows the signal resulting from ions created solely by the ablation laser. This was monitored by adjusting the reflectron voltages so that ions created at the sample surface would be optimally transmitted.

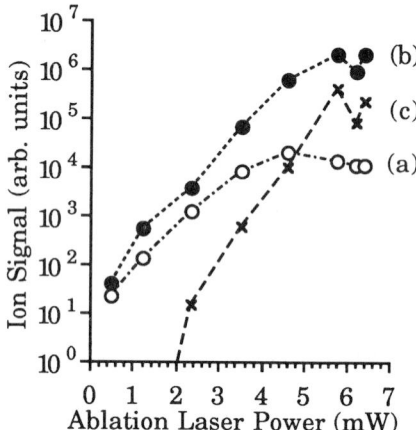

Fig. 1. The effect of increasing the power of the 532nm ablation beam on the signal resulting from (a) ionised neutral atoms (b) ablation laser created ions and (c) the 'leakage' ions.

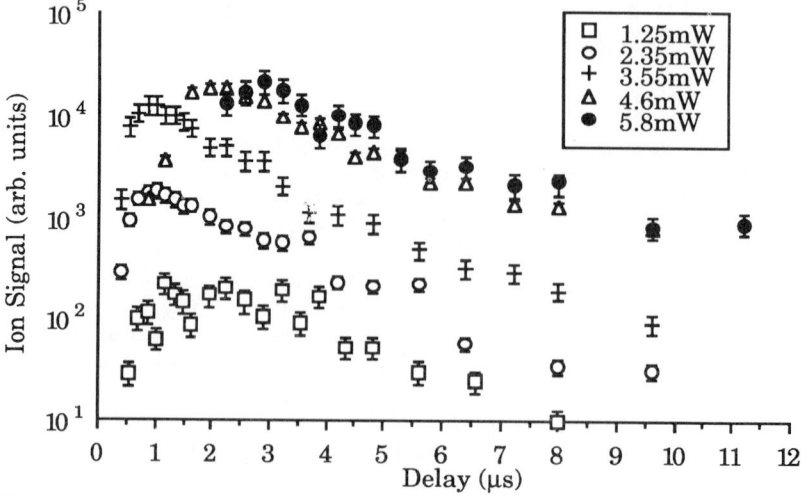

Fig.2 Aluminium ionised neutral atom signal as a function of the delay between the ablating and ionising lasers with a 532nm ablation laser wavelength for different powers.

Ions created by the second laser system several millimetres from the sample surface are subjected to a smaller extraction field and the usual operating arrangement is to set the reflectron voltages so that these ions are preferentially transmitted and the ions created by the ablation laser are suppressed. Line (c) shows the signal resulting from the ablation laser created ions with the reflectron voltages set thus. These ions, the 'leakage' ions, have acquired energies several hundred electron-volts less than the energies of the ions recorded in line (b) and cause an unwanted interference to the ionised neutral atoms signal. It can be seen that as the ablation power increases that the number of ions and neutrals liberated increases in a non-linear fashion and that at the higher powers the leakage ions dominate the observed mass spectra.

The temporal development of the neutrals liberated by the ablation laser can be monitored by changing the delay between the two lasers and depends on the power of the ablation laser as shown in Fig.2. The flattening off of the signal at high powers is in part attributed to saturation of the dual channelplate detector but it is also believed to be due to the increasing density of the laser produced plasma. This both shields the sample surface from the incoming beam and ionises an increasing fraction of the ablated neutrals. This could account for the broadening of the temporal distributions and the increase in the leakage current which are observed at the higher powers. The optimum operating power for analysis was ~3mW at which power the temporal distribution of the ablated neutrals can be seen to be relatively narrow and this would allow for efficient sampling of the plume by the pulsed ionising laser. At this power, corresponding to ~10^8W/cm^2 the amount of material liberated per shot has been determined to be ~10^{-2} of a monolayer per shot (Borthwick *et al* 1992), making this regime suitable for surface analysis.

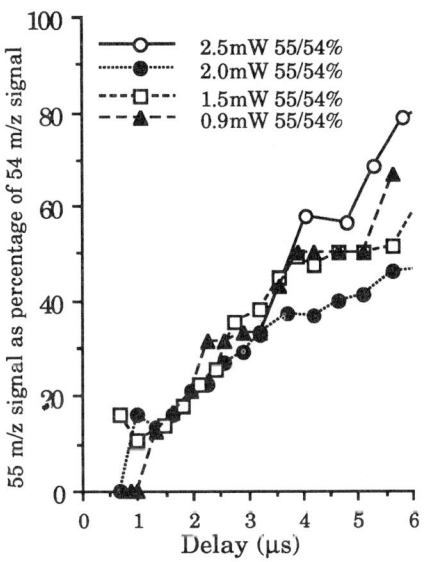

Fig. 3 Temporal variation in the composition of the plume created by the ablation of a steel sample at various powers.

In order to assess the effect of increasing the ablation laser power on the plume composition the ratio of the ^{55}Mn to ^{54}Fe signals obtained using 266nm photo-ionisation was recorded as a function of the delay for various laser powers, as shown in Fig.3. The ablation laser wavelength was 355nm and all the measurements were taken on the same area of a NIST steel sample (1263a). It can been seen that there is no major variation in the ratio between the laser powers, although at the high powers the signals were much larger. It is also important to note that because the different elements have different temporal development the ratio changes markedly with the delay between the lasers.

The effect on the temporal distribution of changing the ablation laser wavelength is shown in Fig.4 .The sample was again aluminium and the ablation fluences were ~10^8W/cm^2. It can be seen that there is little difference between the shapes of the distribution but the signals for the 266nm wavelength were an order of magnitude smaller than those for the other wavelengths. Due to this, and because the third and fourth harmonics of the Nd:YAG laser tended to be the less stable, the second harmonic of 532nm was used for the subsequent analysis.

In order to evaluate the technique using the parameters which have been determined in the preceding sections, a series of aluminium samples containing quantities of magnesium and manganese were analysed. The ionisation was carried out resonantly using one photon to a bound state: 280.2nm for manganese and 285.3nm for magnesium and 355nm from the Nd:YAG to saturate the ionisation step. Fig 5 shows the ion signals as functions of concentrations of Mn and Mg. The results must only be considered preliminary at this stage but the linearity over 4 orders of magnitude is reasonably good and the Mg and Mn signals are of similar size indicating that they are present in similar quantities in the plume Procedures to eliminate sample contamination have still to be optimised and it expected that more linear and sensitive analyses can be carried out in the future.

4. CONCLUSIONS

Some of the important factors affecting the applicability of laser ablation as a microprobe atomisation source for RIMS of surface monolayers have been illustrated and using these criteria trace analysis has been demonstrated.

5. ACKNOWLEDGEMENTS

It is a pleasure to thank Alusuisse-Lonza Services AG for the samples. ISB and CTJS would like to thank the SERC (UK) for studentships.

5. REFERENCES

Beekman DW and Thonnard N 1989 Resonance Ionisation Spectroscopy 1988 Inst Phys Conf Ser **94** pp163-166

Borthwick IS, Ledingham KWD, Sander J and Singhal RP 1992 manuscript in preparation.

Mamyrin BA, Karataev VI, Shmikk DV and Zagulin VA 1973 Sov. Phys. JETP **37** pp45-48

McCombes PT, Borthwick IS, Jennings R, Ledingham KWD and Singhal RP 1991 Optogalvanic Spectroscopy, Inst Phys Conf Ser **113** pp163-168

Towrie M, Drysdale SLT, Jennings R, Land AP, Ledingham KWD, McCombes PT, Singhal RP, Smyth MHC and McLean CJ 1990 Int. J. Mass Spectrom. Ion Proc. **96** pp309-320

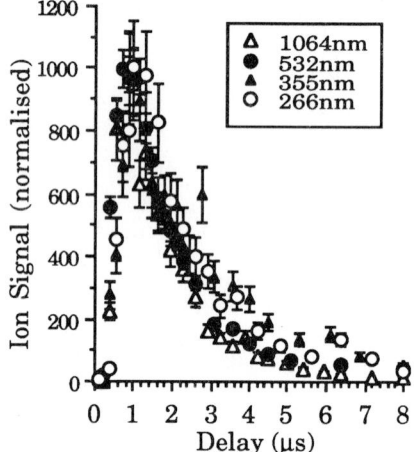

Fig. 4   Effect of changing the ablation laser wavelength on the temporal distribution of neutral atoms from an aluminium sample.

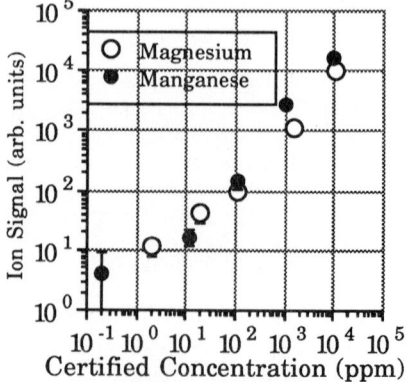

Fig, 5   LARIMS signal as a function of concentration for Mn and Mg in an series of aluminium samples.

Inst. Phys. Conf. Ser. No 128: Section 7
Paper presented at RIS 92, Santa Fe, NM, USA, 24–29 May 1992

283

Development of a resonant ionization mass spectrometer for surface analysis

L Johann, R Stuck, Ph Kern, B Sipp*, P Siffert

Centre de Recherches Nucléaires (IN2P3), Laboratoire PHASE (UPR du CNRS n°292), BP 20, 67037 Strasbourg Cedex 2 (France)
*IPCMS-Groupe d'Optique Non Linéaire, 5 rue de l'Université, 67084 Strasbourg Cedex (France)

ABSTRACT: A resonant ionization mass spectrometer based on a quadrupole SIMS coupled to a tunable dye laser has been developed for the detection of trace amounts of metallic impurities in silicon. The instrument has an hemispherical energy filter which discriminates efficiently the photoions against the "true" secondary ions. The detection limit of the apparatus was measured using standards with known concentrations of impurities and a value of 0.2 ppm was found.

1. INTRODUCTION

It is well known that traces of metallic elements present in semiconductor material at concentrations in the ppb range may affect drastically the performance of electronic or photovoltaic devices. Since sputter induced resonant ionization mass spectrometry has been demonstrated to be a highly sensitive technique [Winograd et al (1982) ; Parks et al (1983) ; Pellin et al (1984) ; Gobert et al (1987)], it appeared interesting to develop such an instrument for the analysis of semiconductor bulk material.

This apparatus, dubed SIRAS (Spectromètre d'Ionisation Résonante pour l'Analyse de Surface) has been described in a previous paper [Stuck et al (1989)] but has been modified to improve its sensitivity. One of the problems in the first version of the instrument was the presence of "true" secondary ions producing a background increasing the detection limits in the case of mass interference, for example when detecting iron in silicon. Therefore an hemispherical velocity filter has been added to eliminate these secondary ions.

2. EXPERIMENTAL

In the instrument, the sample is continuously bombarded under UHV conditions with a 3 keV argon beam at a 45° incidence angle. The maximum intensity is 2 μA and the beam can be focused on a spot with a diameter of 300 μm. Neutrals are eliminated by a 5° bent introduced in the column axis. The samples are fixed on a discus in order to keep the same geometry whatever the thickness of the target (figure 1).

The sputtered neutrals are irradiated by a photon pulse delivered by a dye laser pumped after frequency doubling or mixing with a Nd-Yag laser (Datachrom 5000 from Quantel). The laser is operated at 10 Hz repetition rate and the pulse duration is 10 ns. After frequency doubling of the output of the dye laser in a servo-controlled KDP crystal, a laser beam of about 10 mJ per pulse is obtained in the 280-287 and 292-299 nm ranges with the Rhodamine 590 and 610 dyes, respectively. The spectral width is smaller than 0.003 nm.

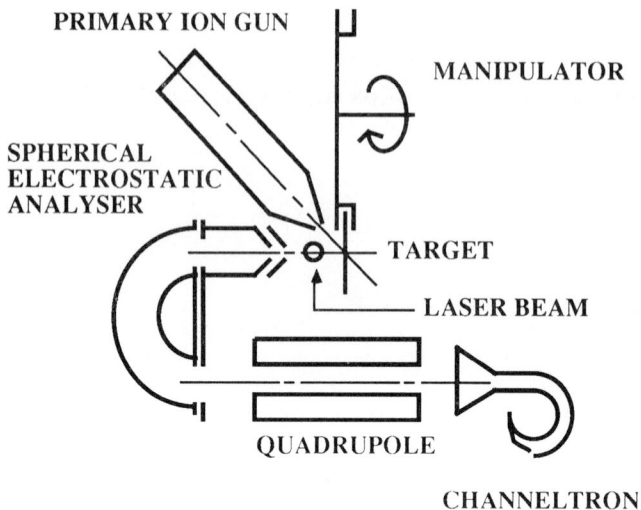

PRIMARY ION GUN

MANIPULATOR

SPHERICAL
ELECTROSTATIC
ANALYSER

TARGET

LASER BEAM

QUADRUPOLE

CHANNELTRON

Fig. 1. Schematic diagram of the experimental set-up

The photon beam is focused to a 2 mm spot to increase the laser beam power density for saturating ionization. The beam axis is parallel to the sample at a distance of about 2 mm. The photoions are detected by a secondary ion optics similar to that described by Magee et al (1978). An immersion lens collects the ions perpendicularly to the surface sample and projects a real magnified image into the entrance aperture of a 180° spherical electrostatic analyser. This analyser has an energy resolution of approximately 4 % which is equivalent to a pass band of 1.4 eV for an extraction potential of 35 Volts. The energy filtered ions are then mass analysed in a 160 x 8 mm quadrupole (Balzers QMA 150). A low resolution ($\Delta M/M$ = 100) is generally chosen to obtain the highest possible transmission. The ions are then detected with a channeltron. Electronic gating allows to count the ions during a time window, opened for 15 µs about 25 µs after the firing of the laser for taking into account the time of flight of the ions in the detection system.

3. RESULTS

3.1 Choice of the wavelengths

Most metallic elements can be resonantly ionized with a one color, one resonance (1 + 1) process and moderate laser power is required to saturate ionization [Gobert et al (1991)].

Table 1 shows the wavelengths which have been used as well as the corresponding transitions according to Moore (1971). To our knowledge, some of these transitions (Mn, Ta, Zr, V) have not yet been used in RIS.

Elements	Fe	Cr	Ti	Co	Zr	Ta	V	Mn	Mg
Wavelength (nm)	293.77	295.72	293.44	292.97	296.17	294.11	294.33	280.19	285.2

Table 1. Transitions used for the detection of metallic impurities

3.2 Energy discrimination

To check the efficiency of the velocity filter we have sputtered an iron sample and measured the iron ion current at m/e = 56 as a function of the potential applied to the sample, with and without laser post-ionization. The primary ion current was reduced in order to avoid saturation of the channeltron. Figure 2 displays the curves obtained which are related to the energy distributions of the photoions and secondary ions when the laser is switched on and to the energy distribution of the "true" secondary (SIMS) ions when the laser is switched off.

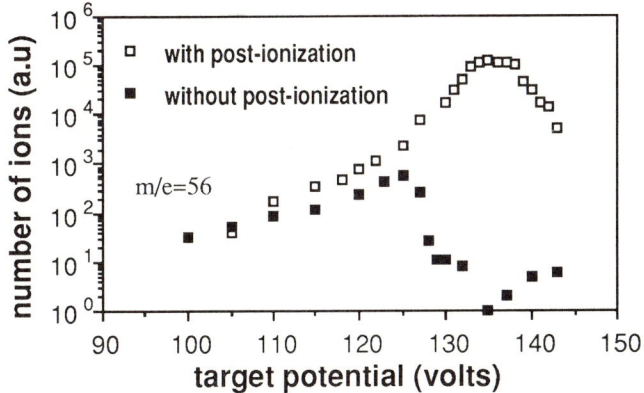

Fig. 2. Fe ion intensity as a function of target potential for 10 mn counting time.

Clearly the mean energy of the SIMS ions is higher than that of the photoions since a lower sample bias is necessary to allow them to pass the energy filter. This effect is due to the fact that the photoions are created at some distance from the surface and are therefore less accelerated than the secondary ions. Furthermore the figure shows also that for the bias resulting in optimal photoion collection, almost no secondary ions are detected since the background is lower than 2 counts for the 10 mn counting time generally used in trace analysis.

Another test was performed by detecting Fe at mass 56 in pure silicon. The efficiency of the energy filter was confirmed since virtually no Si_2 ions were counted in 10 mn when the laser was switched off.

3.3 Detection limits

Since we had not enough silicon standards containing known amounts of metals to perform a calibration, we have first used steel standards to evaluate the detection limits. Figure 3 shows such a calibration plot obtained with chromium doped samples. A detection limit of about 0.2 ppm can be extrapolated from this curve, since only 2 counts are recorded during the same counting time in a chromium free sample.

In silicon, similar values of the detection limits were found. For example, for a silicon standard containing 15 ppm of iron, 400 counts were recorded in 10 mn whereas the background was only 2 counts for the same time. Assuming that the detection limit corresponds to a signal to noise ratio of 2, results in a value equal to 0.15 ppm.

It should be noticed that such low concentrations can only be detected if the surface has been carefully sputter cleaned over a relatively large area in order to avoid any background due to the detection of surface contaminants by edge effects.

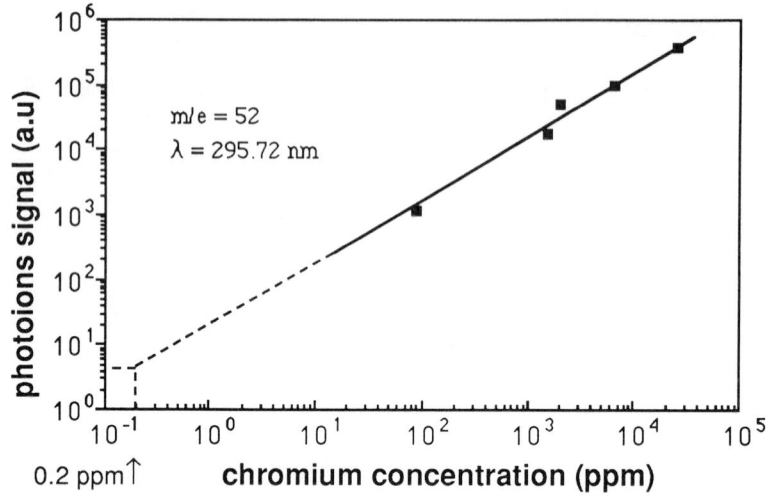

Fig. 3. Calibration plot for chromium in steel standards. Laser energy is 2 mJ/pulse. Primary ion current is 1 μA and counting time for the photoions is 600 s.

4. CONCLUSION

Further improvements of the sensitivity of the instruments are expected from the two following modifications :
- replacement of the actual ion gun by a new column using a more brilliant duoplasmatron source which should allow to increase the number of sputtered neutrals interacting with the laser beam.
- optimization of the immersion lens. The actual lens is designed for operation in the SIMS mode and has a limited field of view (about 300 μm). A computer simulation program will be used to design a lens able to collect more efficiently the photoions which are created in a much larger volume since the laser beam has a diameter of 2 mm. With these changes, the SIRAS should become an useful analytical tool to detect metallic traces in silicon.

ACKNOWLEDGMENTS

We acknowledge the ADEME (French Agence de l'Environnement et de la Maîtrise de l'Energie) for financial support. Many thanks are due to B. Dubreuil (Orléans) for stimulating discussions.

REFERENCES

Gobert O, Dubreuil B, Gelin P and Inglebert RL 1987 *Proc. of the VIth Int. Conf. on SIMS* ed A. Benninghoven, A M Huber and H W Werner (New York : Wiley) .
Gobert O, Gibert T, Dubreuil B, Gelin P and Debrun J L 1991 *J. Appl. Phys.* **70** 12.
Magee C W, Harrington W L and Honig R E 1978 *Rev. Scient. Instrum.* **49** 477.
Moore C E 1971 *Atomic Energy Levels, NSRDS-NBS* **35** / Vol I, II, III (Washington DC : National Bureau of Standards)
Parks J E, Schmitt HW, Hurst GS and Fairbank W M 1983 *Thin Solid Films* **108** 69.
Pellin MJ, Young CE, Calaway WF and Gruen DM 1984 *Surf. Sci.* **144** 619.
Stuck R, Sipp B 1989 *Proc. of the VIIth Int. Conf. on SIMS* eds A Benninghoven, C A Evans, K D Mc Keegan, H A Stroms and H W Werner (New York : Wiley) pp. 263-266.
Winograd N, Baxter J P and Kimock F M 1982 *Chem. Phys. Lett.* **88** 581.

Inst. Phys. Conf. Ser. No 128: Section 7
Paper presented at RIS 92, Santa Fe, NM, USA, 24–29 May 1992

Studies of ion–surface interactions using multiphoton ionization

L.Wang, R.Nor, S.P.Mouncey, and W.G.Graham

Department of Pure & Applied Physics, Queen's University of Belfast, Belfast, UK

ABSTRACT: A experimental setup has been built for the study of low energy reactive ion interactions with surfaces. Particular emphasis is placed on the detection of the neutral material leaving the surface using nonresonant multiphoton ionization. Some preliminary measurements show the suitability of the system for the proposed study.

1. INTRODUCTION

An understanding of basic ion-surface processes is of fundamental importance if an insight is to be gained into how a plasma interacts with its surroundings. This is particularly true in the semiconductor industry, where reactive ion etching is used to manufacture microelectronic circuits. Using a well-defined ion beam to bombard a surface characterized under UHV conditions is an ideal method to elucidate the etching mechanisms because of the ease with which one can control the ion-surface environment. It is well known that the majority of material sputtered from a surface leaves via the neutral channel. This much larger neutral fraction is more likely to be representative of the chemical and sputtering processes than the smaller ionized fraction, due to the fact that it is much less sensitive to the chemical properties of the surface. A non-selective ionization process, such as non-resonant multiphoton ionization which has been successfully used in surface analysis (Becker et al 1984), together with mass and energy analysis, of the sputtered neutral material would a powerful tool to study ion-surface interactions. In this study a three-stage process has been developed. A probe beam of ions causes the sputtering of material which is subsquently ionized using non-resonant multiphoton ionization. The photoions are then analysed using a TOF spectrometer.

2. EXPERIMENTAL ARRANGEMENT

A schematic diagram of the experimental configuration is shown in Fig.1.The target is bombarded by a pulsed ion beam at 45° to the surface normal. Sputtered neutrals are then ionized by a high density, pulsed UV laser focused close to the surface. The photoions are then accelerated by a pulsed voltage applied to the sample and focused using an electrostatic lens assembly. They are then allowed to drift in a field-free region for TOF mass spectral analysis before reaching a channeltron.

The operating pressure of the chamber is typically 1×10^{-8} Torr. A modified Colutron ion source is

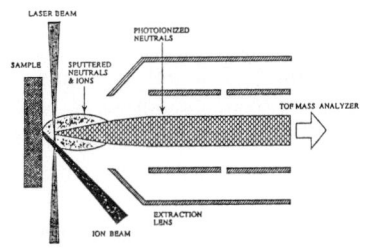

Fig.1 A schematic diagram of the analysis of ion-surface interaction using combined sputtering initiated nonresonant multiphoton ionization and TOF mass spectrometry.

differentially pumped and produces current densities of up to $0.1mAcm^{-2}$ at energies ranging from a few hundred to 5keV. A Lambda Physics LPX 200 excimer laser is used for ionizing the neutrals ejected from the surface, producing UV output at 248nm with pulse duration 20ns and energy/shot up to 500mJ. A pulse counting data acquisition system is used which consists essentially of amplifiers, a constant fraction discriminator, a time-to-amplitude converter and a multichannel analyzer interfaced to an IBM PC. For preliminary measurements, Ar^+ was used to bombard a Si surface, but there is the facility to use beams of reactive ions, such as CF_x (x=1 to 4), as well as using different target materials.Si has been chosen because of the large amount of interest in the reactive ion etching of this material.

3. PRELIMINARY MEASUREMENTS

Fig.2 shows the TOF mass spectrum for photoionized neutrals emitted from a contaminated Si surface under bombardment with 3keV Ar^+ at a current density of $0.2\mu Acm^{-2}$. The photoionized Si^+ at mass 28 is seen clearly, as well as the various hydrocarbon ions which have been observed previously in some typical secondary ion mass spectra from untreated Si wafer (Johnson et al 1992). The relatively poor mass resolution is believed to be due to the large accelerating field gradient in the fairly large ionizing region very near the surface(≤0.5mm), where the laser is focused so as to maximize the number of neutrals within the ionizing volume of the laser beam. The number of neutrals produced, and hence the mass resolution will be improved by a better incident beam quality. A TOF mass spectrum for secondary ions emitted from the same surface by 3keV Ar^+ is shown in Fig.3, and indicates a similar structure but with greater fragmentation of the hydrocarbons.

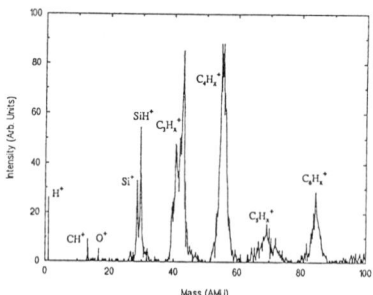

Fig.2 TOF mass spectrum of sputtering of a contaminated Si wafer by 3keV Ar^+ at 45° incidence, with nonresonant multiphoton ionization of the sputtered atoms and molecules by a 248nm, 20ns laser pulse with a power density of about $1\times10^{10}Wcm^{-2}$.

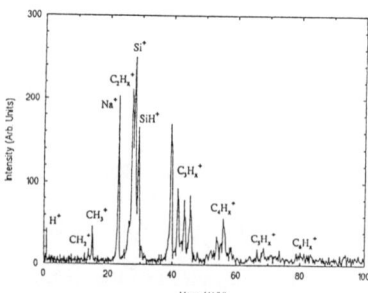

Fig.3 TOF mass spectrum of secondary ions from the Si sample under the same condition as in Fig.2 except a shorter primary ion beam width of $0.4\mu s$.

4. CONCLUSIONS

An experimental setup has been built for the study of ion-surface interactions using non-resonant multiphoton ionization, especially for the interaction of low energy reactive molecular ions with semiconductor surfaces. This will simulate the plasma etching processes in a well controlled environment. Preliminary results are encouraging and some modifications are in progress to improve the performance of the system.

REFERENCES

[1] Becker C H and Gillen K T 1984 Appl. Phys. Lett. **45** 1063.
[2] Johnson D and Hibbert H 1992 Semicond. Sci. Technol. **7** A180.

Inst. Phys. Conf. Ser. No 128: Section 8
Paper presented at RIS 92, Santa Fe, NM, USA, 24–29 May 1992

289

Laser mass spectrometry for biopolymers

K. Tang†, S. L. Allman, and C. H. Chen

Oak Ridge National Laboratory, P.O. Box 2008, Oak Ridge, TN 37831-6378 USA

ABSTRACT: Various matrix materials were used for laser desorption of biological molecules which include large polypeptides and deoxyadenosine oligonucleotides. Both matrix assisted laser desorption ionization (MALDI) and matrix assisted desorption with post-ionization (MADPI) have been used. Detection sensitivity of a few femtomole of both oligomer and protein has been achieved. Both positive and negative ions were observed with little fragmentation.

1. INTRODUCTION

During the past decade, tremendous effort has been expended on applying mass spectrometry to biopolymer measurements for biological and medical applications. However, two major difficulties are developing a reliable method to deliver these nonvolatile biopolymers to the mass spectrometer's ionization zone and achieving ion production without significant fragmentation in either the desorption or ionization steps. Electrospray and laser desorption ionization (LDI) have been considered to have good potential to overcome these difficulties. Hillenkamp and his co-workers (1987) first reported the discovery that large polypeptide molecules could be put into the gas phase and ionized by a laser in the presence of a large excess of nicotinic acid molecules serving as a matrix. Since then, Karas and Hillenkamp (1988) and Beavis *et al* (1989) have reported measurements of protein ions by the matrix assisted laser desorption ionization method.

Matrix assisted UV desorption and ionization has also been applied to DNA segments. However, the success has been limited to very small segments. Spengler *et al* (1990) reported the observation of 8 mer single-stranded DNA ions with detection sensitivity of 10 to 100 pmole. Nelson *et al* (1990) obtained positive DNA ion spectra by laser ablation and ionization from a frozen aqueous matrix with laser wavelength tuned to the electronic resonant excitation of sodium or copper atoms, and detected $p(dA)_8$ single-stranded DNA oligomer.

In this work, we report time-of-flight mass spectra from matrix assisted laser desorption of mixtures of single-stranded oligomers and protein by various laser wavelengths and different matrix materials. The effects of matrix to analyte ratio on the sensitivity and resolution of time-of-flight mass spectrometer are studied. Potential applications and future development for DNA sequencing are also briefly discussed.

2. EXPERIMENTAL

A schematic of the experimental appartus is shown in Figure 1. A Nd-YAG laser capable of delivering four different wavelengths (i.e., 1064 nm, 532 nm, 355 nm, and 266 nm) was used for laser ablation. The maximum energy per laser pulse for 1064 nm, 532 nm, 355 nm, and 266 nm is 700 mj, 400 mj, 200 mj, and 100 mj, respectively. The pulse duration for 1064 nm, 532 nm, 355 nm, and 266 nm is 10 ns, 7 ns, 5 ns, and 5 ns, respectively. However,

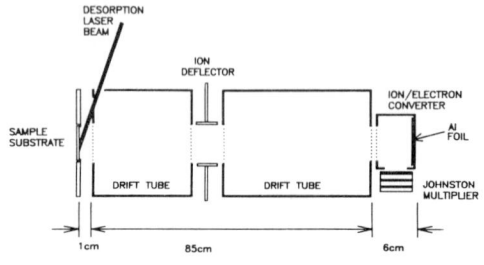

Figure 1 Schematic of experimental setup.

typical laser fluence used in this work was less than 200 mj/cm^2 to prevent any possible production of plasma or fragmentation. No focusing was attempted. A conversion box was used to receive ions and emit secondary electrons. An electron multiplier was used to detect secondary electrons from an aluminum foil in the conversion box. Signals from the multiplier went through a preamplifier and subsequently to a fast digital oscilloscope. The maximum resolution (M/ΔM) of the TOF is only about 200. However, the time-of-flight resolution of heavy biological molecules is somewhat lower. This is probably due to the matrix materials, since the initial velocity distribution of desorbed oligomers is expected to be similar to the velocity distribution of the matrix material. The detection senstivity is in the range of a few femtomole.

3. RESULTS AND DISCUSSION

A laser desorption positive ion spectrum of protein is shown in Figure 2. A laser desorption negative ion mass spectrum of mixtures of 3, 4, 5, 8, 11, 15, 20, and 34 mer of deoxyadenosine oligonucleotides with 3-methylsalicylic acid and 3-hydroxy-4-methoxybenzaldehyde mixtures as matrixes is shown in Figure 3. The wavelength of the laser was 355 nm and the laser fluence was 160 mj/cm^2. Parent ions

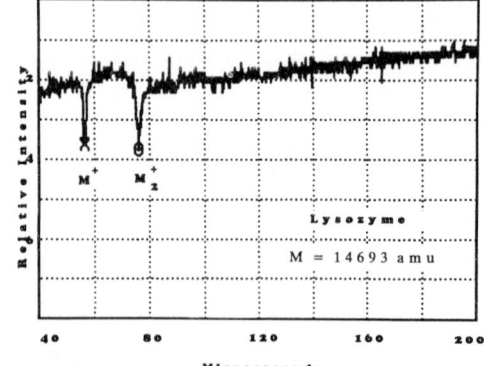

Figure 2 Laser desorption positive ion spectrum of protein.

of each oligomer were observed. No obvious fragmented or dimer ions appeared.

Background signals in the low mass region below 300 a.m.u. are due to the matrix compounds. Some of the minor peaks were from impurities in the samples. The mole ratio of matrix materials to analyte was 15,000 to 1. Similar results were obtained with a 266 nm laser beam for a few different matrix materials which include sinapinic acid, caffeic acid, 3-methylsalicyclic acid, 3-hydroxy-4-methoxybenzaldehyde, and nicotinic acid.

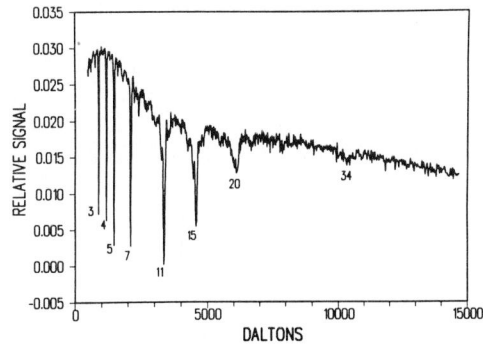

Parent negative ions of oligomers were also observed without significant fragmentation for Rhodamine B dye as the matrix. Experimental results are shown in Figure 4. It was found that the 532 nm laser beam is more efficient in producing DNA ions than UV beams (i.e., 355 nm and 266 nm).

Figure 3 Negative ion mass spectrum obtained from mixtures of deoxyadenosine oligonucleotides which include 3 mer, 4 mer, 5 mer, 8 mer, 11 mer, 15 mer, 20 mer, and 34 mer.

Although most matrix materials tested up to now are organic acids, we found some aldehydes and dye compounds such as 3-hydroxy-4-methoxybenzaldehyde and Rhodamine B are quite efficient for obtaining ions from biological molecules. Positive ions of oligomers from laser desorption and ionization were also obtained, and no significant breakup was observed. When DNA salt was used for positive ion spectra, fragmented ions were clearly observed. This may indicate that DNA molecules with sodium content tend to produce fragmentation from laser ablation. The presence of Na ions may lead to the need for high laser fluence to desorb DNA such that bond breaking becomes possible. Proposing a detailed mechanism will definitely require more study.

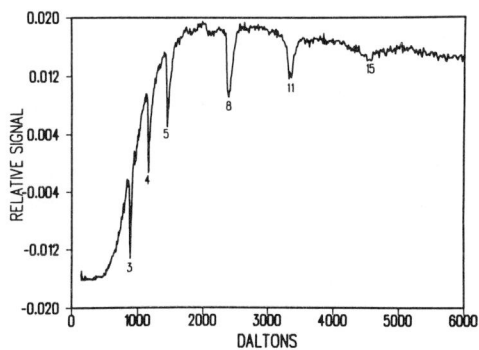

Figure 4 Negative ion mass spectrum obtained from mixtures of deoxyadenosine oligonucleotides which include 3 mer, 4 mer, 5 mer, 8 mer, 11 mer, 15 mer.

Since the discovery of matrix assisted UV desorption, a wide range of mole ratios of matrix compounds to analytes has been tried (from 100:1 to 10,000:1). We found that the resolution defined as $M/\Delta M$ tends to decrease as the ratio increases. However, for a selected matrix, a high ratio of matrix compounds to analytes is more efficient for producing larger oligomer ions. In Figure 3, it can be seen that overall detection efficiency stays roughly constant for oligomers up to 11 mer. Detection efficiency starts to drop about a factor of 2 to 3 for each increase of 5 mer of DNA size. If this general trend can be extended to much larger oligomers, it would be extremely difficult to detect any single-stranded oligonucleotide longer than 100 mer for most commonly used matrix materials. Considering the gain of electron multiplier and the transmission of the TOF, the number of ions produced for 20 mer is estimated to be between 10 ~ 100 ions per laser pulse. The signal observed for 34 mer ions is less than 10 ions per laser pulse. With the typical laser energy per pulse of 400 μj in this work, the number of desorbed matrix molecules is estimated about 10^{11} ~ 10^{13} per laser pulse, assuming a significant fraction of the laser energy contributed to vaporization of matrix material. Thus, the desorbed DNA molecules should be on the order of 10^7 ~ 10^9 per laser pulse. This indicates that the number of negative ions of large oligomers produced compared to the total number of desorbed neutral DNA molecules is less than 1 in 10^6. Recently, Romano and Levis (1991) used gel electrophoresis for identification and observed the desorption of very large single-stranded DNA molecules (~ 1000 mer) with dye as the matrix. It may indicate that a significant amount of neutral oligonucleotide molecules are desorbed during the laser ablation without producing a significant quantity of fragmentation. Thus, we consider there is a decent probability of success using post ionization of a neutral single-stranded DNA molecule or protein molecules after laser desorption to detect very large oligonucleotides by a TOF mass spectrometer.

4. REFERENCES

Karas M, Backmann D, Bahn U, and Hillenkamp F, *Int. J. Mass. Spect. Ion, Proc.* 78, 53 (1987)
Karas M, and Hillekamp F, *Anal. Chem.* 60, 2299 (1988)
Beavis R C, and Chait B T, *Rapid Comm. Mass. Spectrom.* 3, 436 (1989)
Spengler B, Pan Y, Cotter R J, and Kan L, *Rapid Comm. Mass. Spectrom.* 4, 99 (1990)
Nelson R W, Thomas R M, and Williams P, *Rapid Comm. Mass. Spectrom.* 4, 348 (1990)
Romano L J and Levis R J, *J. Am. Chem. Soc.* 113, 9665 (1991)

†Graduate student, Vanderbilt University, Nashville, Tennessee

Inst. Phys. Conf. Ser. No 128: Section 8
Paper presented at RIS 92, Santa Fe, NM, USA, 24–29 May 1992

Resonance ionization mass spectrometry in biochemical analysis

J. E. Anderson, B. L. Perez-Lopez and N. S. Nogar
MS J565, Los Alamos National Laboratories, Los Alamos, NM 87545
R. C. Estler, Dept. of Chemistry, Fort Lewis College, Durango, CO 87103
J. E. Conia, Cell Robotics, 2309 Renard Place, SE, #100, Albuquerque, NM 87106

1. INTRODUCTION

Resonance Ionization Mass Spectrometry (RIMS) is a form of laser-based mass spectrometry that is finding increasing applications in elemental and isotopic analysis, particularly in biochemical systems[Arlinghaus, et al,1991; Moore et al,1986]. Significant attributes include: 1) excellent sensitivity— sub-fg, absolute and sub-ppb, relative abundance sensitivity [Estler and Nogar,1992]; 2) information on isotopic distributions[Fassett and Murphy,1990]; 3) reduction of background (relative to conventional mass spectrometry) due to the sample matrix[Hurst et al,1979]. The high sensitivity allows the use of very small sample sizes; this is potentially important in trace element analysis for infants, neonates, or other hard-to-obtain samples. Isotopic measurements can provide direct absolute signal calibration, as well as being applicable to kinetic analysis, such as trace metal uptake studies. The selectivity of RIMS provides for accurate measurements with little or no sample preparation. This, in turn, improves the speed and accuracy of the analysis. Furthermore, resonance ionization can be adapted for use with many pre-existing mass spectrometric systems. We report here on three applications of RIMS: the detection of trace levels of chromium; the quantification of trace chromium in urine; and, the uptake of copper by single cells.

2. EXPERIMENTAL

2.1 RIMS Apparatus
The measurements described below utilized a tunable dye laser for excitation, and a 0.4 m time-of-flight (TOF) mass spectrometer for ion sorting and detection. This instrument has been described in greater detail previously[Downey et al, 1984; Nogar et al, 1985]; only pertinent aspects are presented here. Pulses, at 20 Hz, from an excimer laser-pumped dye laser (Lambda Physik Model 101/2002) were focused to a spot ≈0.7 mm in diameter in the source region of the mass spectrometer, immediately adjacent to (in some cases, grazing) the sample filament, and retro-reflected with a dielectric mirror. For laser desorption experiments,, the Q-switch synch-out from the Nd:YAG laser (Quanta Ray/Spectra Physics Model DCR 1A) was used to master the timing sequence. This laser was used as the desorption source and operated at the third harmonic, 355 nm. The output of this laser was ~10 nsec in duration and was smooth on the time scale of the detection electronics (~2 nsec). The laser output was focused to a measured spot size of ~1mm diameter onto to a disk of (PMMA) polymer onto which the sample had been placed.

2.2 Sample Preparation
The samples for trace detection of chromium were prepared from known dilute solutions of $Cr(NO_3)_3.9H_2O$ using ethanol as a solvent. Two 10 µl aliquots of each solution were placed onto the surface of a 2.54-cm diameter PMMA disk (0.5 cm thick) by spin coating. For the isotope dilution studies, samples were prepared by dissolution of $Cr(NO_3)_3 \cdot 9H_2O$ in water, dissolution of enriched stable isotopes (^{50}Cr in Cr_2O_3) in an aqueous acid solution, and the reconstitution of a freeze-dried NIST standard urine standard (SRM 2670). The urine was digested and concentrated in Teflon labware using HNO_3/H_2O_2 treatments. An EM Science (EP6739A-1) multi-element ICP standard was used as the primary standard for spike calibration. Determination of Cr in the urine sample was made using standard isotope dilution

techniques[Fassett and Paulsen,1989]. Here, twenty microliter aliquots of the calibrated isotope spike, standard urine and the spiked urine were air dried onto Ta ribbon filaments (0.5 cm x 0.075 cm x 0.0025 cm) for subsequent mass spectral analysis.

In the case of single cell determination, the Datura *innoxia* (Mill.) cell line was grown in the dark at 30°C with agitation. The cell suspension was maintained as an exponentially growing culture by regular transfer at two day intervals into fresh culture medium supplemented with 0.01-1mM $CuSO_4$ (final concentration). Protoplasts were isolated from *D. innoxia* cell cultures by mixing an equal volume of cell suspension with a solution containing 600 mM KCl, 70 mM $CaCl_2$, 0.5% (w/v) Cellulase, 0.05% (w/v) Pectolyase, 0.08% (w/v) dithiotreitol, and 0.6% (w/v) polyethylene glycol 8000. The enzyme solution was sterilized by passage through a 0.2 mm nylon membrane prior to use. Cells were incubated in the presence of the enzymes with agitation for 40 min at 30°C. After digestion of the cell wall, the protoplast suspension was filtered through a 65 mm nylon mesh filter. The protoplasts were rinsed twice with culture medium by successive centrifugations (100 g's, 2 min). This culture medium was similar to the one used for cell suspension culture but was without added copper. The protoplasts were then suspended in this medium at an initial density of about 30,000/ml. For the detection of copper concentration within small populations, aliquots of protoplasts were further diluted in the culture medium lacking copper.

3. RESULTS AND DISCUSSION

3.1 Chromium Desorption

Figure 1 shows the ionization spectrum of Cr near the $(3d^55s^1)$ $^7S_3 \leftarrow$ $(3d^54s^1)$ $^7S_3$ two-photon resonance[Estler and Nogar,1992].   This curve was obtained with from a heated filament with Cr air-dried onto the surface. Several similar spectra were obtained as a function of incident laser energy, and the resulting signal vs. intensity saturation curve fit to a rate-equations formalism. This fit yielded values of $\sigma_{2h\nu} = (5\pm2)$ x 10^{-44} cm^4-sec, $\sigma_I = (7\pm4)$ x 10^{-18} cm^2, and $S_p = 1$ x 10^7 sec^{-1}, where $\sigma_{2h\nu}$ is the effective two-photon excitation cross section, σ_I is the ionization cross section, and S_p is the rate of loss of analyte. The error limits on the two-photon and photoionization cross sections are two-sigma values, while the fit was insensitive to the value of S_p in the range 10^5-10^8 sec^{-1}. We include these values to show that they fall in the range reported previously for similar processes.

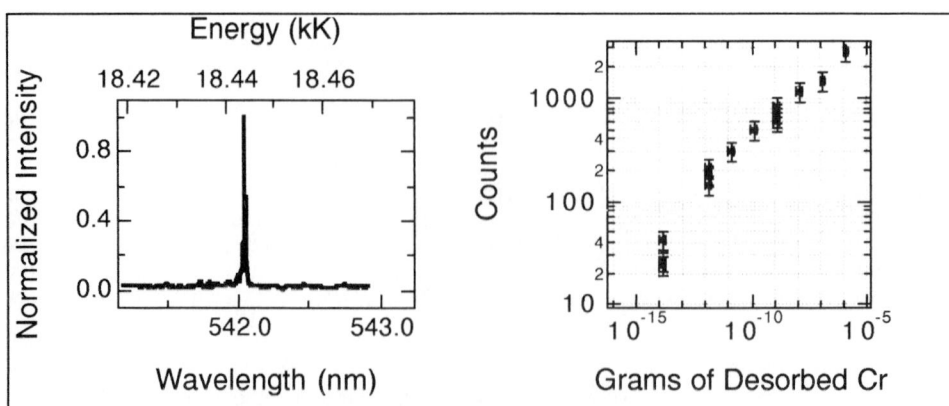

Figure 1 (left) shows the optical spectrum for "2+1" ionization of chromium, while the right side of the figure shows the analytical working curve.

Also shown in Figure 1 is an analytical working curve, showing a log-log plot of the Cr ion signal (in counts) as a function of desorbed sample size. For these experiments, the arrival time distributions of Cr ground state atoms in the ionization region (relative to the desorption laser pulse) were found to be roughly thermal with a temperature of 10^3 K. The detection limit observed here, 10 fg, compares favorably with previous reports for laser-based analysis[Niemax Sdorra,1990]. In addition, the use of mass spectral interrogation allows the potential for isotopic analysis. It should be noted also that the RIMS system used for these experiments must serve multiple purposes, and is not optimized for best sensitivity. In particular, the relatively long distance between the sample surface and the ionization laser, 3.2 cm, dictates that the spatial overlap of the laser desorbed plume with the ionization laser is small. Alteration of our source geometry to reduce this distance could result in a sensitivity increase of perhaps four orders of magnitude.

3.2 Chromium Analysis in Urine

Each of the spectra shown in Fig. 2 represent the accumulation of over 1000 individual mass spectra. These spectra are from samples containing ≈1 ng of chromium. For each spectrum, baselines and data peaks were identified and fit to gaussian functions. The peak areas obtained from the best fits were used for the subsequent concentration determination in order to reduce any possible errors due to the slight electronic ringing present in the spectra. Due to the manner in which the data is collected and analyzed it is not possible to use Poisson-counting statistics to estimate the error in the isotope ratio. Instead, we have used the reproducibility of multiple mass spectra accumulations to estimate this error.

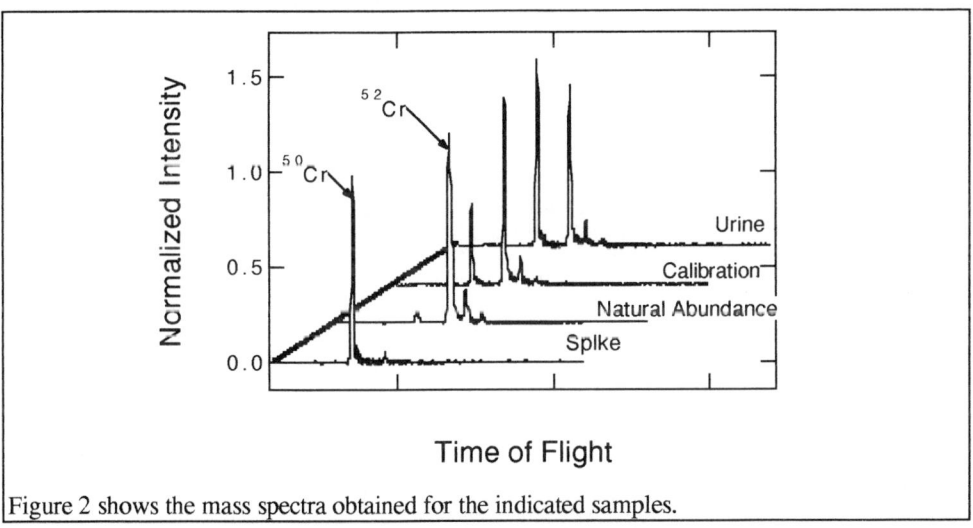

Figure 2 shows the mass spectra obtained for the indicated samples.

From the pertinent isotopic abundances observed in the spectra, we determine the Cr concentration in the NIST standard to be 0.093 ± 0.008 µg/mL in good agreement with the given 0.085 ± 0.006 µg/mL certified value. We believe our results indicate the utility and selectivity of RIMS in isotope dilution analysis of biological samples of complex matrix and history. The application of RIMS to other specific elemental cases, avoiding those elements possessing isotopes of known varying laser-driven ionization efficiencies, should be straightforward. In those cases, RIMS will certainly provide a viable alternative to more complex analyses.

3.3 Copper Detection in Single Plant Cells.

The temporal evolution of the copper signal is displayed in Figure 3, for a sample loading of ≈10 protoplasts and for a sample of culture medium with no added copper. The signal from the copper can be seen to rise and fall in a period of about 2-3 min, while the blank exhibits a minimal signal over a similar period. The rising edge of the waveform corresponds to the heating rate for the filament, and equilibration of the copper to this temperature. The decay of the signal is due to exhaustion of the sample, and is roughly consistent with the calculated decay rates for samples heated to this temperature.   The transient signal from the blank is likely due to radiative evaporation of copper deposited on   nearby parts of the time-of-flight ionization region by past samples, and removed by flash heating during the warm-up cycle of the sample filament.

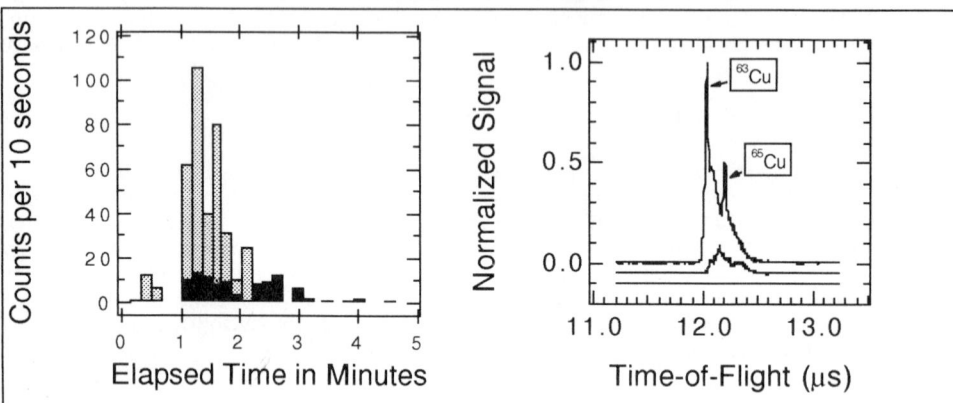

Figure 3 displays (left) the temporal evolution of signal from a blank (solid) and a sample containing 10 protoplast (gray), while on the right, we display mass spectra (from bottom to top) of a blank, 10 protoplasts and 100 protoplasts.

Also in Figure 3 we display the mass spectra due to a loading of ≈100 protoplasts, 10 protoplasts, and a blank. Protoplasts were counted with an optical microscope for the smaller sample, estimated by optical microscopy for the larger sample, and verified absent for the blank. The total signal for the 100 cell sample generated several thousand counts, in rough agreement with the signal level anticipated by calculation.   The sample containing 10 cells produced 530 detected ions, while the blank produced 30 counts. If we assume that the copper was evenly distributed among the cells, this implies a signal to noise ratio for a single cell of $53/(30)^{1/2} \approx 10$. This indicates that we have the ability to detect transition metals in single cells at the millimolar and below level.   Significant improvements in sensitivity may be possible through the use of a multichannel plate detector in place of the electron multiplier, and by the use of a higher repetition rate laser.

5. REFERENCES

Arlinghaus H F, Thonnard N, Spaar M T, Sachleben R A, Larimer F W, Foote R S,
    Woychik R P, Brown G M, Sloop F V Jacobson K B 1991 *Anal. Chem.* **6 3** 402.
Downey S W, Nogar N S Miller C M 1984 *Int. J. Mass Spectrom. Ion Processes* **6 1**
    337.
Estler R C Nogar N S 1992 *Anal. Chem.* **6 4** 465.
Fassett J D Murphy T J 1990 *Anal. Chem.* **6 2** 386.
Fassett J D Paulsen P J 1989 *Anal. Chem.* **6 1** 643A-649A.
Hurst G S, Payne M G, Kramer S D Young J P 1979 *Rev. Mod. Phys.* **5 1** 767.
Moore L J, Parks J E, Taylor E H, Beekman D W Spaar M T In *Inst. Phys. Conf. Ser.*;
    Inst. Phys.: 1986; pp 239.
Niemax K Sdorra W 1990 *Appl. Opt.* **2 9** 5000.
Nogar N S, Estler R C Miller C M 1985 *Anal. Chem.* **5 7** 2441.

Inst. Phys. Conf. Ser. No 128: Section 9
Paper presented at RIS 92, Santa Fe, NM, USA, 24–29 May 1992

297

Resonance ionization of rubidium using sequential diode laser-driven transitions

R. W. Shaw, J. P. Young, and J. M. Ramsey

Analytical Chemistry Division, Oak Ridge National Laboratory, P. O. Box 2008, Oak Ridge TN 37831-6142

ABSTRACT: Diode laser excitation of rubidium atoms at the D2 line followed by [1 + 1] ionization using a pulsed dye laser was examined. Mass-resolved D2 spectra were recorded for both natural isotopes. Excitation using two sequential diode laser photons was accomplished by counterpropagating the beam from the 780 nm laser and one from a 775.8 nm laser; the latter excites Rb from the $5P_{3/2}$ level to the $5D_{5/2}$ level at 25704 cm^{-1}. A photon from the R6G laser then ionized the atom. Spectra resulting from scanning the second diode laser -- with the first at an off-resonance fixed frequency -- revealed both two-photon and sequential [1 + 1] processes.

1. INTRODUCTION

Tunable, semiconductor diode lasers are practical sources for high resolution atomic spectroscopy[1,2]. These lasers oscillate in a single longitudinal mode without any special arrangements, and thus exhibit very narrow spectral bandwidths (approximately 25 MHz). They are conveniently tunable, with no moving parts, by adjusting their temperature and/or operating current. We have previously capitalized on their narrow linewidth for atomic hyperfine spectroscopy and isotopically selective ionization in resonance ionization mass spectrometry (RIMS) experiments[3]. A single diode laser-promoted excitation step was used to create an excited state population, followed by a pulsed dye laser-driven [1 + 1] resonance ionization process. We now report resonance ionization processes that utilize two sequential diode laser excitations. With the involvement of two narrow band lasers, different spectroscopic experiments become possible; the first step wavelength can be held fixed while the second step wavelength is scanned, and vice versa. This two-diode laser excitation demonstration is the next step in a progression toward a resonance ionization process that uses only diode lasers for the optical steps.

2. RUBIDIUM RIMS

RIMS experiments were conducted using rubidium atoms from a heated rhenium filament loaded with $RbNO_3$. Ionization was accomplished using one or more diode lasers in conjunction with a high repetition rate, pulsed dye laser. Figure 1 is an energy level diagram showing the pertinent levels. Our apparatus has been described previously[4] and will be only highlighted here. The diode laser employed for the 780.0 nm rubidium D2 line was a 3 mW Mitsubishi ML-4402 laser. Two different lasers were used for the subsequent excitation step at 775.8 nm: (a) Mitsubishi ML-4102A (3 mW) and (b) Mitsubishi ML-64110N-01 (30 mW).

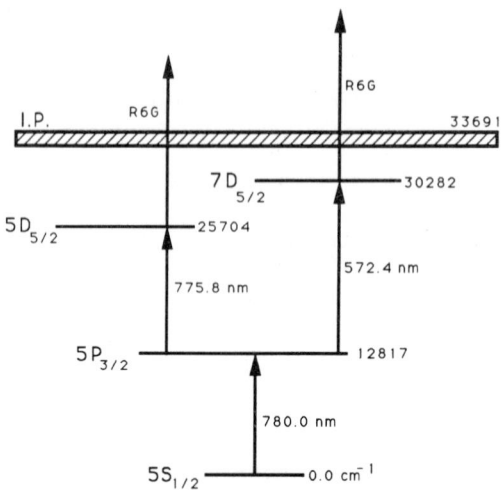

Figure 1. Pertinent rubidium energy levels

The pulsed dye laser was a copper vapor laser-pumped R6G laser producing 50 μJ, 20 nS pulses at 6000 pps. The diode lasers were arranged to counter-propagate, and were crossed at 90° by the dye laser beam. Each of the three beams was focused to approximately 1 mm diameter. Laser-generated ions were mass analyzed using a single-stage magnetic sector mass spectrometer. Spectra were acquired by tuning the mass spectrometer to one or the other of the two natural rubidium isotopes (m/z 85 and 87) and ramping the diode laser drive current repetitively in a sawtooth fashion, while holding the device temperature constant. The frequency extent of these diode laser scans was determined by following the laser mode motion using a scanning Fabry-Perot spectrum analyzer of known free spectral range.

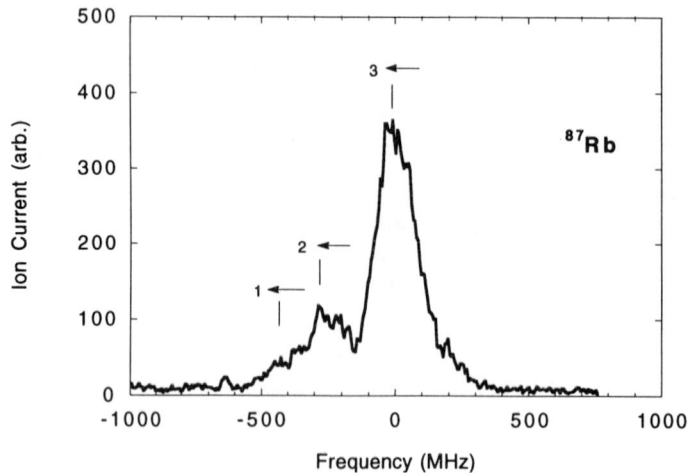

Figure 2. Rb-87 D2 spectrum as detected via a [diode laser + resonant dye laser + dye laser] ionization process.

The initial experiment was to excite at the D2 line (780.0 nm) to populate the $5P_{3/2}$ level from the ground state; [1 + 1] ionization using the dye laser at 572.4 nm created ions via the $7D_{5/2}$ state at 30282 cm^{-1}. By scanning the diode laser over a 3 GHz range, mass-resolved D2 spectra were recorded for both natural isotopes. A scan corresponding to the mass 87 isotope is shown in Figure 2. The extent of the scan is such that only lines from the F=2 component of the ground state are shown. The line labels indicate the F quantum numbers of the $5P_{3/2}$ sublevels. The spectral linewidth is 175 MHz, and the excited state hyperfine splittings are evident. In order to achieve this resolution, the diode laser power was reduced to 50 μW. While we have not made quantitative isotope ratio measurements, the 1 GHz spectral isotope shift of rubidium results in excellent isotopic selectivity for this ionization process.

Excitation using two sequential diode laser photons was accomplished by counterpropagating the beam from the 780 nm laser and one from a 775.8 nm laser; the latter excites Rb from the $5P_{3/2}$ level to the $5D_{5/2}$ level at 25704 cm^{-1}. An arbitrary wavelength photon from the R6G laser then ionized the atom. In order to establish operating conditions for the second diode laser the following process was used. First, the single diode laser, [1 + 1 + 1] process described above was established and optimized. The dye laser was then de-tuned to the red by approximately 2 nm, extinguishing the ionization signal. The ion current was then re-established by admitting the second diode laser and adjusting its drive current and spatial position. An alternate method we attempted was to establish the single diode laser [1 + 1 + 1] process, and then admit and scan the second diode laser, looking for an increase or decrease in the continuous ion current. This method was not as satisfactory because it represents a non-zero background technique. However, when we were able to see the effect, what we observed was a signal increase.

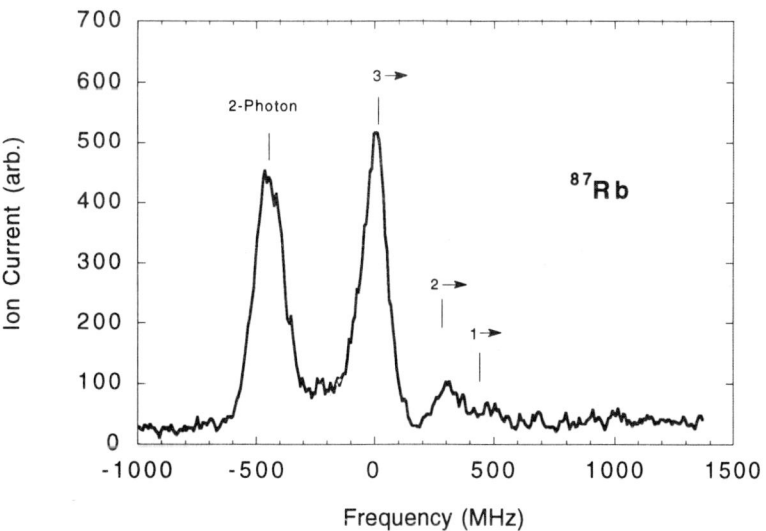

Figure 3. Two-photon and [1 + 1] rubidium-87 ionization lines
for two-diode laser excitation.

The wavelength of either the initial diode laser step or the second could be scanned, and both experiments were accomplished. Figure 3 shows a scan of the diode laser 2 wavelength, with diode laser 1 fixed at a wavelength within the 1 GHz interval that corresponds to the mass 85/87 isotope shift; that is, diode laser 1 is tuned slightly blue of the mass 87 resonance. The ion current signal was at m/z 87, and the counterpropagating diode laser powers were a few milliwatts each. Two distinct excitation processes appear in the spectrum: a concerted two-photon process and a stepwise [1 + 1] process. A similar effect has been reported by Axner[5] for laser enhanced ionization of strontium in a flame. Our lowest energy line is due to a concerted two-photon process and occurs at a frequency where the step 1 and step 2 photon energies sum to the energy of the upper level. The three remaining, higher energy lines are due to sequential [1 + 1] excitation. In the first step, three (F= 1,2,3) of the four hyperfine components of the intermediate $5P_{3/2}$ level are populated ($\Delta F = 0, \pm 1$) within the Doppler width of the experiment; however, the F = 2 to F = 3 transition is the most intense, as confirmed by 6j angular momentum calculations. As the diode laser 2 wavelength was scanned, each of the three populated intermediate sublevels was excited in turn, yielding the three observed lines. The final state hyperfine splitting is small enough (< 100 MHz) that it is inconsequential for the interpretation of this data. The labels associated with the [1 + 1] lines in Figure 4 are scaled to represent the accurately known Rb-87 $5P_{3/2}$ hyperfine splittings; our agreement with the splittings is within the experimental uncertainty of our frequency axis calibration (5%). The FWHM of the F=3 line is 110 MHz.

3. SUMMARY

Semiconductor diode lasers are convenient and practical sources for resonance ionization spectroscopy. We accomplished rubidium ionizations that comprise both one- and two-diode laser excitation steps; they required a pulsed dye laser for the bound-to-continuum step. For the two-diode laser scheme, both concerted two-photon and sequential [1 + 1] processes were observed. We are currently attempting to demonstrate a resonance ionization process wherein only diode lasers are utilized for the optical steps.

4. ACKNOWLEDGEMENTS

Oak Ridge National Laboratory is managed by Martin Marietta Energy Systems, Inc. under contract DE-AC05-84OR21400 with the Department of Energy.

5. REFERENCES

1.   Lawrenz J and Niemax K 1989 *Spectrochim. Acta* **44B** 155

2.   Wieman C E and Hollberg L 1991 *Rev. Sci. Instrum.* **62** 1

3.   Shaw R W, Young J P, Smith D H, Bonanno A S, and Dale J M 1990 *Phys. Rev. A* **41** 2566

4.   Shaw R W, Young J P, and Smith D H 1989 *Anal. Chem.* **61** 695

5.   Axner O and Sjöström S 1992 *Spectrochim. Acta* **47B** 245

Inst. Phys. Conf. Ser. No 128: Section 9
Paper presented at RIS 92, Santa Fe, NM, USA, 24–29 May 1992

Status and future development of resonant laser ion-sources for on-line mass separators

H. L. Ravn

CERN-ISOLDE, CH-1211 Geneva 23, Switzerland

ABSTRACT: The widespread use of resonant laser ionization for detection and study of rare species have now led to its successful use also in the ion-source region of on-line mass separators used for production of intense beams of radioactive ions. The laser technique in many cases solves the principal technical problems in this domain, i.e. the chemical selectivity, ionization efficiency and desorption. Bunched beams with efficiencies of 10-20 % and selectivities of 100-10000 have now been obtained in on-line experiments. Recent developments and some future possibilities of this technique are discussed.

1. INTRODUCTION

Collinear laser spectroscopy and resonance ionization spectroscopy of rare short-lived radioactive isotopes have been particularly fruitful at on-line mass separators as seen in the reviews by Otten(1989) and Kluge (1992). A continuous development of new methods adapted to radioactive ion-beams has produced spin offs in other areas of precision spectroscopy and trace analysis and now also in preparation of radioactive ion-beams.

At on-line mass separators these beams of radioactive isotopes are produced by a nuclear target from which the complex mixture of short-lived nuclear reaction products are transferred to the ion source of the mass spectrometer. A more general discussion of the requirements to such target and ion source configurations are found in Ravn and Allardyce (1989) and Ravn (1992). The development of techniques in which tunable lasers are used in a resonant process to ionize atoms and molecules have leed to new methods for their detection or study. These techniques offer considerable advantages in terms of selectivity and sensitivity for detection of rare species over conventional mass spectrometers. The availability of high repetition rate (10 kHz) copper-vapor lasers for pumping dye lasers provided the break through to reach efficiencies that made it realistic to use the laser ionization technique for on-line preparation of radioactive ion-beams as demonstrated by Andreev et. al. (1987). In fact laser resonant ionization sources may replace the conventional ion sources used in the acceleration of radioactive nuclei. It should be noted that the efficiency of this method, like other techniques for on-line mass separators, depend strongly on the chemistry or atomic levels of the individual elements. The major problem is to find the most efficient excitation scheme and sufficient laser intensity to drive each step into saturation.

As will be discussed in detail below, the non-thermal nature of the resonance process not only gives such ion-sources an exceptional high chemical selectivity but also reaches efficiencies of the order of tens of percent, often exceeding those of conventional sources. In addition this new technique provides bunched beams which open up new possibilities for laser spectroscopy of radioactive atoms.

2. TARGET AND ION-SOURCE CONFIGURATIONS

A number of different configurations for bringing the radioactive nuclei from the nuclear-reaction target into the laser beam have been proposed or tested. This choice is most often dictated by the nuclear reaction type used for production, in particular whether a thick or thin target technique is used. If the radioactivity is accumulated by adsorption on a cold spot it can be desorbed in a pulsed mode by the light from powerful Nd yag, CO_2 or excimer lasers at low laser power (10^6-10^8 W/cm^2 in 10 ns pulses). Ionization of the desorbed atoms may now follow by means of non-resonant or preferably multi-colour multi-step resonant excitation to an autoionizing state or to a Rydberg state followed by field ionization.

For use in mass spectrographic analysis of short-lived nuclei, Eloy and Zirnheld (1976) suggested to use the high speed and chemically unselective He gas-jet transport technique in conjunction with laser desorption and non resonant laser ionization. By desorbing and ionizing the radioactive atoms deposited by the He-jet, limitations in coupling of gas jets to conventional ion sources could possibly be overcome at least in connection with thin targets. Fairbank Jr and Carter (1987) further developed this idea by proposing to use resonant laser ionization of the radioactive atoms deposited by the He-jet and to inject the obtained beam into an on-line mass separator. This technique of resonant ionization of laser desorbed atoms, which seems to be a promising technique for making beams of a number of refractory transition group metals, was, however, first demonstrated in a specific experiment at an on-line mass separator by Krönert et al. (1987). They studied the isobar separation and laser spectroscopy of Au and Pt isotopes implanted into a solid substrate as their Hg precursors.

It is only from the extensive study of Lee et al (1988) and Krönert et. al.(1991) that a detailed description of the method and its performance is available.

A scheme for the application of laser ionization to the complex mixture of radioactive atoms which effuse out of thick spallation reaction targets was proposed by Kluge et al. (1985). In the geometry he proposed, the atoms diffuse into an cylindrical ionization chamber with insulated end caps that allowed to apply an electrical field for extraction of the ions. This method has the disadvantage that it carries over one of the problems of conventional plasma discharge ion-sources, the large insulator surface which acts as an adsorption trap for many elements. For this reason the groups of Alkhazov et al. (1989), Scheerer et al. (1992) and Mishin et al. (1992) based the starting point of their studies on a geometry similar to the

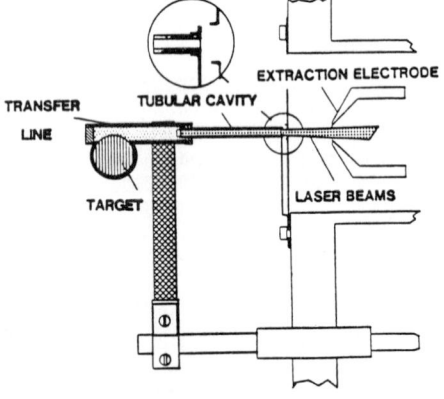

Fig. 1 ISOLDE type target and laser ion-source configuration

well known high temperature and insulator free tubular surface-ionizers often used at on-line mass separators as discussed by Ravn (1989). Figure 1 shows the version used at ISOLDE.

3. RECENT RESULTS

A configuration like the one shown in Figure 1 was first used on-line by Alkhazov et al. (1991) who found ionization efficiencies for the rare earths Yb and Nd of 35% and 20%

respectively and a selectivity of 15 (i. e. laser-ionized ions / surface ionized ions). Mishin et al. (1992) studied in further detail the properties of the source shown in Figure 1. Since there is no field gradient suitable to remove Rydberg electrons they chose stepwise resonant excitation and photoionization in the last step. In a series of off-line and on-line studies the ionization of Sn (E_i = 7.3eV), Tm (E_i =6.2eV), Yb (E_i = 6.2eV) and Li (E_i = 5.4eV) was investigated.The ratio of the laser-ionized and surface-ionized ion currents was measured as function of temperature for different materials (W, Ta, Nb and TaC) of the hot tube. It was shown that this ratio, i.e. the selectivity, rises for Tm from 10 to 10000 with falling temperature strongly depending on the material. Since the lasers are pulsed the ion beam is also bunched with a pulse width of about 10-50 μs as shown in Figure 2. This width is strongly dependent on the potential drop along the tube (caused by the electric current used for heating the tube) and on the alignment of the laser beams with respect to the tube axis.

The selectivity could be further improved by a factor of 10 using a gated detection of the bunched ion beam. Selectivities of up to 10^5 and ionization efficiencies of 15% were achieved generally for the rare earths. An efficient excitation scheme was identified for the ionization of the element tin. The wavelengths λ_1=317 nm, λ_2=812 nm and λ_3=824 nm lead to an auto ionization state in the third transition. The excitation steps can easily be saturated with the following laser powers: I_1=0.6 mW/mm^2, I_2=10 mW/mm^2 and I_3=100 mW/mm^2 leading to an ionization efficiency of 10 %. Interesting information on the production and release of the element Li from a Ta target was obtained by the same group in a study of its laser ionization.

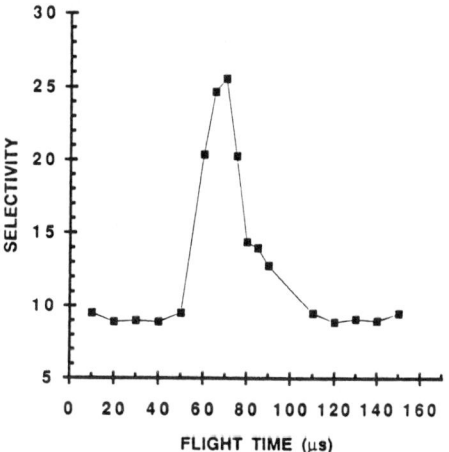

Fig. 2 Pulse shape of the photo-ion current of stable Tin

4. FUTURE DEVELOPMENTS

The measurements performed under on-line conditions by Alkhazov et al. (1991) and Mishin et al. (1992) have shown the strength of the laser ion-source principle for efficient and selective ionization of short-lived isotopes. A chemical selectivity factor of 100 combined with efficiencies of the order of 10-20 % enables the study of a number of exotic, short-lived nuclei far from ß-stability, where the use of conventional ion sources is hindered due to isobaric background. If frequency doubling is applied in the first excitation step, also efficient photo-ionization of elements with ionization potentials as high as 7.3 - 7.6 eV like manganese, silver and tin can be achieved, [Mishin et al. (1992)]. For these elements very strong UV-transitions can be used in the first excitation step, which allows to reach saturation with the limited power available in the UV-wavelength range. By means of transitions to autoionizing states, saturation can be achieved in all excitation steps.

In the case of elements which do not react with oxide or nitrides, as for example silver, a shorter ion bunch and consequently an increased selectivity may be achieved by using a ceramic tube as laser ionizer. This material allows use of an extraction voltage applied across the tube of up to 60 V and leads to corresponding shorter pulse width. An additional increase in selectivity may be achieved by an electrode system placed between the ionizer tube and the extraction electrode.

In the studies of isobar separation of laser desorbed and ionized Au and Pt nuclei Krönert et al.(1991) found an overall efficiency of 10^{-4}.

Further development of this technique indicate that efficiencies of the order of 10^{-3} may be reached. Replacing the implanted mass separated atoms with the skimmed deposit of refractory atoms from a He-jet, as suggested by Fairbank Jr and Carter (1987), may be the best way to study the short-lived isotopes of those elements. The thin target and the low efficiency may be offset by the low decay losses of this fast process.

An interesting application of the laser desorption is seen in combination with conventional ion sources. By creating a cold spot in these sources bunched ion beams could be produced as illustrated in Figure 3.

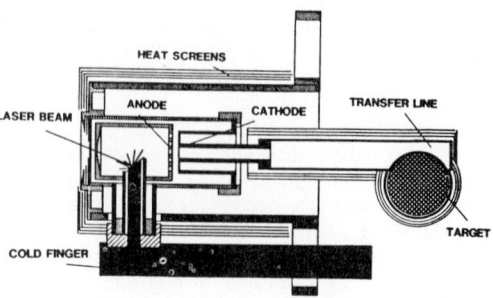

Fig. 3 Layout of a target and high temperature FEBIAD plasma discharge ion-source with pulsed laser desorption

Presently Oshima et al. (1992) and Qamhieh et al. (1992), using the ion guide technique to bring reaction recoils thermalized in a gas cell directly into a mass separator, are studying the possibilities to compensate the gas neutralization losses by means of resonance ionization in the gas cell.

5. REFERNCES

Alkhazov G D E 1989 Berlovich Ye and Panteleyev V N Nucl. Instr. and Meth. **A 280** 141

Alkhazov G D 1991 Batist L Kh Bykov A A Vitman V D Letokhov V S Mishin V I Panteleyev V N Sekatsky S K and Fedoseyev V N Nucl. Instr. and Meth. **A 306** 400

Andreev S V 1986 Mishin V I and Lethokov V S Opt. Commun. **57** 317

Eloy J F 1976 Zirnheld J P Nucl. Instrum. and Methods **135** 111

Fairbank Jr. W M 1987 Carter H K Nucl. Instrum. and Methods **B26** 357

Kluge H -J 1985 Ames F Ruster W and Wallmeroth K in Proc. of Accelerated Beams Workshop Parksville , Canada 1985 eds. Buchmann L and D'Auria J M TRIUMF report TRI-85-1 p 119

Kluge H -J 1992 To be published in Hyperfine Interactions

Krönert U 1987 Becker St Hilberath Th Kluge H -J and Schulz C Appl Phys **A 44 339**

Krönert U 1991 Becker St Bollen G Gerber M Hilberath Th Kluge H -J and Passler G Nucl. Instr. and Meth **A 300** 522

Lee J K P 1988 Savard G Crawford J E Thekkadath G Duong H T Pinard J Liberman S Le Blanc F Kilcher P Obert J Putaux J C Roussière B and Sauvage J Nucl.Instr. and Meth. **B34** 252

Le Blanc F 1991 Pinard J. Arianer J Crawford J E Dautet H Duong H T Kilcher P Thekkadath IPNO-DRE 91-25

Mishin V I 1992 Fedoseev V N Kluge H -J Letokhov V S Ravn H L Scheerer F Shirakabe Y Sundell S and Tengblad O To be submitted to Nucl.Instr. and Meth. **B**

Oshima M 1992 Sekine S Ichikawa S Hatsukawa Y Morikawa T and Nishinaka Proc 12th Int. Conf. on Electromagnetic Isotope Separators and Techniques Related to their Applications Sendai Japan Sept. 1991 To be published in Nucl. Instrum. & Methods **B**

Otten E W 1989 in Treatise On Heavy Ion Science Vol. 8: Nuclei far from stability ed. D. A. Bromley (New York: Plenum Press) p 515

Ravn H L 1989 in Treatise On Heavy Ion Science Vol. 8: Nuclei far from stability ed. D. A. Bromley (New York: Plenum Press) p 363

Ravn H L 1992 To be Published in Nucl. Instrum.& Methods **B**

Scheerer F 1992 Fedoseyev V N Kluge H -J Mishin V I Letokhov V S Ravn H L Shirakabe Y Sundell S and Tengblad O Proc. 4th Int. Conf. on Ion Sources Bensheim-Germany Sept 1991 To be published in the Rev. Sci. Instrum.

Quamhieh Z N 1992 Huyse M Silverans R E Vandeweert E Van Duppen P and Vermeeren Proc 12th Int. Conf. on Electromagnetic Isotope Separators and Techniques Related to their Applications Sendai Japan Sept. 1991 To be published in Nucl. Instrum. & Methods **B**

Inst. Phys. Conf. Ser. No 128: Section 9
Paper presented at RIS 92, Santa Fe, NM, USA, 24–29 May 1992

305

Laser desorption sources for RFQ traps

L. Davey[1], F. Buchinger[1], J.E. Crawford[1], Y.Ji[1], J.K.P. Lee[1], J. Pinard[2], J.L. Vialle[3], W.Z. Zhao[1]

[1]Foster Radiation Laboratory, McGill University, Montréal, Canada
[2]Laboratoire Aimé Cotton, Orsay, France
[3]Départment de Physique, Université de Lyon, Lyon, France

ABSTRACT: Pulsed laser desorption sources for RFQ ion traps have been devised and tested using two techniques. In one method, ions are desorbed from an external target by pulses from a Nd:YAG laser and injected through the trap endcap electrode at a particular phase of the RF voltage. A retarding pulse on the entrance endcap reduces the momentum of the injected ion cloud, permitting capture by the trap fields. Time-of-flight techniques may be used to select and trap particular masses. For internal ion production, material located at the trap ring electrode is desorbed by YAG pulses and ionized by RIS laser beams.

1. INTRODUCTION

Since its invention (Paul 1956) the Radiofrequency Quadrupole (RFQ) or Paul ion trap has seen increasing application as a containment device for laser spectroscopic studies. Trap techniques may be particularly advantageous if the available sample is very small. Detection at the single particle level has been achieved, and the feasibility of high resolution spectroscopy at this level has been demonstrated (Madej 1992). More commonly, studies of hyperfine structure and isotope shift, in which the incident laser beam is scanned, have been carried out with ~10^3 ions in the trap, and fairly conventional techniques (e.g., resonance fluorescence) may be used for detection (Ifflander 1977). Two particularly attractive features of such traps for spectroscopy are the relatively long ion confinement times (many hours, in some cases) and the small perturbation of the confining fields at the trap centre. However, the creation of ions in the trap, or ion injection from outside poses a number of special problems. For internal production the technique that is most frequently used is electron impact ionization—the atoms may be evaporated from a filament and simultaneously ionized by a beam of electrons. This ionization is not selective, however: other evaporated material from the electrode may be ionized, and produce background contamination in the experiment. Nevertheless, critical tuning of the trap may allow ions of unwanted mass to leak from the trap, while desired ions are retained. If sample purity is important, it seems desirable to consider some sort of mass pre-selection outside the trap, followed by injection and capture. Such techniques must somehow address the problem of subsequent escape of the captured ion. A particle that has enough kinetic energy to enter the trap and fall into its potential well can (and will) escape unless its energy can be reduced internally. This can be done by introducing buffer gas into the cell to create collisional loss. External beam injection into RFQ traps has been demonstrated for continuous beams from on-line mass separators (Moore 1992) and for laser-desorbed pulses (Louris 1989). In the latter experiment, the laser desorption source was located close to the trap endcap. The sample selection was only partially effective, excluding only masses below a certain m/Z cutoff, and not highly efficient, since it relied only on buffer gas collisions as its energy-reducing capture mechanism. Another laser-desorption injection technique (Kwong 1990) uses crossed beam collisions to reduce the ion energy.

In the present work, we have built a system to study the capture of ions produced by laser desorption both inside an RFQ trap and by external injection.

2. EXPERIMENT

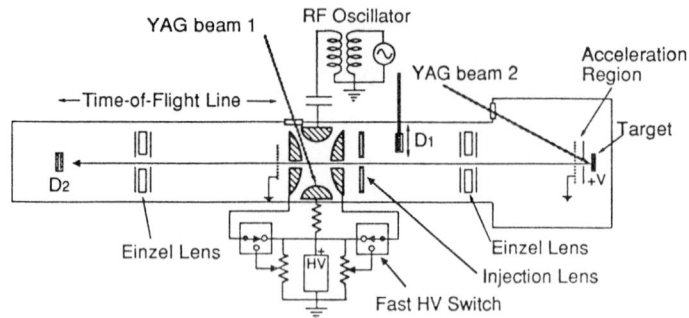

Fig. 1 Schematic of the trap mounted in the TOF test system

Fig. 1 shows schematically the elements of the trap system. The trap is of the usual design: a ring electrode of radius $r_0 = 2.03$ cm with a hyperbolic face, and two hyperboloidal endcaps separated by $2z_0$ ($z_0 = 1.44$ cm). An oscillator drives the ring electrode at a frequency $\nu = 650$ kHz, and amplitude V, relative to the endcaps; the system is designed so that the whole trap assembly may be raised above ground to some DC potential. With the external source, it is possible to select a desired mass group by time-of-flight (TOF) separation, to capture only these ions, and to eject them later from the trap. The target irradiated by the YAG laser and the associated accelerating plates are located 50 cm from the entrance endcap. Normally, the target assembly is operated at +1 kV. After their flight through the input TOF line the ions produced by the YAG pulse are slowed to energy 90eV before injection by setting the trap assembly potential to +910 V. Two lenses (an einzel lens and an injection lens in front of the trap entrance endcap) provide the focusing necessary as the ions decelerate. This first section of the system provides the time separation necessary for selection of the mass to be trapped.

Ion motion within traps is described by Mathieu equations for the r and z motions, and is usually expressed in terms of dimensionless parameters $q_{r,z}$ (related to the AC voltage amplitude) and $a_{r,z}$ (related to a DC bias voltage that may be applied between ring and endcap electrodes). The regions of stability are shown on the 'Mathieu stability diagram'(Morand 1991). In the present experiment, the trap was operated with no DC, so the only parameters affecting the masses that may be trapped are the AC voltage amplitude V and frequency ν. For this simple case, the stability diagram predicts that in a trap made with this standard geometry, a mass A (amu) will be trapped if $A \geq 5.3 \ V/z_0^2 \ \nu^2$, with V in volts, z_0 in m, ν in kHz. To study the selective capture of two separate masses simultaneously produced in a graphite laser desorption source, we chose C_2^+ (carbon-2) and C_3^+ clusters. The trap was operated at V=360v, and $\nu = 650$ kHz; according to the equation above, this permits confinement of ions with $A \geq 22$ amu. The depth D of this trap's pseudopotential well may also be calculated using these parameters. For motion in the z and r directions of a C_3^+ ion, $D_z \approx 30$ eV, and $D_r \approx 15$ eV. For efficient trapping of incoming particles from the external beam it is therefore necessary to inject ions at the appropriate phase of the RF voltage, and to switch on a retarding pulse just after the ions have entered the trap. It is convenient to do this by connecting a fast pulse generator to the entrance endcap, and to trigger it at a particular phase of the trap's RF voltage. We obtained best results by applying a pulse of -165V, τ=660 ns (i.e., about 1/2 cycle of the RF) just after the ions enter the trap. A similar pulsing system is used to extract the ions after the desired storage time, and to direct them into the exit TOF line. Two channelplate detectors D_1 and D_2 are installed in the system so that measurements can be made on ions entering the trap, and ions ejected from it.

Fig. 2 shows how mass selection may be performed by varying the timing of the retarding pulse. The 3 oscilloscope traces show signals at D_2. The small initial pulse is caused by the laser flash. With no retarding pulse, ions simply pass directly through the trap— the two groups in the neighbourhood of 20 μs represent the arrival of C_2^+ and C_3^+ ions. Fig. 2b shows the effect of switching on the retarding pulse 7.6 μs after the YAG. Here, the group of C_2^+ ions at 16 μs has been reduced, since they are captured by the trap; they are subsequently ejected, and arrive at 85 μs. In fig. 2c, the YAG-retardation delay has been increased to 11.0 μs, to correspond to the arrival of C_3^+ ions at the trap.

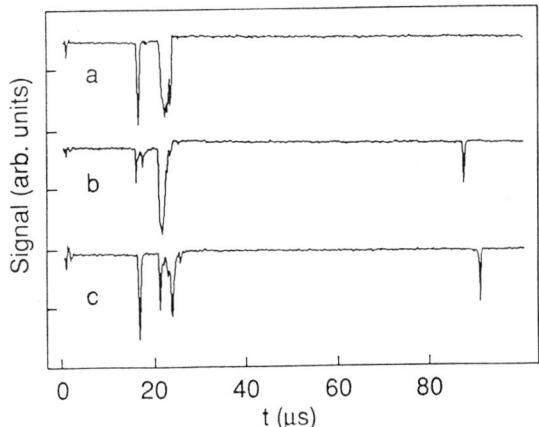

Fig. 2 TOF mass selection by the trap with (a) no retarding pulse (b) YAG-retardation delay= 7,6 μs (c) delay=11.0 μs

Here, the signal from the untrapped group at 20 μs has been reduced, and the C_3^+ group is observed to arrive later than the C_2^+ group of fig. 2b. The spectra show no contamination of the extracted mass groups.

We have investigated the effect of a number of parameters on capture efficiency of the trap. The phasing of both the injection and extraction pulses with respect to the RF excitation is important, and is shown in Fig. 3.

For injection, the YAG-retarding pulse delay is kept constant, but the YAG is triggered to fire at different points on the RF cycle; the ion pulses therefore arrive at the trap at selected phase angles. To reduce the effect of pulse-to-pulse variation of the YAG, the signal is averaged over 16 pulses. The FWHM of the peaks in Fig. 3 is 90°±10°. The signal strength and pulse shape change as the extraction phase is varied; the best TOF resolution occurs close to the signal maximum. Such effects been seen in similar studies using TOF measurements of extracted ions (Lunney 1992) and have been ascribed to the changing shape of the phase-space volume of the trapped ions over the RF cycle.

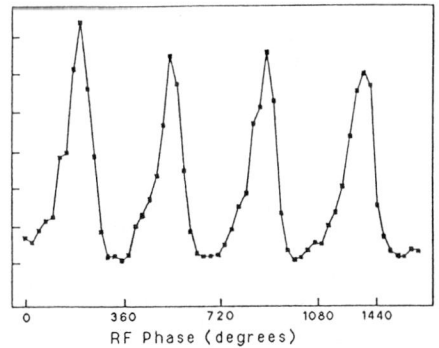

Fig. 3 Variation of extraction signal with trap RF phase at injection

We estimate the trapping efficiency for external injection by comparing the pulse heights of C_2^+ ions arriving at detector D_2 without any retarding pulse (i.e., passing directly through the trap), to those observed with the correct retarding pulse, after a storage time of 150 μs. For this C_2^+ group, the capture efficiency is of the order of 25%. It is likely that the main reason for these losses is the finite spatial extent of the injected ion cloud. Because the ions are produced simply by the heating of the graphite, they are relatively wide—300 ns for C_2^+. For 1 keV ions, this translates into a spatial width of 2.7 cm, which is comparable to the size of the trap before entry into the cavity. Inspection of the direct arrival signals for the C_2^+ and

C_3^+ clusters in fig. 3b and 3c indicates that the signal is both diminished *and* broadened by the trap fields to pulses with widths > 1 μs.

If ions are to be produced within the trap, a Nd:YAG pulse may be focused on a target placed at the edge of the ring electrode. The YAG energy is kept below the threshold for plasma production so that neutral atoms are desorbed; then suitably tuned pulsed dye laser beams can selectively ionize the desired element. In one experiment calcium carbonate ($CaCO_3$) was used as a target and an excitation scheme using two-colour RIS beams of 273.6 nm and 547.1 nm was chosen to ionize the calcium. Tuning of the incident wavelength and TOF analysis confirmed that only Ca^+ ions produced by the RIS beams were stored in the trap. Sweeping the position of the RIS beams in both the r and z directions by ~3 mm had little effect on trapping efficiency. Since the ions are created with thermal velocity, they occupy a small phase-space volume, and the overall trapping efficiency should not be a strong function of the RF phase timing: this was verified experimentally. In a trap with relatively large volume such as this, internal production at thermal velocities should yield trapping efficiencies close to 100%.

To test the selectivity of the method, we changed the RF frequency and voltage to favour the storage of ions with M=24 amu. When the RIS UV laser beam was tuned to 285.3 nm, corresponding to the excitation of the 3s3p state of Mg, only Mg^+ ions (from impurities in the sample) were observed in the TOF spectrum, with no trace of Ca. With the high trapping efficiency and the generous trap phase space volume, this should be a suitable technique for the accumulation of ions from trace elements by repeated desorption-RIS pulses. Unfortunately, the present trap was not suitable for storage times > 1s, and it was not possible to observe this accumulation experimentally.

3. CONCLUSION

Ions produced by laser-desorption sources may be mass-selected by a TOF injection line, and with a correctly timed retarding pulse, it is possible to trap a single mass efficiently. The present experiment used a simple source with desorption, cluster formation, and ionization performed by the same YAG pulse. With narrower ion pulses (produced, for example, by resonant ionization of desorbed atoms in front of the graphite) and careful attention to ion transport, it should be possible to obtain nearly 100% capture efficiency with this type of system. No buffer gas is required for capture, since the retarding pulse provides the energy reducing mechanism. The present system was not designed with ultra-high vacuum, bakable components, and it is not suitable for long storage times. A new system under construction should store ions for much longer periods; this will be used for tests of the cumulative storage of successive laser-desorbed pulses, and for laser spectroscopic experiments. Versions of this type of system should provide a useful method for spectroscopic studies of isotopes produced by accelerators or isotope separators. In some cases this could be done by direct trapping of a pulsed beam; alternatively the isotopes could be accumulated in the graphite reservoir and transferred to the trap by this laser-desorption method. If external injection is not necessary, RIS beams may be combined with a desorbing beam for element-selective ionization within the trap. In this case, some mass selection is possible by trap RF tuning.

4. REFERENCES

Paul W and Steinwedel H Ger. Patent 944900 (1956)
Iffländer R and Werth G 1977 *Metrologia* **13** 167
Madej A 1992 *Physics in Canada* **48** 17 and references therein.
Kwong V H S *et al* 1990 *Rev. Sci. Inst.* **61** 1931
Louris J N *et al* 1989 *Int J. Mass Spectrom. Ion Phys.* **88** 97
Moore RB and Rouleau G 1992 *Journal of Mod. Optics* **39** 361
Morand KL *et al* 1991 *Int J. Mass Spectrom. Ion Phys.* **105** 13
Lunney MDN *et al* 1992 *Jour.of Mod. Optics* **39** 349

Inst. Phys. Conf. Ser. No 128: Section 9
Paper presented at RIS 92, Santa Fe, NM, USA, 24–29 May 1992

Graphite furnace atomization with laser enhanced ionization detection

B W Smith, G A Petrucci, R G Badini and J D Winefordner

University of Florida, Department of Chemistry, Gainesville, FL 32611

ABSTRACT: A graphite furnace is used to atomize samples which are transferred in an argon carrier gas to a low noise miniflame laser enhanced ionization detection system. Two step excitation is provided by dual Nd:YAG pumped dye lasers operating at 30 Hz. The limiting noise is 6 fC and the ionization yields near 0.8. Detection limits for Mg, Tl and In, limited by temporal probing inefficiencies and losses in sample transport, are ca. 10 fg.

1. INTRODUCTION

Laser enhanced ionization in flames (LEI) has been studied as an analytical technique for nearly fifteen years. The collisionally assisted, resonantly pumped ionization which takes place in combustion flames has been used to detect at least 34 elements, generally with exceptionally good detection limits (Axner and Rubinsztein-Dunlop 1989). The mechanism is now well understood and studies have been made concerning the analysis of real samples. When two laser-pumped excitation steps are used, the selectivity is extremely good, although for complex, and especially easily ionized matrices, the detection capability suffers due in large part to modifications in the charge collection process caused by the matrix. The problems associated with measurements in complex matrices have limited applications of LEI mostly to simple aqueous systems such as lake and river waters. The generally accepted theoretical limit of detection for LEI is on the order of 1 pg ml^{-1} or about 10^4 atoms cm^{-3} (Travis et al 1982). This corresponds to an absolute detection limit of only about 10^3 atoms for a laser beam volume of 0.1 cm^3. This comes about from considering the limiting noise to be due to the fluctuation in the several microamps of current carried by the native charged species present in the flame. For several elements (e.g. Pb, Cs, Na, In, Li, Tl and Mg), this theoretical limit has been reached. This implies an ionization yield (defined as the number of charge pairs produced per atom in the probe volume) near unity and, indeed, ionization yields of this magnitude have been measured, (Smith et al 1986) and have been predicted theoretically (Omenetto et al 1986). The collisional rate in the flame is sufficient to provide complete ionization for atoms within about 1 eV of the ionization limit. The efficiency of charge collection has also been shown to be unity (Schenck et al 1981). In general, when this ultimate detection capability has not been achieved, it has been due to a poor choice of excitation wavelengths (to levels farther than 1 eV from the ionization limit), low laser pulse energy (insufficient to saturate the transitions involved) or to unusual sources of limiting noise, particularly radiofrequency noise from the firing of the pump laser. In properly designed experiments, all of these difficulties can be overcome. Because of the problems encountered in dealing with real samples in LEI, and the atomization inefficiencies associated with flames, attempts have been made to observe LEI in alternate atom reservoirs. The graphite furnace is an attractive possibility because of the great improvement in atomization efficiency, especially due to the confinement of the atomized sample to a small volume which can be closely matched to the optical probe volume. Several studies have been reported where detection was carried out by inserting a thin electrode wire into the graphite furnace tube (Magnusson et al 1986, 1987). Two difficult problems have been encountered. The high current (ca. 200 A) used to electrically heat the graphite furnace induces a large background current during the sample atomization. Thus, the limiting noise is orders of magnitude higher than in a flame. Despite the added noise, the absolute detection limits are quite good, indicating that an improvement in sampling and probing efficiency has been achieved. However, the ionization yield in the graphite furnace is poor, typically only about 10^{-5} because of the low collisional rate available in the furnace environment. The use of a Tee-shaped furnace which separated the atomization and ionization/detection processes reduced the current induced noise greatly and provided a substantial improvement (Sjostrom et al 1988). Detection limits of 1 and 2 fg were obtained for Mn and Sr, respectively. However, these might improve by 10^3 X if the ionization yield approached unity. Detection limits near single atom might then be possible. In the case of graphite furnace LEI, the benefit observed in the use of the TEE furnace has been largely one of reduced noise while the ionization yield has remained very low. In this work we have combined a graphite furnace atomization step with a flame LEI detection step in order to acquire the advantages of both graphite furnace atomization and flame ionization detection.

The graphite furnace has been interfaced successfully to microwave and radiofrequency plasma sources for analysis of small volume samples. Samples which have been vaporized in the furnace can be transported in an argon carrier gas and subsequently detected by emission, absorption or, in this case, by laser enhanced ionization. The key to designing an efficient interface is to maintain a dense atomic vapor with a minimum of dilution or sample loss. This was done using a miniflame LEI system which we have developed for use as a resonance ionization detector (Petrucci et al 1992). This device couples with very little dilution to the argon carrier gas from the furnace and presents an active volume which can be very efficiently spatially illuminated with the pump lasers. Detection of the LEI is done in the usual manner with a water-cooled stainless steel electrode immersed in the flame.

2. EXPERIMENTAL

Figure 1 shows schematically the experimental system. The Nd-YAG-dual dye laser system is that used by Petrucci and co-workers (1990) and the modified furnace design has been described previously (Crabi et al 1982). The major modification is a chamber consisting of a ring machined from boron nitride and tightly fitted in a specially shaped stainless steel housing firmly secured with O-rings between the two holders carrying the original left-hand graphite contact cylinders of the HGA-400 (Perkin-Elmer, Norwalk, CT) graphite furnace system. Two holes appropriately positioned in the housing allow sample introduction and operation of the conventional temperature sensor device. A mechanical clamp was fashioned to ensure good electrical contact between the electrodes and graphite tube. One of the window holders was replaced with a PTFE piece designed to accept a PTFE tube (0.64 cm o.d.) which was used to transfer the vaporized sample to the flame. A stainless steel ball valve, inserted in the transfer line near the furnace, was closed during sample introduction into the graphite furnace to prevent flashback of the flame. A PTFE stopper was used to block gas flow from the sample introduction hole in the graphite furnace during atomization. The small flows of transport gas (Ar) were provided from a gas tank and metering system independent of the original furnace gas supply. Carrier gas was introduced by way of the original internal and external gas inlet tubes which feed into the furnace and had been disconnected from the conventional furnace gas supply. Typical Ar carrier gas flow rates were in the range 50-400 mL min^{-1}.

The miniature flame has been described by Petrucci and co-workers (1992) and consisted of a single 0.32 cm o.d. stainless steel capillary tube pressure fitted in the center of a 2.54 cm o.d. phenolic cylinder which in turn fits into the burner sleeve of a commercial premix burner chamber. Argon carrier gas

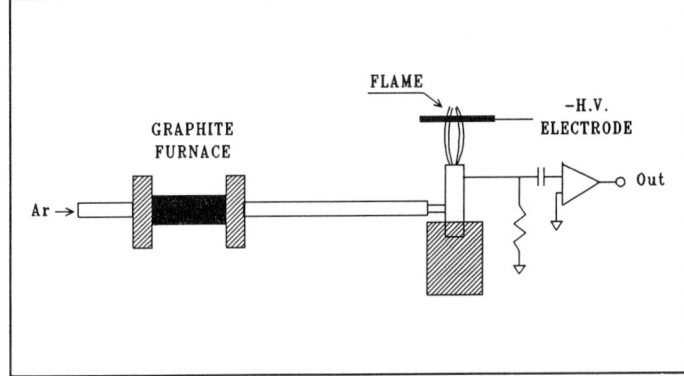

Figure 1. ETA-flame LEI experimental setup.

from the graphite furnace was introduced through another 0.32 cm o.d. tube pressure fitted through the side of the phenolic cylinder. The temperature programs used were those recommended by the furnace manufacturer.

The charge collection method used in conventional flame LEI was employed and has been described (Petrucci et al 1992). The electric field (ca. -1200 V) was applied across a 0.32 cm o.d. stainless steel water cooled electrode and the tip of the stainless steel capillary burner. The pulsed signals were collected at the burner by an ac coupled transimpedance amplifier and recorded through a boxcar averager. The system was calibrated to provide measurements of total absolute charge produced at the flame.

Studies of LEI signals as a function of gas flow rate were carried out using a burner of different design because the single capillary burner was incapable of supporting a stable flame over a wide range of gas flow rates. This burner consisted of a single capillary (0.32 cm o.d.) through which passed the entire Ar flow from the furnace surrounded by an array of capillary tubes which independently carried the premixed combustion gases. Both burners produced identical LEI signals. The single capillary version described above had lower noise characteristics and was therefore used for the analytical studies.

3. RESULTS

A study of S/N and peak shape vs carrier gas flow rate gave an optimum peak shape and magnitude for an argon flow of about 100 cc/min. An estimate of mass transport losses was made by observing the thermally produced dc flame current produced by atomizing aqueous samples of CsCl. It was found that for samples of ca. 200 ng Cs, a minimum of 15% of the introduced mass arrived at the flame. Figure 2 shows several typical ionization signals resulting from the atomization of 40 pg samples of Mg. Excellent linearity was observed for all three elements, extending over at least 3 orders of magnitude. Table 1 summarizes the analytical results. The discrepancy between the observed sensitivity and the theoretical sensitivity (96487 C/mol) is attributable to a temporal probing efficiency of 0.026 for the 30 Hz laser and to the losses in mass transport. Previous work with this LEI system has proven that the ionization yield for Mg is >0.8.

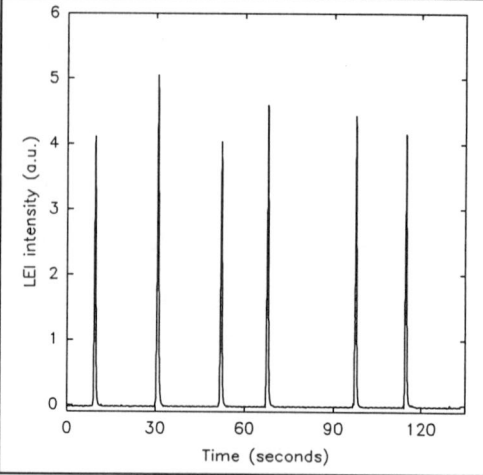

Figure 2 GFLEI signals for 40 pg Mg

Table 1

Element	ΔE (eV)	$\lambda_1 - \lambda_2$	pC/pM	LOD* (fg)
Mg	0.1	285.2 - 435.2	25.3	17
Tl	1	377.6 - 655.6	20.4	118
In	<0.1	303.9 - 786.4	6.8	260

The limits of detection were observed to improve by 10X when both laser beams are expanded in the flame in order to increase the probing efficiency.

4. REFERENCES

Axner A B and Rubinsztein-Dunlop H 1989 *Spectrochim. Acta* **44B** 835
Crabi G, Cavalli P, Achilli M, Rossi G and Omenetto N 1982 *Atom. Spec.* **3** 81
Magnusson I, Axner O, Lindgren I and Rubinsztein-Dunlop H 1986 *Appl. Spectrosc.* **40** 968
Magnusson I, Sjostrom L, Lejon M and Rubinsztein-Dunlop H 1987 *Spectrochem. Acta,* **42B** 713
Omenetto N, Smith B W and Hart L P 1986 *Fresenius Z. Anal. Chem.* **324** 683
Petrucci G A, Stevenson C L, Smith B W, Winefordner J D and Omenetto N 1990 *Spectrochim. Acta* **46B** 975
Petrucci G A, Badini R G and Winefordner J D 1992 *J. Anal. Atom. Spec.* in press
Smith B W, Hart L P and Omenetto N 1986 *Anal. Chem.* **58** 2147
Schenck P K, Travis J C, Turk G C and O'Haver T C 1981 *J. Phys. Chem.* **85** 2547
Sjostrom L, Magnusson I, Lejon M and Rubinsztein-Dunlop H 1988 *Anal. Chem.* **10** 1631
Travis J C, Turk G L and Green R B 1982 *Anal. Chem.* **54** 1006A

Inst. Phys. Conf. Ser. No 128: Section 9
Paper presented at RIS 92, Santa Fe, NM, USA, 24–29 May 1992

A highly-efficient and selective laser ion source by resonance ionization spectroscopy

F. Albus, F. Ames, H.-J. Kluge, S. Kraß, F. Scheerer, B.M. Suri*, A. Venugopalan*
Institut für Physik, Universität Mainz, D-6500 Mainz, Fed. Rep. Germany
R. Deißenberger, S. Köhler, J. Riegel, N. Trautmann, F.-J. Urban
Institut für Kernchemie, Universität Mainz, D-6500 Mainz, Fed. Rep. Germany
R. Kirchner
Gesellschaft für Schwerionenforschung, D-6100 Darmstadt, Fed. Rep. Germany
* Permanent address: Bhabha Atomic Research Centre, Bombay-400085, India

ABSTRACT: Resonance ionization of atoms confined in a hot cylindrical cavity is a very efficient and selective technique for trace analysis. Several applications of this method have been tested or are presently investigated. An ionization efficiency of 14% was obtained for trace analysis of technetium. An efficient path for resonance ionization of tin was found, leading to an autoionizing state at 59375.9 cm^{-1}. The high efficiency makes the laser ion source suitable for trace analysis of actinides for environmental studies. In all cases surface-ionized background has to be suppressed to avoid isobaric interferences. Therefore a new laser ion source has been developed with a cavity made of extremely pure pyrolytically coated graphite.

1. INTRODUCTION

In conventional resonance ionization mass spectroscopy (RIMS) the efficiency is mainly limited by the poor spatial and temporal overlap of the atomic beam with the pulsed laser beams. Hence a laser ion source (LIS) is used consisting of a hot cavity with a small hole to inject the laser beams and to extract the photoions (Kluge 1985, Andreev 1986). The atoms are confined inside the cavity and can interact several times with the laser light. As a result the ionization efficiency is significantly enhanced. In such a cavity a single atom released from the hot walls has three possibilities for escape: it can be photoionized, it can diffuse as an atom out of the cavity, or it can be surface-ionized. Whereas the first mentioned process is desired, the latter two processes limit the total efficiency of the LIS. In most cases of interest surface-ionization of the sample under investigation can be neglected. Then, the efficiency of resonant photoionization of a LIS is given by:

$$\epsilon_{LIS} = \frac{\epsilon_{photo} \cdot \nu_{rep}}{\epsilon_{photo} \cdot \nu_{rep} + \frac{v}{4 \cdot l}}. \tag{1}$$

Assuming saturation in all excitation steps one calculates an efficiency of the LIS of ϵ_{LIS}=17% for technetium. In this case, the photoionization probability of an technetium atom interacting with the laser pulse is taken to be $\epsilon_{photo} = 0.6$. The repetition rate of the used copper vapour laser is $\nu_{rep} = 6.5$ kHz, the mean velocity of the atoms $v = 750$

LASER SYSTEM   SOURCE ION OPTICS   MASS SPECTROMETER

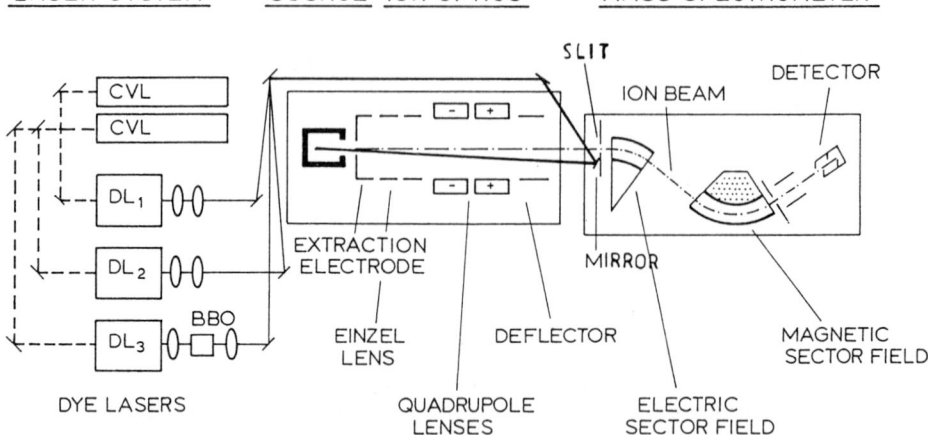

Figure 1: Experimental set-up of the laser ion source with laser system and Mattauch-Herzog mass spectrometer. CVL: copper vapour laser, DL: dye laser.

m/s, and the cavity has a length of $l = 10$ mm. Clearly, the use of lasers with a high repetition rate is essential for achieving high efficiency. A schematic drawing of the experimental set-up used is given in Fig. 1.

Such a laser ion source was tested in Mainz for the detection of trace amounts of technetium. Efficiencies up to 14% were obtained for ^{99}Tc by use of a three-step, three-colour resonant excitation scheme leading to an autoionizing state ($\lambda_1 = 312.12$ nm, $\lambda_2 = 821.13$ nm , $\lambda_3 = 670.74$ nm). However, the determination of 10^8 atoms of ^{97,98}Tc, as required for the measurement of the solar neutrino flux, was hampered by a strong interference of surface-ionized isobaric molybdenum (Ames 1991). This molybdenum is present as impurity in the wall of the tungsten cavity. In order to avoid the interference by molybdenum extremely pure pyrolytically coated graphite will be used for the construction of the hot cell as frequently applied in atomic absorption spectroscopy. Besides its extremely high purity (ash content lower than 10 ppm, no specification for impurities of molybdenum) its very hard and dense surface reduces diffusion of the sample atoms into the bulk material. Because of the high resistivity of graphite ($\rho = 24\mu\Omega\cdot m$) it is possible to heat the cavity directly by a high current ($I_{max} = 500$ A).

2. APPARATUS

A very homogeneous temperature distribution along the cavity can be reached by heating it with an electrical current applied transversely to the direction of flight of the ions. In this way, adsorption and condensation of sample atoms at colder spots can be avoided. Additionally, fast heating-up rates are possible which are essential for a quantitative and controlled volatilization of the sample atoms. The design is shown in Fig. 2. The pyrolytically coated graphite cavity ($l = 11.9$ mm, $\emptyset_{cav} = 9.4$ mm, $\emptyset_{hole} = 2.9$ mm) is placed between two spring-loaded graphite rods serving as electrical contacts. Cooling is provided by two massive water-cooled copper blocks. The whole structure is mounted on a flange for easy dismantling and sample exchange. The photoions are extracted and

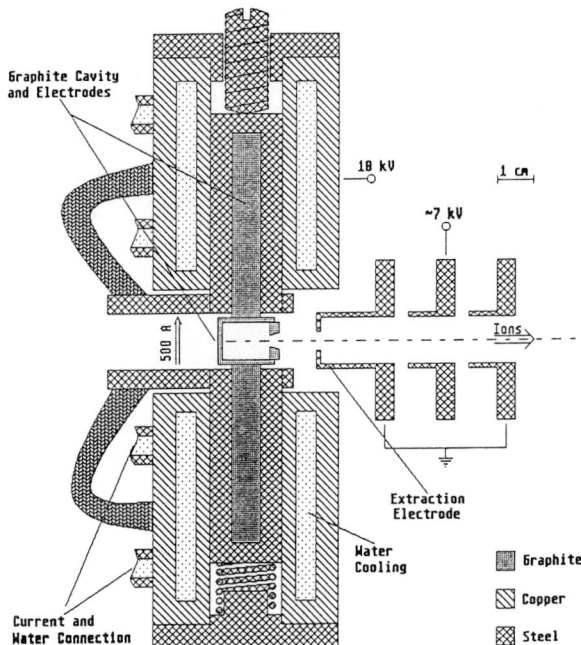

Figure 2: Sectional view of the transversely heated LIS made of pyrolytically coated graphite.

mass-separated by means of a conventional Mattauch-Herzog mass spectrometer and detected at its exit.

3. APPLICATIONS

3.1. Technetium

Trace analysis of technetium has several applications. Most exciting is the determination of the integral solar neutrino flux over the past several million years by measuring the isotopic ratio of ^{97,98}Tc/ ^{99}Tc (Cowan 1982). ^{97,98}Tc is produced by inverse β-decay from ^{97,98}Mo via the neutrino reaction

$$^{97,98}Mo + \nu_e \rightarrow {}^{97,98}Tc + e^-. \tag{2}$$

According to the standard solar model a technetium sample chemically isolated from 10000 tons of molybdenum ore (Henderson Rock mine) should contain about 10^8 atoms of ^{97,98}Tc. Furthermore, 10^{11} atoms of ^{99}Tc from the spontaneous fission of ^{238}U and about 10^{11}-10^{12} atoms of molybdenum are expected in the sample. Therefore, besides a high efficiency of the LIS, an element selectivity of at least 10^4 is required for discrimination against molybdenum contamination.

3.2. Tin

Efficient resonance ionization of tin in a laser ion source is required for a planned experiment at GSI to study the decay properties of very neutron-deficient nuclei near the double-magic ^{100}Sn, e.g. ^{102}Sn. Because of the low production rate (100 atoms/s) and interferences by isobaric indium an efficiency of at least 10% together with high element selectivity is required. We have found a three-colour, three-step resonant excitation scheme (Scheerer 1992) with $\lambda_1 = 317.51$ nm, $\lambda_2 = 811.40$ nm and $\lambda_3 = 823.49$ nm leading to an autoionizing state at 59375.9 cm^{-1}. This excitation scheme is shown in Fig. 3. Saturation in all steps was reached with the power available from the laser system used (Ruster 1989).

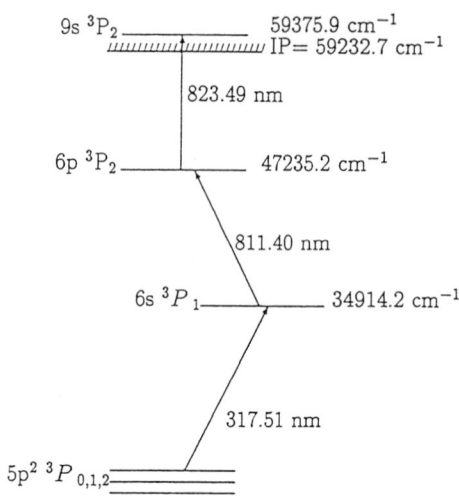

Figure 3: Three-step, three-colour resonant photoionization scheme for tin.

3.3. Trace analysis of actinides in environmental samples

Most frequently α-spectroscopy is applied to trace analysis of long-lived actinides in environmental samples. This detection method suffers from the low specific activity of the investigated isotope. For example, the detection limit for ^{239}Pu ($T_{1/2} = 2.4 \cdot 10^4$ a) by α-spectroscopy is $4 \cdot 10^8$ atoms and that for ^{237}Np ($T_{1/2} = 2.1 \cdot 10^6$ a) is even worse, namely $4 \cdot 10^{10}$ atoms. Confinement of the atoms in the hot cell of a laser ion source, providing high efficiency, together with the use of resonance ionization spectroscopy, yielding high element selectivity as well as efficiency, should improve these limits by several orders of magnitude.

This work was funded by the Deutsche Forschungsgemeinschaft.

References:

Ames F et al 1991 Inst. Phys. Conf. Ser. **114** 289
Andreev S V, Mishin V I, Letokhov V S 1986 Opt. Com. **57** 317
Cowan G A, Haxton W C 1982 Science **216** 51
Kluge H J et al 1985 Proc. on Accelerated Radioactive Beams
    Workshop ed Buchmann L and D'Auria J M TRIUMF Proc. TRI-85-1 119
Ruster W et al 1989 Nucl. Instr. and Methods **A281** 547
Scheerer F et al 1992 Spectrochimica Acta, in press

Inst. Phys. Conf. Ser. No 128: Section 9
Paper presented at RIS 92, Santa Fe, NM, USA, 24–29 May 1992

Trace analysis of oil pollution by time-resolved laser spectroscopy—first results of field measurements

W Schade and J Bublitz

Institut für Experimentalphysik, Universität Kiel, Olshausenstr. 40, D-2300 Kiel, Germany

ABSTRACT: A new application of time-resolved laser-induced fluorescence spectroscopy for detecting oil-pollution in water and in the ground is described. Hydrocarbon concentrations down to 0.5 mg/liter in hydrophobic media and in the ground down to 0.5 mg/kg soil have been measured. The method has also been tested successfully in field experiments.

1. INTRODUCTION

The trace analysis of various pollutants by measuring atomic or molecular fluorescence is superior in sensitivity to other spectroscopic methods, e.g. Raman scattering, because of the large cross sections (Measures (1974)). The results from time-resolved fluorescence measurements contain more information than those associated with the observation of pure fluorescence intensities. Different types of oil products show significantly different fluorescence decay spectra between 400 and 500 nm after excitation with UV-light (Measures et al. (1974), Rayner and Szabo (1978), Camagni et al. (1991)). In addition the decay spectra of water are quite different compared to those of oil. When measuring only intensity spectra it is almost impossible to distinguish between water and oil fluorescence.

The practical application of the time-resolved method was limited up to now because in general it requires complicated techniques for data processing and evaluation. However, the reduction of the time-resolved observation of the fluorescence in an "early" and a "late" time-window with respect to the excitation pulse simplifies this method so that it becomes very attractive for practical applications (Schade et al. (1992)).

The capability of this method for environmental trace analysis of oil-pollution is demonstrated by field measurements in an industrial sewer and by analyzing drillings which have been taken from the ground below a gasoline station.

2. THE EXPERIMENTAL METHOD

In the present set-up the excitation is performed by a sealed-off nitrogen laser with a maximum repetition rate of 30 Hz and a typical pulse duration of 400 ps. A fiber optic system is used for excitation and observation of the fluorescence. A monochromator selects the wavelength and the fluorescence signal is detected by a photomultiplier. The data processing is performed by two gated photon counters. One gate activates the first counter for the time interval 0-100 ns and the other one the second counter for the time interval 100-200 ns with respect to the laser pulse.

In figure 1 fluorescence decay curves are shown for three types of oils (a) and also for pure water (Baltic sea water) and two mixtures with different concentrations of oil in water (b). These curves show that both the pure oil samples and the oil water mixtures emit fluorescence even for times longer than 100 ns after the excitation pulse while for pure water fluorescence is only observed up to 80 ns after the laser pulse. Therefore the first counter measures the "early" fluorescence corresponding to the water, while the "late" signal of the second counter is mainly connected with the oil fluorescence. This enables us to separate even very low oil fluorescence signals from the always present water fluorescence.

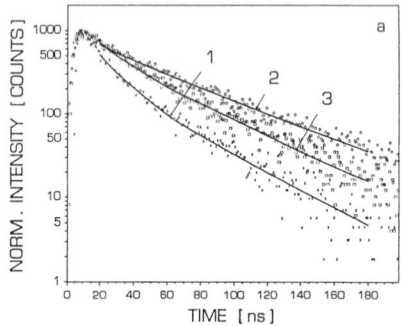 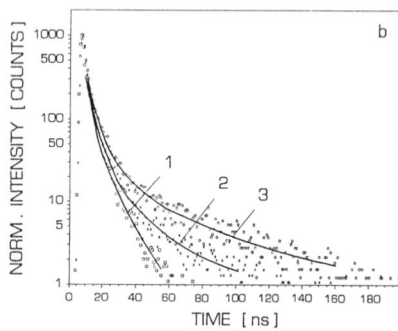

Fig. 1. (a) Fluorescence decay curves of engine oil (1), diesel fuel (2) and light fuel oil (3). (b) Fluorescence decay spectra of pure Baltic sea water (1) and two mixtures of engine oil and Baltic sea water with the concentrations 1 mg/l (2) and 10 mg/l (3). All spectra have been excited by the N_2-laser line of 337 nm and the fluorescence was observed at 400 nm.

An illustration of the integral detection of the fluorescence in an "early" and a "late" time-window is shown in figure 2a. In the present set-up the photons counted in the two time intervals I_1 and I_2 are collected over 1800 laser shots. In the case of low concentrations ($c < 100$ mg/l) the ratio I_2/I_1 can then be used for a quantitative trace analysis of oil in water.

When increasing the oil concentration in the water the ratio I_2/I_1 becomes also larger. This is explained by a more pronounced increase of the fluorescence intensity in the "late" time window when changing the portion of oil in the mixture. The intensity I_1 is here also increasing but the ratio I_2/I_1 is dominated by the increase of the intensity I_2. The results of such measurements which can be used for calibration are given in figure 2b.

Disturbance of the time evolution of the fluorescence signals due to interferences with other organic or anorganic molecules which are present in the water, such as algas, ferric-oxides etc. have also been investigated. These measurements have shown that such mixtures decay similarily to samples which contain only pure sea water. The time constants are always shorter than 10 ns. Therefore the fluorescence that is emitted by these particles is only detected by the "early" time-window. The "late" oil fluorescence is not influenced.

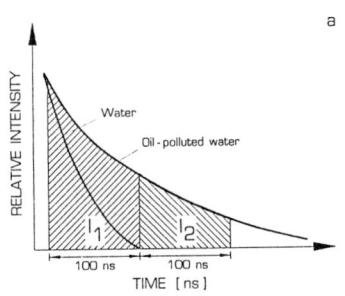

a

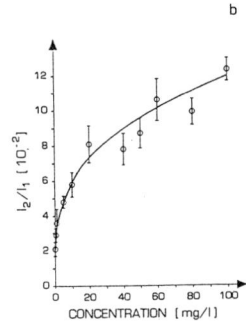

b

Fig. 2. (a) Time-integrated detection of the fluorescence signal in an "early" and a "late" time-window. (b) Ratio I_2/I_1 of an oil-water mixture for different concentrations of oil. The excitation was performed at 337 nm and the fluorescence was observed at 400 nm.

3. RESULTS AND DISCUSSION

The capability of the described method for environmental trace analysis of oil-pollution is demonstrated by field measurements in an industrial sewer and by measurements on drillings which have been taken from the ground below a gasoline station. A first example is given in figure 3 where a scheme of the sewer including the points where the measurements have been performed is given.

The pollution of the water by oil spills was analysed at ten different points. The ratio I_2/I_1 was measured on-line by averaging the fluorescence intensities in the two time intervals I_1 and I_2 for one minute. The results of these measurements are summarized in table 1. The relative values I_2/I_1 are then transformed into absolute concentrations by using the calibration curve shown in figure 2b.

An average density of oil between 1 and 5 mg/l is observed in the sewer. This is in very good agreement with the chemical analysis of the same water samples. The laboratory analysis gives an average oil concentration of 5 mg/l.

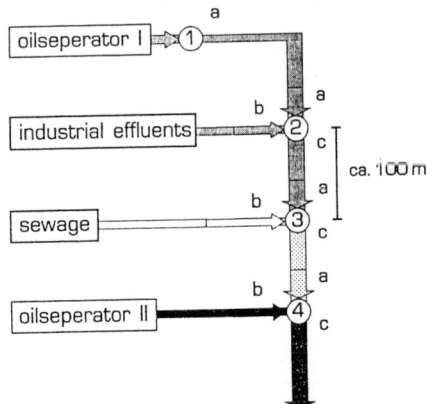

Fig. 3. Scheme for the position of the measuring points in the industrial sewer.

It is important to notice that in contrast to the chemical laboratory analysis this new laser spectroscopic method offers the possibility to detect time dependent concentration fluctuations of pollutants because of the on-line technique. The measurements at the position 4b in figure 3 show a short-term increase of the oil concentration up to 60 mg/l which was observed first by this new diagnostic. The detection limit of oil in hydrophobic media with this method is 0.5 mg/l.

The time-resolved laser-induced fluorescence spectroscopy is also suitable for the trace analysis of oil pollutions in the ground. The drillings are taken from the ground and can be analysed immediately after sampling by measuring the intensities I_2 and I_1. The ratios of the fluorescence intensities I_2/I_1 give then the relative profile of the oil concentrations which are also free from disturbances of the surrounding medium (e.g. fluorescence of the ground water). In figure 4 the evolution of the relative vertical concentration of diesel-oil in a drilling is shown. The sample is taken 1 m beside a diesel gas pump. The excitation was performed with the N_2-laser at 337 nm and the fluorescence was observered at 400 nm. From the earth's surface down to a depth of 0.4 m significant oil-pollutions are observed in the samples. In a separate chemical laboratory analysis the integral oil-concentration of the same samples has been estimated to be 2370 mg/kg soil for the depth 0 to 0.5 m. For depths below 0.5 m the concentration was below 20 mg/kg soil. The minimum oil concentration that can be detected with this method is about 0.5 mg/kg soil. The advantage of the spectroscopic method compared to the chemical laboratory analysis of samples is the on-line diagnostic and the improved spatial resolution.

	I_2/I_1 $[10^{-2}]$			Concentration $\left[\frac{mg}{l}\right]$		
	a	b	c	a	b	c
1	3.1	—	—	1	—	—
2	3.1	3.8	3.9	1	4	5
3	3.9	2.5	3.4	5	<1	2
4	1.3	9.6	3.0	<0.5	60	1

Table 1. Summary of measured intensities I_2/I_1 and absolute concentrations in the sewer.

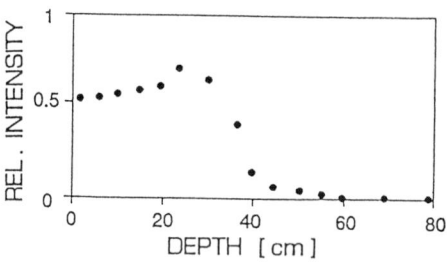

Fig. 4. Relative vertical oil fluorescence intensities in the soil 1 m beside a diesel gas pump.

4. REFERENCES

Camagni P, Colombo A, Koechler C. Omenetto N, Pan Qi and Rossi G 1991 Appl. Opt. **30** 26

Measures R M, Houston W R and Stephenson D G 1974 Opt. Eng. **13** 494

Measures R M 1979 Chemical Analysis Vol. 50, Analytical Laser Spectroscopy ed N Omenetto (New York: Wiley) pp 295

Rayner D M and Szabo A G 1978 Opt. Eng. **17** 1624

Schade W, Bublitz J, Nick K P and Helbig V in press, Laser in der Umweltmeßtechnik ed V Klein, V Klein and C Werner (Berlin: Springer)

Inst. Phys. Conf. Ser. No 128: Section 9
Paper presented at RIS 92, Santa Fe, NM, USA, 24-29 May 1992

Investigation of energy-transfer-processes in photosensitive molecules used for medical applications

W Schade[1], J A Werner[2], S Gottschlich[2], B Lippert[2] and V Helbig[1]

[1](Institut für Experimentalphysik, Universität Kiel, Germany)
[2](Klinik für Hals-, Nasen-, Ohrenheilkunde, Kopf- und Halschirurgie, Universität Kiel, Germany)

ABSTRACT: In this paper time-resolved laser-induced fluorescence spectroscopy is used to investigate energy-transfer-processes of two hematoporphyrin-derivates dissolved in aqueous solution, namely Photosan-3 and Photofrin II. The interpretation of the results by application of the Förster theory shows that the rates for these processes are in the case of Photosan-3 about 30 times larger than those for Photofrin II. Further rotational energy distributions of the excited level have been measured by time-resolved registration of the emitted fluorescence for the two extreme polarization angles $ß=0°$ and $ß=90°$.

1. INTRODUCTION

Hematoporphyrin molecules have become very important in medical applications as photosensitizers for photodynamic cancer therapy (Dougherty et al. (1978)) and in laser spectroscopic applications in the diagnostics of cancer cells (Svanberg (1989)). From clinical studies it is known that special aggregate fractions of these molecules like Photofrin II or Photosan-3 are superior to normal hematoporphyrin solutions concerning the properties of the selective accumulation and the side-effects (Moan and Sommer (1981)).
Time-resolved laser-induced fluorescence measurements give the possibility to study single aggregate fractions of these molecules. Depending on the concentration of hematoporphyrin molecules in the solution the number of decay constants in the time evolution of the fluorescence signal varies and they can be attributed to the monomeric (slow decay) and the dimeric (fast decay) component of the molecule (Andreoni et al. (1982)). Beside this the results of time-resolved fluorescence measurements can be used to discuss resonant-energy-transfer processes between the monomers and the dimers which has been described for the Photofrin II molecule by Yamashita et al. (1984).
If the excitation is performed with linearly polarized laser light and a polarizer is used in the observation channel not only the decay of the lowest singlet state S_1 but also the rotational energy distribution of the exited level in the S_1-state can be measured. This is discussed for aqueous solutions of Photosan-3 and Photofrin II.

2. THE EXPERIMENTAL METHOD

In the present investigation all samples for the fluorescence measurements are solved in aqueous solution which is phosphate-buffered at pH = 7.4. The excitation was performed with a quenched-dye-laser which is pumped by a XeCl-excimer laser (Szatmari and Schäfer (1984), Schade et al. (1987)). The wavelength of the dye laser was 500 nm. The pulse duration was about 400 ps, the spectral band width about 1 nm and the repetition rate 200 Hz. In figure 1 the basic concept of this laser system is shown.

A polarizer in the excitation and in the observation channel was used for all measurements. The relative angle ß for transmission between these two polarizers was ß = 54.7° to be free of laser-induced alignment effects on the decay curves. Only when the rotational energy distribution of the excited S_1-level was measured the decay curves were taken for ß = 0° and ß = 90°.

The reabsorption effect of the fluorescence at high concentrations was carefully avoided by measuring the fluorescence only from the front surface of the sample-cuvette. Therefore the cell was pumped at some angle with respect to its surface. The fluorescence light was focussed on the entrance slit of a monochromator and detected by a photomultiplier.

The time-resolved observation of the fluorescence was performed by a photon counting system. The subnanosecond time resolution of the detection system was achieved by the use of two constant fraction discriminators for the stop- and start-pulses of the time-to-amplitude-converter. A time resolution below 400 ps was observed. The advantages of this method are the high sensitivity and the wide range of intensities and decay times that can be measured.

3. RESULTS AND DISCUSSION

In figure 2 the results of the time-resolved fluorescence measurements for two different concentrations of Photosan-3 in aqueous solution (c = 50 mg/l (a-c), c = 100 mg/l (d-e) and different observation wavelengths (615 nm (a,d), 640 nm (b,e), 670 nm (c,f)) are shown. All decay curves have been measured for the polarization angle ß = 54.7°.

The relative amplitudes a_i and the decay times τ_i have been estimated by fitting a sum of exponentials to the decay curves. This was done by using the formalism developed by Wiscombe and Evans (1977). Up to three decay times in the range between 0.4 and 14.5 ns are observed. However, when decreasing the Photosan-3 concentrations down to c < 10 mg/l in aqueous solution a monoexponential decay with τ_0 = 14.5 ns is observed for the wavelengths 615 nm and 670 nm. This emission can be interpreted as the fluorescence of the Photosan-3 monomer which has two emission peaks at 615 nm and 670 nm, respectively. This is similar to the results of Yamashita et al. (1984) for the Photofrin II molecule. When increasing the concentration of Photosan-3 molecules in the solution up to c = 100 mg/l also fluorescence at 640 nm is observed. But then the fast component of the decay dominates and the slow component is shortened down to 8.5 ns (figure 2e).

On the other hand the comparison of absorption measurements for samples with different Photosan-3 concentrations and laser-induced fluorescence measurements has shown that similar to the results of Yamashita et al. (1984) a part of the

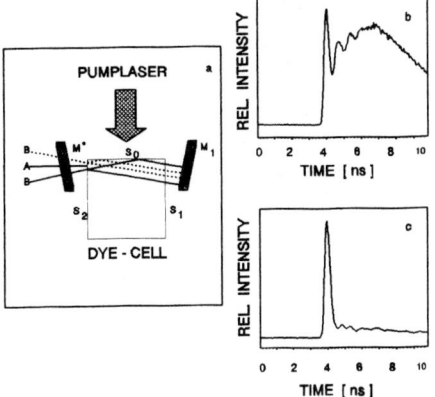

Fig.1. (a) Optical arrangement of the quenched-dye-laser. M_1: High reflectivity mirror; M^*: Quarz-plate; S_1, S_2, S_0: Windows of the quartz dye-cell. (b) The mirrors M_1 and M^* are not aligned. (c) Single pulse operation when M_1 and M^* are aligned.

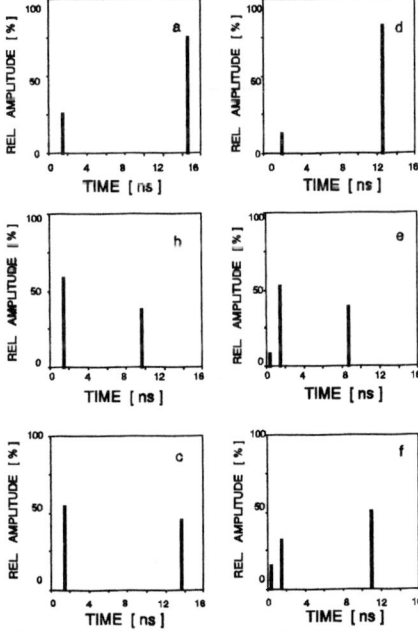

Fig.2. Summary of decay times for Photosan-3 molecules in aqueous solution at pH = 7.4 for c = 50 mg/l (a-c) and 100 mg/l (d-f). The excitation was at 500 nm and the observation of the fluorescence was performed at 615 nm (a, d), 640 nm (b, e) and 670 nm (c, f).

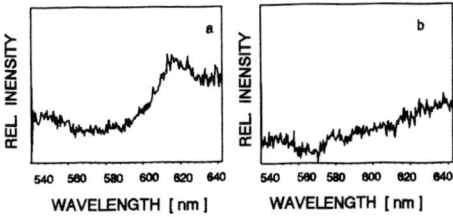

Fig.3. Laser-induced fluorescence spectra of Photofrin II (a) and Photosan-3 molecules in aqueous solution at pH = 7.4 and c = 100 mg/l. The excitation was performed at 337 nm and the fluorescence was observed between 540 and 640 nm.

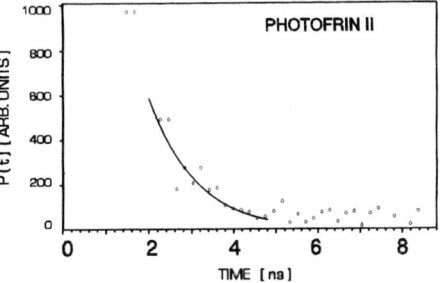

Fig.4. Decay of the laser-induced anisotropy of Photofrin II molecules in aqueous solution at pH = 7.4 and c = 100 mg/l. The excitation was at 500 nm and the observation of the fluorescence at 615 nm. A decay time of $\tau = 1.1$ ns is observed.

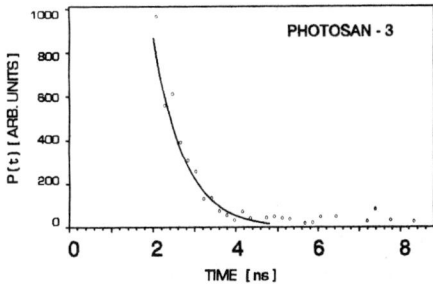

Fig.5. Decay of the laser-induced anisotropy of Photosan-3 molecules in aqueous solution at pH = 7.4 and c = 100 mg/l. The excitation was at 500 nm and the observation of the fluorescence at 615 nm. A decay time of $\tau = 0.7$ ns is observed.

fluorescence spectrum of the Photosan-3 monomer overlaps with the absorption spectrum of the Photosan-3 dimers. Therefore the shortening of the slow decay times at 640 nm in the case of increasing Photosan-3 concentrations in the solution can be interpreted by resonant-energy-transfer from the monomer molecule to the dimer. These processes can be explained quantitatively by the Förster theory (Förster (1948)). The evaluation of the time-resolved measurements by application of the Förster theory shows, that in the case of Photosan-3 the rate for these processes is about 30 times larger than the one for Photofrin II and the same experimental conditions.

This interpretation is confirmed by the laser-induced fluorescence measurements of aqueous solutions of Photofrin II and Photosan-3 molecules for $c = 100$ mg/l (refer to figure 3). While for Photofrin II a significant fluorescence signal is observed at the wavelength 615 nm (a) for Photosan-3 the monomer fluorescence is almost complete quenched at 615 nm (b).

The excitation with linearily polarized laser light induces an alignment in the excited molecular energy level, which is destroyed by collisions when the laser pulse stops. The time-resolved registration of the emitted fluorescence $I(t,ß)$ for two extreme polarization angles (e.g. $ß = 0°$ and $ß = 90°$) gives the possibility to investigate rotational energy distributions in the excited singulet state.

The time evolution of the function $P(t) = (I(0°)-I(90°))/(I(0°)+I(90°))$ gives an information about the destruction of laser-induced alignment. Results of such measurements are shown for Photofrin II and Photosan-3 in the figures 4 and 5. The data points are fitted by a single exponential and decay times of $\tau = 1.1$ ns and $\tau = 0.7$ ns are observed.

These results agree with the interpretation of the time-resolved fluorescence measurements for $ß = 54.7°$ in so far as rotational energy distribution seems to be a faster process for Photosan-3 than for Photofrin II.

4. ACKNOWLEDGEMENTS

We wish to thank Professor Müller von der Haegen for providing the Photosan-3 samples for these measurements.

5. REFERENCES

Andreoni A, Cubeddu R, De Silvestri, Jori G, Laporta P and Reddi E 1982 Z. Naturforsch. **38c** 83

Dougherty T J, Kaufman J E, Goldfarb A, Weishaupt K R, Boyle D G and Mittelman A 1978 Cancer Res. **38** 2628

Förster Th 1948 Ann. Physik **6** 55

Moan J and Sommer S 1981 Photobiochem. Photobiophys. **3** 93

Schade W, Langhans G and Helbig V 1987 Phys. Scr. **36** 890

Svanberg S 1989 Phys. Scr. **T26** 90

Szatmari S and Schäfer F P 1984 Appl. Phys. **B33** 95

Wiscombe W J and Evans J W 1977 J. Comput. Phys. 24 416

Yamashita M, Nomura M, Kobayashi S, Sato T and Aizawa K 1984 IEEE J. Quant. Electr. **QE-20** 1363

Inst. Phys. Conf. Ser. No 128: Section 9
Paper presented at RIS 92, Santa Fe, NM, USA, 24–29 May 1992

325

Cavity-enhanced photothermal measurements of chemisorbed iron on silicate dusts

E. C. Benck and H. A. Schuessler

Department of Physics, Texas A&M University, College Station, TX 77843

ABSTRACT: Cavity-enhanced photothermal spectroscopy is used to measure the relative amounts of iron chemisorbed onto the surface of silicate dust.

1. INTRODUCTION

Measurement of minute amounts of adsorbed surface contaminants is of great importance in many practical applications. We are using the novel technique of cavity-enhanced photothermal spectroscopy (Schuessler 1991, Benck 1991) to demonstrate its use for the measurement of adsorbed surface material. The technique detects the adsorbed material by measuring small changes in the optical absorption of the surface. The absorbed energy changes the index of refraction of the adjacent medium which when probed with a laser yields the photothermal signal. Measurements of optical absorptions as low as 10^{-6} are possible using cavity-enhanced photothermal spectroscopy. An advantage of the photothermal measurements is that they can be done in ambient air. Other techniques for measuring surface composition such as x-ray photoelectron spectroscopy (XPS) or energy dispersive spectroscopy (EDS) need to operate in a high vacuum.

The iron loading of silicate dust surfaces is of interest due to its potential medical application. It has been postulated that the toxicity of dust in the lung is due to chemisorbed iron on its surface (Nolan 1981, Weissner 1988). Once dust is deposited in the lung, iron from the body becomes chemisorbed onto the particles. The chemisorbed iron in turn will produce the hydroxyl radical (OH^-), which is then thought to cause the actual tissue damage. Accurate measurements of the ability of various dusts to chemisorb iron as well as the examination of dust removed from lung tissue will help to demonstrate the validity of this hypothesis.

2. EXPERIMENTAL SETUP

The cavity-enhanced photothermal apparatus shown in Fig. 1 has been previously described in detail. A cw argon ion laser is focused onto a sample surface with a microscope objective. The surface at the focal point of the laser is heated slightly by the absorption of the laser light and in turn heats the air above the sample. A probe beam from a low power HeNe laser is collimated parallel to the surface and when passing above the argon ion laser focal spot is used to detect the heated air. The effects of the heated air are greatly amplified by having the probe beam within a resonant confocal optical cavity. The optical cavity is actively stabilized by a second HeNe laser beam to eliminate the noise from ambient vibrations.

The silicate dust is composed of relatively flat planar crystals with a maximum dimension of about one micron. The samples were cleaned and placed in solutions containing a known concentration of iron ranging from 1×10^{-6}M to 1×10^{-3}M. The samples were then rinsed and dried. These dried samples were lightly pressed onto a glass slide forming an opaque layer which was used for the photothermal measurements.

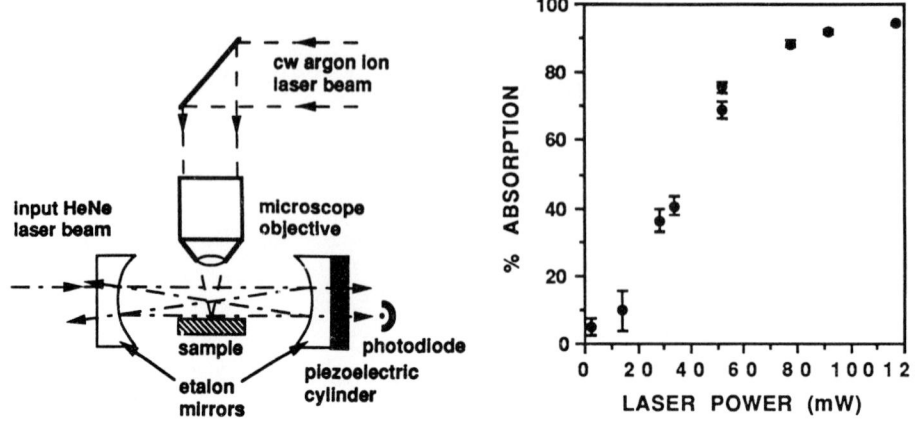

Fig. 1. Photothermal apparatus.

Fig. 2. Photothermal signal versus laser power for the silicate dust sample exposed to the 1.0mM iron solution. The laser was operating at a wavelength of λ=514.7nm.

3. RESULTS

Since clean silicate dust has a relatively high optical absorption, pump laser powers of only 60mW or less were needed. Higher laser powers result in a saturation of the signals as shown in Fig. 2. The signal due to the adsorbed iron had to be observed on top of the already large signal from the silicate substrate. Spatial variations in dust sample thickness caused signal intensities to vary by about 10% with changing position. It was therefore necessary to average measurements from several different locations on a sample.

The results of the photothermal measurements versus the initial iron solution concentrations are shown in Fig. 3. To increase the dynamic range of these measurements, the 1.0, 0.1 and 0.01 mM samples were measured at 32mW and 0.001 mM sample at 62mW. Since signals with less than 80% total absorption were measured to be essentially linear with respect to the laser power, the high laser power measurement of the 0.001mM sample could be scaled down to a value appropriate for comparison with the other samples. We also attempted to measure the iron concentrations using the more traditional techniques of EDS and XPS. We were unable to measure any iron on the dust particles using EDS on a Tracor Northern Series II x-ray microanalysis system. Using XPS, which is more sensitive to surface composition, the surface iron concentrations could be measured with a Kratos SAM800 system as shown in Fig. 4. The iron measurement of the 0.001 mM sample was just above background noise of the XPS system.

Since the required laser power was so small, observations of the photothermal signal could be made with most of the available argon ion laser lines. The results of the signal intensity versus wavelength are shown in Fig. 5. In the future, it should be possible to differentiate contributions to the photothermal signal due to iron and other potential surface adsorbates by spectrally resolving the photothermal signal over the entire visible range by employing a tunable dye laser.

This work is supported by DOE, the Teledyne Research Assistance Program and the Center for Energy and Mineral Resources at Texas A&M University.

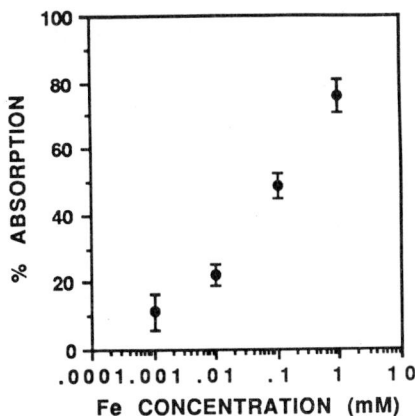

Fig. 3. Photothermal signal versus the iron concentration to which the silicate dust samples were exposed. The laser operated with a wavelength of 514.7nm and a power of 32mW. The 0.001 mM sample was actually measured with 62 mW of laser power and the result was scaled to match the other samples.

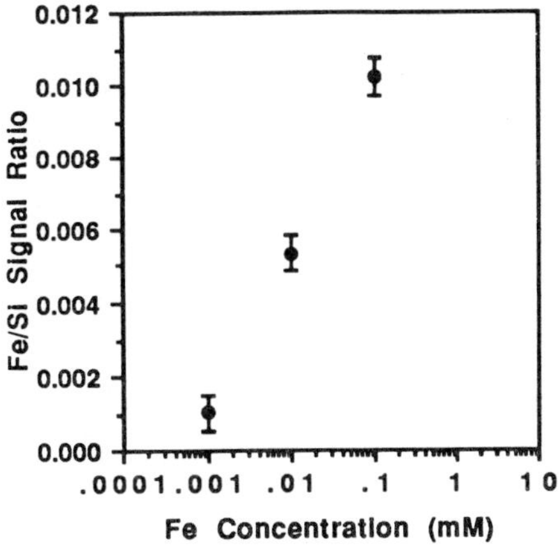

Fig. 4. Ratio of the iron to silicon signals from the XPS measurements versus the iron concentration to which the silicate dust samples were exposed.

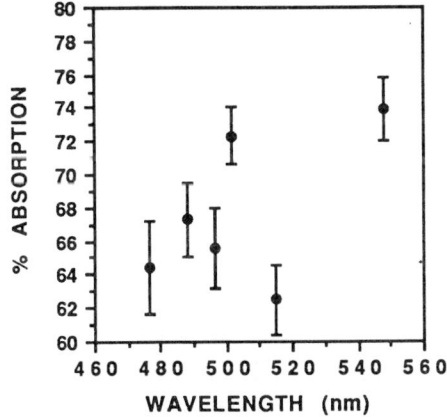

Fig. 5. Photothermal signal versus the laser wavelength. The silicate sample was exposed to the 1.0mM iron solution and measured with a laser power of 56 mW.

REFERENCES

Benck E C, Rong Z, Chen S H, Tang Z C and Schuessler H A (1991)   Appl. Phys Lett. *58* pp 1476-78

Nolan R P, Langer A M, Harington J S, Oster G and Selikoff I J (1981) Environ. Res. *26* pp 503-20

Schuessler H A, Tang J and Benck E C, (1991) in *Photoacoustic and Photothermal Phenomena III, Proceedings of the 7th International Topical Meeting* (New York-Heidlberg: Springer -Verlag)

Weissner J H, Henderson J D Jr., Sohnle P G, Mandel N S and Mandel G S (1988) Am. Rev. Respir. Dis. *138* pp 445-50

Inst. Phys. Conf. Ser. No 128: Section 9
Paper presented at RIS 92, Santa Fe, NM, USA, 24–29 May 1992

329

Detection of rare isotopes by collinear-fast-beam photon-burst spectroscopy

E. C. Benck[a], H. A. Schuessler[a], F. Buchinger[b] and K. Carter[c]

[a]Dept. of Physics, Texas A&M University, College Station, TX 77843, USA

[b]Foster Radiation Laboratory, McGill University, Montreal H3A 2B2, Canada

[c]UNISOR, Oak Ridge Associated Universities, Oak Ridge, TN 37831, USA

ABSTRACT: An experiment is presently being set up which uses collinear-fast-beam laser spectroscopy with photon-burst detection. The large artificial kinetic isotope shift characteristic of an accelerated beam, together with the Doppler-free linewidth of the fluorescence from the practically mono-kinetic fast atoms, lead to a selectivity against neighboring isotopes exceeding parts in 10^9. In addition, photon-burst detection based on photon correlations in the resonance fluorescence is employed to increase the sensitivity. The sensitivity is expected to reach the few ten atom/s level. The technique is being tested using a 10 to 50 KeV beam of krypton atoms and light from a single mode Ti-Sapphire laser at $\lambda=811$ nm.

I. INTRODUCTION

There exist a wealth of applications which require the quantitative measurement of a minute fraction ($< 10^{-9}$) of a rare noble gas isotope in the presence of the more abundant stable isotopes of the same element. They range on the one hand from basic nuclear structure studies (Otten 1988, Kluge 1991) yielding the sizes and deformations of nuclei far off stability with half lives of a few seconds to, on the other hand, determining the age and ground water mixing rates in hydrological problems (Lehmann 1988) which occurred over hundreds of thousands of years.

In this paper we describe an experimental setup based on the combination of collinear-fast-beam laser-spectroscopy (Kaufman 1976) with photon-burst detection (Letokov 1980). Conventional collinear-fast-beam laser-spectroscopy on the noble gases (Schuessler 1990) uses photon counting to detect the laser resonance and is limited by the directly scattered laser light and the finite solid angle of detection to about the 10^5 atoms/s range. Stepwise photoionization (Kudryavtsev 1988) requires ultra high vacuum (10^{-10} torr) to minimize collisional ionization by residual gas molecules. This technique is expected to be ultra sensitive but has still to be tested for the noble gases in the optimal case that narrow band cw lasers are used. Comparatively the present method needs only moderate vacuum (10^{-6} torr) and relies on photon-burst statistics to differentiate against laser stray light and other background noise.

II. THEORETICAL CONSIDERATIONS

The total isotopic selectivity of collinear-fast-beam photon-burst spectroscopy depends on the selectivity of the mass separator, and the selectivities of the laser excitation and the optical photon burst detection. These selectivities are evaluated in the following.

The selectivity of our high current mass separator was measured and is for neighboring isotopes one part in 10^5 ($S=10^5$).

The selectivity of the collinear laser excitation is determined by the width of the signals and their separations. For collinear-fast-beam laser-spectroscopy the sharpness of the Doppler-free optical signal is due to velocity bunching during acceleration to a high energy and approaches for 50 KeV noble gases atoms the natural linewidth of a few MHz. The selectivity against neighboring isotopes depends in addition on the size of the isotope shift (IS) and hyperfine structure (HFS). It is well known that the probability for off-resonance optical scattering is determined by $(2\Delta v/\Gamma)^2$, where Δv is the separation from resonance and Γ is the linewidth. For a thermal noble gas beam there is therefore only a modest selectivity due to the small IS and HFS of the states of interest. However, an additional advantage of the fast beam arises due to the artificial kinetic isotope shift which for 50 KeV noble gas atoms is several GHz in size. This guarantees a single step excitation selectivity between $S = 10^5$ - 10^6.

Photon-burst detection will also increase the selectivity, since for high photon burst rates signals with subnatural linewidths are possible.

Another important parameter required for an ultrasensitive trace element detection system is the attainable sensitivity related to the suppression of the background. Photon-burst detection achieves this by making use of the fact that for a fast recyclable transition the rate of resonantly scattered light is typically 10 MHz and therefore the instantaneous signal from an atom is about a thousand times larger than the background. This high rate during the transit time of the atom in the interaction region is the basis for photon-burst detection. Unfortunately experimental limitations arise due to the light collection efficiency and the short transit time of the fast atom beam. In the present system these factors are reduced by employing a long fiber optical array detector with a 4π light collection geometry and a length corresponding to a transit time of about 1 μsec. All the noble gases have a particular transition which is fast and recylable. In this case the statistics of resonance fluorescence of a two-level system apply and are used for a more complete analysis. In the two extreme cases of low and high exciting laser powers the probability P_n that an atom scatters a burst of n photons is a Poisson distribution. In the intermediate case it is sub-Poissonian. The following discusses only the Poissonian distribution which has the form

$$P_n = (RT)^n e^{-RT} / n!$$ (1)

Here R is the detected rate of photons scattered per atom and T is the transit time. The total photon count rate for a flux of atoms $\dot{N}$ is related to the n-fold burst probability according to

$$R_n^{atoms} = \dot{N} P_n.$$ (2)

The photon correlation described by Eq. (2) has to be compared to the corresponding one for the background pulses which is given by

$$R_n^{back} = R_b e^{-R_b T} (R_b T)^{n-1} / (1 + R_b T)(n-1)!$$ (3)

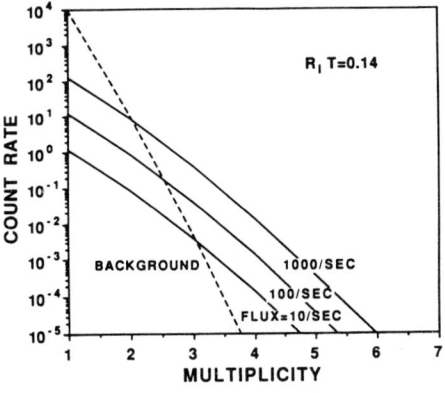

Fig. 1 displays such a comparison between the rate of photon bursts and the rate of background pulses versus their multiplicity with the atom flux as the parameter. It is shown that the photon burst method works, since the background count rate R_n^{back} decreases faster with increasing multiplicity than the total fluorescence photon count rate R_n^{atoms}.

Fig. 1. Rate of photon bursts and rate of background counts versus the multiplicity. The atom flux is the parameter.

III. EXPERIMENTAL ARRANGEMENT

In the following the example of detecting rare krypton isotopes is presented. The method works similarly for all the other noble gases. Fig. 2 depicts schematically the experimental set up. A fast Kr ion beam is produced in a Febiad ion source (efficiency $\approx 10\%$), then accelerated to 10 to 50 KeV and subsequently mass separated. The fast Kr ion beam is converted into a fast neutral beam by near resonant charge exchange in a charge-exchange cell containing an alkali metal vapor (Rb or Cs). This process preferentially populates the desired metastable $5s\,[3/2]_2^0$ level from which laser spectroscopy starts.

Fig. 3 shows in a partial energy level diagram of Kr, the energy levels of interest. A laser beam at $\lambda = 811$ nm is copropagating with the fast neutral beam. The laser frequency is kept fixed and the atoms are Doppler-tuned into resonance by applying a variable potential to the charge exchange cell. The observed fluorescence signal in the first interaction region is Doppler free. In this region we have already performed isotope ratio measurements at the 10^5 atoms/sec level and have

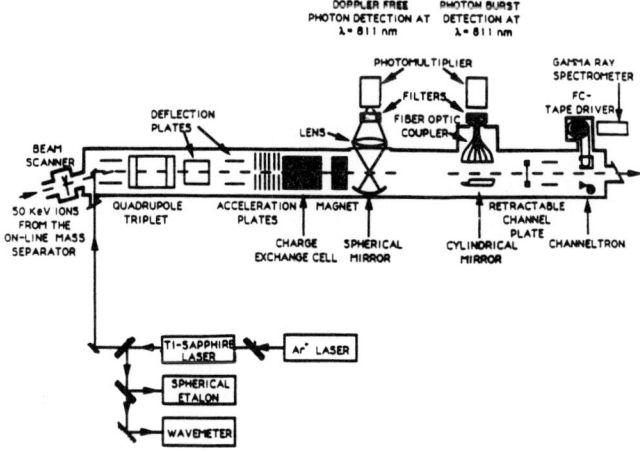

Fig. 2. Experimental setup for collinear-fast-beam photon-burst detection.

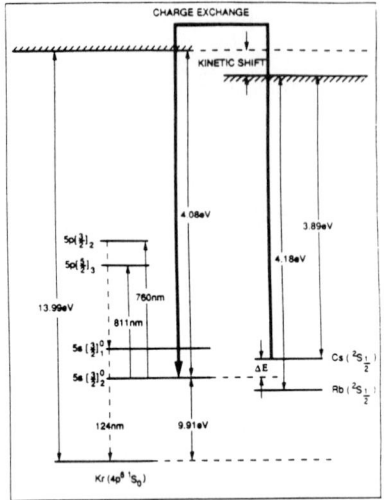

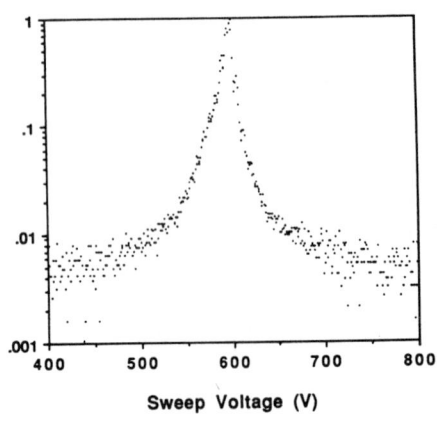

Fig. 3. Partial energy level diagram of Kr showing the population of the 5s [3/2]$_2^0$ metastable state by near reso- nant charge exchange with Cs and Rb.

Fig. 4. Fluorescence signal from a fast beam of ^{84}Kr atoms at λ=760 nm with signal integration of 10 ms/point. The S/N ratio is about 200.

obtained a laser excitation selectivity of S ≈10^4. Fig. 4 displays a conventional fast beam fluorescence signal at λ=760 nm plotted on a logarithmic scale. A signal-to-noise ratio of 10^4 will be reached when integrating for 25 s/point. This demonstrates that so far the combined selectivity for our mass separated collinear fast beam laser spectroscopy apparatus is already about 10^9. However, at λ=811 the transition is recyclable and no optical pumping occurs in the first interaction region so that subsequently in the second interaction region in addition the more sensitive photon-burst detection method can be implemented. This should increase the sensitivity to the few ten atom/s range.

This work is supported by the Teledyne Research Assistance Program, DOE, the Center for Energy and Mineral Resources at Texas A&M University, and the Canadian Natural Science and Engineering Research Council.

REEFERENCES

Kaufman S L (1976) Opt. Comm. 17, pp 309.
Kluge H J (1991), in Resonance Ionization Spectroscopy 90, Inst. Phys. Conf. Ser.114, Bristol pp 439.
Kudryavtsev (1988) Opt. Comm. 68, pp 25.
Lehman B E (1988), in Resonance Ionization Spectroscopy 88, Inst. Phys. Conf. Ser. 94, Bristol pp 213.
Letokov V S (1980) Appl. Phys. 22, pp 245-48.
Otten E W (1988), in Nuclei far from Stability, Treatise in Heavy Ion Physics, Vol.8, Plenum, New York pp 515.
Schuessler H A (1990) Phys. Rev. Lett. 65, pp 1332-34.

Inst. Phys. Conf. Ser. No 128: Section 9
Paper presented at RIS 92, Santa Fe, NM, USA, 24–29 May 1992

333

Resonance detection of photons by atomic ionization

G A Petrucci, R G Badini and J D Winefordner

Department of Chemistry, University of Florida, Gainesville, FL 32611-2046

ABSTRACT: A resonance ionization detector (RID) based on the two-step laser-enhanced ionization of Mg in a miniature acetylene/air flame is described. The detector utilizes the 285.213 nm resonance absorption of Mg as the signal transition. The minimum detectable number of photons (MDP), obtained for the excitation scheme $3s^2\,{}^1S$ (285.2 nm) → $3p\,{}^1P°$ (435.2 nm) → $6d\,{}^1D$, was determined to be 1×10^3 (7×10^{-16} J). The quantum efficiency of this excitation scheme was found to be 0.75. The detector had a stray light rejection (SLR) ratio of approximately 10^{-5} at 100 cm^{-1} displacement from the absorption maximum of the RID at 285.213 nm. The RID is used to record the Raman spectrum of dimethylsulfoxide (DMSO).

1. INTRODUCTION

An interesting variation of the atomic ionization technique is the utilization of the laser induced ionization process as a sensitive photon detector. This concept, first proposed several years ago (Matveev 1979), has received considerable attention (Ganeev 1987; Omenetto 1989) owing to its potential applications. In a resonance ionization detector an atom excited by absorption of a photon resonant with a transition (1→2) of the atom M is further excited by a laser tuned to the excited-state transition (2→3) of that atom, λ_2, promoting M* into a higher energy state (E_3), yielding a highly excited atom. These highly excited atoms are then collisionally ionized or photoionized. The magnitude of the ionization signal is directly proportional to the number of incident signal photons absorbed and hence, the RID behaves as a spectrally selective photon detector, centered at the signal transition, $h\nu_{sig}$. The attractiveness of such an approach is mainly based in its high light throughput, very narrow frequency response bandwidth and subsequently high stray light rejection.

Different general optical schemes have been proposed for the RID (Okada 1989; Bloom 1990). The RID element must possess several favorable characteristics: (i) The number density of atoms in level 1 of the signal transition (usually ground state) must be high enough to ensure complete absorption of all incident signal photons. The signal transition should also possess a favourable oscillator strength (>0.5); (ii) the thermal population of RID atoms in the first excited state (level 2) of the RID scheme must be minimal to reduce the number of thermal ions; (iii) the excited state transition, 2→3, should be of low energy to minimize multiphoton ionization of the RID element (from level 1) or flame species. The oscillator strength of this transition should also be relatively high (>0.05) so that optical saturation can be effected at modest laser powers; (iv) the RID element must possess a possible excitation scheme such that the energy difference, ΔE, between the uppermost laser-populated level and collisionally coupled levels and the IC, be no more than 3 kT to ensure a high probability of collisional ionization (Axner 1989). For the LEI approach to be used as a photon detector, it is imperative that atoms are ionized with 100% efficiency from level 2.

Both the resolution and stray light rejection (SLR) of the RID are fixed by choice of RID element, wavelength scheme and atom cell and are given directly by the absorption profile of the signal transition in the RID. In a flame-based RID, Lorentzian-dominated profiles of the RID element signal transition can yield resolutions on the order of picometers with corresponding SLRs of 10^{-5} at 100 cm^{-1} displacement from λ_0.

An RID based on the two-step LEI of Mg in a miniature air/acetylene flame is described. In all cases, the 285.213 nm resonance transition of Mg was considered as the first (signal) transition. From this first excited state, several different excitation/ionization schemes were studied. Table 1 lists each scheme with corresponding transitions and associated energy levels.

Table 1. Transitions and energy levels considered.

Scheme	λ_2 (nm)	Transition	E_3 (cm^{-1})	ΔE (cm^{-1})
A	571.1	3p ^{1}P$^\circ$ → 5s ^{1}S	52556	- 9113
B	300.9	3p ^{1}P$^\circ$ → continuum	68285	+ 6616
C	435.8	3p ^{1}P$^\circ$ → 6d ^{1}D	58023	- 3646

A - (+) indicates a final energy level below (above) the ionization limit of the atom.

2. EXPERIMENTAL

The experimental set-up is shown in Figure 1 and has been described elsewhere (Petrucci 1992). The burner used to support the air/acetylene flame consisted of a 1.9 mm i.d., 2 cm long stainless steel tube pressure fitted into a phenolic base and was fitted into a conventional atomic absorption premix spray chamber. Due to the high back pressure from such a small flame orifice, Mg solution was introduced with a laboratory-constructed ultrasonic nebulizer and was carried into the flame in the air stream. The concentration of the Mg solution in the nebulizer was 100 ppm, and the nebulization rate was increased until >

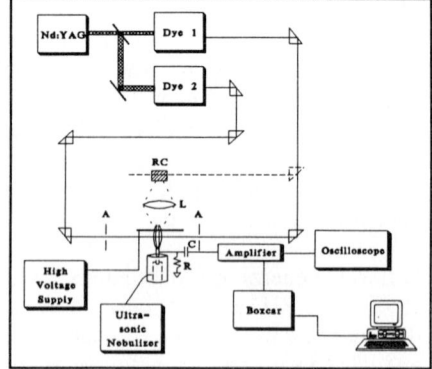

Figure 1. Experimental setup.

95% absorption of the 285.2 nm line from a Mg hollow cathode lamp was obtained.

The wavelength response of the RID was obtained by placing an aluminum plate at the Raman sample position (RC in Fig. 3) to act as a diffuse scatterer, imaging the scatter into the RID. To increase the signal-to-noise of these measurements, the profile determination was performed with the second laser also incident on the flame and tuned to the excited state transition at 435.2 nm.

The Stokes shifted Raman spectra were obtained by scanning the exciting laser at a fixed rate toward the Mg transition maximum at 285.2 nm. The scattered photons were collected and imaged into the flame with a doublet lens. The second laser was tuned to one of the excited-state transitions of the RID. The energy flux of the detection laser was maintained at ~ 1 mJ cm^{-2} to optically saturate the excited-state transition. The energy flux of the exciting (Raman) laser, over the wavelength range scanned (279-286 nm), was approximately 10 mJ cm^{-2}.

The Raman shift is given by $\Delta\tilde{\nu} = \lambda_1^{-1} - \lambda_{12}^{-1}$, where λ_1 is the wavelength of the exciting laser and λ_{12} is the wavelength of the first (signal) transition of the RID element (i.e. 285.2 nm). Since λ_{12} is fixed by the choice of RID element and signal transition, a Raman spectrum is obtained by scanning the exciting laser. When the energy difference between the exciting photons and the energy of the signal transition is equal to a Raman shift of the sample, an enhanced ionization rate is observed in the RID.

3. RESULTS AND DISCUSSION

In the first scheme studied (A), the excited state transition was to the 5s ^{1}S state, occurring at 571.2 nm. In this scheme, the Mg atoms are excited to a final level well below the IC. The quantum efficiency, η, and MDP obtained were 0.069 and 2.4x10^4 photons (17 fJ at 285.2 nm), respectively. Since the excited state transition was optically saturated, η is less than unity most likely because of the large energy defect and correspondingly low ionization probability from level 3. The limiting noise for this scheme was determined to be that of the transimpedance amplifier used.

A potential approach for increasing η is by state-specific ionization of Mg atoms in the 3p^{1}P$^\circ$ state *via* an autoionizing level, a.i. An atom excited to a <u>bound level above</u> the IC must necessarily decay into the continuum with unity probability (Scheme B). The autoionizing transition from the 3p^{1}P$^\circ$ level occurs at 300.9 nm and has a cross-section of 3x10^{-16} cm^2. This is considerably larger than the cross-section for non-specific photoionization. The MDP and η for this scheme were 3x10^5 photons and 0.0049, respectively. Both figures of merit are poor compared to Scheme A.

Due to the still modest (autoionizing) cross-section, the transition could not be saturated with available photon irradiances. As a result, η will be linearly dependent on the energy of the ionizing laser. Therefore, to maximize η, the largest available photon irradiance at 300.9 nm was used. This however, led to a substantial increase in the multiphoton ionization background, since ground state Mg atoms could be ionized by the simultaneous absorption of two photons at 300.9 nm.

The third scheme studied promoted Mg atoms from the 3s^{1}P$^\circ$ state to the 6d^{1}D state by absorption of photons at 435.2 nm. This excited state transition has a relatively large cross-section of 4x10^{-13} cm^2 and promoted the Mg atoms to within 3600 cm^{-1} (~ 2 kT @

2500 K) of the IC. The high probability of ionization from an energy state with such a low energy defect is reflected in the high η of 0.75 obtained with this scheme. A LDR of over 5 orders of magnitude is obtained. Multiphoton ionization is negligible since three photons at 435.2 must be absorbed to ionize ground state Mg atoms directly. This serves to decrease the noise, which, together with the high η, yielded an MDP of 1000 photons. The normalized spectral response profile is shown in Figure 2. The SLR at 100 cm^{-1} displacement from $\tilde{\nu}_0$ is ~3x10^{-5}. The curve is a best-fit using a Lorentzian equation and a FWHM of the resonance absorption line of 0.43 cm^{-1}.

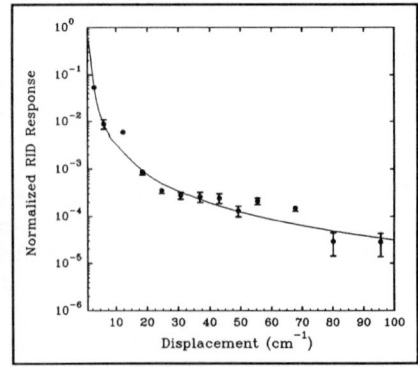

Figure 2. Response profile of RID.

The Raman spectrum of DMSO was recorded with the flame RID operating with Scheme C. The simple lens system used allowed for collection of only ~1.5 % of the signal photons. Clearly, for the RID, which has no limiting aperture, Ω_{exc} could be increased to greater than 2π sr. The Raman spectrum shown in Figure 3 was obtained with a single scan of the excitation laser. A one-second time constant was used. The scan took approximately 15 minutes.

Acknowledgements--Research supported by DOE-DEOFGO5-88ER13881. The authors gratefully acknowledge Dr. N. Omenetto and Dr. B.W. Smith for their many helpful discussions. One of the authors (RGB) would like to thank the Consejo Nacional de Investigaciones Científicas y Técnicas de la República Argentina for the fellowship which made possible his stay at the University of Florida.

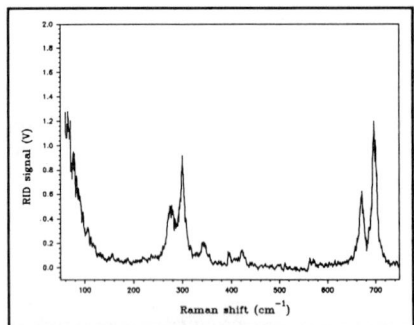

Figure 3. Raman spectrum of DMSO obtained with RID.

4. REFERENCES

Axner O and Berglind T 1989 *Appl. Spec.* **43** 940-952
Bloom S H, Korevaar E, Rivers M and Liu C S 1990 *Opt. Lett.* **15** 294-6
Ganeev A A, Matveev O I and Sholupov S E 1987 *J. Anal. Chem. USSR* **43** 1424-9
Matveev O I, Zorov N B and Kuzyakov Y Y 1979 *J. Anal. Chem. USSR* **34** 654-662
Okada T, Andou H, Moriyama Y and Maeda M 1989 *Opt. Lett.* **14** 987-9
Omenetto N, Smith B W and Winefordner J D 1989 *Spectrochim. Acta B* **Special Supplement on Proceedings of the Symposium held at Pontifical Academy of Sciences 27-28 June** 101-111
Petrucci G A, Badini R G and Winefordner J D 1992 *J. Anal. At. Spectrom.* in press

Inst. Phys. Conf. Ser. No 128: Section 9
Paper presented at RIS 92, Santa Fe, NM, USA, 24–29 May 1992

Laser-based studies with an ion-trap mass spectrometer: ion tomography and analytical applications

M. L. Alexander, M. E. Cisper, P. H. Hemberger, N. S. Nogar
MS J565, Los Alamos National Laboratories, Los Alamos, N. M. 87545
J. D. Williams, Dept. of Chemistry, Purdue University, West Lafayette, In. 47907
J. E. P. Syka, Finnigan MAT, San Jose, Ca. 95134

1. INTRODUCTION

The ion trap mass spectrometer (ITMS) is an ion storage device which consists of two hyperbolic endcaps and a hyperbolic ring electrode[March and Hughes,1989]. This forms a trapping cavity having a volume of several cm^3. An rf potential applied to the ring electrode produces a time-varying potential which can be used to trap and/or manipulate ions under controlled conditions. This device has been used in ion trapping studies for a number of years. More recently, a commercial version has been produced and sold which allows for mass-selective ejection of trapped ions, with subsequent detection by an electron multiplier. In this mode, it operates as a compact, high efficiency, high resolution mass spectrometer. The instrument has found applications in GC/MS, in tandem mass spectrometry and in portable mass spectral analysis Hemberger et. al., 1991; Todd,1991].

In this manuscript, we present a survey of recent results incorporating laser desorption, ionization, or photodissociation with ITMS. In one instance, we describe the use of laser photodissociation to map the spatial distribution of trapped ions in the ITMS. In this tomographic study, we have parameterized the effects of trapping potential, buffer gas pressure, supplementary rf-potential, and laser intensity. In separate studies, laser desorption was used to generate gas phase ions in the ITMS from a solid probe, by irradiation of both neat and matrix-dissolved samples. The latter experiment produced both high molecular weight ions, and significant numbers of negative ions.

2. EXPERIMENTAL

A unique ion trap mass spectrometer system that incorporates an ion optical/laser optical platform within a large (1 m diam) vacuum chamber was developed and built for these experiments. This system was designed specifically to support the development of novel instrumental configurations and combined laser-ion trap studies. The apparatus is based on a Finnigan ITMS and a Questek XeCl excimer laser, as shown below. Details of this experiment may be found elsewhere[Hemberger, et al, 1992]. Ion trap and control electronics were used as supplied by Finnigan.

For tomography experiments, the ring electrode had a pair of slots machined parallel to the z-axis on opposite sides of the electrode, to allow for optical entry and exit; the electrode assembly was otherwise intact. The laser beam was apertured and focused through the slots in the ion trap. The final turning mirror and focusing lens were mounted on a precision translation stage which allowed translation of the beam along the length of the electrode slot. Typical incident laser energies for tomography were $\approx$ 1 mJ, focused to $\approx 10^{-2}$ cm ^{2} in a15 nsec pulse.

In laser desorption experiments, the beam is focused near the surface of a sample probe which is inserted though the ring electrode and is adjacent to the inner wall of the electrode. Both stainless steel and machinable glass sample platforms were used. In this case, typical pulse energies were $\approx$ 0.5-1.0 mJ.

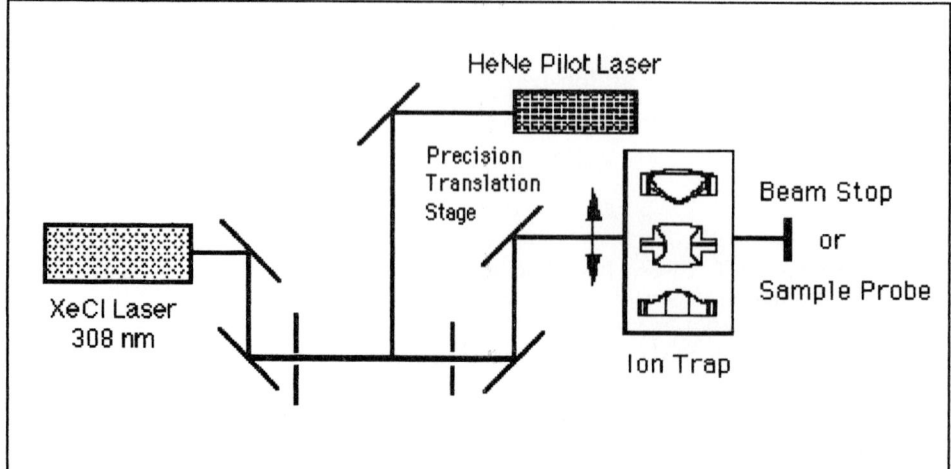

Figure 1 depicts a chematic of the experimental apparatus. For tomography experiments, the beam is translated along the z-axis of the ion trap, with spatial integration along the r-axis.

3. RESULTS AND DISCUSSION

3.1 Ion Tomography

The electrodes used in ion traps are necessarily imperfect. In addition to the inherent difficulties in fabricating hyperbolic surfaces, intentional distortions must be introduced. The endcaps have a number of perforations, in order to allow reagent gases and electrons in, and ions out. Because of these imperfections, it is impossible to calculate with high precision the properties (velocity, position and their time dependence) of the stored ions. It is therefore difficult to model the performance of the trap. We have undertaken experimental studies to verify theoretical models and calculations. In preliminary work, $C_6H_5CO^+$ was generated by electron impact from acetophenone. This ion was photodissociated by the incident 308 nm radiation, and the fragment ion at m/z=77 Da monitored by ion ejection. By monitoring the signal at m/z=77 as a function of laser beam position, we were able to map the spatial distribution of the $C_6H_5CO^+$ parent.

The number density of stored ions was measured as a function of position (z-resolved, radially and phase averaged), and controllable trap parameters such as potential well depth, supplemental rf voltage, and gas pressure. The intrinsic parameters that govern motion of ions in a quadrupole ion trap are the mass-to-charge ratio and velocity of the ion, the radial and axial dimensions of the ion trap electrodes (r_o and z_o, respectively), the magnitude and frequency of the rf voltage (V and Ω, respectively), and the magnitude of any dc component, U, applied to the electrodes. These variables define the dimensionless Mathieu parameters a_z and q_z, where $a_z = -8eU/mr_o^2\Omega^2$ and $q_z = 4eV/mr_o^2\Omega^2$. Figure 2 shows the dependence of the daughter ion signal as a function of buffer gas pressure (left) and of trapping well depth (right).

Clearly, the addition of helium results in a more sharply peaked, and greater intensity signal; this is due to stabilization of ion trajectories, and greater trapping efficiency, through collisional relaxation. The distribution variation with welldepth shows that ions are trapped more

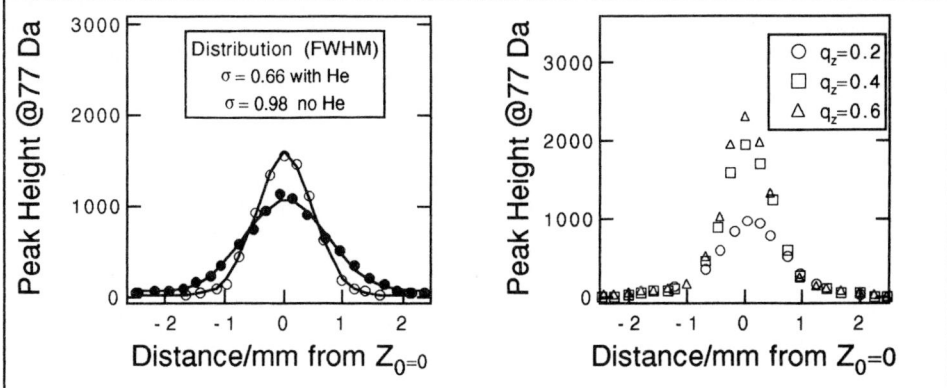

Figure 2 (left) Shows the distribution of ions as a function of He pressure (open circles with He, closed without) for $q_z=0.4$ while the right side shows the distribution as a function of trapping well depth.

efficiently, and in smaller volumes, in relatively high fields. In further studies, we have measured distributions for a number ions, examined the effect of supplemental rf (tickle) voltage, and quantified the effect of additional ions on the measured distribution.

3.2 Negative Ion Laser Mass Spectrometry

The detection, identification, and characterization of polycyclic aromatic hydrocarbons (PAH's) are problems of immediate significance. PAH's are a class of toxic and potentially carcinogenic compounds often resulting from hydrocarbon processing. Conventional mass spectrometry is often able to provide significant information concerning the molecular weight of structurally significant units, but may have difficulty in distinguishing one species from another[Oehme,1983; Stemmler and Buchanan, 1988]. We present here data on mass spectrometry of species generated from laser ablation of a mixture of about a dozen PAH's (composed of from two to four aromatic rings, and ranging in size from $C_{10}H_8$ to $C_{22}H_{12}$).

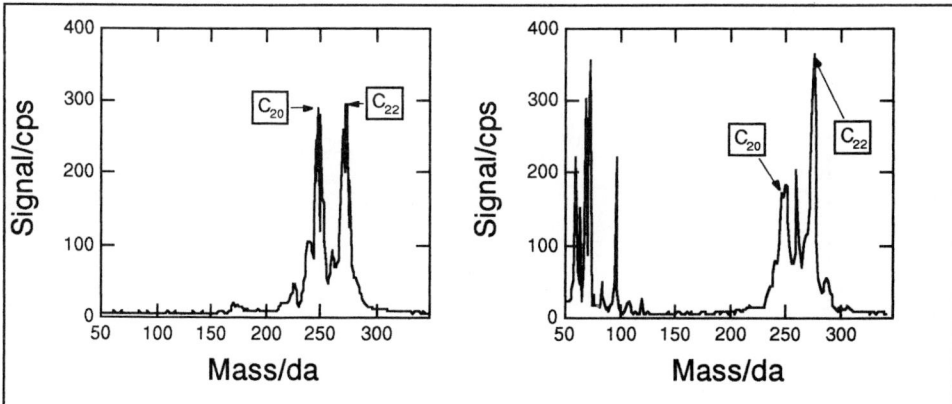

Figure 3 (left) shows the positive-ion mass spectrum generated by laser desorption/ionization of a sample of mixed PAH's; (right) negative-ion mass spectrum generated under similar conditions from the same sample.

Several features should be noted from these spectra. First, there is a very pronounced intensity altering (at high masses) for the positive ion spectrum, which appears to be absent from the negative ion spectrum. This likely reflects "magic number" structures of high stability in the positive species, structures which are apparently lacking in the negative ions. Secondly, the negative ion spectrum clearly exhibits more information at low molecular weights than does the positive ion spectrum. In order to explore this second feature in more detail, we present below an expanded view of the low mass region for the negative ion spectra.

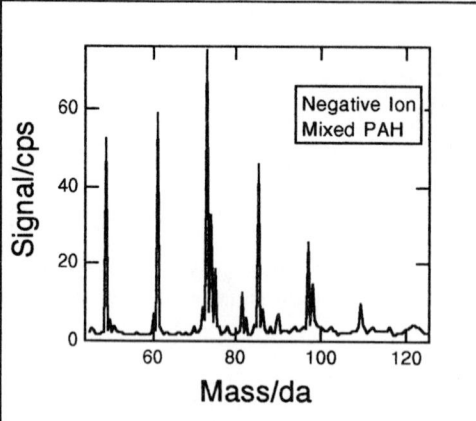

Figure 4 shows detail of the low-mass spectrum for negative ions generated from laser desorption/ionization of a sample containing mixed PAH's.

The highly structured spectra observed at both high and low mass in the laser desorption negative ion mass spectrometry suggest that this may be a promising method for the detection of mutagenic and(or) carcinogenic polycyclic aromatic hydrocarbons (PAH's).

In addition, we observed unusual behavior near the (energy) threshold for production of negative ions from the PAH mixture described above. The figure below shows the total ion intensity as a function of incident laser pulse energy for two chemical systems.

The behavior shown in the left figure suggests the presence of several thresholds, perhaps indicating different thresholds for different compounds, or classes of compounds. For the pure compound on the right, the signal for both positive and negative ions shows a smooth onset as a function of laser energy, and no evidence of secondary threshold(s). The data on the PAH mixture suggests a possible method for distinguishing similar compounds, or classes of compounds based on their threshold behavior for laser-induced negative ion formation.

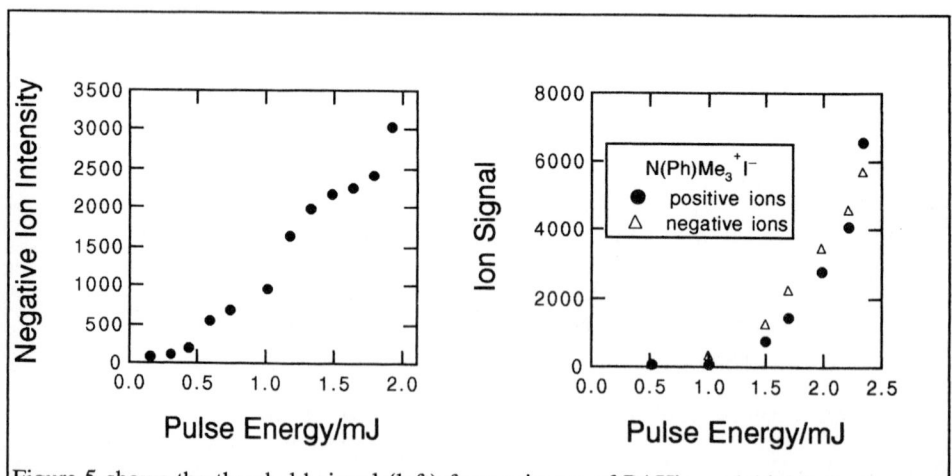

Figure 5 shows the threshold signal (left) for a mixture of PAH's, and (right) for the neat sample of trimethyl-phenyl ammonium iodide.

4. CONCLUSIONS

The advantages of laser-based ITMS, relative to conventional mass spectral strategies, are numerous. First, the ITMS is a relatively simple apparatus: the design and operation of the ITMS make it much more compact, and as much as an order of magnitude less expensive than the multistage (triple quad or tandem double focussing) instruments traditionally used for structural determinations. Second, we have demonstrated the sensitivity and potential selectivity of ITMS in combination with laser desorption techniques for the analysis of complex, nonvolatile organic molecules. The combination of easy optical access and multiplex detection capabilities should routinely allow sub-picomole sample sizes for large molecules. Third, numerous capabilities are available for structural analysis: collision-induced dissociation (CID), surface-induced dissociation(SID) and photodissociation (PD) can all be used to fragment large molecules for structural analysis. Again, this contrasts with conventional instruments, where only one mode of operation is easily accessed. Lastly, by allowing successive PD or CID processes on a single ionic species, sequential-in-time MS/MS can be extended for several "generations" in ITMS to increase structural specificity or provide sequencing information.

5. REFERENCES

Hemberger P H,Alarid J E, Cameron D, Liebman C P, Cannon T M, Wolf M A, Kaiser R E, 1991 *Int. J. Mass Spec. Ion Proc.* **106** 299-313.

Hemberger P H, Nogar N S, Williams J D, Cooks R G Syka J E P 1992 *Chemical Physics Letters* **191** 405-410.

March R E Hughes R J 1989 *Quadrupole Storage Mass Spectrometry* **102** (New York :Wiley-Interscience) 471.

Oehme M 1983 *Anal. Chem.* **55** 2290-5.

Stemmler E A Buchanan M V 1988 *Rapid Commun. Mass Spectrom* **2** 184-8.

Todd J F 1991 *Mass Spec. Rev.* **10** 3-52.

Inst. Phys. Conf. Ser. No 128: Section 9
Paper presented at RIS 92, Santa Fe, NM, USA, 24–29 May 1992

343

Plasma wave spectroscopy: a novel technique in resonance ionization spectroscopy

J Franzke, D Veza and K Niemax

Institut für Spektrochemie und Angewandte Spektroskopie (ISAS), Dortmund, Germany

ABSTRACT: The measurement of frequency changes of plasma waves in low-pressure discharges due to ion density variation by resonant laser excitation allows sensitive and selective detection of impurities in the plasma. The analytical applicability of this novel technique has been studied in argon and neon discharges.

1. INTRODUCTION

It is well known that dc low pressure discharges may generate plasma waves of very stable amplitude and frequency (Boyd and Sanderson 1969). Different types of waves can simultaneously exist in plasmas (Denisse and Delcroix 1963). In the case of low-pressure discharges, pseudosonic waves (PSW) of relatively low frequency (12-250 kHz) can be easily separated and monitored. The frequency of PSW depends on the geometry of the tube, the gas pressure, the voltage applied, the type of gas composition and the gas flux. If one of these parameters is varied, for example the ion density, the wave amplitude and frequency will be different. Recently, a PSW-arrangement has been successfully applied as a very sensitive non-selective detector in gas chromatography by Kuzuya and Piepmeier (1991). About 20 fmol of propanol in argon could be measured. However, PSW-detectors can also be used for selective measurement of analytes if resonance ionization spectroscopy is applied. It is known from optogalvanic spectroscopy that resonant laser excitation of species changes the charge densities in discharge plasmas. Changes of the charge density effect the PSW-frequency and spectra can be measured (Popescu et al 1991). As frequency can be measured with much better precision in comparison with amplitude, a technique which combines the measurement of plasma waves with selective laser excitation is expected to be a very sensitive detection method.
In preliminary studies of the potential of plasma wave spectroscopy we have investigated pure and mixed noble gas discharges (Ar, Ne) inducing one- and two-photon transitions from metastable levels. In order to test the PSW detection power, we have compared the PSW-spectra with absorption as well as optogalvanic signals.

2. EXPERIMENT

The experimental arrangement is shown in Fig. 1. The discharge (DC mode) was burning between two exchangeable flat stainless steel electrodes (distance: 15 cm) in a glass tube (diameter: 20mm) with end-on windows. The noble gas flow through the tube was about 5 cm^3/s and the pressure about 0.5 mbar. The discharge current density of about 0.3 mA/cm^2 was regulated by a current sink. The discharge was probed by laser beams in axially (through small holes in the electrodes) or in transverse direction (through the glass walls). Single mode diode lasers as well as dye lasers were used for excitation. The PSW were detected by a small, movable copper ring around the discharge tube. The laser induced frequency was converted into voltage signals by a frequency to voltage converter and processed by a lock-in amplifier. The amplified signal was recorded by a personal computer (PC) or a strip chart recorder.

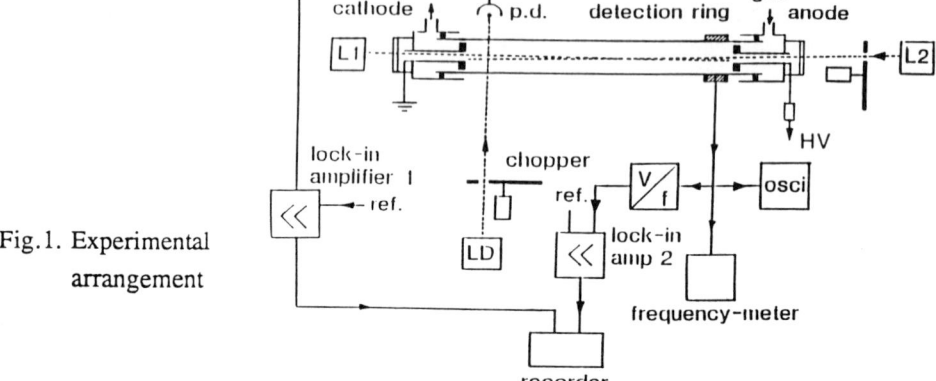

Fig.1. Experimental
arrangement

3. RESULTS AND DISCUSSION

Plasma wave spectroscopy was performed in Ne as well as in Ar glow discharges starting with transitions from the $J=2$ component of the metastable states. The spatial distribution of the metastable atoms was not uniform in the discharge. This is shown in Fig. 2 where the dependence of the transverse absorption of Ar metastable atoms on the distance from the cathode is displayed. The corresponding PSW-signal is shown in Fig. 2b. The largest metastable densities were detected in the positive column. The periodic structure in the metastable density followed the striations in the positive column which can be seen with the naked eye. In argon, the maximum absorption (per 1 cm) was about 60% which corresponds to a metastable density of about 1×10^{11} cm^{-3}. In neon, the absorption (per 1 cm) was about 3%, indicating a neon metastable density of about 1×10^9 cm^{-3}. For one fixed fundamental PSW frequency the laser induced frequency change was found to be linearly dependent on the laser intensity up to about 50 mW/cm^2. The relative change of the PSW frequency is proportional to the fundamental PSW frequency. It follows that measurements should be made with fundamental PSW frequencies as high as possible. The S/N ratio of the Ar 811.75 nm line induced by a diode laser was at least comparable to the absorption and optogalvanic signals recorded simultaneously. For example, the typical absolute frequency difference of PSW was about 75 Hz for Ar at our experimental conditions.

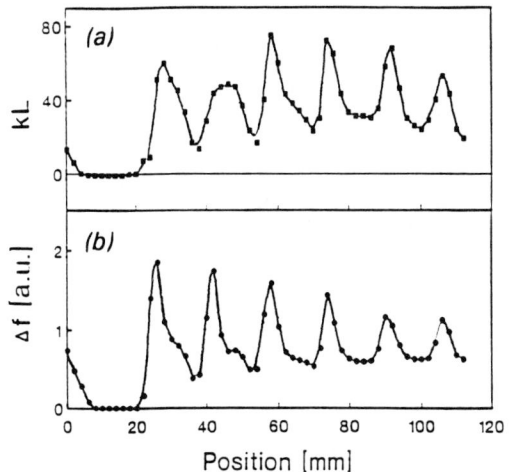

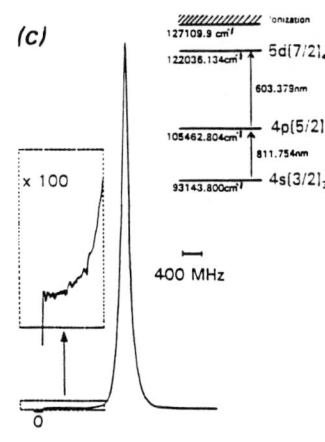

Fig. 2 Absorption by Ar metastables at 811.75 nm measured in transverse direction plotted against the distance from the cathode (a) and the corresponding plasma wave signals (b).(c) PSW-spectrum of a Doppler-free two-photon transition in argon.

The typical S/N ratio was about 300. This value could be improved significantly by excitation of the argon atom to a higher lying level with a larger ionization probability (see Fig. 2c).

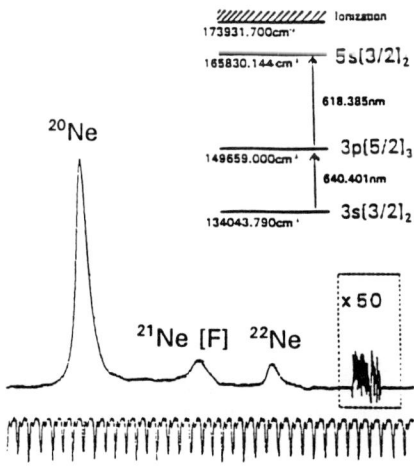

Fig. 3. Doppler-free two-photon PSW-spectrum in Ne. The first laser is slightly detuned to the high frequency side of the ^{20}Ne isotope line and the second laser is scanned. The free spectral range of the Fabry-Perot interferometer fringes (bottom) is 149MHz.

Since the beam of the second laser was transmitted in counterpropagating direction to the first laser beam we obtained a Doppler-free spectrum. We have found a signal enhancement of about one order of magnitude and a better S/N ratio compared with one-photon excitation (see Fig. 2c). Here, the PSW-frequency shift was about 800 Hz and the S/N factor about 3×10^3. Similar results with one- and two-photon excitation from the metastable level were found in Ne. The high detection power of this new technique is demonstrated in Fig. 3, where a two-photon spectrum of neon is shown. The spectrum was measured with two counterpropagating laser beams in axial direction. The wavelength of the first laser was fixed in the Doppler wing of the ^{20}Ne transition. One of the hyperfine structure components of the ^{21}Ne isotope is clearly visible. Taking into account the measured total neon metastable density (1×10^9 cm^{-3}), the ^{21}Ne abundance (0.3%), the line intensity and the measured S/N ratio of about 50, a detection limit of about 5×10^5 metastable ^{21}Ne*($3s[3/2]_2$) atoms/cm^3 could be estimated.

Here one has to note that the ^{21}Ne metastable level splits into nine hf-components. Therefore, the actual number density of probed atoms is even smaller (about 5×10^4 cm^{-3}).

It is interesting to study the Fourier spectrum (FS) of the PSW. This was done by the use of the spectrum analyser mode of our lock-in amplifier. Depending on the purity of the PSW mode (pure sine wave mode or slightly "modulated", but still sinusoidal wave), one line corresponding to the fundamental frequency with a few (usually two) very weak side-components appeared in the FS. In the case of a pure sine wave and laser excitation, the fundamental component shifted corresponding to the frequency changes measured by the frequency-to-voltage converter. If there are side-modes, one can additionally observe that there is a change of the separation of the side-modes. However, because of better stability and better signal-to-noise ratio we operate the discharge always in pure sine wave mode.

4. CONCLUSION

Preliminary measurements applying plasma wave laser spectroscopy have revealed that it is a promising technique for applied spectroscopy. Systematic investigations and further optimization of the experimental parameters, in particular, the dimensions of the discharge tube, will certainly improve the power of this technique.

REFERENCES

T J M Boyd and J J Sanderson 1969 *'Plasma Dynamics'* (London: Nelson)
J F Denisse and J L Delcroix 1963 *'Plasma Waves'* (London: Interscience)
M Kuzuya and E H Piepmeier 1991 *Anal.Chem.* **63**, 1763
I I Popescu, M A Bratescu and M P Dinca 1991 Spectrochim.Acta **46B**, 547

Inst. Phys. Conf. Ser. No 128: Section 9
Paper presented at RIS 92, Santa Fe, NM, USA, 24–29 May 1992

A miniature carbon furnace for mass spectrometry

J. P. Young and R. W. Shaw

Analytical Chemistry Division, Oak Ridge National Laboratory, Oak Ridge, TN 37831-6142

ABSTRACT: A miniature carbon (graphite) disposable tube furnace has been designed and built for use in RIMS. This furnace is an excellent source for generating atoms to be subsequently ionized by optical or thermal means. The complete furnace assembly is very small, less than 4 cm². Electrical connections to the carbon are simple, and no water cooling is required.

1.   INTRODUCTION

Ion sources for mass spectrometers many times consist of a ribbon filament made of rhenium, tantalum, or some other high-melting metal. They are shaped in some fashion depending upon the desired use. Samples of various kinds, such as evaporated solutions, metals, electroplated metals, resin beads, etc., are loaded onto the filament. Carbon is sometimes placed over the sample by vapor deposition or as an aqueous suspension that is subsequently baked. Depending on the chemical preparation of these filaments, heating causes the evolution of neutral and/or ionic species. An idealized concept of a mass spectrometer source for inorganic use would be one that generates a point source of atoms at some location to be coupled with an ionization step that efficiently ionizes this point source of atoms. The heated metal filament arrangement described above only poorly approaches this ideal situation. The sample may be the closest to a point source only if it is a resin bead or a carefully electroplated sample. The effusate consists of neutral species and ions generally spreading out from the filament. The ratio of ions to neutrals normally greatly favors neutrals, and even here many of the neutrals may be molecular. There are two-filament designs wherein neutrals are formed at a sample filament, held at a lower temperature, and ionized by contact with a second, very hot, filament. This design counters some problems arising from overheating the sample, but it can exacerbate the ion spreading problem.

A carbon furnace offers distinct advantages as a mass spectrometer sample source for both thermal ionization mass spectrometry (TIMS) and, particularly, resonance ionization mass spectrometry (RIMS). Carbon is an effective reducing agent; it has a relatively low work function compared to metal filaments normally used as sources. The process of effusing out of the carbon furnace, i.e., collisions, etc., can aid the formation of neutral atoms as opposed to molecules. These attributes favor the release of atoms from heated carbon; further, the tube arrangement has been shown by others (Bushaw and Gerke 1988) to effectively directionalize the effused atoms into a narrow cone. There have been several carbon furnaces described for use as mass spectrometer atom sources (Bekov et al 1986, Bekov et al 1987, Bushaw and Gerke 1988). These furnaces have been shown to be useful for RIMS studies. They require water-cooling for their operation and therefore have a somewhat complex design. We describe here a miniature carbon furnace that requires no elaborate electrical connections and no external water cooling.

2.   CARBON FURNACE

The assembled carbon furnace source is shown schematically in Figure 1, and the components are detailed in Figure 2. In Figure 1, the furnace is mounted on a plate, case plate, that is a standard part of our mass spectrometer. The device consists of an easily handled, sturdy, small carbon furnace 2 cm long by 0.16 cm diameter with a 0.03 cm diameter opening drilled halfway into the furnace. The electrical connection to the carbon is made by compressing the tube between two metal plates. It is necessary to use carbon (graphite) of relatively

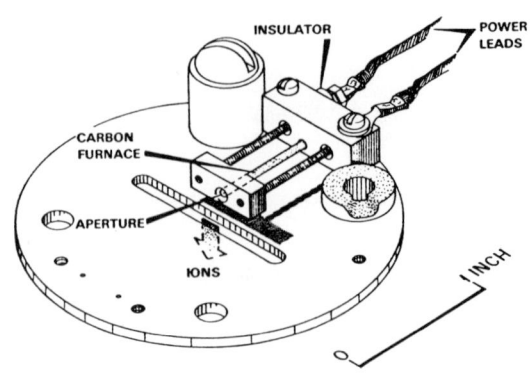

Figure 1.   Assembled Graphite Furnace.

high resistivity, > 0.02 ohms inches. The tube is inserted into two metal end plates which are the case block and the cylindrical part of the suspended block shown in Figure 2. These plates are held against the ends of the furnace by two 0-80 screws which screw into the suspended block and are insulated from the case block. Electrical connections are made to the case block and either of the two insulated screws. The insulation of each screw is accomplished by means of a ceramic sleeve that fits around the screw and into the hole and a ceramic washer that fits between the head of the screw and the case block. As shown in Figure 1, the case block electrical connection is made with a lug that covers one of the 0.086" dia. holes; screws go through these holes into the case plate that is the first surface of our ion lens. It should be pointed out that the critical arrangement and dimensions of the parts shown in Figure 2 have to do with the carbon furnace, carbon furnace inserts, and screw openings for holding the furnace assembly in place. Exterior dimensions and positions and size of the 0.086" dia. holes would be modified to fit the atom source for which this furnace is intended. There is no need for water cooling or elaborate spring loading to maintain electrical contact during heating. The complete furnace assembly is very small, e.g., less than 4 cm². The carbon furnace tube, shown in Figure 2, is only open at one end to provide the greatest number of quasi-collimated effused species. The open end of the furnace is located at the opening into the ion lens. By drilling a hole completely through the furnace and matching that with a hole drilled in the case block, one could have a completely open tube furnace for counterpropagating laser beams.

3.   OPERATION AND RESULTS

The carbon furnace is loaded most easily by pipetting a solution of a sample into the furnace opening. The furnace tube is then dried, or if already in the furnace assembly, a low current can be applied to evaporate the liquid. If desired, small amounts of solid samples, or resin beads could also be loaded into the furnace.

It has been found that heating the furnace to approximately 1100° C causes the effusion of neutral lanthanides from a furnace loaded with an aqueous solution of salts of these elements. To obtain this temperature, approximately 9 amps at 2 volts are passed through the furnace.

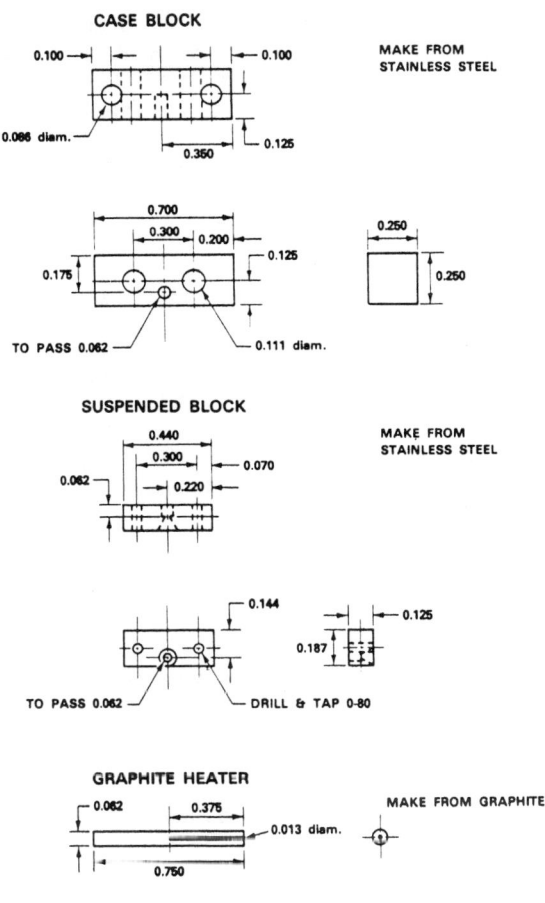

Figure 2. Miniature Graphite Furnace.

Rubidium atoms are effused at much lower temperatures, probably less than 500° C, from aqueous RbNO$_3$ loaded in the furnace. For TIMS we use the two-filament approach described in the introduction; in addition to the carbon furnace, a rhenium filament is placed on the case plate of the ion lens so that it is essentially at a right angle from the carbon furnace opening and the slot in the case plate that permits the extraction of ions into the ion lens. With the carbon furnace heated and the rhenium filament heated, ions are thermally generated at the filament and extracted into the mass spectrometer. In this fashion, TIMS signals have been seen for La, Gd, and Rb. In fact, this setup can be used, with the filament slightly out of line with the furnace opening, to obtain a preliminary focus of the ion lens and verify proper operation of the mass spectrometer for samples to be subsequently studied by RIMS.

We have demonstrated RIMS with RbNO$_3$ loaded in the furnace. In this case a CW diode laser beam was directed along the furnace axis to the opening in the furnace. The diode laser was softly focused on this opening. This laser beam was intersected by a beam from a 6 KHz pulsed copper vapor laser - pumped dye laser (R6G dye solution) that was softly focused and directed parallel to the case plate in a direction along the length of the slot. The diode laser when tuned to 780.0 nm excited ground state ($^2S_{\frac{1}{2}}$) Rb to the $^2P_{3/2}$ level. When the CVL-dye laser was then tuned to 572.4 nm, ionization of Rb occurs by a 1+1 process; i.e., excitation from the $^2P_{3/2}$ level to the $^2D_{5/2}$ level at 30282 cm^{-1} followed by absorption of a second photon to cause ionization. By tuning the wavelength of the single mode diode laser either Rb-85 or Rb-87 could be ionized selectively. A further discussion of the spectroscopy of Rb by this optical process is given in this volume by Shaw *et al.*

4.   DISCUSSION

In our studies to date, the furnace performed well. As noted above, we can generate an atom plume at reasonably low temperatures, < 1200°C. The furnace as presently designed may be limited to this temperature, but simple modifications would increase the temperature range. We are forced to limit the current applied to the furnace to 10 amps because of our power supply. In our experimental set-up, we are concerned with heat transfer from the furnace to the case plate and the ion lens. A temperature of 1200° C is permissible in this respect. The case block could be redesigned to minimize metal contact to the case plate if desired. One could envision other design features, such as heat shields, insulators, etc., which could also be employed if higher temperatures are required.

5.   ACKNOWLEDGEMENTS

Research sponsored by the Office of Energy Research, U.S. Department of Energy, under contract DE-AC05-84OR21400 with Martin Marietta Energy Systems, Inc.

REFERENCES

1.   Bekov G I, Kudryavtsen Y A, Auterinen I and Likonen J. 1986 *Inst. Phys. Conf. Ser.* **84** 97

2.   Bekov G I, Radaev V, Likonen J, Zilliacus R, Auterinen I and Lakomaa E-L, 1987 *Anal. Chem.* **59** 2472

3.   Bushaw B A and Gerke G K 1988 *Inst. Phys. Conf. Ser.* **94** 277

Subject Index

Author Index